The Statistical Theory of Linear Systems

The Statistical Theory of Linear Systems

E. J. HANNAN

The Australian National University
Canberra, Australia

MANFRED DEISTLER

The Technical University of Vienna
Vienna, Austria

JOHN WILEY & SONS

New York · Chichester · Brisbane · Toronto · Singapore

Library of Congress Cataloging-in-Publication Data:

Hannan, E. J. (Edward James), 1921–
The statistical theory of linear systems / E. J. Hannan, M. Deistler.
p. cm.—(Wiley series in probability and mathematical statistics. Probability and mathematical statistics, ISSN 0271-6232)
Bibliography: p.
Includes index of notations.
ISBN 0-471-80777-X
1. System identification. 2. Linear systems. 3. Discrete-time systems. I. Deistler, M. (Manfred) II. Title. III. Series.
QA402.H333 1987
003—dc 19 87-19863
CIP

Printed in the United States of America

10 9 8 7 6 5 4 3 2 1

Probability and Mathematical Statistics (Continued)

MUIRHEAD • Aspects of Multivariate Statistical Theory
PURI and SEN • Nonparametric Methods in General Linear Models
PURI and SEN • Nonparametric Methods in Multivariate Analysis
PURI, VILAPLANA, and WERTZ • New Perspectives in Theoretical and Applied Statistics
RANDLES and WOLFE • Introduction to the Theory of Nonparametric Statistics
RAO • Linear Statistical Inference and Its Applications, *Second Edition*
RAO • Real and Stochastic Analysis
RAO and SEDRANSK • W.G. Cochran's Impact on Statistics
RAO • Asymptotic Theory of Statistical Inference
ROBERTSON, WRIGHT and DYKSTRA • Order Restricted Statistical Inference
ROGERS and WILLIAMS • Diffusions, Markov Processes, and Martingales, Volume II: Îto Calculus
ROHATGI • An Introduction to Probability Theory and Mathematical Statistics
ROHATGI • Statistical Inference
ROSS • Stochastic Processes
RUBINSTEIN • Simulation and The Monte Carlo Method
SCHEFFE • The Analysis of Variance
SEBER • Linear Regression Analysis
SEBER • Multivariate Observations
SEN • Sequential Nonparametrics: Invariance Principles and Statistical Inference
SERFLING • Approximation Theorems of Mathematical Statistics
SHORACK and WELLNER • Empirical Processes with Applications to Statistics
STOYANOV • Counterexamples in Probability

Applied Probability and Statistics

ABRAHAM and LEDOLTER • Statistical Methods for Forecasting
AGRESTI • Analysis of Ordinal Categorical Data
AICKIN • Linear Statistical Analysis of Discrete Data
ANDERSON and LOYNES • The Teaching of Practical Statistics
ANDERSON, AUQUIER, HAUCK, OAKES, VANDAELE, and WEISBERG • Statistical Methods for Comparative Studies
ARTHANARI and DODGE • Mathematical Programming in Statistics
ASMUSSEN • Applied Probability and Queues
BAILEY • The Elements of Stochastic Processes with Applications to the Natural Sciences
BAILEY • Mathematics, Statistics and Systems for Health
BARNETT • Interpreting Multivariate Data
BARNETT and LEWIS • Outliers in Statistical Data, *Second Edition*
BARTHOLOMEW • Stochastic Models for Social Processes, *Third Edition*
BARTHOLOMEW and FORBES • Statistical Techniques for Manpower Planning
BECK and ARNOLD • Parameter Estimation in Engineering and Science
BELSLEY, KUH, and WELSCH • Regression Diagnostics: Identifying Influential Data and Sources of Collinearity
BHAT • Elements of Applied Stochastic Processes, *Second Edition*
BLOOMFIELD • Fourier Analysis of Time Series: An Introduction
BOX • R. A. Fisher, The Life of a Scientist
BOX and DRAPER • Empirical Model-Building and Response Surfaces
BOX and DRAPER • Evolutionary Operation: A Statistical Method for Process Improvement
BOX, HUNTER, and HUNTER • Statistics for Experimenters: An Introduction to Design, Data Analysis, and Model Building
BROWN and HOLLANDER • Statistics: A Biomedical Introduction
BUNKE and BUNKE • Statistical Inference in Linear Models, Volume I

Applied Probability and Statistics (Continued)

CHAMBERS • Computational Methods for Data Analysis
CHATTERJEE and HADI • Sensitivity Analysis in Linear Regression
CHATTERJEE and PRICE • Regression Analysis by Example
CHOW • Econometric Analysis by Control Methods
CLARKE and DISNEY • Probability and Random Processes: A First Course with Applications, *Second Edition*
COCHRAN • Sampling Techniques, *Third Edition*
COCHRAN and COX • Experimental Designs, *Second Edition*
CONOVER • Practical Nonparametric Statistics, *Second Edition*
CONOVER and IMAN • Introduction to Modern Business Statistics
CORNELL • Experiments with Mixtures: Designs, Models and The Analysis of Mixture Data
COX • Planning of Experiments
COX • A Handbook of Introductory Statistical Methods
DANIEL • Biostatistics: A Foundation for Analysis in the Health Sciences, *Fourth Edition*
DANIEL • Applications of Statistics to Industrial Experimentation
DANIEL and WOOD • Fitting Equations to Data: Computer Analysis of Multifactor Data, *Second Edition*
DAVID • Order Statistics, *Second Edition*
DAVISON • Multidimensional Scaling
DEGROOT, FIENBERG and KADANE • Statistics and the Law
DEMING • Sample Design in Business Research
DILLON and GOLDSTEIN • Multivariate Analysis: Methods and Applications
DODGE • Analysis of Experiments with Missing Data
DODGE and ROMIG • Sampling Inspection Tables, *Second Edition*
DOWDY and WEARDEN • Statistics for Research
DRAPER and SMITH • Applied Regression Analysis, *Second Edition*
DUNN • Basic Statistics: A Primer for the Biomedical Sciences, *Second Edition*
DUNN and CLARK • Applied Statistics: Analysis of Variance and Regression, *Second Edition*
ELANDT-JOHNSON and JOHNSON • Survival Models and Data Analysis
FLEISS • Statistical Methods for Rates and Proportions, *Second Edition*
FLEISS • The Design and Analysis of Clinical Experiments
FOX • Linear Statistical Models and Related Methods
FRANKEN, KÖNIG, ARNDT, and SCHMIDT • Queues and Point Processes
GALLANT • Nonlinear Statistical Models
GIBBONS, OLKIN, and SOBEL • Selecting and Ordering Populations: A New Statistical Methodology
GNANADESIKAN • Methods for Statistical Data Analysis of Multivariate Observations
GREENBERG and WEBSTER • Advanced Econometrics: A Bridge to the Literature
GROSS and HARRIS • Fundamentals of Queueing Theory, *Second Edition*
GUPTA and PANCHAPAKESAN • Multiple Decision Procedures: Theory and Methodology of Selecting and Ranking Populations
GUTTMAN, WILKS, and HUNTER • Introductory Engineering Statistics, *Third Edition*
HAHN and SHAPIRO • Statistical Models in Engineering
HALD • Statistical Tables and Formulas
HALD • Statistical Theory with Engineering Applications
HAND • Discrimination and Classification
HOAGLIN, MOSTELLER and TUKEY • Exploring Data Tables, Trends and Shapes
HOAGLIN, MOSTELLER, and TUKEY • Understanding Robust and Exploratory Data Analysis

(*continued on back*)

Preface

One of the central issues of theoretical science is how to create good models from observations. Technically speaking, this is the problem of system identification, the principal topic of this book. More specifically, we treat linear dynamic systems evolving in discrete time.

The past twenty years have witnessed major developments in the theory and application of systems, work that was heavily influenced by the results of Kalman in the early 1960s. On the one hand, certain essential advances have been made in understanding the algebraic and topological structure of linear dynamic systems, with much of this work originating in the fields of systems and control engineering. At the same time, these systems, especially for the scalar output case, have been widely used to model and statistically treat data arising in signal processing (e.g., speech processing). In particular, engineers have developed algorithms for on-line and real time calculations. Finally, statistical time-series analysts, motivated by applications coming from a wide variety of fields, have developed theory and algorithms primarily for off-line calculation. The book *Time Series Analysis, Forecasting and Control* by Box and Jenkins (1970) played a major part in this development. Statisticians have also developed the required asymptotic theory associated with estimation procedures. Of special note in this regard is the early work of Whittle (1951).

There is a requirement for a book bringing all of the foregoing developments together. To be sure, parts are already treated in book form. However, there is no book that treats the above topics together, showing their interconnections; it is this need that we seek to satisfy. This appears of importance since the topics covered in this book are known to different degrees by workers in widely separated areas. For example, some parts of the general theory, such as the structure associated with Kronecker indices, is little known to statisticians, but may prove important since it offers one way of handling the "curse of dimensionality" (too many parameters) plaguing estimation in the vector case. There are many different aspects of the overall problem to be treated and it is not easy to give appropriate emphasis to each in a book of limited size. First, there is the relation between linear systems and the structure of time series in general, especially stationary time series. Then there is structure theory, referred to above. A description of sets of systems in local coordinates must be available because computations are carried out in coordinates, as is the central limit theorem. All systems must be uniquely described. The somewhat delicate topological and geometrical structure must be understood, for these factors relate closely to such things as the stability of algorithms and limiting properties of estimation procedures. Asymptotic theories need to be developed in as general a fashion as possible. In particular, the integer paremeters specifying (local) coordinates need to be estimated. Here work by Akaike (1969) and Rissanen (1978) has been important. There are a great variety of off-line algorithms that need to be discussed, as well as important on-line algorithms. Although all of these topics are treated here to some degree, perhaps we have not given them as extensive treatment as we have on-line algorithms, for example. However, such algorithms have recently been surveyed by Ljung and Söderström (1983) and Young (1984). Finally, we must emphasize the fact that the statistical procedures provide only approximations to the unknown, and possibly unknowable, structures generating the data.

We have attempted to cover, to some extent at least, the topics noted above. However, we have restricted ourselves essentially to the stationary case and the case of noise-free inputs. We have not discussed those theories, developed particularly in relation to economic applications, allowing systems to be nonstationary though on the edge of stationarity, so that they may be brought to stationary form by suitable linear filtering—typically differencing, of the output. Such filters might have to be determined from the data. Our excuse for omitting this important topic is the usual limitation of space and the relative completeness of the stationary theory.

The theory of statistical systems and their treatment is a large and important topic. Even apart from that, and from the subjects listed above, it may be hoped that some rather general theory will be useful to researchers [e.g. that given in subsection 6.6(ii)]. We have almost certainly failed to report some significant developments, through oversight or inadequate knowledge. We ask the forgiveness of those responsible for such developments. The best that can be hoped is that this book will be of value to those who will further advance the study of its subject matter and related topics.

The typing for this book was done by Ann Milligan. Were it not for her skill and assiduity, the work would never have been produced. The authors wish to thank D. S. Poskitt, B. M. Pötscher, and W. Scherrer for their help. Though the book is not formally dedicated, all who know the first author will know to whom his heart is given so that such a formal dedication is superfluous for him.

Canberra, Australia
Vienna, Austria
January 1988

E. J. HANNAN
M. DEISTLER

Contents

Index of Notations

Greek letters follow corresponding Roman ones, with *phi* after *f*, *psi* after π after *p*, *chi* after *xi* after *x*, *eta* after *e* before *epsilon*, and *omega* last. Capitals precede lowercase letters. Subscripted follow unsubscripted and precede superscripted symbols.

The Statistical Theory of Linear Systems

CHAPTER 1

Linear Systems and Stationary Processes

The purpose of this book is to survey part of a large subject, with the emphasis on statistics (i.e., data analysis) for linear systems. The data will be comprised of observed sequences, $y(t)$, $z(t)$, $t = 1, \ldots, T$, of vectors of, respectively, s and m components. The analysis of the data may be for various purposes, of which the most obvious, perhaps, are prediction and to serve as a basis of understanding in order to control a system to which $z(t)$ is an observed input and from which $y(t)$ is an observed output. The estimation of spectra may also be a main purpose. However, there are many other purposes, among which would be investigative work to give understanding and preparatory analysis to form a basis for other analyses of data, for example to form a basis for filtering data so as to allow more effective spectrum and cross-spectrum estimation procedures. However, before such statistical analysis is presented it is necessary to understand the structure of linear systems and to relate such system theory to more general stochastic process theory. This is done in the first two chapters.

In this chapter we review some basic facts about linear dynamic systems and stationary processes. Thus we consider the sequence $y(t)$ as the vector output of some system. In economics the system would be some part of the overall economic system, such as the banking sector, and $y(t)$ would consist of measurements on that sector such as the levels of deposits and advances of various kinds. These are viewed as being jointly determined by the system and in economics would be called endogenous variables. The external forces or inputs that are to be considered and have been measured constitute $z(t)$ and, for example, might include measurements on variables external to the economy being considered: for example, for a "small" country, measures of the condition of the U.S. economy. In economics $z(t)$ would be said to be composed of exogenous variables. In addition to $z(t)$ there would be other unobserved, "residual" influences causing variation in $y(t)$, for no matter how careful the measurement process may be there will always be such additional influences, including errors of measurement.

The relation between $y(t)$ and $z(t)$ will be dynamic in the sense that $y(t)$ is influenced by $z(s)$, $s < t$. Herein, however, we shall restrict ourselves to the linear case so that $y(t)$ will be viewed as determined by $z(t)$ and the residual influences in a linear manner. The justification for this is that experience shows that such linear models work well in a wide range of circumstances and they lead to reasonable calculations and a tractable theory. Such a linear relation can be written as

$$\sum_{j=0}^{p} A(j)y(t-j) = \sum_{j=1}^{r} D(j)z(t-j) + \sum_{0}^{q} B(j)\varepsilon(t-j).$$

Here $A(j)$, $B(j)$, and $D(j)$ are matrices of dimensions, respectively, $s \times s$, $s \times s$, and $s \times m$. Chance, stochastic forces enter the system through the "residual," $\varepsilon(t)$ [and perhaps also the $z(t)$], which are regarded as random vectors. It is not unreasonable to put $E\{\varepsilon(s)\varepsilon(t)'\} = 0$, $s \neq t$, since one can hope to replace any initial set of such unobserved influences by a new set with this desirable property. There are now many questions to be answered in connection with this, so-called, ARMAX representation, the acronym standing for "autoregressive-moving average with exogenous variables." The AR part is the left side and the MA part that involving $\varepsilon(t)$. These questions relate, for example, to the uniqueness of the description of the system and its stability [i.e., loosely whether $y(t)$ will vary in a stable fashion or will oscillate and grow wildly]. These questions are discussed in Section 1 (and also in Chapter 2). There is also the relation to stationary process theory and the conditions under which the $\varepsilon(t)$ are the "innovations" or prediction errors, which are discussed in Section 2.

There is another way in which a linear system may be prescribed, which is also discussed in Section 1. This alternative, so-called, state-space representation arises particularly in connection with problems of control in engineering. The output, $y(t)$, might now be considered to consist of a few measurements on a system whose state might require many more variables for a faithful description. We shall indicate the full state vector of measurements by $x(t)$. Now linearity requires that

$$y(t) = Hx(t) + \eta(t)$$

where $\eta(t)$ is a vector of measurement errors, which might resonably be assumed to satisfy $E\{\eta(t)\eta(s)'\} = 0$, $s \neq t$. This is called the observational equation. As an example, the system might be a power plant and $y(t)$ constituted by a few measurements in which we are interested, for example the temperature of the water coming from the cooling system. However, the full state of the system might require many measurements for a faithful description. In addition, there will be other measurements on variables such as meteorological conditions or fuel use that will constitute part of $z(t)$. To complete the description a state equation prescribing the temporal behavior of $x(t)$ in relation to $z(t)$ must be prescribed. This state equation is assumed to be of the simple form

$$x(t+1) = Fx(t) + Lz(t) + \xi(t),$$

where $\xi(t)$ is assumed to satisfy $E\{\xi(s)\xi(t)'\} = 0$, $s \neq t$. This equation is more general than might appear, as will be apparent from the discussions below. One reason is that if there were further lagged values of $x(t)$ and $z(t)$ on the right side, a redefinition of $x(t)$ and $z(t)$ would bring the representation back to the required form. Thus if $F_1 x(t-1)$ and $L_1 z(t-1)$ did occur, put $\tilde{x}(t)' = (x(t)', x(t-1)')$, $\tilde{z}(t)' = (z(t)', z(t-1)')$, and

$$\tilde{F} = \begin{bmatrix} F & F_1 \\ I & 0 \end{bmatrix}, \qquad \tilde{L} = \begin{bmatrix} L & L_1 \\ 0 & 0 \end{bmatrix}, \qquad \tilde{\xi}(t) = \begin{bmatrix} \xi(t) \\ 0 \end{bmatrix},$$

in which case

$$\tilde{x}(t+1) = \tilde{F}\tilde{x}(t) + \tilde{L}\tilde{z}(t) + \tilde{\xi}(t).$$

Lagged $\varepsilon(t)$ can be handled similarly.

In Section 2 the very complete correspondence between the ARMAX and state-space representations will be discussed. In particular it will be shown how to transform to a new state vector $x(t|t-1)$ so that $\xi(t)$ becomes $K\varepsilon(t)$ and $\eta(t)$ becomes $\varepsilon(t)$. This new $x(t|t-1)$ will consist of predictions of $y(t+j)$ from data to time $t-1$ and the transformation relates to the case where the $\varepsilon(t)$ are the innovations.

Stationarity of the stochastic processes involved will play a major part in the developments, so that, for example, $y(t)$ and $z(t)$ might be regarded as jointly stationary. There is thus a need to relate these ARMAX and state-space representations to the stationarity aspects and in particular to the spectral theory and the prediction theory for stationarity processes. This is done in Section 3.

In addition to the ARMAX and state-space representations there is, at least in the stationary case, a direct representation of $y(t)$ in terms of an infinite sequence of lagged $\varepsilon(t)$, $z(t)$, of the form

$$y(t) = \sum_{0}^{\infty} K(j)\varepsilon(t-j) + \sum_{1}^{\infty} L(j)z(t-j).$$

This can be obtained, for example, from the ARMAX representation by repeatedly substituting for lagged $y(t)$ [i.e., $y(t-j)$ for $j > 0$], expressed in terms of larger lags obtained from the ARMAX representation. Of course, without further restrictions such a representation may not be possible and this question is discussed in Section 1. It is fairly clear that when it exists, such an input-output representation may more easily be prescribed so as to be unique. This is discussed in Section 3. What turns out to be more difficult to prescribe uniquely is the representation of the $K(j)$, $L(j)$ in terms of the $A(j)$, $B(j)$, $D(j)$, or via a state-space representation. This is discussed extensively in Chapter 2.

In Section 4 some technical details concerning the structure of (strictly) stationary processes are given.

1. LINEAR SYSTEMS: INPUT-OUTPUT, ARMAX, AND STATE-SPACE REPRESENTATIONS

For us a system is always given by a concrete representation: for example, as input-output, an ARMAX, or a state-space representation. We do not intend to give an abstract or the most general definition of a linear dynamic system here; for this see, for example, Blomberg and Ylinen (1983), Kailath (1980), Kalman, Falb, and Arbib (1969), and Willems (1986). We consider only *discrete-time* systems, where time is running over the integers $\mathbf{Z}$, or if explicitly stated, over the natural numbers $\mathbf{N}$. Unless the contrary has been stated explicitly, we consider only the case for *time-invariant* and *causal* systems. We explain these terms below.

The *input-output representation* of a *linear time-invariant* system is given by

$$y(t) = \sum_{j=-\infty}^{\infty} L(j)w(t-j), \tag{1.1.1}$$

where $y(t)$ are the outputs, $w(t)$ are the inputs, and $L(j) \in \mathbf{R}^{s \times m}$ are the coefficients of the transfer function

$$l(z) = \sum_{j=-\infty}^{\infty} L(j)z^j, \qquad z \in \mathbf{C}. \tag{1.1.2}$$

$(L(j) | j \in \mathbf{Z})$ is called the weighting function of (1.1.1). The weighting function is also called the impulse response function since $y(t) = L(t - \tau)e(\tau)$ is the time history of an output resulting from an "impulse," $e(\tau)$, at time τ. *Time invariance* means that the $L(j)$ do not depend on t or in general terms that if $(y(t) | t \in \mathbf{Z})$ is the output corresponding to $(w(t) | t \in \mathbf{Z})$, then $(y(t + s) | t \in \mathbf{Z})$ is the output corresponding to $(w(t + s) | t \in \mathbf{Z})$, for any fixed $s \in \mathbf{Z}$. We deal almost exclusively with this case here.

Throughout the book we assume that $y(t)$ and $w(t)$ are random vectors, defined over an underlying probability space $(\Omega, \mathscr{F}, P)$, $y(t)\colon \Omega \to \mathbf{R}^s$, $w(t)\colon \Omega \to \mathbf{R}^m$. In addition, we assume throughout that $y(t)$ and $w(t)$ have finite second moments, that is,

$$\| y(t) \|^2 = E\{y(t)'y(t)\} < \infty, \qquad E\{w(t)'w(t)\} < \infty.$$

Then the elements $y_i(t)$ and $w_i(t)$ of $y(t) = (y_i(t))$ $i = 1, \ldots, s$ and $w(t) = (w_i(t))$ $i = 1, \ldots, m$, respectively, can be considered as elements of the Hilbert space $L^2(\Omega, \mathscr{F}, P)$ of square-integrable complex random variables $x\colon \Omega \to \mathbf{C}$ over $(\Omega, \mathscr{F}, P)$. The limits, as in (1.1.1), are understood in the mean-squares sense if the contrary has not been stated explicitly. The actual conditions for convergence in (1.1.1) and also in (1.1.2) will be discussed later.

A system is called *causal* if $y(t)$ depends only on $w(s)$, $s \le t$. For the input-output representation (and a suitably rich space of inputs) this is equivalent to $L(j) = 0, j < 0$. In this sense also a transfer function may be called causal if it has a power series expansion in a certain neighborhood of zero and if it is clear that the solution is defined by the weighting function corresponding to this power series. A system is called *finite dimensional* if its transfer function is rational; that is, $l = a^{-1}d$, where a and d are polynomial matrices and where $\det a(z) \not\equiv 0$. To model uncertainty, we distinguish between *observed inputs* (or exogenous variables), which we call $z(t)$, and unobserved inputs, or *errors*, $\varepsilon(t)$. With this distinction we write (1.1.1) as

$$y(t) = \sum_{j=-\infty}^{\infty} L(j)z(t-j) + \sum_{j=-\infty}^{\infty} K(j)\varepsilon(t-j), \tag{1.1.3}$$

where

$$L(j) \in \mathbf{R}^{s \times m}, \qquad K(j) \in \mathbf{R}^{s \times s}$$

and

$$l(z) = \sum_{j=-\infty}^{\infty} L(j)z^j, \qquad k(z) = \sum_{j=-\infty}^{\infty} K(j)z^j \tag{1.1.4}$$

are the (observed) inputs to output and the errors to output transfer function, respectively. Note that we have assumed that $y(t)$ and $\varepsilon(t)$ have the same dimension.

In all cases we consider, $\varepsilon(t)$ will be white noise, that is,

$$E\{\varepsilon(t)\} = 0, \qquad E\{\varepsilon(s)\varepsilon(t)'\} = \delta_{st}\Sigma \tag{1.1.5}$$

and we assume, for all s, t,

$$E\{z(s)\varepsilon(t)'\} = 0. \tag{1.1.6}$$

It is in some ways preferable to require (1.1.6) only for $s < t$, so that $z(s)$ can involve feedback from earlier $\varepsilon(t)$, and many of the results below would hold for that case.

Another representation is the *vector difference equation (VDE) representation* of a linear system

$$\sum_{j=0}^{p} A(j)y(t-j) = \sum_{j=1}^{r} D(j)z(t-j) + \sum_{j=0}^{q} B(j)\varepsilon(t-j), \tag{1.1.7}$$

where again $y(t)$, $z(t)$, and $\varepsilon(t)$ are the outputs, the (observed) inputs, and the errors, respectively, and where $A(j) \in \mathbf{R}^{s\times s}$, $D(j) \in \mathbf{R}^{s\times m}$, and $B(j) \in \mathbf{R}^{s\times s}$ are the associated parameter matrices. Starting the summation in the first sum of the right-hand side in (1.1.7) with 1 rather than with zero turns out to be more convenient and is no restriction of generality in the sense that $z(t)$ can always be redefined as $z(t+1)$.

We use z also to denote the backward-shift operator on $\mathbf{Z}$: $z(y(t)|t \in \mathbf{Z}) = (y(t-1)|t \in \mathbf{Z})$; then (1.1.7) (when considered on $\mathbf{Z}$) can be written as

$$a(z)y(t) = d(z)z(t) + b(z)\varepsilon(t), \tag{1.1.8}$$

where

$$a(z) = \sum_{j=0}^{p} A(j)z^j; \qquad d(z) = \sum_{j=1}^{r} D(j)z^j; \qquad b(z) = \sum_{j=0}^{q} B(j)z^j \tag{1.1.9}$$

are the associated polynomial matrices. The assumption

$$\det a(z) \not\equiv 0$$

will always be imposed and is part of the definition of a (complete) VDE system.

Alternatively, sometimes a representation of the form

$$\sum_{j=0}^{p} A(j)y(t+j) = \sum_{j=0}^{r} D(j)z(t+j) + \sum_{j=0}^{q} B(j)\varepsilon(t+j) \tag{1.1.10}$$

is used. Again using the backward-shift operator z (note that the backward shift on $\mathbf{Z}$ is bijective) this can be written as

$$a(z^{-1})y(t) = d(z^{-1})z(t) + b(z^{-1})\varepsilon(t). \tag{1.1.11}$$

Clearly, for identical processes under consideration the parameter matrices $A(i)$, $B(i)$, $D(i)$ in (1.1.7) and in (1.1.10) will not be the same.

Another alternative is to give two separate difference equation systems, one describing the influence of the (observed) inputs, the second describing the influence of the errors on the outputs.

$$\sum_{j=0}^{p} A(j)v(t-j) = \sum_{j=1}^{r} D(j)z(t-j) \tag{1.1.12}$$

$$\sum_{j=0}^{l} C(j)u(t-j) = \sum_{j=0}^{q} B(j)\varepsilon(t-j) \tag{1.1.13}$$

$$y(t) = u(t) + v(t). \tag{1.1.14}$$

A VDE system (1.1.7), where in addition it is assumed that $\varepsilon(t)$ is white noise and that

$$E\{z(s)\varepsilon(t)'\} = 0, \tag{1.1.15}$$

is called an ARMAX (autoregressive-moving average system with exogenous variables) system; if there are no exogenous variables in (1.1.7)

$$\sum_{j=0}^{p} A(j)y(t-j) = \sum_{j=0}^{q} B(j)\varepsilon(t-j) \tag{1.1.16}$$

[i.e., $D(j) = 0$], it is called an ARMA system. As special cases, then, we have AR (autoregressive) systems for $q = 0$ and MA (moving average) systems for $p = 0$.

Sometimes systems of the form

$$a(z)y(t) = d(z)z(t) + u(t) \tag{1.1.17}$$

$$b(z)u(t) = \varepsilon(t), \tag{1.1.18}$$

where $\varepsilon(t)$ is white noise, are used. These system representations are sometimes called dynamic adjustment forms.

An ARMAX system can equivalently be defined as a triple $(a(z), b(z), d(z))$ when a, b, and d are $s \times s$, $s \times s$, and $s \times m$, respectively, polynomial matrices and where $\det a(z) \not\equiv 0$. Analogously, we define an ARMA system as (a, b). Sometimes we call $(a(z^{-1}), b(z^{-1}), d(z^{-1}))$ [where $\det a(z) \not\equiv 0$] corresponding to (1.1.11) [if (1.1.5) and (1.1.6) hold] an ARMAX system.

The third representation we consider is the *state-space representation*

$$x(t+1) = Fx(t) + Lz(t) + \xi(t) \tag{1.1.19}$$

$$y(t) = Hx(t) + \eta(t). \tag{1.1.20}$$

Here $x(t)$ is the n-dimensional state vector, which is not directly observed in general, $F \in \mathbf{R}^{n\times n}$, $L \in \mathbf{R}^{n\times m}$, and $H \in \mathbf{R}^{s\times n}$ are parameter matrices, $\xi(t)$ and $\eta(t)$ are errors satisfying

$$E(\xi(t)', \eta(t)') = 0$$

$$E\begin{pmatrix}\xi(s)\\ \eta(s)\end{pmatrix}(\xi(t)', \eta(t)') = \delta_{st}\begin{pmatrix}Q & S\\ S' & R\end{pmatrix}$$

[thus $(\xi'(t), \eta'(t))'$ is white noise] and $y(t)$ and $z(t)$ again are the outputs and observed inputs, respectively. We also assume that

$$E\{z(s)\xi(t)'\} = 0, \qquad E\{z(s)\eta(t)'\} = 0. \tag{1.1.21}$$

The importance of state-space models has been pointed out by Kalman (see, e.g., Kalman, Falb, and Arbib, 1969) and, indeed, it was his recognition of the generality and simplicity of that representation in Kalman (1960) that enabled him to institute the celebrated Kalman filter, which is dealt with in Chapter 3. These models allow for a description of the internal and external characteristics of the system. The state contains all information from the past of the inputs which is relevant for future outputs. A system (1.1.19), (1.1.20) could arise from "physical" theories; (1.1.19) then describes the unobserved states and $\xi(t)$ are the system's disturbances; (1.1.20) is the measurement equation and $\eta(t)$ are the measurement errors.

As will be shown in the next section, every stable state-space representation (1.1.19), (1.1.20) can be transformed into a state-space system

$$x(t+1) = Fx(t) + Lz(t) + K\varepsilon(t) \tag{1.1.22}$$

$$y(t) = Hx(t) + \varepsilon(t), \tag{1.1.23}$$

where $\varepsilon(t)$ is white noise [and where in addition $\varepsilon(t)$ is the prediction error of the best linear, least-squares prediction of $y(t)$ given $y(s)$, $s < t$ and $z(t)$, $t \in \mathbf{Z}$] and where $F \in \mathbf{R}^{n\times n}$, $L \in \mathbf{R}^{n\times m}$, $K \in \mathbf{R}^{n\times s}$, $H \in \mathbf{R}^{s\times n}$ are parameter matrices.

Of course, every stable system (1.1.22), (1.1.23) also has a steady-state solution [see (1.2.25), (1.2.26) below]

(1.1.24)
$$y(t) = \sum_0^\infty K(j)\varepsilon(t-j) + \sum_1^\infty L(j)z(t-j),$$

where the $K(j)$, $L(j)$ are linked to (1.1.22), (1.1.23) in (1.2.23) below. In the case $L = 0$ this becomes

(1.1.25)
$$y(t) = \sum_0^\infty K(j)\varepsilon(t-j).$$

These representations will be used repeatedly below.

2. RELATION BETWEEN VARIOUS SYSTEM REPRESENTATIONS

Here the relation between input-output, ARMAX, and state-space representations is considered. This problem is also investigated in Chapter 2. Here it will be shown that under certain conditions every rational transfer function can correspond to both an ARMAX or a state-space system, and conversely that every ARMAX or (finite-dimensional) state-space system has rational transfer functions. First we consider the solutions of VDEs, and thus in particular of ARMAX and state-space systems; in this way we relate these system representations to the input-output representation. In a second step, the relation among input-output, ARMAX, and state-space systems is investigated.

Now, for notational simplicity, consider a VDE

(1.2.1)
$$a(z)y(t) = d(z)z(t); \qquad a(z) = \sum_{j=0}^{p} A(j)z^j, \qquad d(z) = \sum_{j=0}^{r} D(j)z^j.$$

Although the discussion is in terms of observed inputs, it will easily be seen to generalize to the general case, (1.1.7). *A solution* (on $\mathbf{Z}$) of the VDE (1.2.1) is a process $(y(t)|t \in \mathbf{Z})$ satisfying (1.2.1) for given $(z(t)|t \in \mathbf{Z})$ and given $a(z)$, $d(z)$. A solution of (1.2.1) always exists. This is clear for the case $s = 1$, since here the first and the last coefficient of $a(z)$ can be assumed to be nonzero, and we can start from arbitrary initial values to substitute forward and then we can substitute backward. For the case $s > 1$, we write (1.2.1), using the Smith form (Lemma 2.2.3) for $a(z)$ as

$$u(z)\Lambda(z)v(z)y(t) = d(z)z(t),$$

where u, u^{-1}, v, and v^{-1} are polynomial matrices, and where Λ is a diagonal polynomial matrix. [We have here used $u(z)$, $v(z)$ for unimodular matrices. See Section 2.2 for the definition of a unimodular matrix. There should be no confusion with the random processes $u(t)$, $v(t)$ introduced earlier because of the different context and because of the occurrence of z in place of t.] Since the s single-output systems

$$\Lambda(z)w(t) = u^{-1}(z)d(z)z(t)$$

always have a solution $w(t)$, we obtain $y(t) = v^{-1}(z)w(t)$ as a solution of (1.2.1).

Analogously, if, in addition, initial values are prescribed, solutions on **N** are defined if $\det A(0) \neq 0$, the initial values are $y(0), \ldots, y(-p+1), z(0), \ldots, z(-r+1)$. If the contrary is not stated explicitly, we consider solutions on **Z** only.

As is well known and can easily be seen, the set of all solutions is the sum of one particular solution of (1.2.1) and the set of all solutions of the homogeneous VDE

$$a(z)y(t) = 0. \tag{1.2.2}$$

The set of homogeneous solutions is described, for example, in Hannan (1970, p. 14). Again this can be reduced to the scalar case since we must have $\Lambda(z)v(z)y(t) = 0$; that is, putting $w(t) = v(z)y(t)$ we have scalar equations for $w(t)$ which convert easily into equations for $y(t) = v(z)^{-1}w(t)$, remembering that $v(z)^{-1}$ is also a matrix of polynomials. It is well known that the solution of a scalar equation $\lambda_i(z)w_i(t) = 0$ is a linear combination of solutions of the form $t^a\rho^t \cos \omega t$, $t^a\rho^t \sin \omega t$ for ρ, ω determined by the zeros of $\lambda_i(z)$ and with $a = 0, 1, \ldots, m-1$, where m is the multiplicity of the zeros.

In most cases, the *stability condition*

$$\det a(z) \neq 0 \qquad |z| \leq 1 \tag{1.2.3}$$

is imposed. Then $(\det a(z))^{-1}$ has a convergent power series expansion within a disk containing the closed unit disk, and thus $a^{-1}(z) = (\det a(z))^{-1} \operatorname{adj} a(z)$ (where $\operatorname{adj} a$ is the adjoint of a) and $l(z) = a^{-1}(z)d(z)$ also have a convergent power series expansion in the same region; in particular we have

$$l(z) = a^{-1}(z)d(z) = \sum_{j=0}^{\infty} L(j)z^j. \tag{1.2.4}$$

Then from (1.2.4) we obtain a causal solution

$$y(t) = \sum_{j=0}^{\infty} L(j)z(t-j) \tag{1.2.5}$$

provided that $z(t)$ is such that the infinite sum on the right-hand side (r.h.s.) of (1.2.5) exists. Since $\|L(j)\| \leq c\rho_0^{-j}$, $\rho_0 > 1$, where ρ_0 is smaller than the modulus of all zeros of $\det a(z)$, where $\|L(j)\|$ is the Euclidean norm. Using $\|x\|$ for the Euclidean length of a vector x, if

$$E\{\|z(t-j)\|\} \leq a_j \rho_0^j, \qquad \sum a_j < \infty, \tag{1.2.6}$$

then (1.2.5) converges almost surely (a.s.) (Gihman and Skorohod, 1974, p. 66). This will be true for stationary $z(t)$, with finite mean, when $y(t)$ is also stationary. If $z(t)$ is stationary with $E\{\|z(t)\|^2\} < \infty$, then (1.2.5) also converges in mean square, as is easily seen, or indeed if $[E\{\|z(t-j)\|^2\}]^{1/2} \leq a\rho_0^j$.

Every solution $y(t)$ of the homogeneous VDE (1.2.2) under (1.2.3) satisfies $\lim y(t) = 0$ for $t \to \infty$. Therefore, (1.2.5) is often called the steady-state solution; in particular, if $z(t)$ is stationary, (1.2.5) is the only stationary solution of (1.2.1).

If we introduce a norm $\|z\| = \sup_t \|z(t)\|$ for stochastic processes $z(t)$, then (1.2.3) implies that (1.2.5) defines a bounded linear operator $z(t) \to y(t)$ and thus "small" inputs imply "small" outputs. It is for this reason that (1.2.3) is called the stability condition. As has been stated already, in most cases we impose (1.2.3) and then we consider only the solution (1.2.5). However, let us also discuss other solutions now.

If instead of (1.2.3) we require only

$$\det A(0) \equiv \det a(0) \neq 0, \tag{1.2.7}$$

then $(\det a(z))^{-1}$ has a convergent power series expansion in some neighborhood of zero which defines a causal solution

$$y(t) = \sum_{j=0}^{\infty} L(j)z(t-j) \tag{1.2.8}$$

by

$$l(z) = a^{-1}(z)d(z) = (\det a(z))^{-1} \operatorname{adj} a(z)d(z) = \sum_{j=0}^{\infty} L(j)z^j$$

provided that the r.h.s. of (1.2.8) exists. In general, of course, this solution is not stable and the r.h.s. of (1.2.8) in general does not exist for stationary processes $(z(t)|t \in \mathbf{Z})$. If (1.2.7) is not fulfilled, we write $\det a = z^{\delta_0}p(z)$, where $\delta_0 > 0$, and $p(z)$ is a polynomial with $p(0) \neq 0$. Then, expanding p^{-1} into a power series about zero we obtain a solution

$$y(t) = z^{-\delta_0}p^{-1}(z)\,\mathrm{adj}\,a(z)d(z)z(t) \tag{1.2.9}$$

again provided that $z(t)$ is such that the sum exists. In this solution $y(t)$ is influenced by a finite number of future inputs. Again, in general, the sum in (1.2.9) will not exist for stationary inputs $(z(t)|t \in \mathbf{Z})$. If, however, arbitrary (or in particular stationary) processes $(z(t)|t \in \mathbf{N})$ are considered, then (1.2.8) and (1.2.9) clearly always define solutions if $z(t)$ is extended to $\mathbf{Z}$ by setting $z(t) = 0$ for $t \leq 0$. In the case (1.2.8) they are the solutions on $\mathbf{N}$ for zero initial values $y(0) = \cdots y(-p+1) = 0, z(0) = \cdots = z(-r+1) = 0$.

On the other hand, if we assume that

$$\det a(z) \neq 0, \qquad |z| = 1, \tag{1.2.10}$$

then $(\det a)^{-1}$ can be expanded into a Laurent series converging on a certain annulus containing the unit circle, thus giving a solution of the form

$$y(t) = \sum_{j=-\infty}^{\infty} L(j)z(t-j); \qquad l(z) = a^{-1}(z)d(z) = \sum_{j=-\infty}^{\infty} L(j)z^j. \tag{1.2.11}$$

The infinite sum in (1.2.11) exists for all $z(t)$ satisfying a condition analogous to (1.2.6) and in particular for every stationary $z(t)$, in which case $y(t)$ is also stationary. As easily seen, under (1.2.10), the solution (1.2.11) is the only stable solution in the sense described above (or in the sense that stationary inputs imply stationary outputs); thus under (1.2.10) and an additional stability requirement, we have a unique solution (1.2.11).

In all these cases (for given inputs) the solution is completely determined from the power series or the Laurent series coefficients of the transfer function l; in this sense, then, the respective representations of the solutions [as (1.2.5) or (1.2.9) or (1.2.11) (or the weighting functions)] define the input-output representation corresponding to the VDE system. Note that a given transfer function may be expanded in different ways and thus may correspond to different solutions; however, here it will always be clear from the context which expansion is meant.

Sometimes, solutions of a VDE system that have an "autonomous" component and thus cannot be described by the (power or Laurent series coefficients of the) transfer functions are also of interest. Consider a system

$$y(t) - y(t-1) = \varepsilon(t), \tag{1.2.12}$$

where $\varepsilon(t)$ is white noise and where the system starts at $t = 1$ with a (non-stochastic, say) initial value $y(0) \neq 0$. Then the solution is a random walk of the form

$$y(t) = \sum_{j=1}^{t} \varepsilon(j) + y(0). \tag{1.2.13}$$

Note that for $\varepsilon(t) \not\equiv 0$ (a.s.) the system has no stationary solution.

The system

$$y(t) - y(t-1) = \varepsilon(t) - \varepsilon(t-1), \tag{1.2.14}$$

where again $\varepsilon(t)$ is white noise, has stationary solutions (on $\mathbf{Z}$) of the form

$$y(t) = \varepsilon(t) + z_0, \tag{1.2.15}$$

where z_0 is arbitrary; (1.2.14) thus has the same transfer function $k = 1$ as the system $y(t) = \varepsilon(t)$, but may have different solutions. Analogously, the system

$$(1 - e^{i\lambda_0}z)(1 - e^{-i\lambda_0}z)y(t) = (1 - e^{i\lambda_0}z)(1 - e^{-i\lambda_0}z)\varepsilon(t), \tag{1.2.16}$$

which again has a transfer function $k = 1$, has stationary solutions of the form

$$y(t) = \varepsilon(t) + c_0 \cos(\lambda_0 t + \phi_0), \tag{1.2.17}$$

where c_0 and ϕ_0 are random variables.

For $c_0 \cos(\lambda_0 t + \phi_0)$, in any situation where only (a part of) *one* realization of $y(t)$ is available, there is no point in taking c_0, ϕ_0 as other than constants. More generally, if $\det a$ only has zeros of modulus equal to and greater than 1, and if stationary solutions exist, then singular components of the form $\sum_{j=1}^{n} c_j \cos(\lambda_j t + \phi_j)$ occur where $e^{-i\lambda_j}, e^{i\lambda_j}, j = 1, \ldots, n$ are zeros of $\det a(z)$. If the λ_j are known, then by including $\cos \lambda_j t$, $\sin \lambda_j t$ among the components of the exogenous variables $z(t)$, this singular component can be accounted for. We will always assume that this is done.

Every rational matrix l can be written as

$$l = c^{-1}N, \tag{1.2.18}$$

where c is a least common denominator polynomial of the entries of l (i.e., a least common multiple of all denominators after numerators and denominators are made prime) and where N is a polynomial matrix; for $l = 0$ we take $c = 1$, $N = 0$.

If in a VDE (1.2.1) for the corresponding transfer function $l = a^{-1}d$, the polynomial c in (1.2.18) satisfies

$$c(z) \neq 0, \qquad |z| \leq 1; \tag{1.2.19}$$

then expanding c^{-1} into a power series in $|z| \leq 1$ gives a causal and stable solution

$$y(t) = c^{-1}(z)N(z)z(t) = \sum_{j=0}^{\infty} L(j)z(t-j) \tag{1.2.20}$$

if the sum exists, which is the case for stationary inputs $z(t)$. As the examples above show, for stationary inputs this solution is not necessarily the unique stationary solution. Note that c divides $\det a$. Analogously, if $c(z) \neq 0$, $|z| = 1$ we have $l(z) = c^{-1}(z)N(z) = \sum_{j=-\infty}^{\infty} L(j)z^j$ (converging on an annulus containing the unit circle) and thus, in the stationary case, a not necessarily unique solution of the form

$$y(t) = \sum_{j=-\infty}^{\infty} L(j)z(t-j)$$

exists.

For a state-space system (1.1.22), (1.1.23) we have $\det(I - Fz) = 1$ for $z = 0$ and thus there always exists a causal solution

$$x(t) = (I - Fz)^{-1}z[Lz(t) + K\varepsilon(t)], \tag{1.2.21}$$

where $(I - Fz)^{-1} = \sum_{j=0}^{\infty} F^j z^j$, provided that the sum in (1.2.21) exists. In this sense, causality is automatically built into the definition of state-space systems.

The transfer functions corresponding to (1.1.22), (1.1.23) are

$$\begin{aligned} k(z) &= H(I_n - Fz)^{-1}zK + I_s = I_s + \sum_{j=1}^{\infty} HF^{j-1}Kz^j \\ l(z) &= H(I_n - Fz)^{-1}zL = \sum_{j=1}^{\infty} HF^{j-1}Lz^j. \end{aligned} \tag{1.2.22}$$

Starting the system with given initial values $x(0)$, $z(0)$, $\varepsilon(0)$, at $t = 1$ we have

$$\begin{aligned} x(t) &= \sum_{j=1}^{t} F^{j-1}Lz(t-j) + \sum_{j=1}^{t} F^{j-1}K\varepsilon(t-j) + F^t x(0) \\ y(t) &= \sum_{j=1}^{t} HF^{j-1}Lz(t-j) + \sum_{j=1}^{t} HF^{j-1}K\varepsilon(t-j) + HF^t x(0) + \varepsilon(t). \end{aligned} \tag{1.2.23}$$

The stability condition $\det(I - Fz) \neq 0$, $|z| \leq 1$ is cleary equivalent to $\det(F - \lambda I) \neq 0$, $|\lambda| \geq 1$ and this is equivalent to

$$|\lambda_n(F)| < 1, \tag{1.2.24}$$

where $\lambda_n(F)$ is an eigenvalue of F with maximal modulus. The steady-state solution is then of the form

$$x(t) = \sum_{j=1}^{\infty} F^{j-1} Lz(t-j) + \sum_{j=1}^{\infty} F^{j-1} K\varepsilon(t-j) \tag{1.2.25}$$

$$y(t) = \sum_{j=1}^{\infty} HF^{j-1} Lz(t-j) + \sum_{j=1}^{\infty} HF^{j-1} K\varepsilon(t-j) + \varepsilon(t). \tag{1.2.26}$$

Since any rational $(s \times m)$ matrix can be written as (1.2.18) for any such l there corresponds a VDE with $a = cI$, $d = N$ such that l is its transfer function. Clearly, this VDE is not the only one corresponding to given l. For $s = 1$ such a VDE is quite "natural" for l; for $s > 1$ this VDE has serious disadvantages. The problem of finding a "good" VDE (ARMAX system) for l is one of the central problems in Chapter 2.

Any ARMAX system with $\det A(0) \neq 0$ can immediately be given a state-space realization. For simplicity of notation we show this for the case $A(0) = I$ here. Let us put

$$x(t) = (y(t-1)', \ldots, y(t-p)', z(t-1)', \ldots, z(t-r)', \varepsilon(t-1)', \ldots, \varepsilon(t-q)')'.$$

Then we define

$$F = \begin{bmatrix}
-A(1) & -A(2) & \cdots & -A(p-1) & -A(p) & D(1) & \cdots & D(r-1) & D(r) & B(1) & \cdots & B(q-1) & B(q) \\
I_s & 0 & \cdots & 0 & 0 & 0 & \cdots & 0 & 0 & 0 & \cdots & 0 & 0 \\
0 & I_s & \cdots & 0 & 0 & 0 & 0 & 0 & 0 & \cdots & 0 & 0 & \\
\vdots & \vdots & \vdots & \vdots & \vdots & \vdots & \vdots & \vdots & \vdots & \vdots & \vdots & \vdots & \vdots \\
0 & 0 & \cdots & I_s & 0 & 0 & 0 & 0 & 0 & \cdots & 0 & 0 & \\
0 & 0 & \cdots & 0 & 0 & 0 & 0 & 0 & 0 & 0 & 0 & & \\
0 & 0 & \cdots & 0 & 0 & I_m & \cdots & 0 & 0 & 0 & \cdots & 0 & 0 \\
\vdots & \vdots & \vdots & \vdots & \vdots & \vdots & \vdots & \vdots & \vdots & \vdots & \vdots & \vdots & \vdots \\
0 & 0 & \cdots & 0 & 0 & 0 & I_m & 0 & 0 & \cdots & 0 & 0 & \\
0 & 0 & \cdots & 0 & 0 & 0 & \cdots & 0 & 0 & 0 & \cdots & 0 & 0 \\
0 & 0 & \cdots & 0 & 0 & 0 & \cdots & 0 & 0 & I_s & \cdots & 0 & 0 \\
\vdots & \vdots & \vdots & \vdots & \vdots & \vdots & \vdots & \vdots & \vdots & \vdots & \vdots & \vdots & \vdots \\
0 & 0 & \cdots & 0 & 0 & 0 & \cdots & 0 & 0 & 0 & \cdots & I_s & 0
\end{bmatrix}$$

$$K' = [I_s, 0, \ldots, 0, I_s, 0, \ldots, 0]$$

$$L' = [0, \ldots, I_m, 0, \ldots, 0, 0]$$

$$H = [-A(1), -A(2), \ldots, -A(p), D(1), \ldots, D(r), B(1), \ldots, B(q)].$$

Since without restriction of generality (see Section 3) always $K(0) = I$ and thus $A(0) = B(0)$ is assumed, obviously we also obtain,

$$
\begin{aligned}
x(t+1) &= Fx(t) + Lz(t) + K\varepsilon(t) \\
y(t) &= Hx(t) + \varepsilon(t),
\end{aligned}
\tag{1.2.27}
$$

which is just a reformulation of (1.1.7). Again the question arises whether this transformation is a "good" one [e.g., in the sense that $x(t)$ has dimension as small as possible]; this problem is discussed in Chapter 2. In fact, this construction is useful only to show that (1.2.27) is possible, and is not otherwise useful. For the stable case we indicate another construction below.

From now on, if the contrary is not stated explicitly, we assume that the transfer functions (k, l) are causal, in the sense that they have a convergent power series expansion in a neighbourhood of $z = 0$, and in addition satisfy

$$
(k(0), l(0)) = (I, 0). \tag{1.2.28}
$$

As has already been mentioned, $l(0) = 0$ can always be achieved by reindexing $z(t)$; and (as will be shown in the next section) if $\varepsilon(t)$ is unobserved, white noise $k(0) = I$ is no restriction of generality. For a given transfer function (k, l) an ARMAX (state space) system with this transfer function is called an ARMAX (*state-space*) *realization* of (k, l).

Partially summarizing the results of this section, we have:

Theorem 1.2.1

(i) Every ARMAX system (1.1.7) with $\det A(0) \neq 0$ *and every state-space system (1.1.22), (1.1.23) has a causal, rational transfer function.*

(ii) Conversely, every causal rational transfer function (k, l) *with* $(k(0), l(0)) = (I, 0)$ *has an ARMAX realization (1.1.7) with* $\det A(0) \neq 0$ *and a state-space realization (1.1.22), (1.1.23).*

(iii) If, in addition, (k, l) *is analytic in a disk containing the closed unit disk, then there exists a stable ARMAX realization [satisfying (1.2.3)] and a stable state-space realization [satisfying (1.2.24)]; conversely, every stable ARMAX or state-space system has a transfer function* (k, l) *which is analytic in a disk containing the closed unit disk.*

We now show that every state-space system (1.1.19), (1.1.20) can be transformed into a form (1.1.22), (1.1.23), called prediction error form. Let us assume that the stability condition (1.2.24) holds and that $z(t)$ are such that the steady-state solutions of (1.1.19) [and thus of (1.1.20)] exist:

$$x(t) = \sum_{j=1}^{\infty} F^{j-1}Lz(t-j) + \sum_{j=1}^{\infty} F^{j-1}\xi(t-j)$$

and thus

$$y(t) = \sum_{j=1}^{\infty} HF^{j-1}Lz(t-j) + \sum_{j=1}^{\infty} HF^{j-1}\xi(t-j) + \eta(t)$$

Define

$$w(t) = x(t) - \sum_{j=1}^{\infty} F^{j-1}Lz(t-j)$$

$$u(t) = Hw(t) + \eta(t)$$

and let $H_u(t)$ be the Hilbert space in $L^2(\Omega, \mathscr{F}, P)$ spanned by the elements, $u_i(j)$ $j \leq t$, of $u(j) = (u_i(j))$ $i = 1, \ldots, s$. Let $w(t|s)$ be the vector whose components are the orthogonal projections of the elements of $w(t)$ on $H_u(s)$. By the linearity of the projection, and since $(\xi(t)', \eta(t)')'$ is white noise, satisfying (1.1.21), we obtain from (1.1.19), (1.1.20):

$$w(t+1)|t) = Fw(t|t) = Fw(t|t-1) + F(w(t|t) - w(t|t-1))$$

$$u(t) = Hw(t|t-1) + \varepsilon(t),$$

where $\varepsilon(t)$ is defined by the last relation. Then the elements of $\varepsilon(t)$ are orthogonal to $H_u(t-1)$ since the elements of $Hw(t|t-1)$ are the orthogonal projections of the elements of $u(t)$ on $H_u(t-1)$. $H_u(t)$ is also spanned by $u_i(j)$, $j < t$, $i = 1, \ldots, s$ and by the elements of $\varepsilon(t)$, and therefore $w(t|t) - w(t|t-1)$ is a linear function of $\varepsilon(t)$ and

(1.2.29) $$w(t+1|t) = Fw(t|t-1) + K\varepsilon(t)$$

Put

$$x(t+1|t) = w(t+1|t) + \sum_{j=1}^{\infty} F^{j-1}Lz(t+1-j).$$

Then

(1.2.30) $$x(t+1|t) = Fx(t|t-1) + Lz(t) + K\varepsilon(t)$$

(1.2.31) $$y(t) = Hx(t|t-1) + \varepsilon(t)$$

(1.2.32) $$E\varepsilon(s)\varepsilon(t)' = \delta_{st}\Sigma$$

since

$$\sum_{j=1}^{\infty} F^{j-1}Lz(t+1-j) - F\sum_{j=1}^{\infty} F^{j-1}Lz(t-j) = Lz(t).$$

Equations (1.2.30), (1.2.31) are called the *prediction error forms* of (1.1.19), (1.1.20), since $\varepsilon(t)$ is the one-step-ahead prediction error of $y(t)$ given $y(s)$, $s < t$ and z(t), $t \in \mathbf{Z}$ (for the best linear least-squares prediction).From now on we consider only state-space systems of the form (1.1.22), (1.1.23).

This construction, *for the case where* (1.2.24) *holds*, may be used in relation to the ARMAX representation, (1.1.7), to obtain a simpler state-space form than that given above (1.2.27). We take the case $A(0) = I$ for brevity of notation, but the only change in the form of (1.2.27) given below if that is not true is to replace $A(j)$ in the F matrix by $A(0)^{-1}A(j)$. Put

$$y(t+j|t) = \sum_{u=j}^{\infty} K(u)\varepsilon(t+j-u) + \sum_{u=j}^{\infty} L(u)z(t+j-u).$$

Then

$$y(t+j|t) = y(t+j|t-1) + K(j)\varepsilon(t) + L(j)z(t), \tag{1.2.33}$$

as is easily seen. Also, putting $m = \max(p, q, r)$,

$$\sum_{0}^{p} A(j)y(t+m-j|t) = c(m),$$

where $c(m)$ is null for $m > q$, r, is $B(q)\varepsilon(t)$ for $m = q > r$, $= D(r)z(t)$ for $m = r > q$ and is $B(q)\varepsilon(t) + D(r)z(t)$ for $m = q = r$. This follows from the fact that, for example, $\sum A(j)K(p+k-j)$ is the coefficient of $\varepsilon(t+m-p-k)$ in $\sum A(j)y(t+m-j|t)$. Thus, using (1.2.33),

$$\begin{aligned} y(t+m|t) &= -\sum_{1}^{p} A(j)y(t+m-j|t) + c(m) \\ &= -\sum_{1}^{p} A(j)y(t+m-j|t-1) - \left\{\sum_{1}^{p} A(j)K(m-j)\right\}\varepsilon(t) \\ &\quad - \left\{\sum_{1}^{p} A(j)L(m-j)\right\}z(t) + c(m) \\ &= -\sum_{1}^{p} A(j)y(t+m-j|t-1) + K(m)\varepsilon(t) + L(m)z(t), \end{aligned}$$

since, for example,

$$\sum_1^p A(j)K(m-j) = -K(m), \quad m > q; \qquad = -K(m) + B(q), \quad m = q.$$

Thus we may take in (1.2.30), (1.2.31)

$$F = \begin{bmatrix} 0 & I_s & 0 & \cdots & 0 \\ 0 & 0 & I_s & \cdots & 0 \\ \vdots & \vdots & \vdots & \vdots & \vdots \\ 0 & 0 & 0 & \cdots & I_s \\ -A(m) & -A(m-1) & -A(m-2) & \cdots & -A(1) \end{bmatrix}, \tag{1.2.34}$$

$$K = \begin{bmatrix} K(1) \\ K(2) \\ \vdots \\ K(m) \end{bmatrix}, \qquad L = \begin{bmatrix} L(1) \\ L(2) \\ \vdots \\ L(m) \end{bmatrix}, \qquad H = [I_s, 0, \ldots, 0],$$

$$x(t|t-1)' = (y(t|t-1)', y(t+1|t-1)', \ldots, y(t+m-1|t-1)').$$

Of course, $A(j) = 0$, $j > p$. Again this may not be a best representation but appears often in the literature, especially for the autoregressive case.

We point out that the eigenvalues of the matrix F in (1.2.34) are precisely the zeros of $\det\{\sum_0^m A(j)z^{m-j}\}$, $A(0) = I_s$. Indeed, in $F - zI_{ms}$ we may eliminate the bottom right-hand block by adding $(A(1) + zI_s)$ by the second last block of rows to the last block of rows. Then add $A(2) + z\{A(1) + zI_s\}$ by the third last block of rows to the last block of rows. Continuing in this way all blocks in the last block of rows are made into the null matrix save that in the left-hand bottom corner, which is $-\sum_0^m A(j)z^{m-j}$, so that $\det(F - zI_{ms})$ is $(-)^{ms}\det\{\sum_0^m A(j)z^{m-j}\}$. Thus the eigenvalues of F are the reciprocals of the eigenvalues of $\det\{a(z)\} = \det\{\sum A(j)z^j\}$; that is, if $a(z)$ is stable, all eigenvalues of F are inside $|z| = 1$ and F is also stable. We state this as a theorem since it will be used later.

Theorem 1.2.2. *If $a(z) = \sum_0^m A(j)z^j$, $A(0) = I_s$, is stable, then F [see (1.2.34)] is also stable and indeed the eigenvalues of F are the solutions of* $\det\{z^m a(z^{-1})\} = 0$.

REMARK. Sometimes the condition that all eigenvalues of F lie in $|z| < 1$ is spoken of as asymptotic stability to differentiate that from the case where the eigenvalues are in $|z| \leq 1$.

Note that an analogous construction to obtain a prediction error form can be performed [without assuming (1.2.24)] starting from the initial value solution (on **N**) of (1.1.19):

$$x(t) = \sum_{j=1}^{t-1} F^{j-1}Lz(t-j) + \sum_{j=1}^{\infty} F^{j-1}\xi(t-j) + F^t x(0),$$

where $x(0)$, $z(0)$, and $\varepsilon(0)$ are the initial values. Thereby

$$w(t) = x(t) - \sum_{j=1}^{t-1} F^{j-1}Lz(t-j) + F^t x(0)$$

$$u(t) = Hw(t) + \eta(t)$$

and we project on the Hilbert space spanned by the elements of $u(s)$, $1 \leq s \leq t$. Since then the processes $w(t)$, $u(t)$ will not be stationary in general; the prediction errors $\varepsilon(t)$ will also not be stationary.

3. STATIONARY PROCESSES AND LINEAR SYSTEMS

To a good part models and methods in time-series analysis are based on stationarity. Here we review some of the relevant results from the linear theory of stationary processes. We assume that the reader is familiar with basic facts of this theory. Our account is far from being self-contained and the reader is referred to the books of Hannan (1970) and Rozanov (1967) for further details and proofs.

In this section, unless the contrary is explicitly stated, the word "stationary" without further specification is used for stationary in the wide sense. ARMA systems play an important role in modeling time series because of the fact that, on the one hand, every linearly regular, stationary process can be approximated with arbitrary accuracy by an ARMA process (and this approximation is quite "efficient," as it corresponds to the rational approximation of the spectral density) and, on the other hand, only a finite number of parameters is needed to describe (the second moments of) an ARMA process. Analogously, ARMAX systems are used to approximate linear (causal) infinite-dimensional systems. We discuss such approximations further in Chapter 7.

One of the main purposes in the use of stationary processes $y(t)$, $y(t) = (y_i(t))$ $i = 1,\ldots,s$, is prediction. Throughout we will be concerned only with best approximation in the least squares sense. Let $y(t+\tau|t)$ denote the best linear (least squares) prediction of $y(t+\tau)$, $\tau > 0$, from $y(s)$, $s \leq t$; that is, $y(t+\tau|t)$ is the (unique) random vector that minimizes $E\{(y(t+\tau) - \tilde{y})'(y(t+\tau) - \tilde{y})\}$ [or equivalently $E\{(y(t+\tau) - \tilde{y})(y(t+\tau) - \tilde{y})'\}$, where the order induced by positive definiteness of matrices is used] over vectors $\tilde{y}$ composed from all linear combinations of the $y_i(s)$, $s \leq t$, $i = 1,\ldots,s$ and their limits.

A stationary process $y(t)$ is called *linearly regular* if $Ey(t) = 0$ and if, for the best linear predictor, $y(t+\tau|t)$ of $y(t+\tau)$ $(\tau > 0)$ based on $y(s)$, $s \leq t$, the relation

$$\lim_{\tau \to \infty} y(t + \tau | t) = 0$$

holds, or equivalently if

$$S = \bigcap_{t \in \mathbf{Z}} H_y(t) = \{0\},$$

where $H_y(t)$ is the Hilbert space spanned by $\{y_i(t-j) | i = i, \ldots, s;\ j \geq 0\}$ in $L^2(\Omega, \mathscr{F}, P)$. A stationary process $y(t)$ is called *linearly singular* if

$$y(t + \tau | t) = y(t + \tau), \quad \text{a.s.}$$

holds [for one τ, $t(\tau > 0)$ and thus for all τ, t], or equivalently, if

$$S = H_y,$$

where H_y is the time domain of $y(t)$, that is, the Hilbert space spanned by $\{y_i(t) | i = 1, \ldots, s, t \in \mathbf{Z}\}$. The terms "linearly regular" and "linearly singular" are often replaced, respectively, by "linearly, purely nondeterministic" and "linearly, purely deterministic."

Now we have (see, e.g., Hannan, 1970, p. 137, or Rozanov, 1967, Sec. II.2.3)

Theorem 1.3.1. *(Wold decomposition) Every stationary process $y(t)$ can be represented in a unique way as*

$$y(t) = u(t) + v(t), \tag{1.3.1}$$

where $u(t)$ and $v(t)$ are obtainable as linear transformations of $y(t)$, where $H_u(t) \subset H_y(t)$, $H_v(t) \subset H_y(t)$, where $Eu(t)v(s)' = 0$ and where $u(t)$ is linearly regular and $v(t)$ is linearly singular. Furthermore, every linear regular process $u(t)$ can be represented (Wold representation) as

$$u(t) = \sum_{j=0}^{\infty} K(j)\varepsilon(t-j), \qquad \sum_{j=0}^{\infty} \|K(j)\|^2 < \infty, \tag{1.3.2}$$

where $H_u(t) = H_\varepsilon(t)$ and where $\varepsilon(t)$ is white noise. Thus $\varepsilon(t)$ are linear innovations of $y(t)$.

In particular we can choose $\varepsilon(t) = u(t) - u(t|t-1)$.

The most important linear singular processes are the harmonic processes, which have a spectral representation of the form

$$v(t) = \sum_{j=-h}^{h} z_j e^{i\omega_j t} = \sum_{j=0}^{h} (a_j \cos \omega_j t + b_j \sin \omega_j t), \tag{1.3.3}$$

where z_j are complex-valued random vectors satisfying $Ez_j^* z_j < \infty$, $Ez_j = 0$ for $\omega_j \neq 0$ and $Ez_j z_l^* = 0$, $j \neq l$ and, since $v(t)$ is real $\omega_j = -\omega_{-j}$, $\omega_0 = 0$ and $\bar{z}_j = z_{-j}$ and where $a_j = z_j + \bar{z}_j$, $b_j = -i(z_j - \bar{z}_j)$. These processes can be obtained as solutions of homogeneous VDEs $a(z)v(t) = 0$, where $\det a(z)$ has only (suitably located) zeros of unit modulus.

In any situation where there is only a part of one realization considered, a_j and b_j could be considered as nonstochastic (of course, then the process in general will no longer be stationary) and the corresponding component will be removed, together with the trend from the original data. Therefore we restrict ourselves now to the linear regular component.

We may consider (1.3.2) as an in general not finite-dimensonal (but causal and stable) linear system with white noise inputs and with transfer function

$$k(z) = \sum_{j=0}^{\infty} K(j) z^j.$$

Here $k(z)$ is analytic within the unit disk and $k(e^{i\omega})$ exists as a limit of $\sum_{j=0}^{n} K(j) e^{i\omega j}$ in the mean square sense or as the radial limit of $k(z)$, $|z| < 1$:

$$\lim_{\rho \uparrow 1} \sum_{j=0}^{\infty} K(j) \rho^j e^{i\omega j}.$$

The system (1.3.2) could be written as an infinite-dimensional state-space system

$$x(t+1) = Fx(t) + K\varepsilon(t) \tag{1.3.4}$$

$$y(t) = Hx(t) + \varepsilon(t), \tag{1.3.5}$$

where

$$x(t) = \begin{pmatrix} y(t|t-1) \\ y(t+1|t-1) \\ y(t+2|t-1) \\ \vdots \end{pmatrix} = \begin{pmatrix} K(1) & K(2) & K(3) & \cdots \\ K(2) & K(3) & K(4) & \cdots \\ \cdots & \cdots & \cdots & \cdots \\ \cdots & \cdots & \cdots & \cdots \end{pmatrix} \begin{pmatrix} \varepsilon(t-1) \\ \varepsilon(t-2) \\ \vdots \\ \vdots \end{pmatrix},$$

$$F = \begin{pmatrix} 0 & I & 0 & \cdots & \cdots \\ 0 & 0 & I & 0 & \cdots \\ 0 & 0 & 0 & I & \cdots \\ \cdots & \cdots & \cdots & \cdots & \cdots \end{pmatrix}; \quad K = \begin{pmatrix} K(1) \\ K(2) \\ \vdots \\ \vdots \end{pmatrix}; \quad H = (I, 0, 0, \ldots).$$

The spectral density of a linear regular process (1.3.2) always exists and is given by

(1.3.6) $$f_y(\omega) = (2\pi)^{-1} k(e^{i\omega}) \Sigma k^*(e^{i\omega}),$$

where $\Sigma = E\varepsilon(t)\varepsilon'(t)$ is the covariance matrix of $\varepsilon(t)$. The problem of finding k, Σ such that (1.3.6) holds for given f_y is called the *spectral factorization problem.* The spectral density of a linear regular process has, almost everywhere, constant rank. A complete characterization of the spectral density of a regular process is given in Rozanov (1967).

A stationary process $y(t)$ is linearly regular if and only if it has a spectral density f_y which has constant rank almost everywhere (a.e.) and where f_y admits a factorization (k, Σ) [in (1.3.6)] with the properties

(1.3.7) $$k(z) = \sum_{j=0}^{\infty} K(j) z^j; \qquad \sum_{j=0}^{\infty} \|K(j)\|^2 < \infty$$

[where $k(e^{i\omega})$ is a radial limit in the sense described above].

Beginning with f_y, without restriction of generality one can always assume, *as we do throughout*, that

(1.3.8) $$k(0) = I.$$

Thus the assumption that $y(t)$ and $\varepsilon(t)$ have the same dimension is no restriction of generality.

From Rozanov (1967, p. 60) we see that in addition Σ is maximal (with respect to the order induced by positive definiteness) among all factorizations of a given f_y satisfying (1.3.7), (1.3.8) if and only if such a factorization corresponds to the Wold representation (1.3.2). Then, of course, the $\varepsilon(t)$ are linear innovations of $y(t)$ and [by (1.3.8)] even the one-step-ahead prediction errors of the best linear prediction of $y(t)$ given $y(s)$, $s < t$.

Unless the contrary is stated explicitly, *we will also assume throughout that f_y has full rank in the sense that*

(1.3.9) $$f_y \text{ has rank } s, \text{ a.e.}$$

holds. Clearly [under (1.3.8)] this is equivalent to $\Sigma > 0$. This assumption is a proper restriction of generality. However, in practical situations singular Σ matrices are unlikely to occur and also they would be detected from the data. It can easily be shown that under (1.3.8) and (1.3.9), the factorization (k, Σ) corresponding to the Wold representation (1.3.2) is unique for given f_y.

For the full-rank case we have (see, e.g., Hannan, 1970, p. 162; Rozanov, 1967, pp. 73–75)

Theorem 1.3.2. *A stationary process that has a spectral density f_y of full rank is linearly regular if and only if*

(1.3.10)
$$\int_{-\pi}^{\pi} \log \det f_y(\omega)\, d\omega > -\infty$$

holds. In this case the factorization (k, Σ) *corresponding to the Wold representation (1.3.2) satisfies the relation*

(1.3.11)
$$\det \Sigma = \exp\left\{(2\pi)^{-1} \int_{-\pi}^{\pi} \log \det 2\pi f_y(\omega)\, d\omega\right\}.$$

More generally, if $y(t)$ is a (not necessarily linearly regular) stationary process with spectral distribution function F_y, then F_y may be uniquely decomposed as $F^{(1)} + F^{(2)}$, where $F^{(1)}$ is the absolutely continuous part, $F^{(1)}(\omega) = \int_{-\pi}^{\omega} f^{(1)}(\omega)\, d\omega$, of F_y and $F^{(2)}$ is the singular part of F_y (consisting of a continuous but not absolutely continuous part and a discontinuous part). If

$$\det \Sigma = \exp\left\{(2\pi)^{-1} \int_{-\pi}^{\pi} \log \det 2\pi f^{(1)}(\omega)\, d\omega\right\} > 0$$

holds, then $f^{(1)}$ is the spectral density f_u of the linearly regular process $u(t)$ in the Wold decomposition (1.3.1) and $F^{(2)}$ is the spectral distribution function of the linearly singular part $v(t)$ (see, e.g., Hannan, 1970, p. 162). Note, however, that there are linearly singular processes which have a spectral density.

Consider an ARMA system

(1.3.12)
$$a(z)y(t) = b(z)\varepsilon(t)$$

with transfer function $k = a^{-1}b$ and with $\Sigma = E\varepsilon(t)\varepsilon(t)'$. Since $\varepsilon(t)$ is white noise, we can characterize its solutions in more detail.

(i) From the preceding section we know that if $\det a(z) \neq 0$, $|z| = 1$, there is a stationary solution of the form

(1.3.13)
$$y(t) = \sum_{j=-\infty}^{\infty} K(j)\varepsilon(t - j),$$

where

$$\sum_{j=-\infty}^{\infty} K(j)z^j$$

is the Laurent series expansion of $k(z)$ in an annulus containing the unit circle. This solution is unique among the stationary ones. If, in addition, $\det a(z) \neq 0$ $|z| \leq 1$ holds, then $K(j) = 0, j < 0$.

(ii) Let $k = c^{-1}N$, where c is the least common denominator of the entries of k. If $c(z) \neq 0\ |z| = 1$, stationary solutions exist; exactly one of them is given by (1.3.13), where the Laurent series expansion of k is obtained from the Laurent series expansion of c^{-1} in an annulus containing the unit circle. It can be shown that every other stationary solution of (1.3.12) can be written as the sum of the solution (1.3.13) and an orthogonal (stationary) harmonic process $v(t)$ satisfying $a(z)v(t) = 0$.

(iii) As can be shown, if $\Sigma > 0$ and if $c(z) = 0$ for some $|z| = 1$, no stationary solution of (1.3.12) exists.

(iv) In the case $\det \Sigma = 0$, a stationary solution of (1.3.12) exists if and only if

$$\int \operatorname{tr}\{k(e^{i\omega})\Sigma k(e^{i\omega})^*\}\, d\omega < \infty$$

(where tr denotes "trace of"). Also, in this case exactly one solution is of the form (1.3.13); however, the $K(j)$ are not necessarily defined from the Laurent series expansion of k.

Every solution of an ARMA system of the form (1.3.13) (i.e., every stationary solution which has a spectral density) is called an ARMA process. An ARMA process has a spectral density of the form

$$f_y(\omega) = (2\pi)^{-1}k(e^{i\omega})\Sigma k(e^{i\omega})^*; \qquad k = a^{-1}b, \tag{1.3.14}$$

where f_y is a rational matrix in the sense that each of its entries is the quotient of trigonometric polynomials.

Frequently, in addition to the stability assumption (1.2.3) $\det a(z) \neq 0$ $|z| \leq 1$, the miniphase assumption

$$\det b(z) \neq 0, \qquad |z| < 1 \tag{1.3.15}$$

is imposed. If for an ARMA system (1.3.12) the stability and the miniphase assumption hold, then (see, e.g., Hannan, 1970, Chap. III) $\varepsilon(t)$ are linear innovations of $y(t)$ and thus $a^{-1}b$ corresponds to the Wold representation. Conversely, if in addition only relatively left prime pairs (a, b) (see Section 2.2) are considered, and if either (1.2.3) or (1.3.15) is violated, then $a^{-1}b$ does not correspond to the Wold representation.

Clearly, the stability assumption (1.2.3) implies that $k(z) = a^{-1}(z)b(z)$ is analytic within a circle containing the closed unit disk and the miniphase assumption (1.3.15) implies that $\det k(z) \neq 0\ |z| < 1$. Conversely, if only relatively left prime pairs (a, b) are considered, then k being analytic within a circle containing the unit circle implies (1.2.3) and $\det k(z) \neq 0$, $|z| < 1$ implies (1.3.15) (see Lemma 2.4.4).

Theorem 1.3.3. *Any rational and a.e. nonsingular spectral density matrix f may be uniquely factorized as in (1.3.14), where $k(z)$ is rational in z, analytic within a circle containing the closed unit disk,* $\det k(z) \neq 0$ $|z| < 1$ *and* $k(0) = I$ *and where* $\Sigma > 0$.

Proof. Note that f is a spectral density if and only if f is integrable, $f(\omega) \geq 0$, a.e. and $f(\omega) = f(\omega)^*$; since we are only interested in real (rather than in complex) random variables, we have, in addition,

$$f(\omega) = f(-\omega)'.$$

Consider first the case $s = 1$ and put $f(\omega) = f_1(e^{i\omega})/f_2(e^{i\omega})$, where f_1 and f_2 are trigonometric polynomials, which, without restriction of generality, are assumed to have no common zero and to be nonnegative. Then it must be true that $f_2(e^{i\omega}) > 0$, $\omega \in [-\pi, \pi]$ since otherwise f would not be integrable. Consider, for $\gamma(j)$ real,

$$f_1(z) = \sum_{j=-n}^{n} \gamma(j) z^j.$$

Since f and thus also f_1 (and f_2) are even real functions we have $\gamma(j) = \gamma(-j)$. If $f_1(z_0) = 0$ for $z_0 \neq 0$ [$f_1(0) = 0$ would mean $f_1 = 0$], then

$$f_1(\bar{z}_0^{-1}) = \sum \gamma(j) \bar{z}_0^{-j} = \sum \gamma(j) \bar{z}_0^{j} = f_1(\bar{z}_0) = \overline{f_1(z_0)} = 0$$

and thus $\bar{z}_0^{-1}$ is also a zero of f_1. Now let there be r zeros z_j with $|z_j| > 1$. Then

$$f_1(e^{i\omega}) = \gamma(n) e^{-in\omega} \prod_{1}^{r} (e^{i\omega} - z_j)(e^{i\omega} - \bar{z}_j^{-1}) \prod_{j=1}^{2s} (e^{i\omega} - e^{i\omega_j}), \tag{1.3.16}$$

where the last factor accounts for the $2s = 2n - 2r$ zeros of unit modulus. Clearly, the zeros of modulus 1 must be of even multiplicity. Let us number s of the ω_j so that $z_{r+k} = e^{i\omega_k}$, $k = 1, \dots, s$, each of the remaining s of the ω_j being equal to one of these first s. Inserting $(e^{i\omega} - \bar{z}_j^{-1}) = e^{i\omega} \bar{z}_j^{-1} (\bar{z}_j - e^{-i\omega})$ into (1.3.16) it follows from the nonnegativity of f_1 that $(-1)^n \gamma(n) \prod_{i=1}^{n} \bar{z}_i^{-1}$ must be nonnegative and thus

$$b(e^{i\omega}) = \left\{ (-1)^n \gamma(n) \prod_{j=1}^{n} \bar{z}_i^{-1} \right\}^{1/2} \prod_{j=1}^{n} (e^{i\omega} - z_j)$$

is a polynomial having no zeros inside the unit circle satisfying $f_1 = b\bar{b}$. Since for every z_j, $\bar{z}_j$ is a zero, the coefficients of b are real. Using the same construction for f_2 proves the theorem for $s = 1$. This factorization is clearly unique.

Consider now the case $s > 1$. We may find a rational matrix $H(z)$ so that $H(e^{i\omega})f(\omega)H(e^{i\omega})^*$ is diagonal. Indeed, let H_1 have zeros everywhere except in the main diagonal, where it has units, and in the first column, where it has $-f_{j1}(\omega)/f_{11}(\omega)$ as the $(j, 1)$ entry, $j = 2, \ldots, s$. Then $H_1 f$ has zeros in its first column except for the $(1, 1)$ entry and $H_1 f H_1^*$ evidently has the first row and column consisting of zeros save for the $(1, 1)$ entry. The result now follows by induction. Let c be a least common denominator polynomial of H. Then $c(e^{i\omega})c(e^{-i\omega}) = b(e^{i\omega})b(e^{-i\omega})$, where b is a polynomial with no zeros inside the unit circle, from the case $s = 1$. Thus $(c(z)/b(z))H(z)$ is analytic within the unit circle. Moreover, HfH^*, which is diagonal, can be written as VV^*, where V is diagonal with rational elements analytic within the unit circle. Hence $f = A_1 A_1^*$, where $A_1 = (b^{-1}cH)^{-1}V$ and A_1 is analytic within the unit circle; indeed, A_1 must be analytic within a circle containing the unit circle for $A_1(f) = A_1(A_1 A_1^*)$ is integrable and this could not be so if any diagonal element of A_1 had a pole on the unit circle. If $\det A_1(z_1) = 0$ for $|z_1| < 1$, there is a constant unitary matrix Q_1 so that $A_1(z_1)Q_1$ has its last column equal to zero and thus the last column of $A_1(z)Q_1$ is divisible by $(z - z_1)$. Multiply this last column by $(1 - \bar{z}_1 z)/(z - z_1)$. This factor is of unit modulus on the unit circle, as is easily checked, and replaces the zero at z_1 by one of $\bar{z}_1^{-1}$. Thus there is a rational matrix R_1 which is unitary for $|z| = 1$, so that $A_1(z)R_1(z) = A_2(z)$ has a zero z_1 of $\det A_1$ changed to $\bar{z}_1^{-1}$. Continuing, we find an R so that $A(z) = A_1(z)R(z)$ when $R(z)$ is unitary on the unit circle and $A(z)$ is analytic within a circle containing the unit circle and has a determinant that is never zero inside the unit circle.

It remains to show the uniqueness of the factorization. If $f = AA^* = BB^*$ are two such factorizations, then $|\det A(0)|^2 = |\det B(0)|^2$. Let $U = B^{-1}A$; then $UU^* = I$ and $|\det U(0)| = 1$. Then from the maximum modulus principle, $\det U$ is a nonzero constant. Since $U^*(e^{i\omega}) = U(e^{i\omega})^{-1}$ and since $U(z)^{-1}$ is analytic within the unit circle (because $\det U$ is a nonzero constant), then U^* must contain only nonnegative powers of $e^{i\omega}$, since U also has this property, U is a constant unitary matrix. Thus any two factorizations $f = AA^* = BB^*$ with A, B analytic within the unit circle, $\det A \neq 0$, $\det B \neq 0$, $|z| < 1$ differ only by right multiplication with a constant unitary factor U. If we put $\Sigma = A(0)UU^*A(0)$, then Σ is independent of U and the factorization $f(\omega) = (2\pi)^{-1}k(e^{i\omega})\Sigma k(e^{i\omega})^*$, $k(z) = (2\pi)^{1/2}A(z)A(0)^{-1}$ is unique.

REMARK. The condition $f(\omega) > 0$, a.e. in Theorem 1.3.3 can be replaced by $f(\omega) \geq 0$, a.e. (see Hannan, 1970, pp. 66, 67; Rozanov, 1967, Sec. I.9). Then a rational f still has a factorization where k is rational, analytic within a circle containing the closed unit disk, $\det k(z) \neq 0$ $z| < 1$ and $k(0) = I$; however, k is nonunique, in general; Σ (which is unique) has rank equal to the (a.e. constant) rank of f.

As a consequence of the theorem and the remark above we see that the assumptions (1.2.3), (1.3.15) and $K(0) = I$ imply no restriction of generality as far as the rational spectral density f_y is concerned. We also see that every ARMA process is linearly regular.

In the full-rank case, (1.2.3), (1.3.15) and $K(0) = I$ may be considered as suitable norming conditions because they imply uniqueness of the factorization (k, Σ). In addition, since this factorization corresponds to the Wold representation, it can be used for prediction. Sometimes this factorization is called the *canonical factorization.*

Of course, $k(0) = I$ implies that $A(0) = B(0)$. We may, for example, put $A(0) = B(0) = I$, which is natural, for example, for prediction, but other normalizations are also used.

As mentioned earlier, every ARMA process is linearly regular. On the other hand, every linearly regular stationary process $y(t)$ can be approximated with arbitrary accuracy by an ARMA process $y_N(t)$ such that

$$\sup_{t \in \mathbf{Z}} E(y(t) - y_N(t))'(y(t) - y_N(t)) \tag{1.3.17}$$

can be made arbitrarily small. This is an immediate consequence of the Wold representation (1.3.2), taking $y_N(t) = \sum_{j=0}^{N} K(j)\varepsilon(t-j)$. The problems of approximating general linearly regular processes by ARMA processes are discussed in detail in Chapter 7.

Let us assume that $(y(t'), z(t)')$ is stationary and let $y(t|z(s), s \in \mathbf{Z})$ denote the vector whose elements are the orthogonal projections of the elements of $y(t)$ on H_z, the time domain of $z(t)$. Thus we have a representation

$$y(t|z(s), s \in \mathbf{Z}) = \lim_{N\to\infty} \sum_{j=-N}^{N} L^{(N)}(j) z(t-j) \tag{1.3.18}$$

and (1.3.18) is the *best linear (least-squares) filter* for $y(t)$ based on $z(t)$. Because of stationarity, the $L_j^{(N)}$ in (1.3.18) do not depend on t. Again we assume that the linearly singular components have already been extracted, so that $(y(t)', z(t)')$ is linearly regular.

In addition, we assume that $z(t)$ *causes* $y(t)$ in the sense that the best linear filter (1.3.18) satisfies $L^{(N)}(j) = 0, j \leq 0$; or in other words, that the best linear filters for $y(t)$ based on $(z(t)|t \in \mathbf{Z})$ and on $(z(s)|s < t)$ are identical. Then the (in general, not finite-dimensional) system

$$y(t) = \lim_{N\to\infty} \sum_{j=1}^{N} L^{(N)}(j)z(t-j) + u(t), \tag{1.3.19}$$

where $Ez(s)u(t)' = 0$ holds, can be approximated by an ARMAX system (a, b, d) such that (1.3.17) can be made arbitrarily small, where

$$y_N(t) = a^{-1}(z)d(z)z(t) + a^{-1}(z)b(z)\varepsilon(t). \tag{1.3.20}$$

The transfer function $l = a^{-1}d$, relating observed inputs to outputs, in (1.3.20), is easier to obtain from the second moments of the observed processes than $k = a^{-1}b$. In particular for the ARMAX case we have: If $(y(t)', z(t)')'$ is stationary and has a spectral density

$$f = \begin{pmatrix} f_y & f_{yz} \\ f_{zy} & f_z \end{pmatrix}$$

and if $f_z(\omega)$ is nonsingular, a.e., then l is given by

$$l(e^{i\omega}) = f_{yz}(\omega)f_z(\omega)^{-1}. \tag{1.3.21}$$

Since l is rational this condition can be weakened, as can easily be seen, to:

(1.3.22) *The spectral distribution function F_z of $z(t)$ has an absolutely continuous component f_z that is nonsingular on a nonempty open subset of $[-\pi, \pi]$.*

Conditions of this kind which guarantee that l is uniquely determined from the second moments of the observed processes $y(t)$, $z(t)$ are called *persistently exciting conditions.*

In the ARMAX case, most results can be shown if only asymptotic stationarity of $z(t)$ is assumed, that is, that

$$\lim \frac{1}{T} \sum_{t=1}^{T} z(t)z(t+s)' = \Gamma_z(s), \quad \text{a.e.} \tag{1.3.23}$$

exists (for all s). For $\Gamma_z(s)$, a spectral distribution function, or, if it exists, a spectral density function, can be defined. If $z(t)$ is asymptotically stationary, and if the stability condition (1.2.3) holds, then $y(t)$ is also asymptotic stationary and a condition analogous to (1.3.22) implies that $l(z)$ is uniquely determined from the second moments of the observed processes (see Section 4.1 for more details).

Of course, once l is uniquely determined, k is obtained (assuming that f_z exists) from the spectral density of $y(t) - l(z)z(t)$, that is, from $f_y(\omega) - l(e^{i\omega})f_z(\omega)l(e^{i\omega})^* = f_y(\omega) - f_{yz}(\omega)l(e^{i\omega})^*$ by the spectral factorization described before, which is unique under the assumptions $K(0) = I, \Sigma > 0$, (1.2.3) and (1.3.15). These assumptions together with a persistent exciting condition as, for example, (1.3.22) are called the *standard assumptions.* They guarantee that (k, l, Σ) is uniquely determined from the second moments of the observed processes $y(t)$, $z(t)$.

If $z(t)$ has a spectral density, f_z, and f_z is also rational, then clearly f is rational and thus $(y(t), z(t))$ may be modeled as the output of an ARMA system which of course has more parameters than the original ARMAX system. On the other hand, ARMA modeling of the observed process $(y(t), z(t))$ may be preferred if we want a symmetric treatment of all observed variables, for example, if we are not sure a priori whether there is causality from $z(t)$ to $y(t)$.

In many applications time series show apparent nonstationarity features. Then often the observations are transformed in order to achieve stationarity. If the underlying model is of the form

$$x(t) = m(t) + y(t) \tag{1.3.24}$$

where $m(t) = Ex(t)$ and $y(t)$ is a mean-zero, stationary process, then often a trend regression is performed. In many cases, then $m(t)$ is parameterized by a finite number of parameters; for example, in the case of polynomial trends, linear least-squares methods can be applied (see, e.g., Anderson, 1971, Chap. III; Hannan, 1970, Chap. VIII). In other cases nonlinear least squares have been used (Hannan, 1971a; Jennrich, 1969). Another common transformation to achieve stationarity is repeated differencing (see, e.g., Box and Jenkins, 1976). A process $x(t)$ is called integrated of order d, if d is the smallest integer such that $(1 - z)^d x(t)$ is stationary. If $(1 - z)^d x(t)$ is an ARMA process, $x(t)$ is called an ARIMA process.

4. STRICTLY STATIONARY PROCESSES

In this short section some facts from the theory of strictly stationary processes, which will be used later, are reported. Proofs will not be given in general and the reader is referred to standard works, as, for example, Rozanov (1967) and Ibragimov and Linnik (1971).

The concept of strict stationarity is important, first since it seems to be physically meaningful for mechanisms where the stochastic laws are not changing in time, and second, because most statistical results do require more than wide-sense stationarity. Some additional results are given in Section 4.1.

In this section, the word "stationary" is always used for strictly stationary processes. Unless the contrary is stated explicitly, we in addition assume that the second moments exist (and thus strict stationarity implies wide-sense stationarity). Let $y(t)$ be a random process, where $y(t)$ is a vector of random variables $y_i(t)$ over probability space $(\Omega, \mathscr{F}, P)$. In most cases, without restriction of generality, we consider the case where $\mathscr{F}$ is the σ-algebra generated by $y(t)$, that is, $\mathscr{F}$ is the smallest σ-algebra generated by all events of the form

$$A = \{\omega \in \Omega | (y_{j_1}(t_1) \cdots y_{j_k}(t_k) \in B\}, \tag{1.4.1}$$

where B is a Borel set in $\mathbf{R}^k$. In this case sometimes we call $\mathscr{F}$, $\mathscr{F}_\infty$. It is convenient to assume that $\mathscr{F}$ is complete, that is, contains all sets that differ from a set in $\mathscr{F}$ by a set contained in a set of probability measure zero. Equally, we may define $\mathscr{F}_t$ as the (completed) σ-algebra generated by $y(s)$, $s \le t$. Then $\mathscr{F}_t \subset \mathscr{F}_{t+1} \subset \mathscr{F}_\infty$. Put $\mathscr{F}_{-\infty} = \bigcap_{-\infty}^{\infty} \mathscr{F}_t$. Then $\mathscr{F}_{-\infty}$ is the σ-algebra of events determined "in the far past."

The time translation $y(t) \to y(t+1)$ induces an equivalent transformation in $\mathscr{F}_\infty$, namely that defined for A by $A \to TA$ where $TA = [\omega | \{y_{j_1}(t_1 + 1), \ldots, y_{j_k}(t_k + 1)\} \varepsilon B]$. The σ-algebra $\mathscr{F}_\infty$ contains as a sub-σ-algebra, the *σ-algebra $\mathscr{I}$ of all invariant events* (i.e., all events A such that A and TA differ by a set of probability measure zero). [Since $y(t)$ is stationary, A and TA have the same measure in any case.] If $\mathscr{I}$ is trivial (i.e., contains only sets of probability measure 0 or 1), then $y(t)$ is said to be *ergodic* (or metrically transitive). In the situation studied in this book, where only one realization, or part of one realization, of the process is ever seen, it is costless to assume that a stationary process $y(t)$ is ergodic, for that merely says that we hold as fixed these aspects of the stochastic structure that are fixed over the one realization we shall ever see.

For a general strictly stationary process we have (compare, e.g., Rozanov, 1967, p. 157)

Theorem 1.4.1

(i) If $y(t)$ has finite first moments (i.e., $E|y(t)| < \infty$), then

$$\lim T^{-1} \sum_{t=1}^{T} y(t) = E(y(0)|\mathscr{I}), \quad a.s. \tag{1.4.2}$$

(ii) If $y(t)$ has finite second moments [i.e., $Ey(t)'y(t) < \infty$], then (for $s \ge 0$)

$$\lim T^{-1} \sum_{t=1}^{T-s} y(t+s)y(t)' = E(y(s)y(0)'|\mathscr{I}), \quad a.s. \tag{1.4.3}$$

(iii) If $y(t)$ is ergodic, then

$$E(y(0)|\mathscr{I}) = Ey(0); \qquad E(y(s)y(0)'|\mathscr{I}) = Ey(s)y(0)'. \tag{1.4.4}$$

The statement of Theorem 1.4.1 can be generalized to hold for the time averages corresponding to every $\mathscr{F}_\infty$-measurable and integrable function.

Of course, $\mathscr{I} \subset \mathscr{F}_{-\infty}$; there are cases where $\mathscr{I}$ is trivial and $\mathscr{F}_{-\infty}$ is not. If $\mathscr{F}_{-\infty}$ is trivial, then $y(t)$ is said to be *regular*. By what was said before every regular process is ergodic. Let $\mathscr{H}_y(t)$ denote the subspace of $L^2(\Omega, \mathscr{F}, P)$ consisting of all zero mean $\mathscr{F}_t$-measurable random variables. Then regularity obviously equivalent to

$$\bigcap_{t \in \mathbf{Z}} \mathscr{H}_y(t) = \{0\}$$

and thus since $\mathscr{H}_y(t) \supseteq H_y(t)$, every regular process with $Ey(t) = 0$ is linearly regular (see Section 3).

The best predictor (in the least-squares sense) of $y(t)$ given $y(s)$, $s < t$ is the conditional expectation $E(y(t)|\mathscr{F}_{t-1})$.

Indeed, if α is measurable $\mathscr{F}_{t-1}$, then

$$\begin{aligned} E[\{y(t) - \alpha\}\{y(t) - \alpha\}'] &= E[\{y(t) - E(y(t)|\mathscr{F}_{t-1})\}\{y(t) - E(y(t)|\mathscr{F}_{t-1})\}'] \\ &\quad + E[\{\alpha - E(y(t)|\mathscr{F}_{t-1})\}\{\alpha - E(y(t)|\mathscr{F}_{t-1})\}'] \end{aligned}$$

since for β measurable $\mathscr{F}_{t-1}$ [take β as $E(y(t)|\mathscr{F}_{t-1}) - \alpha$]

$$E[\{y(t) - E(y(t)|\mathscr{F}_{t-1})\}\beta'] = E[E[\{y(t) - E(y(t)|\mathscr{F}_{t-1})]\beta'|\mathscr{F}_{t-1}] = 0.$$

If $y(t)$ is regular and has zero mean, then

$$E(y(t)|\mathscr{F}_{-\infty}) = 0. \tag{1.4.5}$$

Indeed, $x_j = E\{y(t)|\mathscr{F}_{t-j}\}$ is measurable $\mathscr{G}_j \stackrel{d}{=} \mathscr{F}_{t-j}$. Then $\mathscr{G}_j \supset \mathscr{G}_{j+1}$ and $E\{x_j|\mathscr{G}_{j+1}\} = x_{j+1}$. Thus $(x_j, \mathscr{G}_j)$ is a reversed martingale and thus (Neveu, 1975, p. 119), $y(t)$ being integrable,

$$\lim_{\tau \to \infty} E(y(t)|\mathscr{F}_{t-\tau}) = 0 \quad \text{a.s.} \tag{1.4.6}$$

and in the L_1 sense. If $y(t)$ is square integrable, convergence also holds in the mean-squares sense.

Here linear systems are considered. A main purpose for the construction of such models is linear prediction. However, many stationary phenomena are highly nonlinear in nature, so that a linear prediction procedure will not be a good one. This suggests that the use of a linear system in prediction is fully appropriate only if the best linear predictor is the best predictor (both in the least-squares sense) or something near to that is true.

The statement that the best linear predictor equals the best predictor is equivalent to

$$E\{\varepsilon(t)|\mathscr{F}_{t-1}\} = 0, \tag{1.4.7}$$

where $\varepsilon(t) = y(t) - y(t|t-1)$ are the errors of the linear prediction. Indeed, if the condition holds, then $\varepsilon(t) = y(t) - E(y(t)|\mathscr{F}_{t-1})$, a.s., while if (1.4.5) holds, then $E(y(t) - y(t|t-1)|\mathscr{F}_{t-1}) = 0$, so that $E\{y(t)|\mathscr{F}_{t-1}\} = y(t|t-1)$. Thus we have the following simple result.

Theorem 1.4.2. *The necessary and sufficient condition for the best linear predictor of the stationary process $y(t)$, with finite mean square, to be the best predictor is that the linear prediction errors $\varepsilon(t)$ should satisfy (1.4.7).*

Of course, if the $\varepsilon(t)$ are Gaussian, they are independent because of their covariance structure and hence satisfy (1.4.7). It is difficult to find conditions on $y(t)$ that would justify (1.4.7). It is instead the (often not unreasonable) assumption that linear prediction is fully adequate for $y(t)$ that justifies this condition.

If (1.4.5) holds, the stationary process might be said to be purely nondeterministic. A regular process is necessarily purely nondeterministic but also is such that if $w(t)$ is any new process, formed by translating the arguments of some integrable function of the realization to time zero, then necessarily $w(t)$ is also regular.

If (1.4.5) holds, we also have the following decomposition. Put

$$u(t, t-k) = E\{y(t)|\mathscr{F}_{t-k}\} - E\{y(t)|\mathscr{F}_{t-k-1}\}. \tag{1.4.8}$$

Then

$$y(t) = \sum_{0}^{\infty} u(t, t-k). \tag{1.4.9}$$

Indeed, the sum to n terms on the right is $y(t) - E\{y(t)|\mathscr{F}_{t-n}\}$, so that the series converges to $y(t)$ by (1.4.6). Moreover, if $y(t)$ is square integrable and, say, $j > k$, then from (1.4.8)

$$E\{u(t, t-j)u(t, t-k)'\} = E[u(t|t-j)E\{u(t|t-k)'|\mathscr{F}_{t-k-1}\}] = 0,$$

so that the series in (1.4.9) converges also in mean square since the summands are orthogonal. The importance of (1.4.9) arises from the fact that $u(t|t-k)$, for fixed k, are martingale differences with respect to the σ-algebras $\mathscr{F}_{t-k}$.

CHAPTER 2

Realization and Parameterization of Linear Dynamic Systems

In this chapter we are concerned with the structure of the relation between the external behavior of systems (described by their transfer functions) and the corresponding ARMAX or state-space realizations and with the parameterization of these realizations. These problems concern sets of systems rather than single systems. A number of complications in this context arise only in the multioutput ($s > 1$) case and the case of scalar outputs is significantly simpler. Section 1 gives a general formulation of the problems of realization and parameterization. Sections 2 and 3 deal with the characterization of observationally equivalent ARMAX and state-space realizations, respectively. In Section 4 properties of rational transfer functions and of the associated Hankel matrices are investigated. In the remaining sections different realizations and parameterizations for ARMAX and state-space systems are considered: in Section 5 canonical realizations and their parameterization, in Section 6 the "overlapping" parameterization of the set of all transfer functions of order n as a manifold, and in Section 7 systems with prescribed column degrees. In Section 8 the stable miniphase case is investigated.

1. REALIZATION AND PARAMETERIZATION: GENERAL ASPECTS

Under our standard assumptions the observed processes $y(t)$, $z(t)$ or their second moments uniquely determine the transfer functions (k, l) and thus there is a one-to-one relation between the second moments of the observed processes and (k, l, Σ). In this sense (k, l, Σ) represents the *external behavior* and the *noise characteristics* of the system considered.

As has been stated already, the problem here is to investigate for certain sets of systems, the relation between transfer functions (k, l) and the corresponding ARMAX systems (a, b, d) or the corresponding state-space systems (F, H, K, L). For a complete analysis the relation between (k, l, Σ) and (a, b, d, Σ) or (F, H, K, L, Σ) has to be investigated. However, as far as Σ is concerned, no problems arise; Σ is parameterized by the vector $\sigma(\Sigma)$ of its on- and above-diagonal elements (arranged in a certain order). This maps the set of all symmetric positive definite $s \times s$ matrices homeomorphically onto an open subset in $\mathbf{R}^{s(s+1)/2}$. Indeed, for an arbitrary $\Sigma_0 > 0$, there is an $\varepsilon > 0$, such that all Σ, for which

$$\sum_{i,j:i \leq j} (\sigma_{ij} - \sigma_{ij}^0)^2 < \varepsilon; \qquad \Sigma_0 = (\sigma_{ij}^0), \qquad \Sigma = (\sigma_{ij})$$

holds, are positive definite, since all principal minors of Σ are continuous functions of the σ_{ij}'s.

In addition, we always assume in this chapter that there are no additional restrictions on the parameters of Σ and no joint restrictions between the elements of Σ and those of the $A(i)$, $B(i)$, $D(i)$ (or F, H, K, L) matrices. This is the reason that we can restrict ourselves to the investigation of the relation between (k, l) and (a, b, d) or (F, H, K, L), respectively.

In this chapter, unless the contrary is stated explicitly, we neither impose a stability assumption on a nor a miniphase assumption on b [nor do we impose the analogous assumptions on (F, H, K, L)]. This is done mainly for notational convenience. However, we always impose the causality assumptions $K(i) = 0$, $i < 0$, $L(i) = 0$, $i \leq 0$ and thus the assumption $K(0) = I$. Thus the standard assumptions of this chapter are the standard assumptions of Chapter 1, except for the assumptions just mentioned. By T_A we denote the set of all ARMAX systems (a, b, d), satisfying the standard assumptions of this chapter, where s and m are fixed (but p, q, r are arbitrary). Thus T_A is the set of all polynomial matrices (a, b, d), of appropriate dimensions, such that $\det a(z) \not\equiv 0$ and $a^{-1}(b, d)$ satisfies the assumptions mentioned above. By π: $T_A \to U_A$ we denote the mapping defined by $\pi(a, b, d) = a^{-1}(b, d)$, attaching to every ARMAX system the corresponding transfer function and U_A is the image of T_A under π. Thus U_A is the set of all $s \times (s + m)$ rational and causal transfer functions where $(k(0), l(0)) = (I, 0)$. Analogously, by Δ_A we denote the set of all state-space systems (F, H, K, L), where s and m are fixed (but n is arbitrary) and by ρ: $\Delta_A \to U_A$ we denote the mapping defined by

$$\rho(F, H, K, L) = zH(I - Fz)^{-1}(K, L) + (I, 0).$$

Two ARMAX systems having the same transfer function (k, l) are called observationally equivalent. For given (k, l) any ARMAX system with this transfer function is called an ARMAX realization of (k, l). Let $\hat{T} \subset T_A$ be the actual set of ARMAX systems under consideration (the so-called model class) and let $U = \pi(\hat{T})$ be the corresponding transfer functions. $\hat{T}$ may be considered as a (preliminary) parameter space. From the point of view of the input-output behaviour it is natural to consider the quotient set $\hat{T}/\pi =$

$\{\pi^{-1}(k,l) \cap \hat{T} | (k,l) \in U\}$ whose elements $\pi^{-1}(k,l) \cap \hat{T}$ are called the classes of all *observationally equivalent* ARMAX systems corresponding to (k,l) [or the *(k,l)-equivalence classes*] *in* $\hat{T}$. Of course, one would like to have a unique description of U by its parameter space. $\tilde{T} \subset T_A$ is called *identifiable* if π restricted to $\tilde{T}$ is injective, that is, if within $\tilde{T}$ the ARMAX systems (a,b,d) are uniquely determined from the transfer functions $a^{-1}(b,d)$. Starting from $\hat{T}$, identifiable parameter spaces can be obtained by selecting a unique representative from every equivalence class.

If $(a,b,d) \in \tilde{T}$, not all elements in the matrices $A(i)$, $B(i)$, and $D(i)$ may be needed for the description of (a,b,d), as some elements may be a priori known to be constant over $\tilde{T}$ or to be functions of other elements. Let τ denote a vector of elements of the $A(i)$, $B(i)$, and $D(i)$ such that the $(a,b,d) \in \tilde{T}$ are uniquely determined from the corresponding τ and that τ contains a minimal number of elements. We call τ a vector of *free parameters* for $U = \pi(\tilde{T})$ (or for $\tilde{T}$). In addition to identifiability, a desirable property of $\tilde{T}$ is that τ is finite dimensional. In this case there exists a bijective function of ψ: $U \to T$ such that $\psi(k,l)$ is a vector of free parameters for (a,b,d) corresponding to (k,l) and $T = \psi(U) \subset \mathbf{R}^d$, for some d. We call ψ an *ARMAX parameterization* of U (or of $\hat{T}/\pi$ or of $\tilde{T}$) and T is a (final) *parameter space* for U. In the same way, observational equivalence, realizations, identifiability, free parameters, parameter spaces, and parameterizations for state-space systems are defined.

In the general (finite-dimensional) case a *parameterization* is just a bijective mapping ψ: $U \to T \subset \mathbf{R}^d$, where the elements of $\tau = \psi(k,l)$ do not necessarily have an interpretation as ARMAX or state-space parameters (see, e.g., Son and Anderson, 1971; Solo, 1986; and Section 6.2). The actual choice of the parameterization depends on the special problem under consideration. It is assumed throughout that T is of finite dimension.

2. THE STRUCTURE OF ARMAX REALIZATIONS; SOME PROPERTIES OF POLYNOMIAL MATRICES

In this section some properties of polynomial matrices which are relevant for the analysis of ARMAX systems are treated. Of special interest is the relation between observationally equivalent ARMAX systems.

As has already been stated, every rational matrix (k,l) can be written as

$$(k,l) = c^{-1}N, \tag{2.2.1}$$

where c is a least common denominator of the elements of (k,l) and where N is a polynomial matrix. [For the case $(k,l) = 0$ we take $c = 1$, $N = 0$.] Thus every rational matrix (k,l) has a *(left) matrix fraction description* (MFD)

$$(k,l) = a^{-1}(b,d), \tag{2.2.2}$$

where (a, b, d) are polynomial matrices and where

$$\det a \not\equiv 0. \tag{2.2.3}$$

The ARMAX realization (cI, N) defined by (2.2.1), however, is not very convenient, as, for example, too many real parameters are needed for its description. So we have to investigate the problem of finding convenient ARMAX realizations.

First, let us repeat some facts from the algebra of polynomial and rational matrices (see, e.g., Barnett, 1972; MacDuffee, 1956; Newman, 1972; Wolovich, 1974; in particular we draw heavily from Kailath, 1980). We consider only polynomials and rational functions with real coefficients, although the variable may be complex. As the polynomials form a principal ideal domain and the rational functions form a field, from an abstract point of view polynomial matrices are considered as matrices over a principal ideal domain (as are, for example, matrices with integer entries) and rational matrices are considered as matrices over a field (and thus the latter share most properties with "ordinary" matrices with real entries). From this abstract point of view, linear independence, rank, determinant, adjoint (sometimes called the adjugate), and the inverse of a matrix are defined in a straightforward manner. Clearly, every polynomial matrix can be considered as a rational one. Note that a rational square matrix a has an inverse if and only if

$$\det a(z) \not\equiv 0,$$

that is, if $\det a(z) \neq 0$ for at least one and thus for all but a finite number of $z \in \mathbf{C}$. Let (a, b, d) be a polynomial matrix. If there exist polynomial matrices u and $(\hat{a}, \hat{b}, \hat{d})$ such that

$$(a, b, d) = u(\hat{a}, \hat{b}, \hat{d}), \tag{2.2.4}$$

then u is called a *(common) left factor* (or *left divisor*) of (a, b, d), and (a, b, d) is called a *right multiple* of u. A left divisor u is said to be a *greatest (common) left divisor* if it is a right multiple of all left divisors.

A polynomial matrix u is called *unimodular* if $\det u$ is a constant not equal to zero. A polynomial matrix u is unimodular if and only if u^{-1} exists and is a polynomial matrix. If u is unimodular, then from

$$u^{-1} = (\det u)^{-1} \operatorname{adj} u$$

we see that u^{-1} is a polynomial matrix since the adjoint $\operatorname{adj} u$ is a polynomial matrix and $\det u$ is constant. Conversely, if u and u^{-1} are polynomial, so are

$\det u$ and $\det(u^{-1})$. Since $\det u \det(u^{-1}) = 1$ holds, $\det u$ and $\det(u^{-1})$ must be constant. The elementary row (or column) operations for polynomial matrices are of the form:

(i) Interchange any two rows (columns).
(ii) Add to any row (column) a polynomial multiple of any other row (column).
(iii) Multiply any row (column) by a nonzero real number.

As easily seen, elementary row (column) operations correspond to left (right) multiplication by special unimodular matrices. [It can even be shown that every unimodular matrix u is of the form $u = e_1, \ldots, e_N$, where $e_1, \ldots, e_N$ are matrices corresponding to elementary row (column) operations; see, e.g., MacDuffee (1956).]

Greatest (left) divisors for (a, b, d) always exist and they can be constructed as follows: Since (a, b, d) has rank less than or equal to s, we can find elementary column operations and thus a unimodular matrix such that

$$(a, b, d)u = (r, 0), \tag{2.2.5}$$

where r is $s \times s$ and 0 is $s \times (s + m)$. Now using a conformal partitioning

$$u^{-1} = \begin{pmatrix} u_{11} & u_{12} \\ u_{21} & u_{22} \end{pmatrix}^{-1} = \begin{pmatrix} v_{11} & v_{12} \\ v_{21} & v_{22} \end{pmatrix},$$

we obtain from (2.2.5)

$$a = rv_{11}; \qquad (b, d) = rv_{12} \tag{2.2.6}$$

and then r is a (left) divisor of (a, b, d). Note that we have

$$r = au_{11} + (b, d)u_{21}, \tag{2.2.7}$$

so that if r_1 is any other (left) divisor,

$$(a, b, d) = r_1(a_1, b_1, d_1),$$

then

$$r = r_1(a_1 u_{11} + (b_1 d_1)u_{21})$$

so that r is a right multiple of r_1 and thus r is a greatest (left) divisor of (a, b, d).

We remind the reader that we assume that $\det a \not\equiv 0$. Thus greatest left divisors are unique up to right multiplication by unimodular matrices. Indeed, two greatest left divisors r_1, r_2 must be related by

$$r_1 = r_2 p_1, \qquad r_2 = r_1 p_2,$$

where p_1, p_2 are polynomial matrices. Since $\det a \not\equiv 0$ implies that $\det r_2 \not\equiv 0$, we have $p_1 = r_2^{-1} r_1 = p_2^{-1}$ and thus p_2 is unimodular. The converse direction, that every right multiplication of a greatest left divisor by a unimodular matrix gives another greatest left divisor, is evident.

The polynomial matrix (a, b, d) is called *(relatively) left prime* (or *left coprime*) if the only left divisors of (a, b, d) are the unimodular ones. The corresponding MFD then is called *irreducible.* Let p be a polynomial; then by $\nu(p)$ we denote the degree of p. Thereby we use the convention $\nu(0) = -1$.

Lemma 2.2.1. *For the polynomial matrix* (a, b, d) *(where* $\det a \not\equiv 0$*) the following statements are equivalent:*

(a) (a, b, d) *is (relatively) left prime.*

(b) The unit matrix I_s *is a greatest left divisor of* (a, b, d).

(c) $(a(z), b(z), d(z))$ *has rank* s *for all* $z \in \mathbf{C}$.

(d) There exist polynomial matrices g, h *such that*

$$ag + (b, d)h = I_s. \tag{2.2.8}$$

(e) $\nu(\det a)$ *is minimal among all matrices* $(\bar{a}, \bar{b}, \bar{d})$ *such that*

$$\bar{a}^{-1}(\bar{b}, \bar{d}) = a^{-1}(b, d). \tag{2.2.9}$$

Proof

$(a) \Rightarrow (b)$. Since all left divisors are unimodular, all greatest left divisors are also unimodular and thus I_s is a greatest left divisor.

$(b) \Rightarrow (c)$. Suppose that there exists a $z_0 \in \mathbf{C}$ such that $(a(z_0), b(z_0), d(z_0))$ has rank less than s. Then we can find a nonsingular constant matrix U such that $U(a(z_0), b(z_0), d(z_0))$ has a row of zeros as its first row and thus

$$U^{-1} \cdot \begin{pmatrix} z - z_0 & 0 \\ 0 & I_{s-1} \end{pmatrix}.$$

must be a left divisor of (a, b, d). As this left divisor is nonunimodular, I_s cannot be a greatest left divisor.

$(c) \Rightarrow (a)$. If (a, b, d) has nonunimodular left divisor u, there is a z_0 such that $\det u(z_0) = 0$ and then (c) cannot hold.

$(a) \Rightarrow (d)$. If (a, b, d) is left prime, then r in (2.2.6) must be unimodular. Then (2.2.8) is obtained by right multiplication of (2.2.7) with r^{-1}.

$(d) \Rightarrow (e)$. If $(\bar{a}, \bar{b}, \bar{d})$ satisfies (2.2.2), and if (g, h) is a polynomial matrix satisfying (2.2.8), then

$$(a, b, d) = a\bar{a}^{-1}(\bar{a}, \bar{b}, \bar{d}) \tag{2.2.10}$$

and

$$a\bar{a}^{-1}\bar{a}g + a\bar{a}^{-1}(\bar{b}, \bar{d})h = I$$

and thus

$$\bar{a}g + (\bar{b}, \bar{d})h = \bar{a}a^{-1}.$$

Therefore, $u = \bar{a}a^{-1}$ must be a polynomial matrix. Let us write (2.2.10) as

$$(\bar{a}, \bar{b}, \bar{d}) = u(a, b, d). \tag{2.2.11}$$

This implies that

$$\nu(\det \bar{a}) = \nu(\det u) + \nu(\det a) \geq \nu(\det a),$$

and thus $\nu(\det a)$ is minimal for all MFDs of $a^{-1}(b, d)$.

$(e) \Rightarrow (a)$. If (a) did not hold, we could reduce $\nu(\det a)$ by extracting a common nonunimodular left factor.

From the last lemma it is clear that every polynomial matrix (a, b, d) where $\det a \not\equiv 0$ can be transformed to a left prime matrix in a finite number of steps. It is also clear that $\nu(\det a)$ is an invariant for all irreducible MFDs of the same (k, l). A left prime ARMAX realization (a, b, d) has the advantage of being nonredundant in a certain sense. This is easy to see for the case $s = 1$, where left primeness just means that the degrees in a, b, d cannot be reduced; for $s > 1$ the meaning of the statement will become clear later. The following theorem gives a characterization of the equivalence classes.

Theorem 2.2.1. *Two ARMAX systems $(\bar{a}, \bar{b}, \bar{d})$ and (a, b, d), where (a, b, d) is left prime, are observationally equivalent if and only if there exists a (nonsingular) polynomial matrix u such that*

$$(\bar{a}, \bar{b}, \bar{d}) = u(a, b, d). \tag{2.2.12}$$

If $(\bar{a}, \bar{b}, \bar{d})$ is also left prime, then u must be unimodular.

Proof. One direction is evident. Conversely, if $(\bar{a}, \bar{b}, \bar{d})$ and (a, b, d) are observationally equivalent and (a, b, d) is left prime, then as shown in the proof $(d) \Rightarrow (e)$ of Lemma 2.2.1, u in (2.2.12) must be a polynomial matrix. If $(\bar{a}, \bar{b}, \bar{d})$ is also left prime, then u^{-1} must be a polynomial matrix by an analogous argument, and thus u must be unimodular.

Let $c = (c_{ij})$, $i = 1, \ldots, s$, $j = 1, \ldots, n$ be a polynomial matrix. The degree of c, $\nu(c)$ let us say, is defined as the maximum degree of its entries: $\nu(c) = \max_{i,j} \nu(c_{ij})$. In the same way the degree of the ith column (the ith row) of c is defined.

Note that for an ARMAX system (a, b, d) the degree of the ith row is the maximum lag occurring in the ith equation and the degree of the ith column is the maximum lag occurring in the ith component of $(y'(t), z'(t), \varepsilon'(t))$ (over all equations).

Let $c_i = \sum_{u=0}^{\nu(c_i)} c_i(u) z^u$ denote the ith row of c, where the $c_i(u) \in \mathbf{R}^n$ are the coefficients corresponding to the power z^u. The matrix $[c]_r = (c_1(\nu(c_1))', \ldots, c_s(\nu(c_s))')' \in \mathbf{R}^{s \times n}$ is called the *row end matrix* of c. Completely analogous the *column end matrix* $[c]_c$ is defined from the highest degree coefficients of the columns of c. A matrix c is called *row reduced (column reduced)* if $[c]_r$ ($[c]_c$) has rank equal to $\min(s, n)$.

If a is an $s \times s$ polynomial matrix, the degree of $\det a$ cannot exceed the sum of the row degrees $\nu(a_i)$ (or of the column degrees):

$$\nu(\det a) \leq \sum_{i=1}^{s} \nu(a_i).$$

It is easily seen that the coefficient of $\det a$ corresponding to power z^n, $n = \sum \nu(a_i)$ is $\det[a]_r$ (and the analogous statement holds for $\det[a]_c$). Thus a square matrix a is row reduced (column reduced) if and only if the degree of $\det a$ equals the sum of the row (column) degrees of a.

By elementary operations we can transform polynomial matrices to several useful normal forms.

Lemma 2.2.2. *Every $s \times s$ polynomial matrix $\hat{a}$ of rank s can be transformed by left multiplication by a unimodular matrix to a matrix a with the following properties:*

(i) a is lower triangular.

(ii) a_{ii} is monic (i.e., the highest-degree coefficient of a_{ii} equals 1).

(iii) $\nu(a_{ji}) < \nu(a_{ii})$; $j \neq i$.

Proof. Let a be constructed by the following elementary row operations First consider the entry of the last column of $\hat{a}$ which has lowest degree among the nonzero entries of this column and bring it in the (s, s) position by row interchange. Call this entry $\tilde{a}_{ss}$. In the second step, applying the Euclidean division algorithm, subtract from every row except for the last one an appropriate polynomial multiple of the last row such that the new entries in the (j, s) positions, $j < s$, have degrees strictly less than $\nu(\tilde{a}_{ss})$. Now repeat the first two steps with a new (s, s) element of degree lower than $\nu(\tilde{a}_{ss})$ and continue until all elements in the last column, except the (s, s) element, are zero. Now, consider the second last column of the resulting matrix. Ignore the sth row and repeat the foregoing procedure until all entries above the $(s-1, s-1)$ entry are zero. If the $(s, s-1)$ entry does not have degree lower than the $(s-1, s-1)$ entry, then again we apply the division algorithm and the corresponding subtraction of a suitable multiple of the next-to last row from the last row to obtain an $(s, s-1)$ entry having lower degree than the $(s-1, s-1)$ entry. Continuing the procedure for all remaining columns finally gives the desired properties (i)–(iii).

As an immediate consequence of (i)–(iii), a is column reduced. As shown in Section 5, (i)–(iii) guarantee that a is a unique representative for the class $\{u\hat{a} \mid \det u = \text{const} \neq 0\}$ (this is called the left equivalence class of a). The matrix a is called the *Hermite form* of $\hat{a}$.

Next we consider the so-called *Smith form* Λ of a polynomial matrix c, which is obtained by elementary row and column operations.

Lemma 2.2.3. *Every $s \times n$ polynomial matrix c of rank r can be written as*

$$c = u\Lambda v, \tag{2.2.13}$$

where u and v are unimodular, Λ is diagonal of the form

$$\Lambda(z) = \left(\begin{array}{ccc:c} \lambda_1(z) & & & \\ & \ddots & 0 & 0 \\ 0 & & \ddots & \\ & & \lambda_r(z) & \\ \hdashline & 0 & & 0 \end{array}\right),$$

and the λ_i are monic polynomials such that λ_i divides λ_{i+1}, $i = 1, \ldots, r-1$.

Proof. To construct $\Lambda(z)$ we proceed as follows. First, by elementary row and column operations bring the nonzero element of least degree of c in the $(1, 1)$ position. Now, analogous to the construction of Hermite form, by elementary row and column operations make all entries in the first column, except for the first, and then all entries in the first row, except for the first, equal to zero. In making the entries of the first row equal to zero, nonzero entries may reappear in the first column and then we have to repeat the procedure until finally all entries in the first column and in the first row are zero, except for the (1, 1) entry. Next, if this (1, 1) element does not divide every other element in the whole matrix, by using the Euclidean division algorithm and row and column interchanges we can bring a lower degree element to the (1, 1) position and then repeat the steps above to zero all other elements in the first column and first row. Proceeding in this way will give a matrix of the form

$$s-1\{\begin{pmatrix} \lambda_1(z) & \vdots & \overbrace{0}^{n-1} \\ \cdots & \cdots & \cdots \\ 0 & \vdots & r_1(z) \end{pmatrix}$$

where $\lambda_1(z)$ divides every element of $r_1(z)$. Repeating the operations above in $r_1(z)$ finally gives the Smith form $\Lambda(z)$. Note that $\Lambda(z)$ in (2.2.13) is unique, whereas u and v are not necessarily unique.

3. THE STRUCTURE OF STATE-SPACE REALIZATIONS: OBSERVABILITY, REACHABILITY, AND MINIMALITY

We consider a state-space system (F, H, K, L), that is,

$$\begin{aligned} x(t+1) &= Fx(t) + Lz(t) + K\varepsilon(t) \\ y(t) &= Hx(t) + \varepsilon(t), \end{aligned} \tag{2.3.1}$$

where $F \in \mathbf{R}^{n\times n}$, $H \in \mathbf{R}^{s\times n}$, $K \in \mathbf{R}^{n\times s}$, $L \in \mathbf{R}^{n\times m}$. The corresponding transfer function is given by

$$(k(z), l(z)) = zH(I - Fz)^{-1}(K, L) + (I, 0). \tag{2.3.2}$$

As has been shown in Chapter 1, for every ARMAX system satisfying the standard assumptions of Chapter 2 [and thus for every $(k, l) \in U_A$] there is a state-space representation having the same transfer function. In this section we investigate the structure of the set of all observationally equivalent state-space systems corresponding to given (k, l). For references, see Kalman (1960), Kalman (1963), and Kalman, Falb, and Arbib (1969).

The system (2.3.1) (or F, H) is said to be *observable* if

$$\mathcal{O}' := (H', F'H', F'^2H', \ldots, F'^{n-1}H') \tag{2.3.3}$$

is full of rank n. There can be no gain from including further terms to the right in (2.3.3) since any matrix satisfies its own characteristic equation $\det(F - \lambda I) = 0$, which is of degree n, so that the columns of $F'^j H'$, $j \geq n$, are linear combinations of the columns of $\mathcal{O}'$ and their adjunction on the right cannot increase the rank.

If $\mathcal{O}$ is of full rank, we can solve $y(t) = HF^t x(0)$, $t = 0, 1, \ldots, n-1$, for $x(0)$, uniquely, given the $y(t)$. It is from this that the terminology derives; in the absence of $z(t)$, $\varepsilon(t)$ we can determine (observe) from $y(t)$, $t = 0, \ldots, n-1$ the initial value $x(0)$.

If $\mathcal{O}$ has rank $n_1 < n$, we can find a nonsingular matrix T such that $T'^{-1}\mathcal{O}'$ has the last $n - n_1$ rows null and the first n_1 rows of rank n_1. Then partitioning conformally as

$$TFT^{-1} = \begin{pmatrix} F_{11} & F_{12} \\ F_{21} & F_{22} \end{pmatrix}, \qquad HT^{-1} = (H_1, 0),$$

we have

$$0 = (F'_{12}H'_1, F'_{12}F'_{11}H'_1, F'_{12}F'^2_{11}H'_1, \ldots)$$

since for example, the second term is obtained from the last columns (transposed) of $HF^2T^{-1} = (H_1F_{11}, H_1F_{12})TFT^{-1} = (H_1F_{11}, 0)TFT^{-1} = (H_1F_{11}^2, H_1F_{11}F_{12})$ and these last columns are null. However, this implies that $F_{12} = 0$ since it multiplies a matrix of full rank. Hence transforming to TFT^{-1}, HT^{-1}, TK, TL, $Tx(t)$ and then, to conserve notation, calling the new quantities with the old names, F, H, K, L, $x(t)$, we have, after partitioning $x(t) = (x_1(t)', x_2(t)')$,

$$\begin{aligned} x_1(t+1) &= F_{11}x_1(t) + L_1z(t) + K_1\varepsilon(t) \\ x_2(t+1) &= F_{21}x_1(t) + F_{22}x_2(t) + L_2z(t) + K_2\varepsilon(t) \\ y(t) &= H_1x_1(t) + \varepsilon(t). \end{aligned}$$

Here we have also partitioned K, L conformally. Now it is evident that $x_2(t)$ can be ignored since it does not enter the equation for $x_1(t+1)$ nor the equation for $y(t)$. Hence as far as what is observed is concerned, it is no restriction to consider only (F_{11}, H_1, K_1, L_1) and thus to assume that the system is observable.

The system (or (F, K, L)) is said to be *reachable* if

$$\text{(2.3.4)} \qquad \mathscr{C} := ((K, L), F(K, L), F^2(K, L), \ldots, F^{n-1}(K, L))$$

is of full rank n. If that is so, we can write the solution of the first part of (2.3.1) as [see (1.2.23)]

$$x(t+1) = \sum_{j=1}^{t+1} F^{j-1}(L, K)(z(t+1-j)', \varepsilon(t+1-j)')' + F^{t+1}x(0).$$

Thus, for $t \geq n-1$, if $z(t), \varepsilon(t), \ldots, z(0), \varepsilon(0)$ were under our control, we could maneuvre $x(t+1)$ into any state given any initial state—hence the term "reachable."

Now exactly as before, if the system is not reachable, we can find a nonsingular matrix so that, again using the old symbols for the transformed quantities,

$$\begin{aligned} x_1(t+1) &= F_{11}x_1(t) + F_{12}x_2(t) + L_1 z(t) + K_1 \varepsilon(t) \\ x_2(t+1) &= F_{22}x_2(t). \end{aligned}$$

Now (F_{11}, K_1, L_1) is reachable. If (F, H) is observable and $H = (H_1, H_2)$ is a conformal partition, then (H_1, F_{11}) are again observable since the first rows of $\mathcal{O}'$ are $(H_1', F_{11}'H_1', \ldots, F_{11}'^{n-1}H_1')$ and hence this matrix is also of full rank.

The transfer function, (k, l), depends only on (F_{11}, H_1, K_1, L_1). Indeed, from (2.3.2), taking

$$F = \begin{pmatrix} F_{11} & F_{12} \\ 0 & F_{22} \end{pmatrix}, \qquad [K, L] = \begin{bmatrix} K_1 & L_1 \\ 0 & 0 \end{bmatrix},$$

then

$$(k, l) = zH_1(I - zF_{11})^{-1}(K_1, L_1) + (I, 0).$$

However, $y(t)$ is dependent on $x_2(t)$ in a way that cannot be eliminated by a redefinition of an initial condition for $x_1(t)$ since this would require

$$H_1 F_{11}^t x_1(0) = HF^t \begin{pmatrix} x_1(0) \\ x_2(0) \end{pmatrix}$$

and if (F, H) is observable, this is possible only if $x_2(0) = 0$ [i.e., $x_2(t) = F_{22}^t x_2(0) \equiv 0$]. Of course, if F_{22} is asymptotically stable (which will be so if F is asymptotically stable), then $x_2(t)$ decreases to zero at a geometric rate and it could not, asymptotically, affect any of the statistical procedures we discuss below. Moreover, for reasons of the same kind as were given in Section 1.3 it seems reasonable to believe that a term coming from eigenvalues of F of unit modulus [of the form of (1.3.3)] would in any case have been recognized to exist and have been incorporated in $z(t)$ by the model builder.

In any case we have the following result:

Theorem 2.3.1. *Any transfer function* $(k,l) \in U_A$ *can be expressed as* $zH(I - Fz)^{-1}(K,L) + (I,0)$, *where* (F,K,L) *is reachable and* (F,H) *is observable.*

We now introduce an important concept. The state-space system (F,H,K,L) will be said to be *minimal* if F is of minimal possible dimension among all realizations of the transfer function. The concept of minimality of a state-space realization is in some sense analogous to that of irreducibility of an MFD, as we shall see later.

The *(block) Hankel matrices* [a matrix is called a block Hankel matrix if the (i,j) block depends only on $i + j$]

$$\mathscr{H} := \begin{pmatrix} [K(1),L(1)] & [K(2),L(2)] & [K(3),L(3)] & \cdots \\ [K(2),L(2)] & [K(3),L(3)] & [K(4),L(4)] & \cdots \\ \cdots & \cdots & \cdots & \cdots \\ \cdots & \cdots & \cdots & \cdots \end{pmatrix} \tag{2.3.5}$$

are in one-to-one correspondence with the corresponding transfer function $(k(z),l(z)) = (I,0) + \sum_{i=1}^{\infty}(K(i),L(i))z^i$. Furthermore, let

$$\mathscr{H}_i^j := \begin{pmatrix} K(1),L(1) & K(2),L(2) & \cdots & K(j),L(j) \\ \cdots & \cdots & \cdots & \cdots \\ K(i),L(i) & K(i+1),L(i+1) & \cdots & K(i+j-1),L(i+j-1) \end{pmatrix},$$

$i,j \in \mathbf{N} \cup \{\infty\}$.

Since for the state-space system,

$$(K(i),L(i)) = HF^{i-1}(K,L), \tag{2.3.6}$$

we have

$$\mathscr{H} = (H', F'H', F'^2H', \ldots)'((K,L), F(K,L), F^2(K,L), \ldots) \tag{2.3.7}$$

and

$$\mathscr{H}_n^n = \mathscr{O}\mathscr{C}. \tag{2.3.8}$$

Theorem 2.3.2. *Let* (F,H,K,L) *be a state-space system with* $F \in \mathbf{R}^{n \times n}$. *Then* $\mathscr{H}$ *and* $\mathscr{H}_n^n$ *have rank smaller than or equal to* n. *If* (F,H,K,L) *is observable and reachable, equality holds.*

Proof. The first part follows immediately, by what was said before about $\mathcal{O}'$ and $\mathcal{C}$, and also both factors on the right-hand side of (2.3.7) have rank smaller than or equal to n.

If (F, H, K, L) is both observable and reachable, $\mathcal{O}$ and $\mathcal{C}$ have rank n. Thus $\mathcal{O}'\mathcal{O}$ and $\mathcal{C}\mathcal{C}'$ are nonsingular $n \times n$ matrices and so is

$$\mathcal{O}'\mathcal{O}\mathcal{C}\mathcal{C}'.$$

But this implies that $\mathcal{H}_n^n = \mathcal{O}\mathcal{C}$ has rank n and thus also, $\mathcal{H}$ has rank n.

Theorem 2.3.3. *A state-space system (F, H, K, L) is minimal if and only if it is observable and reachable.*

Proof. The "only if" part has been proved since our constructions show that if observability or reachability fails, F may be reduced in dimension. Conversely, if the state-space system is both observable and reachable, then by Theorem 2.3.2, $\mathcal{H}_n^n$ has the rank n. If $(\tilde{F}, \tilde{H}, \tilde{K}, \tilde{L})$ is a realization of the same transfer function with $\tilde{F} \in \mathbf{R}^{\tilde{n} \times \tilde{n}}$, $\tilde{n} \leq n$, then again by Theorem 2.3.2, $\mathcal{H}_n^n$ is of rank less than or equal to $\tilde{n}$ and thus $n = \tilde{n}$.

The next theorem gives a characterization of the equivalence classes for minimal state-space systems. It is the state-space analogue of Theorem 2.2.1.

Theorem 2.3.4. *Two minimal state-space systems $(\bar{F}, \bar{H}, \bar{K}, \bar{L})$ and (F, H, K, L) are realizations of the same transfer functions if and only if there exists a nonsingular matrix T such that*

$$\bar{F} = TFT^{-1}, \qquad \bar{H} = HT^{-1}, \qquad (\bar{K}, \bar{L}) = T(K, L). \tag{2.3.9}$$

Proof. The "if" part is obvious and has already been mentioned. Conversely, if $(\bar{F}, \bar{H}, \bar{K}, \bar{L})$ and (F, H, K, L) realize the same transfer function, then in an obvious notation

$$\mathcal{H}_n^n = \mathcal{O}\mathcal{C} = \bar{\mathcal{O}}\bar{\mathcal{C}}$$

and thus

$$\mathcal{C} = (\mathcal{O}'\mathcal{O})^{-1}(\mathcal{O}'\mathcal{O})\mathcal{C} = (\mathcal{O}'\mathcal{O})^{-1}\mathcal{O}'\bar{\mathcal{O}}\bar{\mathcal{C}}.$$

Similarly,

$$\bar{\mathcal{C}} = (\bar{\mathcal{O}}'\bar{\mathcal{O}})^{-1}\bar{\mathcal{O}}'\mathcal{O}\mathcal{C}. \tag{2.3.10}$$

Let $\bar{T} = (\mathcal{O}'\mathcal{O})^{-1}\mathcal{O}'\bar{\mathcal{O}}$ and $T = (\bar{\mathcal{O}}'\bar{\mathcal{O}})^{-1}\bar{\mathcal{O}}'\mathcal{O}$. Then it follows that $T\bar{T} = I$. From

$$\bar{\mathcal{O}}\bar{F}\bar{\mathcal{C}} = \mathcal{O}F\mathcal{C}$$

we then have $\bar{F} = TFT^{-1}$ and (2.3.10) implies that $(\bar{K}, \bar{L}) = T(K, L)$ and similarly, we obtain $\bar{H} = HT^{-1}$.

Lemma 2.3.1. *The linear dependence relations between the rows of $\mathcal{H}$ and the rows of*

$$(H', F'H', F'^2H', \ldots)' \tag{2.3.11}$$

and between the columns of $\mathcal{H}$ and the columns of

$$((K, L), F(K, L), F^2(K, L), \ldots), \tag{2.3.12}$$

where (F, H, K, L) is any minimal state-space realization of (k, l), are identical.

Proof. If a selection of rows in (2.3.11) is linearly dependent, then as directly seen from (2.3.7), the corresponding rows in $\mathcal{H}$ must also be linearly dependent. Conversely, if a selection of rows in (2.3.11) is linearly independent, then as $\mathcal{C}$ has rank n by Theorem 2.3.3, the corresponding rows in $\mathcal{H}$ are also linearly independent. The proof for the columns is completely analogous.

4. SOME PROPERTIES OF RATIONAL TRANSFER FUNCTIONS

Here some further properties of rational transfer functions and of the related Hankel matrices are investigated. The reader is referred to Kailath (1980) and Kalman, Falb, and Arbib (1969).

In many cases it turns out to be (technically) more convenient to use z^{-1} instead of z for the backward shift operator and to write the transfer functions as functions of the forward-shift operator z; in this way we obtain

$$(\tilde{k}(z), \tilde{l}(z)) = (k(z^{-1}), l(z^{-1})) - (I, 0) = \sum_{j=1}^{\infty} (K(j), L(j))z^{-j}. \tag{2.4.1}$$

We shall continue to use z for the backward shift, but the functions $\tilde{k}$, $\tilde{l}$ will be important. Here, in addition, the coefficient I, corresponding to power zero, has been subtracted from $k(z^{-1})$. Of course, there is a one-to-one relation between the "old" transfer functions (k, l) and the "new" ones $(\tilde{k}, \tilde{l})$ and in this sense we can identify them; for example, we use the notation $(\tilde{k}, \tilde{l}) \in U_A$ when we mean $(\tilde{k}, \tilde{l}) \leftrightarrow (k, l) \in U_A$. Now $(\tilde{k}, \tilde{l})$ is *strictly proper*; that is, all denominator degrees are higher than numerator degrees, since

$$\lim_{|z|\to\infty} (\tilde{k}(z), \tilde{l}(z)) = (0, 0), \tag{2.4.2}$$

and this has a number of (purely technical) advantages, as will be discussed below. The properness of $\tilde{k}, \tilde{l}$ [i.e., that the limit on the r.h.s. of (2.4.2) is finite] is equivalent to causality of (k, l).

Clearly, $(\tilde{k}, \tilde{l})$ is also rational. Let us consider an MFD

$$(\tilde{k}, \tilde{l}) = \tilde{a}^{-1}(\tilde{b}, \tilde{d}) \tag{2.4.3}$$

of $(\tilde{k}, \tilde{l})$, where

$$\tilde{a}(z) = \sum_{j=0}^{\tilde{p}} \tilde{A}(j)z^j; \qquad \tilde{b}(z) = \sum_{j=0}^{\tilde{q}} \tilde{B}(j)z^j; \qquad \tilde{d}(z) = \sum_{j=0}^{\tilde{r}} \tilde{D}(j)z^j.$$

This corresponds to a VDE [i.e., to a realization of $(\tilde{k}, \tilde{l})$ or of (k, l)] of the form

$$\sum_{j=0}^{\tilde{p}} \tilde{A}(j)y(t+j) = \sum_{j=0}^{\tilde{r}} \tilde{D}(j)z(t+j) + \sum_{j=0}^{\max(\tilde{p},\tilde{q})} (\tilde{B}(j) + \tilde{A}(j))\varepsilon(t+j). \tag{2.4.4}$$

Given (2.4.3) the MFD corresponding to (k, l) is given by

$$(a(z), b(z), d(z)) = \operatorname{diag}\{z^{n_i}\}(\tilde{a}(z^{-1}), \tilde{b}(z^{-1}) + \tilde{a}(z^{-1}), \tilde{d}(z^{-1})), \tag{2.4.5}$$

where n_i is the degree of the ith row of $\tilde{a}$, and where in (2.4.5) z denotes the backward shift. It should be stressed again that (2.4.4) has the same input-output behavior as the ARMAX system (a, b, d). Clearly, (2.4.5) defines a one-to-one relation between (a, b, d) and $(\tilde{a}, \tilde{b}, \tilde{d})$. From

$$\tilde{a}(z)\left(\sum_{j=1}^{\infty} (K(j), L(j))z^{-j}\right) = (\tilde{b}(z), \tilde{d}(z)) \tag{2.4.6}$$

we see that the strict properness of $(\tilde{k}, \tilde{l})$ implies that the degree of the ith row of $(\tilde{b}, \tilde{d})$ is strictly less than n_i; thus (a, b, d) defined by (2.4.5) is indeed a polynomial matrix and the row degrees of $\tilde{a}$ and of $(\tilde{a}, \tilde{b}, \tilde{d})$ are the same. The latter is one of the advantages of the new notaion. As an example, consider a scalar MA system $y(t) = \varepsilon(t) + b\varepsilon(t-1)$. Then $k(z) = 1 + bz$, $\tilde{k}(z) = bz^{-1}$, and thus (2.4.4) is of the form $y(t+1) = \varepsilon(t+1) + b\varepsilon(t)$.

Note that even if each row of $(\tilde{b}, \tilde{d})$ has degree strictly less than the degree of the corresponding row of $\tilde{a}$, $\tilde{a}^{-1}(\tilde{b}, \tilde{d})$ is not necessarily strictly proper and the corresponding $a^{-1}(b, d)$ may be not causal, as can easily be shown by an example. However, we have:

Lemma 2.4.1. *If $\tilde{a}$ is row reduced, then $\tilde{a}^{-1}(\tilde{b}, \tilde{d})$ is strictly proper if and only if each row of $(\tilde{b}, \tilde{d})$ has degree strictly less than the degree of the corresponding row of $\tilde{a}$.*

Proof. It only remains to show the 'if' part. Let $\tilde{k}_{ij}$ denote the (i,j) entry of $\tilde{k}$ and consider equation (2.4.3). Then from Cramer's rule we have

$$\tilde{k}_{ij} = (\det \tilde{a})^{-1} \det(\tilde{a}^{ij}), \tag{2.4.7}$$

where $\tilde{a}^{ij}$ is the matrix obtained from $\tilde{a}$ by replacing its jth column by the ith column of $\tilde{b}$. Since any row of $\tilde{b}$ has degree strictly less than the degree of the corresponding row in $\tilde{a}$, $[\tilde{a}^{ij}]_r$ has a column of zeros and thus is singular. Therefore, the degree of the numerator in (2.4.7) is strictly less than the degree of the denominator and thus $\tilde{k}$, and by the same argument $\tilde{l}$, are strictly proper.

Lemma 2.4.2. *If $(\tilde{a}, \tilde{b}, \tilde{d})$ is left prime, then [via (2.4.5)] the corresponding (a, b, d) is left prime if and only if $A(0)$ is nonsingular.*

Proof. From Lemma 2.2.1 we see that if $(\tilde{a}, \tilde{b}, \tilde{d})$ is left prime, then $(\tilde{a}(z^{-1}), \tilde{b}(z^{-1}), \tilde{d}(z^{-1}))$ has rank s for all $z \neq 0$. Therefore, $(\tilde{a}(z^{-1}), \tilde{b}(z^{-1}) + \tilde{a}(z^{-1}), \tilde{d}(z^{-1}))$ has rank s for all $z \neq 0$, since $c(\tilde{a}(z^{-1}), \tilde{b}(z^{-1}) + \tilde{a}(z^{-1}), \tilde{d}(z^{-1})) = 0$ for some $c \in \mathbf{R}^s$, $c \neq 0$ implies that $c\tilde{a}(z^{-1}) = 0$, and thus $c\tilde{b}(z^{-1}) = 0$. Thus $(a(z), b(z), d(z))$ has rank s for all $z \neq 0$. As $(a(0), b(0), d(0)) = (A(0), A(0), 0)$, the statement follows directly from Lemma 2.2.1.

From Lemma 2.2.1 it is clear that $n = \nu(\det \tilde{a})$ is an invariant for the class of all irreducible MFDs $(\tilde{a}, \tilde{b}, \tilde{d})$ of $(\tilde{k}, \tilde{l})$. This n is called the *order* (or the *McMillan degree*) of $(\tilde{k}, \tilde{l})$. In the same way one could define the order $\hat{n}$ of (k, l). The order of $(\tilde{k}, \tilde{l})$ is different in general from the order of the corresponding (k, l), and we consider only the order of $(\tilde{k}, \tilde{l})$, which, as will be shown later, has a nice interpretation as the dimension of the state in the minimal state-space realizations. This is one of the main advantages of $(\tilde{k}, \tilde{l})$ over (k, l).

Let $M(n)$ denote the set of all $(s \times (s + m))$, strictly proper, rational transfer functions $(\tilde{k}, \tilde{l})$ of order n. Identifying $(\tilde{k}, \tilde{l})$ and (k, l) we write $M(n) \subset U_A$. Next we consider some properties of the Hankel matrix $\mathscr{H}$ [see (2.3.5)] associated with the transfer function (k, l). Identifying $(\tilde{k}, \tilde{l})$ with (k, l) we also associate the same $\mathscr{H}$ with $(\tilde{k}, \tilde{l})$.

Let $h(i,j)$ denote the jth row in the ith block of rows of $\mathscr{H}$; thus $h(i,j)$ is the $[(i-1)s + j]$th row of $\mathscr{H}$.

Theorem 2.4.1. *The Hankel matrix $\mathscr{H}$ of the transfer function (k, l) has the following properties:*

(i) If $h(i,j)$ is the linear span of $h(i_1, j_1), \ldots, h(i_K, j_K)$, then *$h(i+1, j)$ is in the linear span of $h(i_1 + 1, j_1), \ldots, h(i_K + 1, j_K)$ (and the analogous statement holds for the columns of $\mathscr{H}$).*

(ii) The rank of $\mathscr{H}$ is finite if and only if (k, l) is rational.

(iii) The order n of $(\tilde{k}, \tilde{l})$ is equal to the number of linearly independent rows (and columns) of $\mathscr{H}$. If $\mathscr{H}$ has rank n, then $\mathscr{H}_n^n$ has rank n.

Proof

(i) is an immediate consequence of the block Hankel structure of $\mathscr{H}$.

(ii) Clearly, (k, l) is rational if and only if $(\tilde{k}, \tilde{l})$ is rational. Let $\tilde{a}^{-1}(\tilde{b}, \tilde{d}) = (\tilde{k}, \tilde{l})$ be an irreducible MFD and assume that $\tilde{a}$ is in Hermite form (Lemma 2.2.2). From a comparison of coefficients (corresponding to the negative powers of z) in

$$\tilde{a}(z)(k(z^{-1}) - I, l(z^{-1})) = (\tilde{b}(z), \tilde{d}(z)) \tag{2.4.8}$$

we obtain

$$(\tilde{A}(0), \tilde{A}(1), \ldots)\mathscr{H} = 0. \tag{2.4.9}$$

From this and from the properties of Hermite's form we see that $h(v(\tilde{a}_{ii}) + 1, i)$, $i = 1, \ldots, s$ are linear combinations of the rows $h(j, i)$, $j \leq v(\tilde{a}_{ii})$, $i = 1, \ldots, s$ and thus, by (i), $\mathscr{H}$ must have finite rank, not exceeding $\sum_{i=1}^{s} v(\tilde{a}_{ii})$.

Conversely, if $\mathscr{H}$ has finite rank, n say, then

$$\mathscr{H}_n^{\infty} = \begin{pmatrix} K(1), L(1) & K(2), L(2) & \cdots \\ \cdots\cdots\cdots & \cdots\cdots\cdots\cdots\cdots & \cdots \\ K(n), L(n) & K(n+1), L(n+1) & \cdots \end{pmatrix},$$

due to (i), also has rank n. Then expressing every row of $(K(n + 1), L(n + 1), K(n + 2), L(n + 2), \ldots)$ as a linear combination of the rows of $\mathscr{H}_{n+1}^{\infty}$ via

$$(\tilde{A}(0), \tilde{A}(1), \ldots, \tilde{A}(n))\mathscr{H}_{n+1}^{\infty} = 0$$

defines a polynomial matrix $\tilde{a}(z) = \sum_0^n \tilde{A}(i)z^i$. Relation (2.4.8) then defines the polynomial matrix $(\tilde{b}, \tilde{d})$ and thus (as $\det \tilde{a} \not\equiv 0$) (k, l) is rational.

(iii) If $(\tilde{a}, \tilde{b}, \tilde{d})$ is left prime and $\tilde{a}$ is in Hermite form, the order n of $(\tilde{k}, \tilde{l})$ is equal to $\sum_{i=1}^{s} v(\tilde{a}_{ii})$. If the rank of $\mathscr{H}$ would be less than n, then by what was said above not all rows $h(j, i)$, $j \leq v(\tilde{a}_{ii})$, $i = 1, \ldots, s$ are linearly independent and by (i) there are $m_i \leq v(\tilde{a}_{ii})$, $i = 1, \ldots, s$ where at least one strict inequality holds such that the $h(j, i)$, $j \leq m_i$, $i = 1, \ldots, s$ form a basis for the row space of $\mathscr{H}$. Expressing the rows $h(m_i + 1, i)$, $i = 1, \ldots, s$ as linear combinations of these basis rows, by (2.4.9) and (2.4.8), gives an MFD $(\tilde{\tilde{a}}, \tilde{\tilde{b}}, \tilde{\tilde{d}})$ of $(\tilde{k}, \tilde{l})$ where $\det \tilde{\tilde{a}} = \sum m_i < n$, in contradiction to the left primeness if $(\tilde{a}, \tilde{b}, \tilde{d})$ (Lemma 2.2.1), and thus the first statement in (iii) has been shown. The second statement is an immediate consequence of (i).

Now we are concerned with the problem of selecting from the rows of $\mathscr{H}$ a basis for its row space. For convenience we consider only *nice selections,*

selections with the property that if $h(i+1, j)$ is in the basis, then $h(i, j)$ is also. [By Theorem 2.4.1 (i) such nice selections exists.] Let n denote the rank of $\mathscr{H}$. A nice selection is described by a *multi-index* $\alpha = (n_1 \cdots n_s)$, indicating that the $n = n_1 + \cdots + n_s$ basis rows selected are $h(1,1), \ldots, h(n_1, 1), \ldots, h(1, s), \ldots, h(n_s, s)$. The symbol $|\alpha|$ is used to denote $n_1 + \cdots + n_s$.

If, for given $\mathscr{H}$, α gives a selection of rows that form a basis for the row space of $\mathscr{H}$, then α is called a *structure index* of $\mathscr{H}$ and of (k, l) [or of $(\tilde{k}, \tilde{l})$]. By $U_\alpha \subset U_A$ we denote the set of all transfer functions where $\alpha = (n_1 \cdots n_s)$ is a structure index, i.e., where $h(1,1), \ldots, h(n_1, 1), \ldots, h(1, s), \ldots, h(n_s, s)$ form a basis for the row space of $\mathscr{H}$. In general, for $s > 1$, the structure index α is not unique for given $\mathscr{H}$, and thus there may exist α, β, $\alpha \neq \beta$ such that $(k, l) \in U_\alpha$, $(k, l) \in U_\beta$.

A special selection is obtained if we seek for the *first* (in natural order) basis rows of $\mathscr{H}$. By Theorem 2.4.1 (i) this gives a nice selection. The corresponding multi-index $\alpha = (n_1 \cdots n_s)$ is called the *Kronecker indices* of $\mathscr{H}$ or of (k, l) (or of $\tilde{k}, \tilde{l}$). Of course, the Kronecker indices are unique for given (k, l). By $V_\alpha \subset U_A$ we denote the set of all transfer functions with Kronecker indices α. Clearly, $V_\alpha \subset U_\alpha \subset M(n)$.

Let h be a rational $s \times n$ matrix and write $h = c^{-1} N$, where c is the monic least common denominator of the entries of h and N is a polynomial matrix. Then as an evident consequence of Lemma 2.2.3, we have

Lemma 2.4.3. *Every $s \times n$ rational matrix h of rank r can be written as*

$$h = u \Lambda v, \tag{2.4.10}$$

where u and v are unimodular (polynomial) matrices, and Λ is diagonal of the form

$$\Lambda(z) = \left(\begin{array}{ccc|c} \lambda_1(z) & & & \\ & \ddots & 0 & 0 \\ & & \ddots & \\ 0 & & \lambda_r(z) & \\ \hline & 0 & & 0 \end{array}\right)$$

with $\lambda_i = \varepsilon_i \cdot \psi_i^{-1}$, where ε_i, ψ_i are relatively prime monic polynomials, ε_i divides ε_{i+1}, $i = 1, \ldots, r-1$, and ψ_{i+1} divides ψ_i, $i = 1, \ldots, r-1$.

Λ in (2.4.10) then is called the *Smith-McMillan form* of h. As can be shown (Kailath, 1980), Λ is unique for given h, whereas u and v are not unique in general.

There are several, partly different ways to define poles and zeros for a rational $s \times n$ matrix h. Following Rosenbrock (1970) and Kailath (1980) we say that the zeros of h are the zeros of the (nonzero) numerator polynomials $\varepsilon_i(z)$ in the Smith-McMillan form of h, and the poles of h are the zeros of denominator polynomials ψ_i in the Smith-McMillan form of h. Note that h may have a pole and a zero at the same z.

Lemma 2.4.4

(i) h has a pole at z_0 if and only if $\det a(z_0) = 0$ for one (and then for every) irreducible MFD, $h = a^{-1} b$ of h or if and only if $c(z_0) = 0$, where c is a least common denominator of h.

(ii) If h is square, then h has a zero at z_0 if and only if $\det b(z_0) = 0$ for one (and then for every) irreducible MFD $h = a^{-1}b$.

Proof

(i) Note that, as is easily seen from Lemmas 2.4.3 and 2.2.1,

$$h(z) = (\psi(z)u^{-1}(z))^{-1}(\varepsilon(z)v(z)),$$

where

$$\psi(z) = \begin{pmatrix} \operatorname{diag}\{\psi_i(z)\} & 0 \\ 0 & I_{s-r} \end{pmatrix}; \qquad \varepsilon(z) = \begin{pmatrix} \operatorname{diag}\{\varepsilon_i(z)\} & 0 \\ 0 & 0 \end{pmatrix} \in \mathbf{C}^{s \times n}$$

is an irreducible MFD of h and thus by Theorem 2.2.1, z_0 is a zero of $\det \psi$ if and only if it is a zero of $\det a$ for any other irreducible MFD $h = a^{-1}b$. The second part of (i) follows from the fact that c divides $\det a$ and that, since $(cI)^{-1}N = h$ holds, $\det(cI) = c^s$ must be a (polynomial) multiple of $\det a$, by Theorem 2.2.1.

(ii) is shown in the same way as the first part of (i).

Now let us introduce a topology for U_A. We identify (k, l) with the sequence $(K(i), L(i)), i \in \mathbf{N}$ of its power series coefficients. Clearly, $(K(i), L(i)) \in (\mathbf{R}^{s \times (s+m)})^{\mathbf{N}}$. Now $(\mathbf{R}^{s \times (s+m)})^{\mathbf{N}}$ is endowed with the product topology of the Euclidean space $\mathbf{R}^{s \times (s+m)}$. $(\mathbf{R}^{s \times (s+m)})^{\mathbf{N}}$, then, is a metrizable space and convergence $(k_t, l_t) \to (k_0, l_0)$ is equivalent to pointwise (in $i \in \mathbf{N}$) convergence of the corresponding power series coefficients, that is,

$$(K_t(i), L_t(i)) \to (K_0(i), L_0(i)) \qquad \text{for all } i \in \mathbf{N}.$$

This is the reason why the corresponding relative topology on U_A is called the *pointwise topology*, T_{pt}. For subsets $U \subset U_A$ we always take the corresponding relative topology.

In general, if A is a set in a topological space, by $\bar{A}$ we denote its closure. Subsets $\Theta \subset \mathbf{R}^d$ are always endowed with the relative topology of $\mathbf{R}^d$.

5. CANONICAL ARMAX AND STATE-SPACE REALIZATIONS

We now consider the case where we have no additional a priori restrictions on the ARMAX or state-space parameters. Thus the model class considered is T_A or Δ_A. In this case two main lines have been followed: canonical realisations, which we treat in this section, and the overlapping parameterization of the manifold of all systems of order n, which is treated in the next section.

A canonical ARMAX (state-space) realization is a value of a mapping $C: T_A/\pi \to T_A$, $(C: \Delta_A/\rho \to \Delta_A)$, attaching to every (k,l)-equivalence class [or alternatively to every $(k,l) \in U_A$] a unique representative.

In many cases we consider only the left prime (the minimal) case; then our problem is to prescribe unique representatives for the classes of left equivalent matrices $\{u(a,b,d) | \det u = \text{const} \neq 0\}$ (or the classes of similarity transforms $\{(TFT^{-1}, HT^{-1}, TK, TL) | \det T \neq 0\}$). Of course, there are many ways of prescribing canonical realizations.

Most of the results presented in this and the following sections have been derived to cope with the genuine problems arising in the multi-output case (i.e., $s > 1$), and in large measure the problems of the single-output case are significantly simpler and concepts that are different for $s > 1$ coincide for $s = 1$. The first canonical realizations (for $s > 1$) have been proposed by Luenberger (1967). The concept has been elaborated in a series of publications, first for the state space, then also for the ARMAX case (see, e.g., Denham, 1974; Dickinson, Kailath, and Morf, 1974; Forney 1975; Glover and Willems, 1974; Guidorzi, 1975; Guidorzi, 1981; Hannan, 1971b; Kalman, 1965; Popov, 1969; Popov, 1972; Rissanen, 1974).

In order to construct canonical ARMAX or state-space realizations, we begin with the (block) Hankel matrix $\mathscr{H}$, by selecting a basis for its row space. As mentioned in Section 4, it is convenient to consider only nice selections, selections with the property that if $h(i+1, j)$ is contained in this selection, then $h(i,j)$ is also. A nice selection is described by a multi-index $\alpha = (n_1 \cdots n_s)$. For the case $s > 1$, clearly nice selections can be defined in several different ways, and as will be seen below, every such selection of a basis defines unique (with respect to this selection) ARMAX or state-space realizations.

First let us consider the case where $y(t)$ is an ARMA process, with Wold representation (1.3.2). Since $\mathscr{H}$ [which now contains only the $K(j)$] has finite rank, by Theorem 2.4.1, we may find a submatrix $\mathscr{H}_\alpha^\infty = S\mathscr{H}$, which consists of basis rows for the row space of $\mathscr{H}$, corresponding to a structure index α.

Thus S is a selector matrix that has each row composed of zeros save for a unit in the column corresponding to the row of $\mathscr{H}$ that this row selects. Put $\varepsilon'_{t-1} = (\varepsilon(t-1)', \varepsilon(t-2)', \ldots)$ and, following Kalman (1963), define the state vector as

$$x(t) = \mathscr{H}_\alpha^\infty \varepsilon_{t-1}. \tag{2.5.1}$$

Then

$$x(t+1) = S\mathscr{H}\varepsilon_t = S\begin{bmatrix} K(2) & K(3) & \cdots \\ K(3) & K(4) & \cdots \\ \vdots & \vdots & \vdots \end{bmatrix}\varepsilon_{t-1} + S\begin{bmatrix} K(1) \\ K(2) \\ \vdots \end{bmatrix}\varepsilon(t). \tag{2.5.2}$$

Since the matrix to which S is applied in the first term on the right side of (2.5.2) is composed of rows of $\mathscr{H}$, its rows are linear combinations of rows of $\mathscr{H}$ and

$$S\begin{bmatrix} K(2) & K(3) & \cdots \\ K(3) & K(4) & \cdots \\ \vdots & \vdots & \vdots \end{bmatrix} = F\mathscr{H}_\alpha^\infty. \tag{2.5.3}$$

Put

$$K = S(K(1)', K(2)', \ldots)'. \tag{2.5.4}$$

Observe that, from (1.1.25), $y(t) = [K(1), K(2), \ldots]\varepsilon_{t-1} + \varepsilon(t)$ and that, for the same reason as gave (2.5.3),

$$[K(1), K(2), \ldots] = H\mathscr{H}_\alpha^\infty. \tag{2.5.5}$$

Thus we have

$$x(t+1) = Fx(t) + K\varepsilon(t), \qquad y(t) = Hx(t) + \varepsilon(t).$$

Note that F and H are uniquely determined from (2.5.3) and (2.5.5), respectively, since $\mathscr{H}_\alpha^\infty$ has rank equal to $n = |\alpha|$ and that (F, H, K) is minimal by Theorems 2.3.2 and 2.3.3. The system (F, H, K) is in prediction error form since the $\varepsilon(t)$ are the innovations. As already stated, the state contains the information from the past of the inputs which is relevant for future outputs; this is clear from (2.5.2) or from (1.3.4).

As first pointed out in Akaike (1974), we can give a nice Hilbert space interpretation to what was said above as follows: It is clear from the expression below (1.3.5) that the linear dependence relations between the rows of $\mathscr{H}$ (note that we always assume that $\Sigma > 0$) and the (one-dimensional) components of the infinite-dimensional state vector $x(t)$ in (1.3.4), considered as elements of $L^2(\Omega, \mathscr{F}, P)$, are identical; that is, if some of the rows of $\mathscr{H}$ are linearly dependent or independent, then so are the corresponding elements of $x(t)$. If $\mathscr{H}(I_\infty \otimes \Sigma^{1/2})$ instead of $\mathscr{H}$ is considered, even the inner products are identical; here

$$I_\infty \otimes \Sigma^{1/2} = \begin{pmatrix} \Sigma^{1/2} & 0 & \cdots \\ 0 & \Sigma^{1/2} & \\ \vdots & & \ddots \\ \vdots & & & \ddots \end{pmatrix}$$

and $\Sigma^{1/2}$ is a symmetric, positive square root of Σ.

It is clear from (2.5.1) and from what was said before that the components of $x(1)$ [in (2.5.1)] form a basis for the *predictor space*, that is, the Hilbert space spanned by the components of the $y(\tau|0)$, $\tau > 0$, in $L^2(\Omega, \mathscr{F}, P)$; or in other words, the projection of the Hilbert space $H_y^+(1)$ spanned by the components of the $y(\tau)$, $\tau > 0$, on the Hilbert space $H_y(0)$ spanned by the components of $y(s)$, $s \leq 0$. By Theorem 2.4.1, k is rational if and only if the predictor space is finite dimensional.

Given (k, l) and the corresponding Hankel matrix $\mathscr{H}$, the corresponding state-space realizations (F, H, K, L) are defined completely analogously. The state is now introduced as

$$x(t) = \mathscr{H}_\alpha^\infty(\varepsilon(t-1)', z(t-1)', \varepsilon(t-2)', z(t-2)', \ldots)' \tag{2.5.6}$$

and the parameter matrices are obtained from

$$F\mathscr{H}_\alpha^\infty = S\begin{pmatrix} K(2), L(2) & K(3), L(3) & \cdots \\ K(3), L(3) & K(4), L(4) & \cdots \\ \cdots & \cdots & \cdots \end{pmatrix} \tag{2.5.7}$$

$$(K, L) = S([K(1), L(1)]', [K(2), L(2)]', \ldots)' \tag{2.5.8}$$

and

$$H\mathscr{H}_\alpha^\infty = (K(1), L(1), K(2), L(2), \ldots) \tag{2.5.9}$$

and again (F, H, K, L) is minimal by Theorems 2.3.2 and 2.3.3.

Since $z(t)$ in general is not white noise, the elements of $x(t)$ do not have the same interpretation as predictors as they had in the ARMA case; also, $(z(t-1)', z(t-2)', \ldots)$ may not necessarily consist only of linearly independent elements.

To construct canonical ARMAX realizations, we begin with equation (2.4.6) [i.e., $\tilde{a}(\tilde{k}, \tilde{l}) = \tilde{b}, \tilde{d}$]. From a comparison of all coefficients in (2.4.6) we obtain

(2.5.10) $$(\tilde{A}(0), \tilde{A}(1), \ldots)\mathscr{H} = 0$$

and

(2.5.11)

$$(\tilde{A}(0), \tilde{A}(1), \ldots)\cdot\begin{pmatrix} 0 & 0 & \cdots \\ K(1), L(1) & 0 & \\ K(2), L(2) & K(1), L(1) & 0 \\ K(3), L(3) & K(2), L(2) & \cdots \\ \vdots & \vdots & \end{pmatrix} = (\tilde{B}(0), \tilde{D}(0), \tilde{B}(1), \tilde{D}(1), \ldots).$$

Thus, given $(\tilde{k}, \tilde{l})$, every $\tilde{a}$ such that $\det \tilde{a} \not\equiv 0$, satisfying (2.5.10) and determining $(\tilde{b}, \tilde{d})$ from (2.5.11), gives a unique MFD of $(\tilde{k}, \tilde{l})$. This leads us to construct canonical ARMAX realizations from $\mathscr{H}$ in a way analogous to canonical state-space realizations. Let us describe this for the, perhaps, most important case, of *echelon* and *reversed echelon (canonical) ARMAX realizations.*

Let $\alpha = (n_1 \cdots n_s)$ be the Kronecker indices of $\mathscr{H}$ and let $h(1,1), \ldots, h(n_1, 1), \ldots, h(1, s), \ldots, h(n_s, s)$ be the corresponding first basis rows of $\mathscr{H}$. If $n_i = 0$, for some i, then no $h(j, i)$, $j = 1, 2, \ldots$ are contained in this basis. Then the equation system

(2.5.12) $$-h(n_i + 1, i) = \sum_{j=1}^{s} \sum_{u=1}^{n_{ij}} \tilde{a}_{ij}(u-1)h(u, j); \qquad i = 1 \cdots s,$$

where

(2.5.13) $$n_{ij} = \begin{cases} \min(n_i + 1, n_j) & \text{for } j < i \\ \min(n_i, n_j) & \text{for } j \geq i \end{cases}$$

defines unique coefficients $\tilde{a}_{ij}(u)$. The equations (2.5.12) give $-h(n_i + 1, i)$ as a linear combination of its *preceding* basis rows. These equations can be considered as special relations of the form (2.5.10), where the $\tilde{a}_{ij}(u)$ are the (i, j) elements in $\tilde{A}(u)$, $\tilde{a}_{ii}(n_i) = 1$, $i = 1 \cdots s$ and all other elements are equal to zero.

In the trivial case $n_i = 0$, $i = 1 \cdots s$ we put $\tilde{a} = I$. In this way we have defined $\tilde{a}$; $(\tilde{b}, \tilde{d})$ is then obtained from (2.5.11). As the Kronecker indices are unique for given $(\tilde{k}, \tilde{l})$ and as (2.5.12) and (2.5.11) in the way described above give a unique MFD $\tilde{a}^{-1}(\tilde{b}, \tilde{d}) = (\tilde{k}, \tilde{l})$, we have defined a unique ARMAX system $(\tilde{a}(z^{-1}), \tilde{b}(z^{-1}) + \tilde{a}(z^{-1}), \tilde{d}(z^{-1}))$ [corresponding to (2.4.4)] called the *echelon ARMAX realization* of (k, l). Using (2.4.5) we obtain from $(\tilde{a}, \tilde{b}, \tilde{d})$ the *reversed echelon ARMAX realization* (a, b, d).

Echelon realizations have been introduced by Popov (1969) and Guidorzi (1975, 1981). For the next theorem see also Dickinson, Kailath, and Morf (1974) and Forney (1975).

Theorem 2.5.1. *(Echelon ARMAX realizations) For every $(k, l) \in U_A$ [with Kronecker indices $\alpha = (n_1 \cdots n_s)$] a unique realization $(\tilde{a}(z^{-1}), \tilde{b}(z^{-1}) + \tilde{a}(z^{-1}), \tilde{d}(z^{-1}))$ is defined by (2.5.12) and (2.5.11). This $(\tilde{a}, \tilde{b}, \tilde{d})$ has the following properties:*

(i) $(\tilde{a}, \tilde{b}, \tilde{d})$ is left prime.

(ii) $\tilde{a}_{ii}$ are monic polynomials and

$$\begin{aligned} &\nu(\tilde{a}_{ij}) \le \nu(\tilde{a}_{ii}) = n_i; && j \le i \\ &\nu(\tilde{a}_{ij}) < \nu(\tilde{a}_{ii}); && j > i \\ &\nu(\tilde{a}_{ji}) < \nu(\tilde{a}_{ii}); && j \ne i \\ &\nu(\tilde{b}_{ij}) < \nu(\tilde{a}_{ii}); && \nu(\tilde{d}_{ij}) < \nu(\tilde{a}_{ii}). \end{aligned}$$

Conversely, every $(\tilde{a}, \tilde{b}, \tilde{d})$ satisfying (i) and (ii) [via (2.4.4)] corresponds to an echelon ARMAX realization.

Proof. The first statement has been proved already. (ii) is a straight forward consequence of (2.5.12) and (2.5.11). Note that by (ii) $\tilde{a}$ is both row reduced and column reduced and thus

$$|\alpha| = \nu(\det \tilde{a})$$

holds. On the other hand, by Theorem 2.4.1, the order n of $(\tilde{k}, \tilde{l})$ is equal to $|\alpha|$ and thus $n = \nu(\det \tilde{a})$ and $(\tilde{a}, \tilde{b}, \tilde{d})$ must be left prime by Lemma 2.2.1.

Conversely, if $(\tilde{a}, \tilde{b}, \tilde{d})$ satisfies (i) and (ii), by the left primeness of $(\tilde{a}, \tilde{b}, \tilde{d})$ and since $\tilde{a}$ is row reduced, $\mathscr{H}$ has rank $n = \nu(\det \tilde{a}) = \sum \nu(\tilde{a}_{ii})$ by Theorem 2.4.1. (ii) together with (2.5.10) imply that the rows $h(j, i), j \le \nu(\tilde{a}_{ii}), i = 1 \cdots s$, span the row space of $\mathscr{H}$ (and thus they are a generating set) where the rows $h(\nu(a_{ii}) + 1, i)$, $i = 1 \cdots s$ are linear combinations of the preceding rows from this generating set. Thus if $\alpha = (m_1 \cdots m_s)$ denotes the Kronecker indices, we have $\nu(\tilde{a}_{ii}) \ge m_i$; but as $n = \sum \nu(\tilde{a}_{ii}) = \sum m_i$, equality must hold and thus

$(\nu(\tilde{a}_{ii}), i = 1 \cdots s)$ are the Kronecker indices and $(\tilde{a}, \tilde{b} + \tilde{a}, \tilde{d})$ is an echelon realization.

By what was said before, we have

$$\tilde{a}(z) = \{\operatorname{diag} z^{n_i}\} + n(z),$$

where the degree of the ith column of $n(z)$ is smaller than n_i, and

$$\tilde{a}(z) = \begin{pmatrix} z^{n_1} & & & & 0 \\ x & z^{n_2} & & & \\ \vdots & & \ddots & & \\ \vdots & & & \ddots & \\ x & x & \cdots & \cdots & z^{n_s} \end{pmatrix} + m(z),$$

where the degree of the ith row of $m(z)$ is smaller than n_i; the last fact implies that for reversed echelon realizations (a, b, d), the matrix $A(0)$ [$= B(0)$] is lower triangular with units in the main diagonal. More generally as a direct consequence of Theorem 2.5.1 and of Lemma 2.4.2, we have the following result for reversed echelon ARMAX realizations. For every $(k, l) \in U_A$ [with Kronecker indices $\alpha = (n_1 \cdots n_s)$] a unique realization (a, b, d) is defined by (2.5.12), (2.5.11), and (2.4.5). (a, b, d) has the following properties:

(i) (a, b, d) is left prime.

(ii) $A(0)$ [$= B(0)$] is lower triangular; all its diagonal elements are equal to 1; $D(0) = 0$; the degree of the ith row of (a, b, d) is n_i; $z^{n_i - n_{ij}}$ [with n_{ij} given by (2.5.13)] is a divisor of $\tilde{a}_{ij}$.

Echelon state-space realizations are defined as follows. Let $(k, l) \in U_A$ be an arbitrary transfer function and let $\mathscr{H} = (h'(1,1), \ldots, h'(1,s), h'(2,1), \ldots, h'(2,s), \ldots)'$ be the corresponding Hankel matrix. For a given multi-index $\alpha = (n_1 \cdots n_s)$ we define

$$\mathscr{H}_\alpha^\infty = (h'(1,1), \ldots, h'(n_1,1), h'(1,2), \ldots, h'(n_2,2), \ldots, h'(1,s), \ldots, h'(n_s,s))'$$

$$\mathscr{H}_{\alpha+s}^\infty = (h'(2,1), \ldots, h'(n_1+1,1), h'(2,2), \ldots, h'(n_2+1,2), \ldots,$$
$$h'(2,s), \ldots, h'(n_s+1,s))'$$

[where it is understood that if $n_i = 0$, then no $h(j,i)$ occurs in $\mathscr{H}_\alpha^\infty$ or in $\mathscr{H}_{\alpha+s}^\infty$] and by $\mathscr{H}_\alpha^1 \in \mathbf{R}^{n \times (s+m)}$ we denote the submatrix of $\mathscr{H}_\alpha^\infty$ consisting of its first $s + m$ columns. We repeat that $\mathscr{H}_1^\infty$ is the matrix consisting of the first row of blocks of $\mathscr{H}$. Note that in $\mathscr{H}_\alpha^\infty$, above, the rows are not in the same order as in $\mathscr{H}$.

Theorem 2.5.2. *(Echelon state-space realizations) Let $(k, l) \in U_A$ and let $\alpha = (n_1 \cdots n_s)$ be its Kronecker indices. Then a unique state-space realization (F, H, K, L) is given by*

$$F\mathscr{H}_\alpha^\infty = \mathscr{H}_{\alpha+s}^\infty \tag{2.5.14}$$

$$H\mathscr{H}_\alpha^\infty = \mathscr{H}_1^\infty \tag{2.5.15}$$

$$(K, L) = \mathscr{H}_\alpha^1. \tag{2.5.16}$$

Here (F, H, K, L) has the following properties:

(i) (F, H, K, L) is minimal.

(ii) $F = (F_{ij})_{i,j=1\dots s} \in \mathbf{R}^{n \times n}$; $F_{ij} \in \mathbf{R}^{n_i \times n_j}$, where

$$F_{ii} = \begin{pmatrix} 0 & & \\ \vdots & I_{n_i-1} & \\ 0 & & \\ -\tilde{a}_{ii}(0), & \dots, & -\tilde{a}_{ii}(n_i - 1) \end{pmatrix};$$

$$F_{ij} = \begin{pmatrix} & & 0 & & & \\ -\tilde{a}_{ij}(0), & \dots, & -\tilde{a}_{ij}(n_{ij}), & 0, & \dots, & 0 \end{pmatrix}, \qquad \text{for } i \neq j$$

where n_{ij} is given by (2.5.13).

(iii) If $n_i > 0$, $i = 1 \cdots s$, then

$$H = \left(\begin{array}{c:c:c:c} \overbrace{1, 0, \dots, 0}^{n_1} & \overbrace{0, 0 \dots, 0}^{n_2} & & \overbrace{0}^{n_s} \\ & 1, 0, \dots 0 & \dots\dots & \\ 0 & 0 & & 1, 0 \dots 0 \end{array}\right).$$

Conversely, every (F, H, K, L) satisfying (i), (ii), (iii) is an echelon state-space realization.

Proof. From (2.5.14) and (ii) it is clear that again the $h(n_i + 1, i)$, $i = 1 \cdots s$ are expressed as linear combinations of the corresponding preceding basis rows and thus the $\tilde{a}_{ij}(u)$ in (2.5.12) and in F are identical. Of course, once F is unique, then H, K, L are uniquely obtained from (2.5.15) and (2.5.16). (ii) and (iii) are straightforward consequences of the construction of (F, H, K, L). Since $\mathscr{H}$ has rank $n = |\alpha|$, (F, H, K, L) is minimal by Theorems 2.3.2 and 2.3.3.

Conversely, if (F, H, K, L) satisfies (i), (ii), and (iii), then the Kronecker indices of $\mathscr{H}$, $m_1 \cdots m_s$ say, must fulfill $m_i \leq n_i$, $i = 1 \cdots s$ and since by (i) and Theorem 2.4.1, $\mathscr{H}$ has rank n, equality must hold.

REMARK. It is straightforward to see that (iii) in Theorem 2.5.2 may be formulated more generally as follows. If $n_i > 0$, the ith row of H is a row of zeros save for a unit in the column numbered

$$\left(\sum_{j=1}^{i-1} n_j\right) + 1.$$

If $n_i = 0$, the ith row of H is a row of zeros save for the columns numbered

$$\left(\sum_{l=0}^{j-1} n_l\right) + 1$$

where j runs over values for which $n_j > 0, j < i$. In these positions the elements $-\tilde{a}_{ij}(0)$ appear.

Now consider the *parameterization* problem associated with echelon realizations. In a first step U_A is partitioned into the subclasses V_α, where V_α, $\alpha = (n_1 \cdots n_s)$ is the set of all $(k, l) \in U_A$ with Kronecker indices α. Every V_α is then parameterized by (2.5.12) and (2.5.11) in the ARMAX case [by (2.5.14)–(2.5.16) in the state-space case].

For given Kronecker index α, in the ARMAX case, we form a parameter vector τ consisting of all entries of the $A(i)$, $B(i)$, $D(i)$, corresponding to reversed echelon canonical realization (arranged in a certain order), which are not explicitly restricted to zero or 1 by (ii) in Theorem 2.5.1. Let $T_\alpha^{(1)}$ denote the set of all such τ corresponding to V_α via (2.5.12) and (2.5.11) and let $\psi_\alpha^{(1)}: V_\alpha \to T_\alpha^{(1)}$ denote the mapping attaching to every $(k, l) \in V_\alpha$ the corresponding parameter vector τ. As is seen from (i) in the next theorem, τ is a vector of free parameters.

Let d_α be the dimension of τ. The one-to-one relation between τ and the corresponding ARMAX realization (a, b, d) obviously can be extended in a one-to-one way to $\mathbf{R}^{d_\alpha}$.

We partition τ as $(\tau_1', \tau_2')'$ where the entries of τ_1 are the $\tilde{a}_{ij}(u)$ in (2.5.12) describing $\tilde{a}$ and τ_2 are the parameters for $\tilde{b}$, $\tilde{d}$. Now consider the case of echelon state-space realizations. Since the $\tilde{a}_{ij}(u)$ in $\tilde{a}$ and in F (and H if $n_i = 0$ for some i) are the same, by Theorem 2.5.2 we can describe (F, H, K, L) by a vector $\vartheta = (\tau_1', \vartheta_2')'$, where the entries of ϑ_2 are the entries of $\mathscr{H}_\alpha^1$.

Note that on $\mathbf{R}^{d_\alpha}$, there is a one-to-one relation between τ and ϑ, since for given (k, l, τ_1) both τ_2 and ϑ_2 are uniquely determined by (2.5.11) and (2.5.16) and as easily seen from (2.5.11), this relation is a homeomorphism (i.e., $\tau \to \vartheta$ and $\vartheta \to \tau$ are continuous) on $\mathbf{R}^{d_\alpha}$.

In this sense we identify τ and ϑ, $T_\alpha^{(1)}$ and the set of all ϑ corresponding to V_α, $\Delta_\alpha^{(1)}$ say, and the ARMAX parameterization $\psi_\alpha^{(1)}: V_\alpha \to T_\alpha^{(1)}$ with the corresponding state-space parameterization $\phi_\alpha^{(1)}: V_\alpha \to \Delta_\alpha^{(1)}$. Let $\beta = (m_1 \cdots m_s)$ and $\alpha = (n_1 \cdots n_s)$. Then $\beta \le \alpha$ means that $m_i \le n_i$, $i = 1 \cdots s$ and $\beta < \alpha$ is used if at least one strict inequality holds.

Theorem 2.5.3. *The ARMAX parameterization $\psi_\alpha^{(1)}: V_\alpha \to T_\alpha^{(1)}$ has the following properties.*

(i) $T_\alpha^{(1)}$ is an open and dense subset of $\mathbf{R}^{d_\alpha}$, where

$$(2.5.17)\quad d_\alpha = |\alpha|(s + m + 1) + \sum_{i,j:\, j<i} \{\min(n_i, n_j) + \min(n_j, n_i + 1)\}; \qquad \alpha = (n_1 \cdots n_s).$$

(ii) $\psi_\alpha^{(1)}: V_\alpha \to T_\alpha^{(1)}$ is a $(T_{pt}$-$)$homeomorphism.

(iii) $\{V_\alpha \,|\, |\alpha| = n\}$ is a disjoint partition of $M(n)$ containing $\binom{n+s-1}{s-1}$ sets.

(iv) $\pi(\overline{T_\alpha^{(1)}}) = \bigcup_{\beta \le \alpha} V_\beta$.

(v) For every $(k, l) \in V_\beta$, $\beta \le \alpha$ the (k, l)-equivalence class in $\overline{T_\alpha^{(1)}}$ is an affine subspace of dimension

$$\sum_{i=1}^{s} \sum_{j=1}^{s} (n_{ij} - n'_{ij}), \; n_{ii} := n_i$$

$$n'_{ij} = \begin{matrix} \min(n_i + 1, m_j), & j < i, \\ \min(n_i, m_j), & j \ge i, \end{matrix}$$

where $\alpha = (n_1 \cdots n_s)$, $\beta = (m_1 \cdots m_s)$ and n_{ij} corresponds to α, via (2.5.13).

(vi) V_α is $(T_{pt}$-$)$open in $\bar{V}_\alpha$.

(vii) $\pi(\overline{T_\alpha^{(1)}}) \subset \bar{V}_\alpha$ and equality holds for $s = 1$.

(viii) $\tau \leftrightarrow \vartheta$ is a homeomorphism on $\mathbf{R}^{d_\alpha}$; the statements (i), (ii), (iv), (v), and (vii) also hold for the state-space parameterization $\phi_\alpha^{(1)}: V_\alpha \to \Delta_\alpha^{(1)}$.

For this theorem, see Deistler and Hannan (1981), Hannan and Kavalieris (1984), and Deistler (1985).

Proof

(viii) The first statement has been shown before and the second statement follows from the respective proofs.

(i) (2.5.17) is easy to see from Theorem 2.5.1 or Theorem 2.5.2. To prove that $\Delta_\alpha^{(1)}$ is open in $\mathbf{R}^{d_\alpha}$, we proceed as follows. Note that the mappings defined on $\mathbf{R}^{d_\alpha}$ by

$$\vartheta \leftrightarrow (F, H, K, L) \mapsto (H', F'H', \ldots, F'^{n-1}H') = \mathcal{O}'$$

$$\vartheta \leftrightarrow (F, H, K, L) \mapsto ((K, L), F(K, L), \ldots, F^{n-1}(K, L)) = \mathcal{C}$$

are continuous. If $\vartheta_0 \in \Delta_\alpha^{(1)}$, then by Theorem 2.5.2, $\vartheta_0 \leftrightarrow (F_0, H_0, K_0, L_0)$ is minimal, and thus (by Theorem 2.3.3) both $\mathcal{O}$ and $\mathcal{C}$ have rank n. But then by the continuity of the mappings defined above, there is a neighborhood of ϑ_0, where $\vartheta \leftrightarrow (F, H, G, K,)$ is still minimal, and thus, by Theorem 2.5.2, this neighborhood is contained in $\Delta_\alpha^{(1)}$ and therefore $\Delta_\alpha^{(1)}$ is open in $\mathbf{R}^{d_\alpha}$. Since $\tau \leftrightarrow \vartheta$ is a homeomorphism on $\mathbf{R}^{d_\alpha}$, $T_\alpha^{(1)}$ is open. By Theorem 2.5.2, $\mathbf{R}^{d_\alpha} - \Delta_\alpha^{(1)}$ corresponds to the set of all (F, H, K, L) satisfying (ii) and (iii) in Theorem 2.5.2 which are not minimal. Let $\vartheta_0 \in \mathbf{R}^{d_\alpha} - \Delta_\alpha^{(1)}$; then in an arbitrarily small neighborhood of ϑ_0 there is a ϑ such that $\mathcal{O}$ and $\mathcal{C}$ both have rank n and thus $\vartheta \in \Delta_\alpha^{(1)}$ and therefore $\Delta_\alpha^{(1)}$ (and $T_\alpha^{(1)}$) is dense in $\mathbf{R}^{d_\alpha}$.

(ii) Cleary, $\psi_\alpha^{(1)}$ is bijective by construction, since π restricted to $\mathbf{R}^{d_\alpha}$ is well defined and π restricted to $T_\alpha^{(1)}$ is injective, as can be seen from (2.5.12) and (2.5.11). From a comparison of coefficients in $a(k, l) = (b, d)$, we get

$$(K(1), L(1)) = A^{-1}(0)[(B(1), D(1)) - A(1)(I, 0)]$$

$$(K(2), L(2)) = A^{-1}(0)[(B(2), D(2)) - A(1)(K(1), L(1)) - A(2)(I, 0)] \tag{2.5.18}$$

$$\cdots\cdots\cdots\cdots\cdots\cdots\cdots\cdots\cdots\cdots\cdots$$

and thus, since $\det A(0) = 1$, π restricted to $\mathbf{R}^{d_\alpha}$ is $(T_{pt}$-)continuous. Conversely, let (k_0, l_0), $(k_t, l_t) \in V_\alpha$ and $(k_t, l_t)' \to (k_0, l_0)$ (in T_{pt}); then, using an obvious notation for the corresponding Hankel matrices and their submatrices $\mathcal{H}_{\alpha,t}^{\infty} \to \mathcal{H}_{\alpha,0}^{\infty}$ and $h_t(n_i + 1, i) \to h_0(n_i + 1, i)$ (elementwise). Since $\mathcal{H}_{\alpha,0}^{\infty}$ has rank n, there is an $n \times n$ submatrix of $\mathcal{H}_{\alpha,0}^{\infty}$ which is nonsingular and (since the regular matrices are an open set) the corresponding submatrices of $\mathcal{H}_{\alpha,t}^{\infty}$ also have rank n from a certain t_0 onward. Thus the coefficients $a_{ij}(u)$ in (2.5.12) continuously depend on the transfer functions. (2.5.11) then implies that also τ_2 continuously depends on (k, l) and thus $\psi_\alpha^{(1)}$: $V_\alpha \to T_\alpha^{(1)}$ is bijective, continuous mapping, and $\psi_\alpha^{(1)-1}$ is also continuous.

(iii) It is straightforward to check that there are $\binom{n+s-1}{s-1}$ indices $\alpha = (n_1 \cdots n_s)$ such that $|\alpha| = n$ and every such α is a Kronecker index for a suitably chosen (k, l). The rest is an obvious consequence of the fact that for every $(k, l) \in M(n)$ the Kronecker indices are uniquely defined.

(iv) If $(k, l) \in V_\beta$, $\beta = (m_1 \cdots m_s) < \alpha = (n_1 \cdots n_s)$, τ can be embedded in $\mathbf{R}^{d_\alpha}$ in a natural way, where in the reversed echelon realization, $(a, b, d) \leftrightarrow \tau \in T_\beta^{(1)}$, which has row degrees $m_1 \cdots m_s$, zero elements as coefficients are added up to the maximum powers described by α. But then [by (i)] $\tau \in \mathbf{R}^{d_\alpha} = \overline{T_\alpha^{(1)}}$. Conversely, let $\tau_0 \in \overline{T_\alpha^{(1)}}$. Then there is a sequence $\tau_t \in T_\alpha^{(1)}$ converging to τ_0, and thus by the continuity of π we have $\mathscr{H}_t \to \mathscr{H}_0$ (elementwise) for the corresponding Hankel matrices. Since $\mathscr{H}_{\gamma,t}^\infty$, for all $\gamma \geq \alpha$, have rank n, by the continuity of the determinant, $\mathscr{H}_{\gamma,0}^\infty$, for all $\gamma \geq \alpha$, has rank less than or equal to n and thus $\pi(\tau_0) \in V_\beta$, $\beta \leq \alpha$.

(v) Let $(k_0, l_0) \in V_\beta$, $\beta = (m_1 \cdots m_s) \leq \alpha = (n_1 \cdots n_s)$. Consider equation (2.5.12), where $h(n_i + 1, i)$ is expressed as a linear combination of the rows $h(u, i)$, $u \leq n_i$ and $h(u,j)$; $u \leq n_{ij}, j \neq i$, where n_{ij} is defined by (2.5.13). But only the rows $h(u, i)$, $u \leq m_i$ and $h(u,j)$, $u \leq \min(n_i + 1, m_j)$, for $j < i$, and $\min(n_i, m_j)$ for $j \geq i$ are linearly independent and thus [for given (k, l)] the set of solutions of (2.5.12), for $i = 1 \cdots s$ is an affine subspace of dimension

$$\sum_{i=1}^{s} \sum_{j=1}^{s} (n_{ij} - n'_{ij})'$$

Equation (2.5.11) then shows that the corresponding solutions for the parameters τ_2 in the $\tilde{B}(i)$, $\tilde{D}(i)$ also form an affine subspace of the same dimension and are unique for given τ_1 and the analogous statement holds for ϑ_2.

(vi) Let $(k_0, l_0) \in V_\alpha$; then $H_{\gamma,0}^\infty$, for all $\gamma \geq \alpha$ has rank $n = |\alpha|$. Thus, as the nonsingular matrices of given dimension form an open set, there is a neighbourhood of (k_0, l_0) where $\mathscr{H}_\alpha^\infty$ still has rank n. By the continuity of the determinant, if every $\mathscr{H}_\beta^\infty$ corresponding to $(k, l) \in V_\alpha$ contains a certain linear dependent row, the same must hold for $\mathscr{H}_\beta^\infty$ corresponding to $(k, l) \in \bar{V}_\alpha$. Thus by Theorem 2.4.1 there is a neighborhood for (k_0, l_0) in $\bar{V}_\alpha$ which is still contained in V_α.

(vii) $\pi(\overline{T_\alpha^{(1)}}) \subset \bar{V}_\alpha$ immediately follows from the continuity of π. For $s = 1$ we have $\alpha = (n)$ $[M(n) = V_{(n)}]$, and the result follows from the fact that for $(k, l) \in \bar{V}_{(n)}$, $\mathscr{H}_{(m)}^\infty$, for $m \geq n$ has rank less than or equal to n and from (iv).

REMARK 1. We now discuss the results of Theorem 2.5.3. By (i) the $\tau \in T_\alpha^{(1)}$ are indeed free parameters. (iv) and (v) show that the boundary $\overline{T_\alpha^{(1)}} - T_\alpha^{(1)}$ of $T_\alpha^{(1)}$ consists of equivalence classes, each of which is an affine subspace, and these equivalence classes correspond to all lower-dimensional systems (i.e., systems such that their Kronecker indices are smaller than α). The latter, by Theorem 2.5.1, means that for the corresponding (a, b, d) either the row degrees are smaller than those prescribed by α or the (a, b, d) are not relatively left prime. Of special interest are the consequences of Theorem 2.5.3 for estimation. (viii) says that, in a certain sense, estimation of state space and of ARMAX parameters is the same thing. As will be shown in Chapter 4 (for maximum-likelihood-type estimators at least), the "natural" consistency result is coordinate free in the sense that only transfer functions are primarily concerned; it says that if $(k_0, l_0) \in \bar{V}_\alpha$ are the true transfer functions, their estimators over V_α, (k_t, l_t), say (which may also attain values in $\bar{V}_\alpha$), converge almost surely in T_{pt} to (k_0, l_0). Now going back into coordinates, we have the following. If $(k_t, l_t) \in \bar{V}_\alpha$ is any sequence (not necessarily the likelihood estimates) converging in T_{pt} to $(k_0, l_0) \in \bar{V}_\alpha$, three cases have to be distinguished.

(i) If $(k_0, l_0) \in V_\alpha$, then by (vi) and (ii), (k_t, l_t) will be in V_α from a certain T_0 onward; thus $\psi_\alpha^{(1)}(k_t, l_t)$ is defined for $t \geq T_0$ and $\psi_\alpha^{(1)}(k_t, l_t) \to \psi_0^{(1)}(k_0, l_0)$. Therefore, in this case, the consistency result mentioned above for the transfer functions implies consistency for the parameters.

(ii) Consider $(k_0, l_0) \in \pi(\overline{T_\alpha^{(1)}}) - V_\alpha$; that is, there is a $\beta < \alpha$ such that $(k_0, l_0) \in V_\beta$ and thus (k_0, l_0) is represented by a nontrivial equivalence class in $\overline{T_\alpha^{(1)}}$. Then we have the following situation. If we put a suitable prior bound on the norm of the elements $\tau \in \overline{T_\alpha^{(1)}}$ $(= \mathbf{R}^{d_\alpha})$ by $\|\tau\| \leq C \in \mathbf{R}^+$, then using an obvious notation $a_t^{-1}(b_t, d_t) \to (k_0, l_0)$ implies that $(b_t, d_t) - a_t(k_0, l_0) \to 0$ and thus the distance of any (not necessarily unique) $\tau_t \in \mathbf{R}^{d_\alpha}$, corresponding to (k_t, l_t), to the (k_0, l_0)-equivalence class converges to zero. Without imposing such a bound, the last statement is not true in general; as can easily be seen from an example, there are sequences τ_t such that $\pi(\tau_t)$ converges to (k_0, l_0), but τ_t does not converge to the corresponding equivalence class.

(iii) The case $(k_0, l_0) \in \bar{V}_\alpha - \pi(\overline{T_\alpha^{(1)}})$ can occur only for $s > 1$. In this case (k_0, l_0) corresponds to the "point of infinity" in $\overline{T_\alpha^{(1)}}$ in the sense that even if $(k_t, l_t) \in V_\alpha$, then $(k_t, l_t) \to (k_0, l_0)$ implies that $\|\psi_\alpha^{(1)}(k_t, l_t)\| \to \infty$.

We illustrate (ii) and (iii) by an example. Take the ARMA case, $s = 2$ and $n_1 = n_2 = 1$. Consider the reversed echelon form with $A_t(0) = B_t(0) = I_2$ and

$$A_t(1) = \begin{bmatrix} -(a_t + \alpha_t) & -a_t \\ a_t & a_t + \gamma_t \end{bmatrix}, \qquad B_t(1) = \begin{bmatrix} -a_t & -a_t \\ a_t & a_t \end{bmatrix}.$$

Assume that $\alpha_t \to 0, \gamma_t \to 0, a_t\alpha_t \to \alpha, a_t\gamma_t \to \gamma$. Unless $0 = \alpha = \gamma$, then, of course, the a_t sequence must diverge with $|a_t| \to \infty$. Now it is easy to see that $k_t(z)$ converges to

$$k_0(z) = \begin{bmatrix} 1 - (\alpha + \gamma)z^2 & 0 \\ 1 & 1 \end{bmatrix}^{-1} \begin{bmatrix} 1 - \gamma z^2 & -\gamma z^2 \\ 1 & 1 \end{bmatrix}.$$

At $\alpha = \gamma = 0$ this is I_2 and we can have points in $T_\alpha^{(1)}$ converging to points in $\overline{T_\alpha^{(1)}}$ whose images under the mapping π are in V_β for $\beta < \alpha$. Of course, we could also have sequences with $|a_t| \to \infty$ for which $k_t(z) \to I_2$, by allowing α_t, γ_t to converge fast to zero. For either or both of α, γ not zero, the Kronecker indices are $n_1 = 2$, $n_2 = 0$ and $k_0(z) \notin \pi(\overline{T_\alpha^{(1)}})$ since it belongs to V_β for $\beta \nleq \alpha$.

This case can arise only because $\|\psi_\alpha^{(1)}(k_t)\| \to \infty$ (i.e., τ_t goes to the 'point at infinity').

REMARK 2. Let us discuss the implications of additionally imposing the stability assumption $\det a(z) \neq 0 |z| \leq 1$ and the miniphase assumption $\det b(z) \neq 0 |z| < 1$, for the results of Theorem 2.5.3. For the moment let T_α denote the subset of $T_\alpha^{(1)}$ where the stability and the miniphase assumption holds. As easily seen, in general, T_α is neither open nor dense in $\mathbf{R}^{d_\alpha} (= \overline{T_\alpha^{(1)}})$; however, T_α contains the open set $\{\tau \in T_\alpha | \det b(z) \neq 0 |z| \leq 1\}$. $\bar{T}_\alpha - T_\alpha$ is the set of all $\tau \in \mathbf{R}^{d_\alpha}$ that are either in $T_\alpha^{(1)}$, and where $\det a(z) \neq 0$, $|z| < 1$, $\det a(z) = 0$ for some $|z| = 1$ and $\det b(z) \neq 0 |z| < 1$ hold, or which are in $\overline{T_\alpha^{(1)}} - T_\alpha^{(1)}$ and where in addition $\det a(z) \neq 0$, $\det b(z) \neq 0$, $|z| < 1$. Thus $\bar{T}_\alpha$ contains also unstable systems where $\det a$ has roots of modulus 1. The (k, l) equivalence classes in $\bar{T}_\alpha$ are subsets of the affine subspaces described in (iv) of Theorem 2.5.3, which contain a relatively open set.

There are many other ways to obtain canonical realizations from $\mathscr{H}$, for example, *Hermite* realizations (see, e.g., Dickinson, Kailath, and Morf, 1974; Kailath, 1980; Rosenbrock, 1970) are obtained analogously to echelon realizations from (2.5.12), (2.5.11) in the ARMAX or from (2.5.14)–(2.5.16) in the state-space case, the only difference being that now the *first* basis rows $h(i, j)$ of $\mathscr{H}$ are defined as the first with respect to lexicographical order of (j, i); that is, we take all $h(i, 1,)$ $i = 1 \cdots \tilde{n}_1$, which are linear independent, next all $h(i, 2)$, $i = 1 \cdots \tilde{n}_2$ such that $h(i, 1)$, $i = 1 \cdots \tilde{n}_1$, $h(i, 2)$, $i = 1 \cdots \tilde{n}_2$ are linear independent, and so on.

With the same argument as in the proof of Theorem 2.5.1 it can be shown that $(\tilde{a}, \tilde{b}, \tilde{d})$ obtained from such a selection $\alpha = (\tilde{n}_1 \cdots \tilde{n}_s)$ from (2.5.12) and (2.5.11) is left prime. Furthermore, it is straightforward to show that $(\tilde{a}, \tilde{b}, \tilde{d})$ is obtained by bringing $\hat{a}$ from an arbitrary left prime MFD $(\hat{a}, \hat{b}, \hat{d})$ of $\tilde{a}^{-1}(\tilde{b}, \tilde{d})$ into Hermite form (see Lemma 2.2.2). One disadvantage of Hermite ARMAX realizations is that they are not necessarily row reduced and thus *reversed Hermite* realizations, defined by (2.4.5), are not necessarily left prime.

6. THE MANIFOLD STRUCTURE OF $M(n)$

In this section we also consider the case where no additional a priori restrictions on the parameters are used. However, we use other parameterizations, taking into account the fact that $M(n)$ is a manifold.

We consider U_α, that is, the set of all transfer functions $(k, l) \in U_A$ where $\alpha = (n_1 \cdots n_s)$ is a structure index, or in other words where the rows of $\mathscr{H}_\alpha^\infty$ form a basis for the row space of $\mathscr{H}$. Clearly, $U_\alpha \subset M(n)$; $n = |\alpha|$.

ARMAX realizations (a, b, d) for $(k, l) \in U_\alpha$ are obtained as follows. For given $\mathscr{H}$, the equation system

$$(2.6.1) \qquad h(n_i+1,i) = -\sum_{j=1}^{s}\sum_{u=1}^{n_j} \tilde{a}_{ij}(u-1)h(u,j); \qquad i = 1\cdots s;$$

defines unique coefficients $\tilde{a}_{ij}(u)$ (since $\mathscr{H}_\alpha^\infty$ has full rank n) and (2.6.1) can be considered as a special relation of the form (2.5.10), where the $\tilde{a}_{ij}(u)$ are the (i,j) entries in $\tilde{A}(u)$, where $\tilde{a}_{ii}(n_i) = 1, i = 1\cdots s$ and where all other elements are equal to zero. Having defined $\tilde{a}(z) = \sum \tilde{A}(u)z^u$ in this way from (2.6.1) we obtain $(\tilde{b},\tilde{d})$ from (2.5.11) and thus an MFD $\tilde{a}^{-1}(\tilde{b},\tilde{d}) = (\tilde{k},\tilde{l})$ and an ARMAX realization (a,b,d) from (2.4.5). This (a,b,d) is unique with respect to a given structure index α. However, since in general a structure index is not unique for given $\mathscr{H}$, we did not define a canonical realization in this way.

A difference between (2.6.1) and (2.5.12) (where Echelon canonical realizations have been defined) is that in (2.6.1) $h(n_i+1,i)$ is expressed as a linear combination of *all* basis rows, whereas in (2.5.12) only the basis rows that *precede* $h(n_i+1,i)$ are used.

Overlapping state-space parameterizations of $M(n)$ have been proposed by Glover and Willems (1974) and Ljung and Rissanen (1976). The manifold structure of $M(n)$ was discovered by Kalman (1974) and Clark (1976); see also Hazewinkel and Kalman (1976) and Byrnes and Martin (1980). Some of the additional properties of the overlapping ARMAX parameterization of $M(n)$ are discussed in Deistler and Hannan (1981).

Theorem 2.6.1. *For every $(\tilde{k},\tilde{l}) \in U_\alpha$ an MFD $\tilde{a}^{-1}(\tilde{b},\tilde{b})$, unique with respect to $\alpha = (n_1\cdots n_s)$, and thus a corresponding ARMAX realization (a,b,d), is defined by (2.6.1), (2.5.11), and (2.4.5). $(\tilde{a},\tilde{b},\tilde{d})$ has the following properties:*

(i) $(\tilde{a},\tilde{b},\tilde{d})$ is left prime.

(ii) $\tilde{a}_{ii}$ are monic.

(iii) $v(\tilde{a}_{ji}) < v(\tilde{a}_{ii}) = n_i; j \neq i$.

Proof. The first statement has been proved already and (ii) and (iii) are straight forward to see from the way $\tilde{a}$ has been defined. $\tilde{a}$ is column reduced due to (ii) and (iii), and thus $v(\det \tilde{a}) = n_1 + \cdots + n_s$. On the other hand, $n = |\alpha| = n_1 + \cdots + n_s$ is the rank of $\mathscr{H}$ and thus, by Theorem 2.4.1 and Lemma 2.2.1, $(\tilde{a},\tilde{b},\tilde{d})$ is left prime.

The row degrees of $\tilde{a}$ may be larger than the corresponding column degrees n_i, and a is not necessarily row reduced. Thus in the corresponding ARMAX realization (a,b,d), $A(0)$ may be singular and thus (see Lemma 2.4.2) (a,b,d) may not be left prime.

As has already been mentioned, $U_\alpha \supset V_\alpha$. If $\alpha = (n_1\cdots n_s)$ is such that $n_1 = n_2 \cdots = n_k = n_{k+1} + 1 \cdots = n_s + 1$ for a suitable k (i.e., the rows of $\mathscr{H}_\alpha^\infty$ are just the first rows of $\mathscr{H}$), we have $U_\alpha = V_\alpha$. This V_α is called the *generic neighborhood* of $M(n)$; clearly in this case, (a,b,d) must be left prime.

A similar result holds for the corresponding state-space realizations.

Theorem 2.6.2. *For every $(k,l) \in U_\alpha$ a state-space realization (F,H,K,L), unique with respect to $\alpha = (n_1 \cdots n_s)$, is defined by*

$$F \cdot \mathscr{H}_\alpha^\infty = \mathscr{H}_{\alpha+s}^\infty \tag{2.6.2}$$

$$H \cdot \mathscr{H}_\alpha^\infty = \mathscr{H}_1^\infty \tag{2.6.3}$$

$$(K,L) = \mathscr{H}_\alpha^1. \tag{2.6.4}$$

(F,H,K,L) has the following properties:

(i) (F,H,K,L) is minimal.

(ii) $F = (F_{ij})_{i,j=1\cdots s} \in \mathbf{R}^{n\times n}$ ($n = |\alpha|$) and $F_{ij} \in \mathbf{R}^{n_i \times n_j}$, where

$$F_{ii} = \begin{pmatrix} 0 & & \\ \vdots & I_{n_i-1} & \\ 0 & & \\ -\tilde{a}_{ii}(0), & \ldots, & -\tilde{a}_{ii}(n_i-1) \end{pmatrix}; \; F_{ij} = \begin{pmatrix} & 0 & \\ -\tilde{a}_{ij}(0), & \ldots, & -\tilde{a}_{ij}(n_j-1) \end{pmatrix}.$$

for $i \neq j$

(iii) (H,K,L) have the same properties as in Theorem 2.5.2(iii) and as in the remark subsequent to this theorem, with the only exception that if $n_i = 0$ in the ith row of H, only the $-\tilde{a}_{ij}(0)$ appear. Conversely, if (F,H,K,L) satisfies (i)–(iii), its transfer function is in U_α, $\alpha = (n_1 \cdots n_s)$.

Proof. Since $(k,l) \in U_\alpha$ implies that $\mathscr{H}_\alpha^\infty$ has rank $n = |\alpha|$, (F,H,K,L) is uniquely defined from (2.6.2)–(2.6.4). (ii) is straightforward to show; the "nontrivial" part of (2.6.2) shows that again the $h(n_i+1,i)$, $n_i > 0$, are expressed as linear combinations of all (basis) rows in $\mathscr{H}_\alpha^\infty$. Thus the $\tilde{a}_{ij}(u)$ in (ii) are the same as in (2.6.1); for $n_i = 0$ they appear in H. (iii) is a direct consequence of (2.6.3) and (2.6.4). Since $\mathscr{H}$ has rank $n = |\alpha|$, (F,H,K,L) is minimal by Theorems 2.3.2 and 2.3.3. Conversely, if (F,H,K,L) satisfies (i)–(iii), then $\mathscr{H}$ has rank n (by Theorem 2.3.2) and $\alpha = (n_1 \cdots n_s)$ is a structure index since (2.6.2) is fulfilled and thus $\mathscr{H}_\gamma^\infty$, for $\gamma \geq \alpha$ has the same rank as $\mathscr{H}_\alpha^\infty$, which then must be n.

A state-space system (F,H,K,L) with the properties (i)–(iii) of Theorem 2.6.2 can be described by a vector $\vartheta \in \mathbf{R}^{n(2s+m)}$ whose elements are the ns entries $\tilde{a}_{ij}(u)$ of F in (ii) (and of H, if $n_i = 0$, for some i) and all $n(s+m)$ entries of (K,L) (arranged in a certain order). Let $\Delta_\alpha^{(2)} \subset \mathbf{R}^{n(2s+m)}$ denote the set of all such vectors ϑ corresponding to U_α and let $\phi_\alpha^{(2)}\colon U_\alpha \to \Delta_\alpha^{(2)}$ denote the corresponding state-space parameterization. Again imposing only (ii) and (iii) of the last theorem (and not minimality), the one-to-one relation $\Delta_\alpha^{(2)} \ni \vartheta \leftrightarrow (F, H, K, L)$ can be extended in the natural way to a one-to-one relation between $\vartheta \in \mathbf{R}^{n(2s+m)}$ and the corresponding (F, H, K, L).

The concept of an analytic manifold plays some part in our consideration, although no deep theory is needed. We digress now to define it. As already said, we will consider sets of transfer functions which are endowed with the pointwise topology. These spaces are metrizable and thus Hausdorff (i.e., for any two different points there are disjoint neighborhoods for each point, which implies the existence of unique limits for convergent series). They are separable (i.e., they have a countable dense subset). These properties follow directly from the fact that the product space $(\mathbf{R}^{s \times (s+m)})^{\mathbf{N}}$ has the same properties.

A *real analytic (coordinate) manifold* of dimension n (see, e.g., Auslander and MacKenzie 1977) is a separable Hausdorff space M together with a family $((U_\alpha, \phi_\alpha), \alpha \in I)$ where U_α is an open subset of M and ϕ_α is a homeomorphism from U_α onto an open subset of $\mathbf{R}^n$ satisfying the following properties:

(i) $((U_\alpha), \alpha \in I)$ is an (open) covering of M.

(ii) If $U_\alpha \cap U_\beta \neq \emptyset$, then $\phi_\beta \circ \phi_\alpha^{-1}\colon \phi_\alpha(U_\alpha \cap U_\beta) \to \phi_\beta(U_\alpha \cap U_\beta)$ is an analytic function [i.e., it is expressible by a convergent power series in some neighborhood of each point in $\phi_\alpha(U_\alpha \cap U_\beta)$].

The integer n is called the dimension of the manifold, U_α a coordinate neighborhood, ϕ_α a coordinate mapping, $((U_\alpha, \phi_\alpha) | \alpha \in I)$ a system of local coordinates, and finally, $\phi_\beta \circ \phi_\alpha^{-1}$ is called a coordinate transformation.

Theorem 2.6.3. *$M(n)$ together with $((U_\alpha, \phi_\alpha^{(2)}) | \, |\alpha| = n)$ is a real analytic manifold of dimension $n(2s + m)$.*

Proof. By Theorem 2.6.2, $\phi_\alpha^{(2)}\colon U_\alpha \to \Delta_\alpha^{(2)}$ is bijective. Since $(K(i), L(i)) = HF^{i-1}(K, L)$; $i > 0$, ρ and thus $(\phi_\alpha^{(2)})^{-1}$ are continuous and $\phi_\alpha^{(2)}$ from (2.6.2) is easily seen to be continuous, since $\mathscr{H}_\alpha^\infty$ has rank n on U_α; thus $\phi_\alpha^{(2)}$ is a $(T_{pt}$-)homeomorphism.

The openness of $\Delta_\alpha^{(2)}$ is shown in the same way as in Theorem 2.5.3. The mapping defined on $\mathbf{R}^{n(2s+m)}$ $(\supset \Delta_\alpha^{(2)})$ by

$$\vartheta \leftrightarrow (F, H, K, L) \to (H', F'H', \ldots, F'^{n-1}H') = \mathcal{O}'$$

is continuous. Thus for any $\vartheta_0 \in \Delta_\alpha^{(2)}$, there is a neighborhood in $\mathbf{R}^{n(2s+m)}$, where $\mathcal{O}$ is still of rank n. With an analogous argument there is also a neighborhood of ϑ_0 where in addition $\mathscr{C} = ((K, L), F(K, L), \ldots, F^{n-1}(K, L))$ has rank n and thus by Theorem 2.3.3, $(F, H, K, L) \leftrightarrow \vartheta$ is minimal over this neighborhood. Then, by the last statement of Theorem 2.6.2, this neighborhood is contained in $\Delta_\alpha^{(2)}$ and thus $\Delta_\alpha^{(2)}$ is open in $\mathbf{R}^{n(2s+m)}$.

Finally, we prove that $\phi_\beta^{(2)} \circ \phi_\alpha^{(2)-1}\colon \phi_\alpha^{(2)}(U_\alpha \cap U_\beta) \to \phi_\beta^{(2)}(U_\alpha \cap U_\beta)$ is an analytic function. Let $(F_\alpha, H_\alpha, K_\alpha, L_\alpha)$ and $(F_\beta, H_\beta, K_\beta, L_\beta)$ $(|\alpha| = |\beta|)$ denote the realizations (2.6.2)–(2.6.4) corresponding to α and β, respectively. As can easily be seen (due to Theorem 2.4.1) $(F_\beta, H_\beta, K_\beta, L_\beta)$ can be obtained using only the entries of $\mathscr{H}_{n+1}^{n+1}$, instead of those of $\mathscr{H}$ in (2.6.2)–(2.6.4). Clearly, the function defined by $(F_\alpha, H_\alpha, K_\alpha, L_\alpha) \mapsto \mathscr{H}_{n+1}^{n+1}$ [given by (2.3.6)] is a finite and rational function, and so (by Cramer's rule) is the function defined by $\mathscr{H}_{n+1}^{n+1} \mapsto (F_\beta, H_\beta, K_\beta, L_\beta)$. Therefore, $\phi_\beta^{(2)} \circ \phi_\alpha^{(2)-1}$ is a finite rational function; as $\phi_\alpha^{(2)}(U_\alpha \cap U_\beta)$ is open, then $\phi_\beta^{(2)} \circ \phi_\alpha^{(2)-1}$ is a real analytic function.

Clearly, in the case $s = 1$ we have $\alpha = (n)$ and $M(n) = U_\alpha$ $(= V_\alpha)$, and thus in this case only one coordinate neighborhood is needed to cover $M(n)$. For general s, in our case $M(n)$ is covered by the $\binom{n+s-1}{s-1}$ coordinate neighborhoods U_α such that $|\alpha| = n$. It can be shown that every such U_α is needed to cover $M(n)$. As has been shown in Hazewinkel (1979) using other systems of local coordinates, the number of coordinate neighborhoods may be reduced, but in general $M(n)$ cannot be described by a single coordinate neighborhood.

Now let us consider the ARMAX parameterization of U_α corresponding to the realizations (2.6.1), (2.5.11). Whereas the restrictions in (F, H, K, L) [given in (ii), (iii) of Theorem 2.6.2] are of a simple form (they are either "zero" or "one" restrictions) and all other entries form the vector ϑ of free parameters (indeed, ϑ is a vector of free parameters by Theorem 2.6.3), not all the entries in the $\tilde{A}(i)$, $\tilde{B}(i)$, $\tilde{D}(i)$ corresponding to $(\tilde{a}, \tilde{b}, \tilde{d})$ defined in Theorem 2.6.1 which are not explicitly restricted by (ii) and (iii) in this theorem are free in general, as there are nonlinear restrictions between them.

We have:

Theorem 2.6.4. *The mapping $\psi_\alpha^{(2)}$ attaching to every $(k, l) \in U_\alpha$,* the vector *$\tau \in \mathbf{R}^{n(2s+m)}$ consisting of the elements*

$$\tilde{a}_{ij}(u), \quad u = 0, 1, \ldots, n_j - 1; \quad j, i = 1 \cdots s \tag{2.6.5}$$

$$\tilde{b}_{ij}(u), \quad u = 0, 1, \ldots, n_i - 1; \quad i, j = 1 \cdots s \tag{2.6.6}$$

$$\tilde{d}_{ij}(u), \quad u = 0, 1, \ldots, n_i - 1; \quad i = 1 \cdots s, \ j = 1 \cdots m \tag{2.6.7}$$

of $(\tilde{a}, \tilde{b}, \tilde{d})$ in Theorem 2.6.1 is an ARMAX parameterization of U_α and a homeomorphism between U_α and $T_\alpha^{(2)} = \psi_\alpha^{(2)}(U_\alpha)$, where $T_\alpha^{(2)}$ is open in $\mathbf{R}^{n(2s+m)}$.

Proof. Let $\tau_1 \in \mathbf{R}^{ns}$ denote the vector made from the $\tilde{a}_{ij}(u)$ in (2.6.5) and write $\vartheta \in \Delta_\alpha^{(2)}$ as $\vartheta = (\tau_1', \vartheta_2')'$, where ϑ_2 consists of the elements of $(K, L) = \mathscr{H}_\alpha^1$, and write $\tau \in T_\alpha^{(2)}$ as $\tau = (\tau_1', \tau_2')'$. Then for given τ_1, τ_2 is a linear function of ϑ_2, as we have from (2.5.11)

$$\tilde{b}_{ij}(u) = \sum_{l=1}^{s} \sum \tilde{a}_{il}(v) k_{lj}(u - v),$$

where $k_{lj}(u)$ is the (l, j) element of $K(u)$ and where the inner summation is ranging over $u \le v \le n_l - 1 + \delta_{il}$ ($\delta_{il} = 0$ for $i \ne 1$, $\delta_{il} = 1$ for $i = l$). Let b_j denote the vector of the $\tilde{b}_{ij}(u)$, $u = 0, 1, \ldots, n_i - 1$, $i = 1 \cdots s$, and let k_j denote the vector of the $k_{ij}(u)$, $u = 1 \cdots n_i$, $i = 1 \cdots s$, then

$$b_j = Ak_j, \tag{2.6.8}$$

where A is a function of the $\tilde{a}_{ij}(u)$ in τ_1 independent of j and arranging the elements of b_j, k_j in a suitable order, A can be made upper triangular with units in the main diagonal. A completely analogous result holds for the $d_{ij}(u)$. Thus, for given τ_1, the relation between τ_2 and ϑ_2 is one-to-one and a homeomorphism. Therefore, also the mapping defined by $\tau' = (\tau_1', \tau_2')' \leftrightarrow \vartheta = (\tau_1', \vartheta_2')$ is a homeomorphism.

As can easily be seen from the last proof, the mapping $\tau \leftrightarrow \vartheta$ defined by the relations of the form (2.6.8) can be extended to a homeomorphism on $\mathbf{R}^{n(2s+m)}$.

Next we consider some additional properties of the parameterizations considered. (The mapping, ρ: $\Delta_\alpha \to U_\alpha$, was defined in Section 1.)

Theorem 2.6.5. *$M(n)$ together with $((U_\alpha, \phi_\alpha^{(2)}) \,|\, |\alpha| = n)$ has the following properties.*

- *(i) $\Delta_\alpha^{(2)}$ ($|\alpha| = n$) is dense in $\mathbf{R}^{n(2s+m)}$.*
- *(ii) $\bar{U}_\alpha = \overline{M(n)} = \bigcup_{i \le n} M(i)$ (for any $|\alpha| = n$).*
- *(iii) $\rho(\overline{\Delta_\alpha^{(2)}}) = \bigcup_{\beta \le \alpha} U_\alpha$.*
- *(iv) For every $(k, l) \in \rho(\overline{\Delta_\alpha^{(2)}})$, the (k, l)-equivalence class in $\overline{\Delta_\alpha^{(2)}}$ is an affine subspace of dimension $s(|\alpha| - n_0)$, where n_0 is the order of $(\tilde{k}, \tilde{l})$.*
- *(v) U_α is (T_{pt}-)open in $\bar{U}_\alpha$.*
- *(vi) $\rho(\Delta_\alpha^{(2)}) \subset \bar{U}_\alpha$ and equality holds for $s = 1$.*
- *(vii) $\tau \leftrightarrow \vartheta$ is a homeomorphism on $\mathbf{R}^{n(2s+m)}$; the statements (i), (iii), (iv), and (vi) also hold for the ARMAX parameterization $\psi_\alpha^{(2)}$: $U_\alpha \to T_\alpha^{(2)}$.*

Proof

(vii) The first statement has been proved before. The remainder will be seen from the corresponding proofs below.

(i) By Theorem 2.6.2, $\mathbf{R}^{n(2s+m)} - \Delta_\alpha^{(2)}$ corresponds to the set of nonminimal (F, H, K, L) satisfying (ii) and (iii) in Theorem 2.6.2. Let $\tau_0 \in \mathbf{R}^{n(2s+m)} - \Delta_\alpha^{(2)}$, then by Theorem 2.3.3, at least one of the matrices $\mathcal{O}$ and $\mathcal{C}$ [defined by (2.3.3), (2.3.4)] corresponding to τ_0 has rank less than n. However, as easily seen, in every neighborhood of τ_0 there is a τ such that the corresponding $\mathcal{O}$ and $\mathcal{C}$ both have rank n and thus (again by Theorem 2.3.3) τ corresponds to a minimal realization and therefore $\tau \in \Delta_\alpha^{(2)}$ and $\Delta_\alpha^{(2)}$ therefore is dense in $\mathbf{R}^{n(2s+m)}$ and the same holds for $T_\alpha^{(2)}$.

(iii) Since $\tau \in \mathbf{R}^{n(2s+m)} - \Delta_\alpha^{(2)}$ corresponds to a nonminimal (F, H, K, L) (by

Theorems 2.3.3 and 2.3.2), $\mathscr{H}$ has rank less than n. From (2.6.2) and (ii) in Theorem 2.6.2, then $\rho(\tau) = (k, l) \in U_\beta$, $\beta < \alpha$, and thus $\rho(\Delta_\alpha^{(2)}) \subset \bigcup_{\beta \le \alpha} U_\beta$ follows. Conversely, if $(k, l) \in U_\beta$, $\beta < \alpha$, then as $h(n_i + 1, i)$ is linear dependent from the rows $\mathscr{H}_\alpha^\infty$ ($\alpha = (n_1 \cdots n_s)$), (2.6.2)–(2.6.4) has a (nonunique) solution $(F, H, K, L) \leftrightarrow \tau \in \mathbf{R}^{|\alpha|(2s+m)}$ [satisfying (ii) and (iii) of Theorem 2.6.2] and thus $\tau \in \Delta_\alpha^{(2)}$ by (i).

(ii) Clearly, $U_\alpha \subset M(n)$ implies that $\bar{U}_\alpha \subset \overline{M(n)}$. Conversely, let $(k_0, l_0) \in M(n)$; then (for arbitrary α such that $|\alpha| = n$) examining the Hankel matrix $\mathscr{H}$ we can always find in any neighborhood of (k_0, l_0) a $(k_t, l_t) \in U_\alpha$ and thus a sequence in U_α converging to (k_0, l_0). Therefore, U_α is dense in $M(n)$ and we have $\bar{U}_\alpha = \overline{M(n)}$. If $(k_0, l_0) \in \overline{M(n)}$, then, by definition, there is a sequence $(k_t, l_t) \in M(n)$ converging to (k_0, l_0) and thus, using an obvious notation, $\mathscr{H}_t \to \mathscr{H}_0$. By Theorem 2.4.1 all minors of $\mathscr{H}_t$ larger than $n \times n$ must be zero, and thus by the continuity of the determinant, this must also hold for $\mathscr{H}_0$. Thus, again by Theorem 2.4.1, (k_0, l_0) has order less than or equal to n and therefore $\overline{M(n)} \subset \bigcup_{i \le n} M(i)$. Conversely, if $(k, l) \in M(i)$, $i < n$, there is a β, $|\beta| = i$ such that $(k, l) \in U_\beta$ and a suitably chosen γ with $\beta < \gamma$ and $|\gamma| = n$. Thus by (iii) and the continuity of ρ, $(k, l) \in \overline{M(n)}$.

(iv) If $(k, l) \in \rho(\overline{\Delta_\alpha^{(2)}})$ and thus by (iii) $(k, l) \in U_\beta$, $\beta \le \alpha$ for a suitably chosen β, we see from (2.6.2) [or (2.6.1)] that the set of solutions F of (2.6.2) is an affine subspace whose dimension is s times the dimension of the left kernel of $\mathscr{H}_\alpha^\infty$. Let $\alpha = (n_1 \cdots n_s)$, $\beta = (m_1 \cdots m_s)$; then clearly the dimension of the left kernel of $\mathscr{H}_\alpha^\infty$ is $\sum n_i - \sum m_i$; as (2.6.3) and (2.6.4) for given F, uniquely determine H, K, L the result follows. For the corresponding ARMAX parameterization the proof is completely analogous.

(v) Let $(k_0, l_0) \in U_\alpha$. Clearly, then, there is a neighborhood of (k_0, l_0) in $\bar{U}_\alpha$, where $\mathscr{H}_\alpha^\infty$ still has rank n. On the other hand, as $\mathscr{H}_\beta^\infty$ has rank n for any $\beta > \alpha$ for any $(k, l) \in U_\alpha$ these ranks are smaller than or equal to n for any $(k, l) \in \bar{U}_\alpha$. Thus there is a neighborhood of $(k_0, l_0) \in U_\alpha$ where α are structure indices and thus U_α is open in $\bar{U}_\alpha$.

(vi) is an immediate consequence of (iii), (ii) and the fact that, for $s = 1$, $M(n)$ is $U_{(n)}$.

REMARK 1. Remarks completely analogous to those in Remarks 1 and 2 below Theorem 2.5.3 for echelon realizations may be made for the overlapping parameterization of $M(n)$. $\overline{M(n)}$ may be partitioned into the three sets $U_\alpha(|\alpha| = n)$, $\pi(\overline{T^{(2)}}) - U_\alpha$ and $\bar{U}_\alpha - \pi(\overline{T^{(2)}})$ and the convergence behavior for parameters corresponding to sequences $(k_t, l_t) \to (k_0, l_0)$ is analogous, as described in Remark 1 below Theorem 2.5.3. The set $\pi(\overline{T^{(2)}}) - U_\alpha$ is the set of all transfer functions for which there is a structure index $\beta < \alpha$. $\bar{U}_\alpha - \pi(\overline{T^{(2)}})$ is the set of all transfer functions of order smaller than or equal to n for which there is no structure index β such that $\beta \le \alpha$.

REMARK 2. Clearly, for a unique description of $(k, l) \in M(n)$ by its parameters, both n and $\alpha(|\alpha| = n)$ such that $(k, l) \in U_\alpha$ have to be known. However, since every U_α, such that $|\alpha| = n$ is open and dense in $M(n)$ [since "almost every" point of $M(n)$ is also in U_α] by the theorem above, once n is known the choice of α is not that critical in general. Moreover, if $(k, l) \in M(i)$, $i < n$, then this (k, l) is on the boundary of $U_\alpha(|\alpha| = n)$ [where U_α is open in $\overline{M(n)}$] and thus (k, l) can be approximated with arbitrary accuracy within U_α. Of course, on the one hand these facts constitute an advantage of the overlapping parameterization of $M(n)$ compared to the parameterization of the V_α (when n is known, at least) since unless V_α is the generic neighborhood, the V_α's $(|\alpha| = n)$ are manifolds of lower dimension and thus $\bar{V}_\alpha$ is a proper (and even "small") subset of $\overline{M(n)}$. On the other hand, a price has to be paid for this advantage, or in other words the use of the V_α has the advantage that in general fewer than $n(2s + m)$ parameters are needed for description.

REMARK 3. As explained below Theorem 2.6.1, there is only one V_α that is also a U_α and this is for $n_1 = n_2 = \cdots = n_q = p + 1, n_{q+1} = \cdots = n_s = p$, where $n = ps + q, 0 \leq q < s$. For this V_α the basic rows for $\mathscr{H}$ are the first n. We shall call this special $U_\alpha = V_\alpha$ just $U(n)$, for brevity. It will be important in Chapter 6.

The ARMAX parameterization of U_α (if U_α is not a generic neighborhood) has another shortcoming, both compared to that of V_α but also, in a sense compared to the state-space representation. From the proof of Theorem 2.6.4 it is seen that there may be nonlinear restrictions between the parameters. Also, $A(0)$ may be singular and (a, b, d) may not be relatively left prime, which is a disadvantage as, for example, in many cases we may wish to transform our system by left multiplication with $A(0)^{-1}$.

7. STRUCTURAL IDENTIFIABILITY

In many applications, especially when the system parameters have direct "physical" interpretation we have a priori information in the form of constraints on these parameters; for example, we might know from existing theories that one particular input variable does not have any influence on another particular output variable, and thus the corresponding parameters are restricted to be zero. In this context the question arises whether the a priori restrictions are sufficient to guarantee identifiability. In econometrics this is called structural identifiability, to distinguish this case from the case where no extra restrictions are known and where we pick out an arbitrary, suitable representative from each equivalence class. Often, the a priori restrictions lead to parameter spaces of relatively small dimension such that estimation of (multivariable) systems is possible also when the sample sizes are rather small (which is the case, for example, in most econometric applications).

At first we consider a case where a certain number of elements in (a, b, d) are a priori known to be zero; that is, we know that certain outputs, inputs, or errors do not occur in a certain equation at any time lag. In this case we do not require left primeness for (a, b, d). Thus the matrix u relating two observationally equivalent MFDs (a, b, d) and $(\bar{a}, \bar{b}, \bar{d})$

$$(\bar{a}, \bar{b}, \bar{d}) = u(a, b, d) \tag{2.7.1}$$

is in general a rational matrix. We have (Hannan, 1971b)

Theorem 2.7.1. *Consider a class of ARMAX systems where:*

(i) In every row of (a, b, d) there are at least $s - 1$ elements prescribed to be zero.

(ii) Call c_i the matrix consisting of those columns of (a, b, d) where we have a zero prescribed in the ith row. Then c_i is assumed to have rank $s - 1$ [as a polynomial matrix, i.e., $c_i(z)$ has rank $s - 1$ for all but a finite number of z's].

(iii) Every row of (a, b, d) is relatively prime (i.e., there is no common polynomial factor of degree greater than zero).

(iv) The diagonal elements of $A(0)$ are equal to 1.

Then this class is identifiable.

Proof. From (2.7.1) we have

$$0 = u_i c_i, \tag{2.7.2}$$

where u_i denotes the ith row of u. As the ith row of c_i is zero and as c_i has rank $s - 1$, we have from (2.7.2) $u_{ij} = 0, j \neq i$ [where u_{ij} is the (i, j) element of u], that is, u must be diagonal. By (iii) the rational functions u_{ii} must be constant and by (iv) $u_{ii} = 1$. Therefore, $u(z) = I_s$; in other words, any two observationally equivalent MFDs satisfying (i)–(iv) must be identical (i.e., the equivalence classes are singletons and the class considered is identifiable).

In a certain sense, identifiability in this case is reduced to the single-equation case, as by (i) and (ii) no observationally equivalent MFDs can be generated by combining different equations. For every equation, due to (iii), the degrees of the polynomials a_{ij}, b_{ij}, and d_{ij} are uniquely determined, and thus to obtain identifiability in this case the maximum degrees in (a, b, d) need not be described a priori. This is of some advantage in situations where there is no strong a priori knowledge about these maximum degrees.

(i) requires a great amount of a priori information about the system which in many cases might not be available. Note, however, that every transfer function (k, l) can be represented by an MFD (a, b, d) satisfying the conditions of the theorem. Let (a^*, b^*, d^*) be an ARMAX system, where a^* is in Hermite form (see Lemma 2.2.2) and thus lower triangular. We can find a lower triangular matrix v, not necessarily unimodular, such that $b := vb^*$ is upper triangular. Indeed, we can always find a unimodular matrix u such that ub^* is upper triangular. Take the last row of u as the last row of v. Now we can find an $(s-1)$-rowed unimodular matrix which does the same for the $(s-1)$-rowed square matrix in the top left-hand corner of b^*. Take the last row of this unimodular matrix followed by a zero as the second last row of v. In this fashion v is constructed. Since $a = va^*$ is lower triangular as the product of two lower trianguar matrices, defining $d = vd^*$, we have an ARMAX system (a, b, d) with $s-1$ prescribed zero elements in every row. Evidently, (ii) is satisfied since a and b are both of rank s. We may eliminate a nonconstant greatest common divisor in each row and satisfy (iv) by left multiplication by a diagonal matrix.

Now we assume that the MFDs are irreducible and that the maximum degrees, say $p_1 \cdots p_s, q_1 \cdots q_s, r_1 \cdots r_m$ of the columns $a_1 \cdots a_s, b_1 \cdots b_s, d_1 \cdots d_m$ of (a, b, d) are a priori prescribed. Let $a_j(i), b_j(i), d_j(i)$ denote the coefficients of z^i in the columns a_j, b_j, d_j. If the column end matrix

$$C_\alpha = (a_1(p_1), \ldots, a_s(p_s), b_1(q_1), \ldots, b_s(q_s), d_1(r_1), \ldots, d_m(r_m)) \tag{2.7.3}$$

corresponding to the prescribed degrees $\alpha = (p_1 \cdots p_s, q_1 \cdots q_s, r_1 \cdots r_m)$, has full rank s, the unimodular matrix u in (2.7.1) must be a constant, since otherwise at least one column in $(\bar{a}, \bar{b}, \bar{d})$ would have a higher degree than the prescribed one. If, in addition, we assume that $A(0) = I$, then by (2.7.1) this constant u must be the unit matrix I. Thus we have (Hannan, 1969, 1971b)

Theorem 2.7.2. *The class of all ARMAX systems with (a, b, d) left prime, with prescribed maximum column degrees $\alpha = (p_1 \cdots p_s, q_1 \cdots q_s, r_1 \cdots r_m)$, where the column end matrix C_α has rank s and where $A(0) = I$, is identifiable.*

The assumption that $A(0) = I$ may be replaced by a more general one to allow for cases where in the same equation several output components occur at the present time (this is the properly "structural" case in econometrics). As u has to be constant by the other assumptions of Theorem 2.7.2, we have from (2.7.1)

$$\bar{C} = uC, \tag{2.7.4}$$

where

$$C := (a_1(0), \ldots, a_1(p_1), \ldots, a_s(p_s), b_1(0), \ldots, b_s(q_s), d_1(1), \ldots, d_m(r_m))$$

and $\bar{C}$ is defined in the same way for $(\bar{a}, \bar{b}, \bar{d})$. Thus, if we assume that in every row of C there are at least $s - 1$ prescribed zeros, we have from (2.7.4)

$$0 = u_i C_i, \tag{2.7.5}$$

where u_i is the ith row of u and C_i is the matrix consisting of those columns of C where in the ith row the zeros have been prescribed. If we assume that C_i has rank $s - 1, i = 1 \cdots s$, then by (2.7.5) u must be diagonal. Thus, imposing the additional norming assumption that the diagonal elements of $A(0)$ are 1, we have obtained identifiability.

For Theorem 2.7.2 a rather large number of integer-valued parameters, namely $p_1 \cdots p_s, q_1 \cdots q_s, r_1 \cdots r_m$ have to be specified. Perhaps the most important special case is when only p, q, r or $\max(p,q,r)$ (let us call this p again) have been specified a priori and when $(A(p), B(q), D(r))$ or $(A(p), B(p), C(p))$ has to be of rank s. For brevity we shall call these sets $M(p,q,r)$ or $M(p,q)$ in the ARMA case. For $k(z)$, $l(z)$ analytic for $|z| \le 1$ and $\det(k) \ne 0$, $|z| \le 1$, this set is mapped homeomorphically onto an open set in Euclidean space of dimension $(p + q)s^2 + rsm$, using the freely varying coefficients in the ARMAX representation as coordinates. There are some disadvantages to this representation. In the first place the same pair (k, l) may have representations in different $M(p,q,r)$ so that these overlap. For example, taking an ARMA example,

$$k(z) = \begin{bmatrix} 1 & z \\ 0 & 1 \end{bmatrix} = \begin{bmatrix} 1 & -z \\ 0 & 1 \end{bmatrix}^{-1}$$

belongs to $M(0, 1)$ and to $M(1, 0)$. Also, there are (k, l) not represented in any $M(p,q,r)$. An ARMA example is $k(z) = a^{-1}b$, $a = I_2 + \frac{1}{2}Bz$, $b = I_2 + Bz$,

$$B = \begin{bmatrix} 1 & 1 \\ 0 & 0 \end{bmatrix}.$$

Now $[\frac{1}{2}B, B]$ is of rank unity and it can easily be seen (see Hannan, 1971b) that this $k(z)$ cannot be brought into the required form for an $M(p,q)$ by any unimodular left multiplication. However, for every ARMAX system column degrees $\alpha = (p_1 \cdots r_m)$ can be found such that the corresponding column end matrix has rank s. The last statement is seen directly if we bring a to Hermite's form, as in this form the column end matrix of a corresponding to the degrees of the diagonal elements has rank s. The prescribed column degrees α may be obtained either by a priori knowledge or they give just another covering for U_A.

Note that the conditions of Theorem 2.7.2 are not necessary. Consider, for example, the ARMA case with $s = 1$ and $p = q = 1$. In this case $a(z) = 1 + A(1)z$, $b(z) = 1 + B(1)z$. If the condition that $C_\alpha = (A(1), B(1))$ has rank 1 is not imposed and the other conditions are fulfilled, this class of systems is still identifiable, as in this case we only add $(a, b) = (1, 1)$. However, in this case problems in estimation would arise (see Deistler, Dunsmuir, and Hannan, 1978; and Section 5.4).

We now deal with a generalization of the preceding results (see Deistler, 1978), where we allow for a more general way of prescribing maximum lag lengths, including, for example, prescriptions of the maximum degrees of the elements a_{ij}, b_{ij}, d_{ij} and for a more general kind of restriction on the parameters. Again we assume left primeness for (a, b, d).

We assume that some upper bounds p, q, and r for the degrees of a, b, and d are a priori known. From (2.7.1) we have $u = \bar{a}a^{-1}$; thus

$$u \det a = \bar{a} \operatorname{adj} a.$$

As $v(\operatorname{adj} a) \le (s-1)p$ we have for the degree of u: $v(u) \le sp$. Let $l = sp$. We can write (2.7.1) as

(2.7.6)

$$(\bar{A}(0) \cdots \bar{A}(p) 0 \cdots 0 \bar{B}(0) \cdots \bar{B}(q) 0 \cdots 0 \bar{D}(1) \cdots \bar{D}(r) 0 \cdots 0) = (U(0) \cdots U(l)) \hat{A},$$

where $\hat{A}$ is defined as follows:

$$\hat{A} = (\tilde{A}, \tilde{B}, \tilde{D}),$$

where

$$\tilde{A} = \begin{pmatrix} A(0) & A(1) & \cdots & A(p) & 0 & \cdots & 0 \\ 0 & A(0) & A(1) & \cdots & A(p) & 0 & \vdots \\ \vdots & \ddots & \ddots & \ddots & & \ddots & 0 \\ 0 & \cdots & 0 & A(0) & A(1) & \cdots & A(p) \end{pmatrix}$$

and $\tilde{B}$ and $\tilde{D}$ are defined in the same way from b and d, respectively. Here $\hat{A}$ is a $(l+1)s \times v$ matrix with $v = s(p+1+l) + s(q+1+l) + m(r+l)$. We have $u = \sum U(j) z^j$ and the bar denotes the parameter matrices of $(\bar{a}, \bar{b}, \bar{d})$.

If D is a matrix with columns d_i, by $\operatorname{vec} D$ we mean the vector $(d_1' \cdots d_p')'$. Defining

$$\bar{\tau} := \operatorname{vec}(\bar{A}(0) \cdots \bar{A}(p) 0 \cdots 0 \bar{B}(0) \cdots \bar{B}(q) 0 \cdots 0 \bar{D}(1) \cdots \bar{D}(r) 0 \cdots 0)$$

$$\tilde{u} := \operatorname{vec}(U(0) \cdots U(l)),$$

we can write (2.7.6) as

$$\bar{\tau} = (\hat{A}' \otimes I_s)\tilde{u}, \tag{2.7.7}$$

where

$$\hat{A}' \otimes I_s$$

is of dimension $sv \times s^2(l+1)$.

To express the "apparent" zeros in τ $(:= \mathrm{vec}(A(0) \cdots A(p)0 \cdots 0B(0) \cdots B(q)0 \cdots 0D(1)0 \cdots 0, D(r)0 \cdots 0))$ as well as the additional restrictions on τ (according to our a priori knowledge), which are assumed to be affine, we write

$$R\tau = r, \tag{2.7.8}$$

where $R \in \mathbf{R}^{w \times sv}$ and $r \in \mathbf{R}^{w \times 1}$ are known and fixed.

Theorem 2.7.3. *The class of all ARMAX systems with* (a, b, d) *left prime (prescribed* p, q, r*), where (2.7.8) is fulfilled and where in addition*

$$R(\hat{A}' \otimes I_s) \quad \textit{has rank} \quad s^2(sp+1) \tag{2.7.9}$$

is identifiable.

Proof. Let τ and $\bar{\tau}$ correspond to two observationally equivalent MFDs. From (2.7.7) and (2.7.8) we have

$$r = R\bar{\tau} = R(\hat{A} \otimes I_s)'\tilde{u}'$$

and

$$r = R\tau = R(\hat{A} \otimes I_s)'(\mathrm{vec}(I_s, 0 \cdots 0))$$

and thus

$$R(\hat{A} \otimes I_s)'(\tilde{u} - \mathrm{vec}(I_s, 0 \cdots 0)) = 0.$$

Then (2.7.9) implies that $\tilde{u} = \mathrm{vec}(I_s, 0 \cdots 0)$ (i.e., $u = I_s$) and thus our theorem has been proved.

If the restrictions (2.7.8) only attach fixed values to single parameters, the rows of R can be chosen as unit vectors having the 1 in the same position as the corresponding restricted element in τ has. In this case $R(\hat{A} \otimes I)'$ is the matrix of all rows of $(\hat{A} \otimes I)'$ in these positions and (2.7.9) means that there have to be $s^2(sp + 1)$ such linearly independent rows. In this way it is straightforward to show that Theorem 2.7.3 as a special case covers Theorem 2.7.2. For nonnecessarily linear restrictions, a result analogous to Theorem 2.7.3 on local identifiability can be obtained (Deistler, 1978).

Now, let us discuss structural identifiability for state-space forms (2.3.1). We refer to Glover and Willems (1974) for this case. If we postulate minimality, two realizations $(\bar{F}, \bar{H}, \bar{K}, \bar{L})$ and (F, H, K, L) of the same transfer functions are related by (2.3.9)

$$\bar{F} = TFT^{-1}, \qquad \bar{H} = HT^{-1}, \qquad (\bar{K}, \bar{L}) = T(K, L).$$

Therefore, we have

$$\bar{F} - F = \bar{F} - F + TF - \bar{F}T = (T - I_n)F - \bar{F}(T - I_n)$$

and thus

$$\text{(2.7.10)} \qquad \operatorname{vec} \bar{F} - \operatorname{vec} F = (F' \otimes I_n) \operatorname{vec}(T - I_n) - (I_n \otimes \bar{F}) \operatorname{vec}(T - I_n).$$

Similarly, we obtain from (2.3.9)

$$\text{(2.7.11)} \qquad \operatorname{vec}(\bar{K}, \bar{L}) - \operatorname{vec}(K, L) = ((K, L)' \otimes I_n) \operatorname{vec}(T - I_n)$$

and

$$\text{(2.7.12)} \qquad \operatorname{vec} \bar{H} - \operatorname{vec} H = -(I_n \otimes \bar{H}).$$

Let

$$\tau := (\operatorname{vec} F, \operatorname{vec}(K, L), \operatorname{vec} H)$$

and let

$$\text{(2.7.13)} \qquad R\tau = r$$

with known R, r express the affine restrictions on the parameters; then we have from (2.7.10)–(2.7.13)

$$0 = R(\bar{\tau} - \tau) = R \begin{bmatrix} (F' \otimes I_n) - (I_n \otimes \bar{F}) \\ ((K, L)' \otimes I_n) \\ -(I_n \otimes \bar{H}) \end{bmatrix} \operatorname{vec}(T - I_n),$$

where $\bar{\tau}$ corresponds to $(\bar{F}, \bar{H}, \bar{K}, \bar{L})$; thus we have

Theorem 2.7.4. *The class of all minimal state-space forms, where (2.7.13) is fulfilled and where in addition*

$$R\begin{bmatrix} (F' \otimes I_n) - (I_n \otimes \bar{F}) \\ ((K, L)' \otimes I_n) \\ -(I_n \otimes \bar{H}) \end{bmatrix} \tag{2.7.14}$$

has rank n^2 for all feasible pairs (F, H, K, L) and $(\bar{F}, \bar{H}, \bar{K}, \bar{L})$ $[\neq (F, H, K, L)]$ is identifiable.

This result has been derived by Glover and Willems (1974) as a consequence of a more general theorem. They also treat local identifiability in the case of general nonlinear restrictions.

In addition to the purely algebraic problem of identifiability, the problem of parameterization again arises and some topological and geometrical properties of the parameter space and of the parameterization are again important for estimation and numerical calculation. We will investigate these properties for classes of ARMAX systems as described in Theorem 2.7.2. Analogous results may be obtained for the other classes discussed; see Deistler and Wang (1987).

Let $\alpha = (p_1 \cdots p_s, q_1 \cdots q_s, r_1 \cdots r_m)$ denote the prescribed column degrees of (a, b, d). If no additional restrictions have been imposed, then, according to Theorem 2.7.2, the *free parameters* in (a, b, d) may be written as a vector

$$\tau = \operatorname{vec}(a_1(1), a_1(2) \cdots a_1(p_1) \cdots a_s(1) \cdots a_s(p_s), b_1(1) \cdots b_1(q_1) \cdots b_s(q_s),$$
$$d_1(1) \cdots d_m(r_m))$$

with dimension $d := s(p_1 + \cdots + p_s + q_1 + \cdots + q_s + r_1 + \cdots + r_m)$. We will identify (a, b, d) and τ. Let $T_\alpha^{(3)} \subset \mathbf{R}^d$ denote the set of all τ's which are left prime and where the column end matrix (for prescribed α) has rank s. Thus $T_\alpha^{(3)}$ is the parameter space under the identifiability assumptions of Theorem 2.7.2 and the mapping π attaching to every $\tau \in \mathbf{R}^d$ the corresponding transfer functions (k, l) when restricted to $T_\alpha^{(3)}$ is injective. Thus $\psi_\alpha^{(3)}$: $W_\alpha \to T_\alpha^{(3)}$ is an ARMAX parameterization of $\pi(T_\alpha^{(3)}) = W_\alpha \subset U_A$. Let

$$C(-i) := (a_1(p_1 - i), \ldots, a_s(p_s - i), b_1(q_1 - i), \ldots, b_s(q_s - i),$$
$$d_1(r_1 - i), \ldots, d_m(r_m - i)),$$

where $a_i(j) = b_i(j) = 0, j < 0$; $d_i(j) = 0, j \leq 0$.

Theorem 2.7.5. *$T_\alpha^{(3)}$ has the following properties:*

(i) $T_\alpha^{(3)}$ is open and dense in $\mathbf{R}^d$ where $d = s(p_1 + \cdots + r_m)$.
(ii) $\psi_\alpha^{(3)}: W_\alpha \to T_\alpha^{(3)}$ is a T_{pt}-homeomorphism.
(iii) $\pi(\overline{T_\alpha^{(3)}}) = \bigcup_{\beta \le \alpha} W_\beta$.
(iv) For every $(k,l) \in W_\beta$, $\beta \le \alpha$, the (k,l)-equivalence class in $\overline{T_\alpha^{(3)}}$ is an affine subspace of dimension equal to

$$(2.7.15)\quad s\left(s^2p - \operatorname{rank}\begin{pmatrix} C(0) & 0 & \cdots & \cdots & 0 \\ C(-1) & C(0) & 0 & \cdots & 0 \\ \cdots\cdots & \cdots & \cdots\cdots & & \cdots \\ C(-s^2p+1) & \cdots & \cdots & \cdots & C(0) \end{pmatrix}\right),$$

where the $C(-i)$ correspond to an arbitrary left prime $\tau \in \pi^{-1}(k,l) \cap T_\alpha^{(3)}$.
(v) W_α is (T_{pt}-)open in $\overline{W_\alpha}$.
(vi) $\pi(\overline{T_\alpha^{(3)}}) \subset \overline{W_\alpha}$ and equality holds for $s = 1$.

For the proof of the theorem, see Deistler, Dunsmuir, and Hannan (1978) and Deistler (1983). Again, analogous results as stated in Remarks 1 and 2 following Theorem 2.5.3 hold.

REMARK. As has already been shown (using Hermite's normal form), for every ARMAX system there is an α such that $(k,l) \in \pi(T_\alpha^{(3)})$. As is easily seen, such an α is not uniquely determined from (k,l), for example, an ARMA system with $s = 1, p = 1, q = 0$, where $A(1) \neq 0$ may also be described with $\alpha = (1,1)$, as then $C(0) = (A(1), 0)$ is still of rank 1. In the case $s = 1$, however, we can always find for a given (k,l) a minimal α, given by the actual degrees of the left prime MFD (a,b,d). For $s > 1$, there may exist α and β such that $\alpha \not\le \beta$ and $\beta \not\le \alpha$ (where $\alpha \le \beta$ means componentwise inequality) and $(k,l) \in \pi(T_\alpha^{(3)})$, $(k,l) \in \pi(T_\beta^{(3)})$. Consider, for example, an ARMA system where

$$k(z) = \begin{pmatrix} 1 & az \\ 0 & 1 \end{pmatrix}^{-1} \begin{pmatrix} 1 & 0 \\ 0 & 1 \end{pmatrix} = \begin{pmatrix} 1 & 0 \\ 0 & 1 \end{pmatrix}^{-1} \begin{pmatrix} 1 & -az \\ 0 & 1 \end{pmatrix}$$

and where $\alpha = (0,1,0,0)$ and $\beta = (0,0,0,1)$.

8. FURTHER DETAILS CONCERNING THE STABLE, MINIPHASE CASE

In this section we assume that $k(z)$, $l(z)$ are rational and analytic for $|z| \le 1$ and that $\det\{k(z)\} \neq 0$, $|z| < 1$. These conditions were not needed in earlier parts of this chapter, although they were referred to in Remark 2 to Theorem 2.5.3. We shall now use some of the apparatus of this chapter to discuss an aspect of the stable, miniphase case. The result is of interest but also illustrates the key part played by echelon forms.

In considering the prediction of $y(t)$,

$$y(t) = \sum_0^\infty K(j)\varepsilon(t-j) + \sum_1^\infty L(j)z(t-j), \tag{2.8.1}$$

from the past, that is, from $y(s)$, $z(s)$, $s < t$, we may reduce ourselves to the consideration of the first component on the right in (2.8.1), which we call $u(t)$. Since $k(z)$ is rational and $\Sigma > 0$, it is not possible precisely to predict $u(t+1)$ from its past since

$$\int_{-\pi}^{\pi} \log\det\{k(e^{i\omega})\Sigma k(e^{i\omega})^*\}\, d\omega > -\infty.$$

However, if there are zeros of $\det\{k(z)\}$ on $|z| = 1$, then (and only then) there are linear functions, $a'y(t)$, that can be exactly interpolated, from $y(s)$, $s \neq t$ (see Hannan, 1970, p. 164). Thus if there are zeros of $\det(k)$ on $|z| = 1$, there is a linear function of the future, $y(s)$, $s > t$ that is, a.s., equal to a linear function of the past, $y(s)$, $s \leq t$. (This might more precisely be called the present and past but we shall retain the simpler terminology.) We wish to discuss this further. Let us call H_F the Hilbert space spanned by all components of $y(s)$, $s > t$, with the inner product defined by the covariance function. Thus H_F is a subspace of $L^2(\Omega, \mathscr{F}, P)$ (see Section 1.1) and indeed of H_y (see Section 1.3). Similarly, call H_P the Hilbert space spanned by the components of $y(s)$, $s \leq t$. This space H_P was called $H_y(t)$ in Section 1.3. We seek for pairs, $\xi^{(j)} \in H_P$, $\eta^{(j)} \in H_F$ so that

$$E\{\xi^{(j)}\xi^{(k)}\} = E\{\eta^{(j)}\eta^{(k)}\} = \delta_{jk}, \qquad E\{\xi^{(j)}\eta^{(j)}\} = \rho_j. \tag{2.8.2}$$

where

$$\rho_1 \geq \rho_2 \cdots \geq \rho_n > 0 \tag{2.8.3}$$

and the ρ_j are as large as possible, subject to the orthonormality conditions, (2.8.2). It is implied by the notation in (2.8.3) that there will be n such ρ_j that are positive, where n is the McMillan degree. Since the perpendicular projection of H_F on H_P is n-dimensional [see Section 2.5, below (2.5.5)], it is evident that there will be only n of the $\rho_j > 0$ and the remainder will be null. The ρ_j are of course called canonical correlations (between future and past) and the $\xi^{(j)}$, $\eta^{(j)}$ are the corresponding "discriminant functions" (see Anderson, 1958, for the classical, finite-dimensional definitions). As we have said it is possible to have $\rho_1 = 1$ if $\det(k)$ is zero on $|z| = 1$. Let us illustrate this by a simple example. Let $s = 1 = n$ and $y(t) = \varepsilon(t) - \varepsilon(t-1)$, $E\{\varepsilon(t)^2\} = 1$. The time-reversed process [i.e., $x(t) = y(-t)$] has precisely the same spectrum and thus we also have a representation $y(t) = \eta(t) - \eta(t+1)$, $E\{\eta(t)^2\} = 1$. We point out that $E\{\varepsilon(t)\eta(t+1)\} = 1$. Indeed, $\varepsilon(t)$ is the limit in mean square of a sequence $\varepsilon_N(t)$, $N = 1, 2, \ldots,$

$$\varepsilon_N(t) = \sum_0^{N-1} \left(1 - \frac{j}{N}\right) y(t-j)$$

(see Hannan, 1970, p. 131). Similarly,

$$\eta_N(t+1) = \sum_0^{N-1} \left(1 - \frac{j}{N}\right) y(t+j+1)$$

converges in mean square to $\eta(t+1)$. Evidently, $E\{\varepsilon_N(t)\eta_N(t-1)\} = 1$, so that we have shown that $E\{\varepsilon(t)\eta(t+1)\} = 1$. Thus $\varepsilon(t) = \eta(t+1)$, a.s. We shall establish the following result, due to Hannan and Poskitt (1987).

Theorem 2.8.1. *Let $2\pi f(\omega) = k(e^{i\omega})\Sigma k^*(e^{i\omega})$, $\Sigma > 0$, be the canonical factorization of a rational spectrum. Then the number of canonical correlations between future and past that are unity is the number of zeros of* $\det(k)$ *of unit modulus, counting these with their multiplicities.*

Proof. Let ξ, η be a pair of discriminant functions for unit canonical correlation. It is known (Hannan, 1970, Chap. II) that there is an isometric isomorphism between H_y and the Hilbert space of all vector functions, $h(e^{i\omega})$, square integrable with respect to $f(\omega)$. In this correspondence $y_j(t)$ maps into $e^{-it\omega}e_j$, where e_j is a vector with unity in the jth place and zeros elsewhere. Thus ξ, η correspond to vector-valued functions

$$\xi(z) = \sum_0^\infty \xi_k z^k, \qquad \eta(z) = \sum_1^\infty \eta_k z^{-k}$$

with

$$1 = \int \xi(e^{i\omega})' f(\omega) \eta(e^{-i\omega})\, d\omega = \int \xi(e^{i\omega})' f(\omega) \xi(e^{-i\omega})\, d\omega$$

$$= \int \eta(e^{i\omega})' f(\omega) \eta(e^{-i\omega})\, d\omega.$$

It follows from a direct generalization of the general form of Schwarz's inequality (Hardy, Littlewood, and Pólya, 1964, p. 132) that we must have $\xi'(e^{i\omega})k(e^{i\omega})\Sigma^{1/2} = c\eta'(e^{i\omega})k(e^{i\omega})\Sigma^{1/2}$, a.e. $d\omega$, where c is a positive constant. Thus, since $\Sigma > 0$, we must have

$$\xi(z)'k(z) = \eta(z)'k(z), \qquad |z| \le 1. \tag{2.8.4}$$

The Hardy space, H_2, is the space of all functions analytic for $|z| < 1$ and square integrable on $|z| = 1$. We shall say that a vector of functions belongs to H_2 if each element does so. Thus from (2.8.4) $\eta(z)'k(z) \in H_2$, since $\xi(z)'k(z)$ clearly does so.

The matrix $f(-\omega)$ is again rational and thus has a factorization (Theorem 1.3.3)

$$2\pi f(-\omega) = \lambda(e^{i\omega})\Omega\lambda(e^{i\omega})^*, \qquad \Omega > 0,$$

where $\lambda(z)$ is analytic for $|z| \le 1$ with $\det \lambda(z) \ne 0, |z| < 1$ and $\lambda(0) = I_s$. Indeed $\det\{f(\omega)\} = \det\{f(-\omega)\}$ so that $|\det\{\lambda(e^{i\omega})\}|^2 \det \Omega = |\det\{k(e^{i\omega})\}|^2 \det \Sigma$ and $\det k$, $\det \lambda$ are analytic for $|z| \le 1$, never zero in $|z| < 1$ and have $\det k(0) = \det \lambda(0) = 1$ so that $\det \Sigma = \det \Omega$ and $\det k(z) = \det \lambda(z)$. Now $z^{-1}\eta'(z^{-1})\lambda(z) \in H_2$, the factor z^{-1} coming from the fact that $\eta_0 = 0$. Thus

$$\eta(z)'k(z) = h_1(z)', \qquad z\eta(z)'\lambda(z^{-1}) = h_2(z^{-1})',$$

where h_1, $h_2 \in H_2$. Put $k(z)^{-1} = m(z)^{-1}r(z)$, $\lambda(z)^{-1} = m(z)^{-1}s(z)$, where $m(z)$ is a polynomial with $m(0) = 1$ and having zeros precisely those of $\det k(z) = \det \lambda(z)$ on $|z| = 1$. Then $r(z)$, $s(z)$ are analytic for $|z| \le 1$. Also, $m(z)/m(z^{-1}) = (-)^a z^\mu$, where μ is the degree of $m(z)$ and a is the number of zeros of $m(z)$ that are at $z = 1$. (We leave the reader to check this simply established fact.) Thus

$$\eta(z)' = m(z)^{-1}h_1(z)'r(z) = z^{-1}m(z^{-1})^{-1}h_2(z^{-1})'s(z^{-1})$$

so that

$$h_1(z)'r(z) = (-)^a z^{\mu-1}h_2(z^{-1})'s(z^{-1}).$$

This is possible only if $h_1(z)'r(z)$ and $h_2(z)'s(z)$ are polynomials of degree $\mu - 1$. Thus

$$\xi(z)' = \eta(z)' = m(z)^{-1}p(z)' = (-)^a z^{-\mu}m(z^{-1})^{-1}p(z)',$$

where $p(z)'$ is a row of polynomials of degree $\mu - 1$. However, $\xi(z)'k(z) =$

$\eta(z)'k(z)$ must also be square integrable on $|z| = 1$ (i.e., each component of the vector must be square integrable). Using Lemma 2.4.3, we put $k(z) = u(z)\Lambda(z)v(z)$. Now $\xi'k$ is square integrable if and only if $\xi'u\Lambda$ is square integrable since v^{-1} is polynomial. Let $m_i(z)$, with $m_i(0) = 1$, be composed of the zeros of $\varepsilon_i(z)$ (see Lemma 2.4.3) that lie on $|z| = 1$. Then $m(z) = m_1(z)m_2(z)\cdots m_s(z)$. Let μ_i be the degree of $m_i(z)$ and $\operatorname{diag}\{m_i(z)\}$ be the diagonal matrix with $m_i(z)$ in the ith place. Then $\varepsilon_i(z)\psi_i(z)^{-1}m_i(z)^{-1}$ (again see Lemma 2.4.3) is bounded and bounded away from zero on $|z| = 1$ so that $\xi'k$ is square integrable if and only if $\xi'u\operatorname{diag}\{m_i(z)\}$ is so. Thus $p(z)'u(z)v(z)^{-1}$ must be square integrable where $v(z)$ is diagonal with $v_i = m(z)/m_i(z)$ in the ith place. Put $j(z) = u(z)v(z)^{-1} = q(z) + t(z)$, where $q(z)$ is the polynomial matrix of quotients obtained by dividing each denominator into its numerator. Then $t(z)$ is strictly proper and may be written in echelon form $t(z) = c(z)^{-1}d(z)$ (Theorem 2.5.1). Thus $j(z) = c(z)^{-1}\{c(z)q(z) + d(z)\}$ is again a left coprime MFD since $(c(z), c(z)q(z) + d(z))$ is of full rank if $(c(z), d(z))$ is of full rank. Thus $c(z)q(z) + d(z)$ is unimodular since $j(z) = \{u(z)^{-1}v(z)\}^{-1}I_s$ is a left coprime MFD (see Theorem 2.2.1). The McMillan degree n of $j(z)$ is μ. Also, the largest Kronecker index is no larger than μ since $m(z)j(z)$ is a matrix of polynomials (i.e., $mc^{-1}d$ must be polynomial). Thus, if $n_1, n_2, \ldots, n_s$ are the Kronecker indices, $\sum'(\mu - n_i) = \mu$, the sum being over all i for which $n_i < \mu$. Now for the square integrability of $p'(z)j(z)$, we must have $p'(z)j(z)$ a matrix of polynomials, for otherwise it must have a pole on $|z| = 1$ and could not be square integrable. Thus $p'(z) = e(z)'c(z)$, where $e(z)$ is a row of polynomials. (Again this follows from Theorem 2.2.1.) We may certainly find a space of such polynomials of dimension μ since the ith row of $c(z)$ is of degree n_i so that the ith element of $e(z)$ may be of degree $\mu - n_i - 1$ and $\sum'(\mu - n_i) = \mu$. Here and below we take a polynomial to be null if its degree is negative. However, these are the actual maximum degrees. Indeed, let $\delta_i = \mu - 1 - n_i + a_i$, $a \geq 0$, be the degree of $e_i(z)$ if $n_i < \mu$ and let the degree be $\delta_i = a_i - 1$, $a_i \geq 0$ if $n_i = \mu$. Let j be the greatest i for which the a_i take their greatest value. Then the degrees, d_i, of the $e_i(z)c_{ij}(z)$ are as follows (see Theorem 2.5.1):

$$
\begin{aligned}
i < j; &\quad d_i \leq \delta_i + n_i - 1 \leq \mu + a_i - 2 < \mu + a_j - 1 \\
i > j; &\quad d_i \leq \delta_i + n_i \leq \mu + a_i - 1 < \mu + a_j - 1 \\
i = j; &\quad d_j = \delta_j + n_j = \mu + a_j - 1.
\end{aligned}
$$

Thus the jth element of $p(z)' = e(z)'c(z)$ is of degree $\mu + a_j - 1$, so that we must have $a_j = 0$, all a_i are zero, and the degrees of the $e_i(z)$ are $\mu - n_i - 1$. Thus the space of $p(z)$ for which $\xi(z)'k(z) = m(z)^{-1}p(z)'k(z)$ is square integrable is of dimension μ and there are precisely μ canonical correlations that are unity. This completes the proof.

The theorem to a certain extent justifies the strengthening of the miniphase assumption, $\det(k) \neq 0$, $|z| < 1$ to $\det(k) \neq 0$, $|z| \leq 1$, since in modeling phenomena one does not expect to be concerned with situations for which some linear function of the future can be perfectly predicted from the past. Also, it is partly for this reason that we have included Theorem 2.8.1, for in a number of situations in later chapters we shall be forced to strengthen the miniphase condition.

CHAPTER 3

The Kalman Filter

1. INTRODUCTION

The apparatus of the Kalman filter, which it is the purpose of this chapter to describe, has proved enormously important in data analysis, over a very wide range. The reason for this will be apparent from what is said below. As a consequence, this apparatus has associated with it a considerable literature, among which we mention Anderson and Moore (1979) and Jazwinski (1970). Because such a large amount of information is available, it would be futile to attempt completeness here and we shall try only to give those essential details that will enable the reader to understand later uses of the method and will thus make this book, to that extent, self-contained.

To some extent the subject matter of this book is built around the topic of this chapter for the papers of Kalman (1960) and Kalman and Bucy (1961) might be regarded as introducing an era, emphasizing state-space representations, in which most of the work discussed in the previous chapters was done and which, consequently, has heavily influenced the work in the chapters to come.

The Kalman filter apparatus is concerned with the problem of constructing the best estimate of $x(t)$ in (1.1.19), (1.1.20) from observations on $y(s)$, $z(s)$, $1 \leq s \leq t + n$. Actually, the term "filter" is reserved for the case where $n = 0$, while for $n < 0$ "predictor" is used, and for $n > 0$, "smoother." Of course, there was an august theory of prediction and signal extraction in existence before 1960. (See the preface to Dym and McKean, 1976, for some history.) The great importance of the Kalman filter is that it is not restricted to the stationary case, so that it will apply, in particular, to the case of a stable system that has recently been initiated but has not yet settled into a stationary state. In addition, the method emphasizes the recursive calculations of the estimates so that the calculations can be effected in real time and even in an off-line situation may result in a reduced burden of calculation. On the other hand, the methods are substantially restricted to the rational transfer function case.

From another point of view the method can be viewed as effecting the Gram-Schmidt orthogonalization of a sequence, $y(t)$, $t = 1, \ldots, T$, on the basis of a special model for the process generating this series, by producing a corresponding innovation sequence. This connects the method with the construction of the Gaussian likelihood, as we shall show.

2. CONSTRUCTION OF THE FILTER

We begin from the model (1.1.19), (1.1.20), which we repeat for convenience.

$$\begin{aligned} x(t+1) &= Fx(t) + Lz(t) + \xi(t), \\ y(t) &= Hx(t) + \eta(t), \end{aligned} \tag{3.2.1}$$

where this is initiated at time $t = 1$. We also assume, as usual, that

$$E\{\xi(s)\xi(t)'\} = \delta_{st}Q, \qquad E\{\eta(s)\eta(t)'\} = \delta_{st}R, \qquad E\{\xi(s)\eta(t)'\} = \delta_{st}S.$$
$$E\{\xi(t)x(0)'\} = E\{\eta(t)x(0)'\} = 0, \qquad t > 0.$$

In this chapter the $z(t)$ will be regarded as a "sure" sequence, that is, a sequence of observed vectors with no statistical properties.

Let us call $\hat{x}(t|s)$ the best estimate, in the least-squares sense, of $x(t)$ from $y(u)$, $z(u)$, $1 \le u \le s$. (See later for precise details when $L \neq 0$.) In general in this book a "hat" has been used to indicate an approximation or an estimate (which is much the same). Here, in $\hat{x}(t|s)$, we are using a hat in a slightly different sense. However, the notation is not altogether inconsistent since $\hat{x}(t|s)$, as we shall see, converges as $t \to \infty$ to $x(t|s)$, so that $\hat{x}(t|s)$ is in that sense an approximation to $x(t|s)$. Correspondingly, we shall put

$$e(t) = y(t) - H\hat{x}(t|t-1).$$

We have inserted a "hat" on $\hat{x}(t|s)$ to differentiate this from the corresponding quantity $x(t|s)$, which would depend on the "infinite" past of $y(u)$, $z(u)$, $u \le s$, in cases where that infinite past has meaning, for example, when the system is stationary. Of course, we would put $\varepsilon(t) = y(t) - Hx(t|t-1)$. The quantities $x(t|s)$, $\varepsilon(t)$ are, however, theoretical and cannot be computed, so it is $\hat{x}(t|s)$, $e(t)$ that are computationally important. We consider first the case $L = 0$ [i.e., when $z(t)$ is not present].

Theorem 3.2.1 *For (3.2.1) with $L = 0$,*

$$\hat{x}(t+n|t) = F\hat{x}(t+n-1|t), \quad n > 1; \qquad e(t) = y(t) - H\hat{x}(t|t-1),$$
$$\hat{x}(t+1|t) = F\hat{x}(t|t-1) + K(t)e(t), \qquad \hat{x}(1|0) = 0,$$
$$\hat{x}(t|t) = \hat{x}(t|t-1) + P(t)H'\Sigma(t)^{-1}e(t).$$

$$K(t) = \{FP(t)H' + S\}\Sigma(t)^{-1}, \qquad \Sigma(t) = \{HP(t)H' + R\},$$
$$P(t+1) = FP(t)F' + Q - K(t)\Sigma(t)K(t)', \qquad P(1) = E\{x(1)x(1)'\}.$$

REMARK 1. *Although we have prescribed these results for the time-homogeneous case, where F, H, L, Q, R, S are independent of t, the results all hold if in (3.2.1) all of these are functions of t.* F in the first line then becomes $F(t+n-1)$, and elsewhere F becomes $F(t)$, S becomes $S(t)$, H becomes $H(t)$, and so on. The proof is almost unchanged.

REMARK 2. We define $\Sigma(t) = E\{e(t)e(t)'\}$, $P(t) = E[\{x(t) - \hat{x}(t|t-1)\} \times \{x(t) - \hat{x}(t|t-1)\}']$ and will show that they are given by the formulas in the theorem. In some ways a better notation would be to write $P(t|t-1)$ in place of $P(t)$ so that we can write $P(t|t)$ for $E[\{x(t) - \hat{x}(t|t)\}\{x(t) - \hat{x}(t|t)\}']$. Although the evaluation of this is not stated in the theorem, which is self-contained without it, this last quantity is $P(t) - P(t)H'\Sigma(t)^{-1}HP(t)$ as easily follows from the description of $\hat{x}(t|t)$ in line 3 of the theorem and the theory below.

REMARK 3. The first two lines of the theorem provide a description of the system in a way closely corresponding to (1.1.22), (1.1.23), namely,

$$\hat{x}(t+1|t) = F\hat{x}(t|t-1) + K(t)e(t), \qquad y(t) = H\hat{x}(t|t-1) + e(t). \tag{3.2.2}$$

REMARK 4. The essential part of the calculation is that giving $P(t)$ (i.e., line 5). This line is often spoken of as the (discrete) Riccati equation because in the continuous-time theory a nonlinear differential equation, which is a matrix form of Riccati's equation, is obtained.

REMARK 5. If (3.2.1) is put in prediction error form, namely

$$\begin{aligned} x(t+1|t) &= Fx(t|t-1) + Lz(t) + K\varepsilon(t) \\ y(t) &= Hx(t|t-1) + \varepsilon(t), \end{aligned} \tag{3.2.3}$$

then $R = \Sigma$, $Q = K\Sigma K'$, $S = K\Sigma$, and

$$P(1) = E\{x(1|0)x(1|0)'\} = F\{x(0|-1)x(0|-1)'\}F' + K\Sigma K',$$

so that, if the system is stationary, then

$$P(1) = FP(1)F' + K\Sigma K', \tag{3.2.4}$$

which is a linear system of equations for $P(1)$. Of course, $P(1)$ prescribes the confidence we place in the prescription $\hat{x}(1|0) = 0$.

Proof. Call $\mathscr{H}(t)$ the space spanned by the components of $y(1), \ldots, y(t)$ considered as square-integrable functions over the measure space in which they are defined. Then $\mathscr{H}(t) = \mathscr{H}(t-1) \oplus \mathscr{T}(t)$, where $\mathscr{T}(t)$ is spanned by the elements of $e(t)$, which are evidently orthogonal to $\mathscr{H}(t-1)$ since, by definition, $H\hat{x}(t|t-1)$ is composed of the projections of the elements of $y(t)$ on $\mathscr{H}(t-1)$. We shall write $e(t) \perp \mathscr{H}(t-1)$, and say "$e(t)$ is orthogonal to $\mathscr{H}(t-1)$," for brevity. Since $x(t+n) = Fx(t+n-1) + \xi(t+n-1)$ and $\xi(t+n-1) \perp \mathscr{H}(t)$, $n > 1$, the first line is immediate. Now

$$E\{e(t)e(t)'\} = R + HP(t)H'$$

follows immediately from $e(t) = \eta(t) + H\{x(t) - \hat{x}(t|t-1)\}$ since $\eta(t) \perp \{\mathscr{H}(t-1), x(t)\}$. Also, $\hat{x}(t+1|t) = F\hat{x}(t|t-1) + E\{x(t+1)e(t)'\}\Sigma(t)^{-1}e(t) = F\hat{x}(t|t-1) + K(t)e(t)$, where $K(t)$ is as stated in the theorem, because

$$\begin{aligned} E\{x(t+1)e(t)'\} &= FE\{x(t)e(t)'\} + E\{\xi(t)e(t)'\} \\ &= FE[x(t)\{x(t) - \hat{x}(t|t-1)\}]H' + FE\{x(t)\eta(t)'\} \\ &\quad + E\{\xi(t)\eta(t)'\} + E[\xi(t)\{x(t) - \hat{x}(t|t-1\}]H' \end{aligned}$$

and the second and fourth terms are null since $x(t) \perp \eta(t)$, $\xi(t) \perp \{x(t), \mathscr{H}(t-1)\}$, while the first term is $FP(t)H'$ and the third is S. Since $\mathscr{H}(t) = \mathscr{H}(t-1) \oplus \mathscr{T}(t)$, then

$$\hat{x}(t|t) = \hat{x}(t|t-1) + E\{x(t)e(t)'\}\Sigma(t)^{-1}e(t),$$

and as we have just seen, $E\{x(t)e(t)'\} = P(t)H'$. It remains therefore to establish line 5 of the theorem. Put

$$\begin{aligned} x(t+1) - \hat{x}(t+1|t) &= \{x(t+1) - \hat{x}(t+1|t-1)\} \\ &\quad - \{\hat{x}(t+1|t) - \hat{x}(t+1|t-1)\} \end{aligned}$$

so that -1 by the second terms is the projection of the first on $\mathscr{H}(t)$. This is the same as the projection on $\mathscr{T}(t)$ since the first term is orthogonal to $\mathscr{H}(t-1)$. Thus, since $e(t) \perp \mathscr{H}(t-1)$,

$$\hat{x}(t+1|t) - \hat{x}(t+1|t-1) = E\{x(t+1)e(t)'\}\Sigma(t)^{-1}e(t) = K(t)e(t). \tag{3.2.5}$$

Then, since

$$x(t+1) - \hat{x}(t+1|t-1) = F\{x(t) - \hat{x}(t|t-1)\} + \xi(t),$$

$$x(t+1) - \hat{x}(t+1|t) = F\{x(t) - \hat{x}(t|t-1)\} + \xi(t) - K(t)e(t). \tag{3.2.6}$$

Thus we obtain from (3.2.6) that

$$P(t+1) = FP(t)F' + Q + K(t)\Sigma(t)K(t)' - 2K(t)\Sigma(t)K(t)'.$$

Indeed, $\xi(t) \perp \{x(t) - \hat{x}(t|t-1)\}$, $e(t) \perp \{x(t) - \hat{x}(t|t-1)\}$ and $E\{\xi(t)e(t)'\} = E\{x(t+1)e(t)'\} = K(t)\Sigma(t)$, because $\xi(t) \perp x(t)$, which establishes what is required and completes the proof of the theorem.

Let us consider evaluating $\hat{x}(t|s)$, $s > t$. First

$$\begin{aligned} &E[x(t)\{x(t+1) - \hat{x}(t+1|t)\}'] \\ &\quad = E[x(t)\{x(t) - \hat{x}(t|t-1)\}'F' + x(t)\{\xi(t) - K(t)e(t)\}'] \\ &\quad = P(t)F' - P(t)H'K(t)'. \end{aligned} \tag{3.2.7}$$

Now put

$$A(t) = \{P(t)F' - P(t)H'K(t)'\}P(t+1)^{-1}$$

where $P(t+1)^{-1}$ is the obvious generalized inverse obtained by replacing all nonzero singular values of $P(t+1)$ by their reciprocals, in the singular-value decomposition. Then

$$\begin{aligned} &A(t)E[x(t+1)\{x(t+1) - \hat{x}(t+1|t)\}'] \\ &\quad = A(t)P(t+1) \\ &\quad = E[x(t)\{x(t+1) - \hat{x}(t+1|t)\}']P(t+1)^{-1}P(t+1) \\ &\quad = E[x(t)\{x(t+1) - \hat{x}(t+1|t)\}'] \end{aligned}$$

from (3.2.7). The last equality holds since the column space of $E[x(t)\{x(t+1) - \hat{x}(t+1|t)\}']$ is contained in the column space of $P(t+1)$. Thus

$$E[\{x(t) - A(t)x(t+1)\}\{x(t+1) - \hat{x}(t+1|t)\}'] = 0. \tag{3.2.8}$$

Moreover, $x(t+2) - \hat{x}(t+2|t+1) = F\{x(t+1) - \hat{x}(t+1|t)\} + \xi(t+1) - K(t+1)e(t+1)$, from (3.2.6), so that

$$E[\{x(t) - A(t)x(t+1)\}\{x(t+2) - \hat{x}(t+2|t+1)\}'] = 0,$$

from (3.2.7), because $\xi(t+1) \perp \{x(t), x(t+1)\}$ and $e(t+1) = y(t+1) - H\hat{x}(t+1|t) = \eta(t+1) + H\{x(t+1) - \hat{x}(t+1|t)\}$ and $\eta(t+1) \perp \{x(t), x(t+1)\}$. Thus, by induction,

(3.2.9)

$$E[\{x(t) - A(t)x(t+1)\}\{x(t+j) - \hat{x}(t+j|t+j-1)\}'] = 0, \qquad j > 0.$$

From (3.2.9), since, as before, $e(t+j) = \eta(t+j) + H\{x(t+j) - \hat{x}(t+j|t+j-1)\}$,

$$E[\{x(t) - A(t)x(t+1)\}e(t+j)'] = 0, \qquad j > 0.$$

Also, for $s > t$,

$$\hat{x}(t|s) - \hat{x}(t|t) = \sum_{t+1}^{s} E\{x(t)e(j)'\}\Sigma(j)^{-1}e(j)$$

because $\mathscr{H}(s) = \mathscr{H}(t) \oplus S$ and S is spanned by $e(j), j = t+1, \dots, s$. Thus

$$\hat{x}(t|s) - \hat{x}(t|t) = A(t)\sum_{t+1}^{s} E\{x(t+1)e(j)'\}\Sigma(j)^{-1}e(j). \tag{3.2.10}$$

Also,

$$\begin{aligned}\hat{x}(t+1|s) - F\hat{x}(t|t) = &\sum_{j=1}^{t} E\{Fx(t)e(j)'\}\Sigma(j)^{-1}e(j)\\ &+ \sum_{t+1}^{s} E\{x(t+1)e(j)'\}\Sigma(j)^{-1}e(j)\\ &- F\sum_{j=1}^{t} E\{x(t)e(j)'\}\Sigma(j)^{-1}e(j) + S\Sigma(t)^{-1}e(t).\end{aligned} \tag{3.2.11}$$

This is because $\{\hat{x}(t+1|s) - x(t+1)\} \perp e(j), j = 1, \dots, s$; $\{\hat{x}(t|t) - x(t)\} \perp e(j), j = 1, \dots, t$, $x(t+1) = Fx(t) + \xi(t)$ and $\xi(t) \perp e(j), j = 1, \dots, t-1$, $E\{\xi(t)e(t)'\} = E\{\xi(t)\eta(t)'\} = S$. Thus, from (3.2.10),

$$\begin{aligned}\hat{x}(t|s) &= \hat{x}(t|t) + A(t)\{\hat{x}(t+1|s) - F\hat{x}(t|t) - S\Sigma(t)^{-1}e(t)\}\\ &= \hat{x}(t|t) + A(t)\{\hat{x}(t+1|s) - \hat{x}(t+1|t)\},\end{aligned}$$

the latter following from $x(t+1) = Fx(t) + \xi(t)$. Thus we have established the following.

Theorem 3.2.2. *For $s > t$, $L = 0$*

$$\begin{aligned}\hat{x}(t|s) &= \hat{x}(t|t) + A(t)\{\hat{x}(t+1|s) - \hat{x}(t+1|t)\},\\ A(t) &= \{P(t)F' - P(t)H'K(t)'\}'P(t+1)^{-1}.\end{aligned}$$

REMARK 1. *Again this result holds, with almost the same proof if F, H, Q, R, S depend on t. It is only necessary to replace F, H, by $F(t)$, $H(t)$ in $A(t)$.*

REMARK 2. To use this result to find $\hat{x}(t|s)$, one must first use Theorem 3.2.1 to find $\hat{x}(s-1|s-1)$, $\hat{x}(s|s)$, $\hat{x}(s|t)$ and then find $\hat{x}(s-1|s)$ from Theorem 3.2.2, and so on down to $\hat{x}(t|s)$. What is here called smoothing is an aspect of what, in Anderson and Moore (1979, p. 157), is called fixed-interval smoothing [i.e., the discovery of $\hat{x}(t|s)$ for all $t \leq s$, s fixed]. Slightly simpler formalus apply for fixed-point smoothing (i.e., t fixed, s varying) or fixed-lag smoothing ($s-t$ fixed, s, t varying.) Although we make no further use of Theorem 3.2.2 in this book, that result has been used by Shumway and Stoffer (1982) to evaluate the log Gaussian likelihood for a time-varying regression model.

REMARK 3. Recalling the notation introduced in Remark 2 below Theorem 3.2.1, we put

$$P(t|s) = E[\{x(t) - \hat{x}(t|s)\}\{x(t) - \hat{x}(t|s)\}'].$$

Of course, $P(t|t-1)$ is what has previously been called $P(t)$. Now it is immediate that

$$P(t|s) = P(t|t) + A(t)\{P(t+1|s) - P(t+1|t)\}A(t)', \qquad s > t.$$

Indeed, $\{x(t) - \hat{x}(t|s)\} + \{\hat{x}(t|s) - \hat{x}(t|t)\} = x(t) - \hat{x}(t|t)$ and the two terms on the left are orthogonal while $\hat{x}(t|s) - \hat{x}(t|t) = A(t)\{x(t+1) - \hat{x}(t+1|t)\} - A(t)\{x(t+1) - \hat{x}(t+1|s)\}$ and the second term on the right is orthogonal to that on the left. Now, $P(t|t) = E[x(t)\{x(t) - \hat{x}(t|t)\}'] = E[x(t)\{x(t) - \hat{x}(t|t-1) - P(t)H'\Sigma(t)^{-1}e(t)\}'] = P(t) - E\{x(t)e(t)'\}\Sigma(t)^{-1}HP(t) = P(t) - P(t)H'\Sigma(t)^{-1}HP(t)$ and this enables $P(t|s)$ to be recursively calculated.

We now consider the case where $L \neq 0$. Put

$$\tilde{x}(t) = x(t) - \zeta(t), \qquad u(t) = y(t) - H\zeta(t)$$

$$\zeta(t) = \sum_{1}^{t-1} F^{j-1}Lz(t-j), \quad t > 0, \qquad \zeta(0) = 0.$$

Then $u(t)$, $\tilde{x}(t)$ obey (3.2.1) for $L = 0$, with $\tilde{x}(1) = x(1)$. Now put

$$\zeta(t+n|t) = \zeta(t+n), \quad n \leq 1;$$

$$\zeta(t+n|t) = \sum_{n}^{t+n-1} F^{j-1}Lz(t+n-j), \quad n > 1.$$

We do this because, at time t, $z(t)$ is available, so that $\zeta(t+n)$ can be formed, for $n \leq 1$, while for $n > 1$ we cannot observe $z(t+n-j)$ for $j < n$. Now *define*

$$\hat{x}(t + n|t) = \hat{\hat{x}}(t + n|t) + \zeta(t + n|t). \tag{3.2.12}$$

In adopting this definition we are taking no account of any ability we might have to predict $z(t)$. To that extent the theory is defective. Of course, if $z(t)$ could be predicted, one procedure would be to extend the model so that $z(t)$ was also described by a rational transfer function system.

We have

$$\zeta(t + n|t) = F^{n-1}\zeta(t + 1|t), \qquad n > 1$$

so that, from (3.2.12), Theorem 3.2.1

$$\hat{x}(t + n|t) = F^{n-1}\hat{x}(t + 1|t), \qquad n > 1.$$

Also, as is easily checked,

$$\zeta(t + 1|t) = F\zeta(t|t - 1) + Lz(t),$$

so that

$$\hat{x}(t + 1|t) = F\hat{x}(t|t - 1) + K(t)e(t) + Lz(t), \qquad \hat{x}(1|0) = 0. \tag{3.2.13}$$

Of course, $e(t) = y(t) - H\hat{x}(t|t - 1)$, as before. Thus all of Theorem 3.2.1 continues to hold, except that line 2 is replaced by (3.2.13).

Finally, from Theorem 3.2.2,

$$\begin{aligned}\hat{x}(t|s) &= \hat{\hat{x}}(t|s) + \zeta(t), \qquad s > t \\ &= \hat{\hat{x}}(t|t) + A(t)\{\hat{\hat{x}}(t + 1|s) - \hat{\hat{x}}(t + 1|t)\} + \zeta(t) \\ &= \hat{x}(t|t) + A(t)\{\hat{x}(t + 1|s) - \hat{x}(t + 1|t)\}\end{aligned}$$

since $\zeta(t + 1|s) - \zeta(t + 1|t) = 0$.

Thus we have the following result.

Theorem 3.2.3. *Define* $\hat{x}(t + n|t)$ *as in (3.2.12). Then all of Theorems 3.2.1 and 3.2.2 hold for* $L \neq 0$ *except that (3.2.13) replaces the second line of Theorem 3.2.1.*

REMARK. *The same remark may again be made as in the first remark below Theorems 3.2.1 and 3.2.2, replacing* F, L *by* $F(t)$, $L(t)$ *in (3.2.13).*

Let us now relate these results to the ARMAX prediction error, state-space form [i.e., to the case where $\tilde{x}(t)$ is stationary]. Now we have

$$x(t+1|t) = Fx(t|t-1) + Lz(t) + K\varepsilon(t)$$
$$(3.2.14) \qquad y(t) = Hx(t|t-1) + \varepsilon(t),$$

so that $x(t|t-1)$ is what was previously $x(t)$. Now

$$\Sigma(t) = E[\{u(t) - H\hat{\hat{x}}(t|t-1)\}\{u(t) - H\hat{\hat{x}}(t|t-1)\}']$$

and $H\hat{\hat{x}}(t|t-1)$ is the projection of $u(t)$ on $\mathscr{H}(t-1)$, which is spanned by the elements of $u(1), \ldots, u(t-1)$. The projection of $u(0)$ on the space spanned by the elements of $u(-1), \ldots, u(-t+1)$ evidently converges in mean square to $u(0|-1)$ [i.e., the prediction of $u(0)$ from its infinite past]. Thus by stationarity $\Sigma(t) \to \Sigma$ and $e(t)$ converges in mean square to $\varepsilon(t)$. Evidently also, $\hat{\hat{x}}(t+1|t)$ converges in mean square to $\tilde{x}(t+1|t)$, so that $P(t) \to 0$ and $K(t) \to S\Sigma^{-1} = E\{K\varepsilon(t)\varepsilon(t)'\}\Sigma^{-1} = K$. Thus we have the following.

Theorem 3.2.4. *If $u(t)$ is stationary then, as $t \to \infty$, $\Sigma(t) \to \Sigma$, $K(t) \to K$, $P(t) \to 0$, $e(t)$ converges in mean square to $\varepsilon(t)$.*

There is some point, partly in view of the discussion in the following section, in examining the stability of the Kalman filter. Of course, this is a big subject and many different stability aspects could be considered. Here we discuss only one aspect for the ARMAX case. From Theorems 3.2.1 and 3.2.3,

$$(3.2.15) \qquad \hat{x}(t+1|t) = (F - K(t)H)\hat{x}(t|t-1) + K(t)y(t) + Lz(t).$$

From Theorem 3.2.4, $(F - K(t)H) \to F - KH$ and thus all eigenvalues of $F - K(t)H$ will eventually be inside the unit circle if and only if that is true for $F - KH$. If that is so, (3.2.15) is "uniformly (or exponentially) asymptotically stable" (see Jazwinski 1970, p. 240) and in particular any errors in initiating the filter will have an effect that decreases to zero at a geometric rate, as would also, for example, the effects from an error in the observation of one $y(t)$ or $z(t)$. Now assume that F, H, K define a minimal prediction error realization of a (stationary) ARMAX system of McMillan degree n. Then F has all eigenvalues less than unity in modulus. Also,

$$\det\{zI_n - (F - KH)\} = \det\{zI_n - F\}\det\{I_n + (zI_n - F)^{-1}KH\}.$$

The first factor is nonzero for $|z| \geq 1$. The second is easily seen to be the same as $\det\{I_s + H(zI_n - F)^{-1}K\} = \det k(z^{-1})$, so that the second factor will be nonzero for $|z| \geq 1$ if $\det\{k(z)\} \neq 0$, $|z| \leq 1$. In this case the ARMAX system is sometimes said to be "stable," although "uniformly asymptotically stable" would be a better term. In Chapter 2 we have spoken of it as a strengthening of the miniphase assumption. The converse is also clearly true; that is,

$(F - KH)$ can have a zero of unit modulus only if *det* $k(z)$ has a zero on the unit circle. [See Section 2.8 for an argument suggesting that $\det\{k(z)\} \neq 0$, $|z| \leq 1$, is not an unreasonable assumption.]

Theorem 3.2.5. *In the (stationary) ARMAX case where (3.2.14) is minimal,* $(F - KH)$ *has all eigenvalues of modulus less than 1 if and only if* $\det\{k(z)\} \neq 0$, $|z| \leq 1$.

It may be that

$$\left\| \prod_{1}^{t} (F - K(j)H) \right\| \to 0, \tag{3.2.16}$$

although not at a geometric rate even if $\det\{k(z)\} = 0$, for some z on $|z| = 1$. For example, if $y(t) = \varepsilon(t) - \varepsilon(t-1)$, then

$$\hat{y}(t|t-1) = \sum_{1}^{t} \{1 - j/(t+1)\} y(t-j)$$

so that

$$\hat{y}(t+1|t) = \frac{t+1}{t+2}\hat{y}(t|t-1) + \frac{t+1}{t+2}y(t)$$

and

$$\prod_{1}^{t} (F - K(j)H) = \prod_{1}^{t} \frac{j+1}{j+2} = \frac{2}{t+2} \to 0.$$

The condition (3.2.16) ensures that an error in initiation will converge to zero. However, the negligibility of a continuing stream of small errors is not ensured. In any case, (3.2.16) does not always hold if $\det\{k(z)\}$ has zeros on $|z| = 1$. This can be seen from the simple example $y(t) = \varepsilon(t) - 2\varepsilon(t-1) + \varepsilon(t-2)$.

3. ALTERNATIVE FORMS OF THE KALMAN FILTER AND THE GAUSSIAN LIKELIHOOD

As we have already indicated, there is much more to the question of stability than is obtained from the simple discussion in Section 2. One important aspect is numerical stability. One procedure that appears to improve performance here is to ensure that $P(t)$ is symmetric, as it should be. Another alternative is to replace the discrete Riccati equation by some more computationally robust alternative. One such procedure is to update $P(t)^{-1}$ (using the matrix inversion lemma) rather than $P(t)$. This leads to what is called the information filter. A second procedure involves update equations not for $P(t|t-1) = P(t)$ and $P(t|t)$ (see the remarks following Theorem 3.2.1) but rather for square roots of these. These square-root algorithms are known to be numerically more stable than the Riccati equation procedure. A third procedure is one giving a recursion for $\Delta(t) = P(t+1) - P(t)$. This involves fewer computations when n is large and again may have numerical advantages. Such procedures are spoken of as Chandrasekhar-type algorithms. All of these questions are discussed in Chapter 6 of Anderson and Moore (1979). We omit any discussion here.

Of course, the Kalman filter may be used to construct the Gaussian likelihood, and this will be emphasized in Chapter 6. Indeed, $e(t)$, $t = 1, \ldots,$ T is just the Gram-Schmidt orthogonalized equivalent of $u(t) = y(t) - \zeta(t)$. The latter quantities, on Gaussian (stationary) assumptions for the ARMAX case, provide the Gaussian likelihood by writing down the joint probability density for $u(t)$, $t = 1, \ldots, T$. Thus it is equivalent to writing down the joint probability density of $e(t)$, $t = 1, \ldots, T$. Of course these quantities themselves and their covariance matrices, $\Sigma(t)$, depend on the parameters since their construction depends on F, H, K, L, Σ. Ignoring a constant, $-2/T$ by the log likelihood is

$$L_T(\theta) = \frac{1}{T}\sum_1^T \log\det\{\Sigma(t)\} + \frac{1}{T}\sum_1^T e(t)'\Sigma(t)^{-1}e(t),$$

using θ to indicate the parameter vector. Of course, the optimization of this is not a trivial matter. Some further discussion is given in Section 6.2.

There are alternative parameterisations of the system, (3.2.3), that will be emphasized in Section 6.2 and that lead to an alternative form of the Kalman filter. These parameterizations (due to Son and Anderson, 1971) replace K, Σ by certain autocovariances. Indeed, Σ is replaced by $\Gamma_u(0)$ [the covariance matrix of $u(t)$], while K, which will be constituted by certain elements of the Hankel matrix $\mathscr{H}$, composed from the $K(j)$ (see Chapter 2), will be replaced by the same elements of a Hankel matrix composed, as was $\mathscr{H}$ from the $K(j)$, from the $\Gamma_u(j) = E\{u(t)u(t+j)'\}$. We leave the details until Section 6.2 (see Theorems 6.2.1 and 6.2.2).

CHAPTER 4

Maximum Likelihood Estimation of ARMAX Systems

In this chapter we consider the situation where certain integer-valued parameters as, for example, n or (p, q, r) or $(n_1 \cdots n_s)$ determining a class of ARMAX systems are prescribed a priori. Of course, in most actual applications these integer-valued parameters are not known a priori and in the next chapter we show how to estimate them. In this chapter we derive the asymptotic properties (consistency and asymptotic normality) of maximum likelihood estimators. The results here are given for rather general parameter spaces, including the cases treated in Chapter 2. In Section 1 some preliminary results are introduced; Section 2 deals with the consistency of the maximum likelihood estimators. In Section 3 a central limit theorem for maximum likelihood estimators is derived.

1. SOME PRELIMINARY RESULTS

We now investigate some properties of the (sample) second moments of the observed processes $y(t)$, $z(t)$ corresponding to a linear system.

The standard assumptions of Chapter 1 are assumed throughout here. In particular it is assumed that the system is an ARMAX system (a, b, d) satisfying $\det a(z) \neq 0$, $|z| \leq 1$; and $\det b(z) \neq 0$, $|z| < 1$ (or equivalently, a stable state-space system in prediction error form). Thus the solution considered is of the form

$$y(t) = \sum_{j=1}^{\infty} L(j)z(t-j) + \sum_{j=0}^{\infty} K(j)\varepsilon(t-j)$$

and $y(t)$ is stationary for stationary $z(t)$. Since the results of this chapter are independent of the form of representation (i.e., ARMAX or state space) we shall formulate all results in terms of the ARMAX representation. We discuss this further in Section 3.

Insofar as $z(t)$ is concerned, it is required that it is asymptotically stationary, that is,

$$\lim_{T\to\infty}\frac{1}{T}\sum_{t=1}^{T} z(t)z(t+s)' = \Gamma_z(s) \tag{4.1.1}$$

holds in the sense that the limit exists [see also (1.3.23)]. A weaker requirement is that

$$z(t)/t^{1/2} \to 0. \tag{4.1.2}$$

Since $T^{-1}\sum z_j(t)^2$ converges, by (4.1.1), then (4.1.2) follows from (4.1.1). Neither (4.1.1) nor (4.1.2) is essential, as may be seen from Fuller, Hasza, and Goebel (1981), who relax it for special cases of the class of structures considered here, but we shall assume (4.1.1). It is a consequence of the existence of the limit in (4.1.1) that

$$\Gamma_z(t) = \int_{-\pi}^{\pi} e^{it\omega}\, dF(\omega), \tag{4.1.3}$$

where $F_z(\omega) = F(\omega)$ is a nonnegative definite Hermitian matrix-valued function with $F(\omega_1) - F(\omega_2) \geq 0$, $\omega_1 \geq \omega_2$ and $F(-\omega) = \Gamma_z(0) - F(\omega)'$, at all points of continuity of $F(\omega)$ (see Hannan, 1970, Chap. II). It has not been required that $F_z(\omega)$ define an absolutely continuous, matrix-valued measure on $[-\pi, \pi]$ because $F_z(\omega)$ might well jump at some ω. For example, among the components of $z(t)$ may be included one that is identically unity, to account for mean correction, and correspondingly $F(\omega)$ will jump at $\omega = 0$. In addition to (4.1.1), a *persistent exciting condition* is required to guarantee that the knowledge of the second moments of the observed processes $y(t)$, $z(t)$ is sufficient to determine uniquely the transfer function $l(z)$. For stationary $z(t)$ we have the following relations between the second moments (see, e.g., Rozanov, 1967, p. 35):

$$\begin{aligned}\Gamma_y(t) &= E(y(s)y(s+t)') \\ &= \sum_{j,k=1}^{\infty} L(j)\Gamma_z(t+j-k)L(k)' + \sum_{j=0}^{\infty} K(j)\Sigma K(j+t)' \\ &= \int_{-\pi}^{\pi} e^{i\omega t} l(e^{i\omega})\, dF(\omega) l(e^{i\omega})^* + (2\pi)^{-1}\int_{-\pi}^{\pi} e^{i\omega t} k(e^{i\omega})\Sigma k(e^{i\omega})^*\, d\omega\end{aligned} \tag{4.1.4}$$

$$\begin{aligned}\Gamma_{yz}(t) &= E(y(s)z(s+t)') = \sum_{j=1}^{\infty} L(j)\Gamma_z(t-j) \\ &= \int_{-\pi}^{\pi} e^{i\omega t} l(e^{i\omega})\, dF(\omega)\end{aligned} \tag{4.1.5}$$

and in this case (1.3.22) is such a persistent exciting condition. Conditions (1.3.22) can be stated analogously for the case of asymptotically stationary $z(t)$. These conditions may be further weakened and we will assume only:

(4.1.6) Equations (4.1.3), (4.1.4), and (4.1.5) uniquely determine k, l, and Σ from the sequences $\Gamma_y(t)$, $\Gamma_z(t)$, $\Gamma_{yx}(t)$, $\forall t \in \mathbf{Z}$.

It is clear that this condition is needed, for the $\Gamma_y(t)$ and so on are all that we can know in the Gaussian case. We call this condition the "persistent exciting condition" because it is a replacement for (1.3.22) or for the weaker condition that $l(z)$ is uniquely determined by (4.1.3) and (4.1.5). This last requirement amounts to the statement that $z(t)$ "excites" sufficiently many frequencies in a persistent fashion, so that (4.1.1) determines $F(\omega)$ as a function increasing at sufficiently many ω to determine $l(z)$ uniquely. However, (4.1.6) in its entirety is needed, as the following simple example shows. Assume that $z(t) \equiv 1$; that is, $m = 1$ and the only exogenous term comes from the constant term. Then assume that the ARMAX system has been reasonably specified so that in a canonically chosen matrix fraction description, $d(z) \equiv D(1)$, a column vector. However, now $F(\omega)$ increases only at $\omega = 0$, where it jumps by unity. Thus from (4.1.5) we know only $l(1)$ and not $l(z)$. Nevertheless, (4.1.6) in its entirely suffices because the first term on the right in (4.1.4) is $l(1)l(1)'$, which is known so that the second term is known for all t and hence $k\Sigma k^*$ and hence k, Σ are known. Thus a, b, d, Σ are known and hence k, l, Σ. However, (4.1.6) is needed. To illustrate this, say $y(t)$, $z(t)$ are observed for $s = m = 1$ and

$$y(t) + \alpha y(t-1) = \delta(1)z(t-1) + \delta(2)z(t-2) + \varepsilon(t) + \beta\varepsilon(t-1),$$

where $|\alpha|, |\beta| < 1$, $\alpha \neq \beta$, $\delta(1), \delta(2) \neq 0$. Say, however, that while $z(t)$ is not identically constant, $\Gamma_z(t)$ [see (4.1.1)] is identically unity. Such a sequence can easily be constructed, for example, as $z(t) = 1 + \sin(\omega/t)$, $t > 0$. Then (4.1.6) will not hold since all we can know is $\delta(1) + \delta(2)$ [i.e., $d(z)$ at $z = 1$]. In this case our theorems below would not hold, for $z(t)$ persistently excites only one frequency, and that is not enough. We cannot consistently estimate $\delta(1)$ and $\delta(2)$.

The condition (4.1.6) has the following consequence that we use below. Say k_0, l_0, and Σ_0 satisfy (4.1.3), (4.1.4), and (4.1.5) for given sequences $\Gamma_y(t)$, $\Gamma_z(t)$, and $\Gamma_{yz}(t)$, say $k = k_0$, $\Sigma = \Sigma_0$, and $l \neq l_0$. Then, writing, $l\,dF$, and so on, for $l(e^{i\omega})\,dF(\omega)$,

$$\int_{-\pi}^{\pi} \operatorname{tr}\{(l - l_0)\,dF(l - l_0)^*\} > 0.$$

Indeed, otherwise, for all t,

$$\int_{-\pi}^{\pi} e^{it\omega} l\, dF(\omega) = \int e^{it\omega} l_0\, dF(\omega),$$

and hence, also,

$$\int_{-\pi}^{\pi} e^{it\omega} l\, dF(\omega) l^* = \int_{-\pi}^{\pi} e^{it\omega} l_0\, dF(\omega) l_0^*$$

and (4.1.6) does not hold, contrary to assumption.

Thus the question now arises as to whether $\Gamma_z(t)$, $\Gamma_y(t)$, and $\Gamma_{yz}(t)$ can be known from data, for all t. The condition (4.1.1) asserts this for $\Gamma_z(t)$ and we turn to the remaining two sequences. Put

$$G_y(s) = T^{-1} \sum_{t=1}^{T-s} y(t)y(t+s)' = G_y(-s)', \qquad s \geq 0, \tag{4.1.7}$$

$$G_{yz}(s) = T^{-1} \sum_{t=1}^{T-s} y(t)z(t+s)' = G_{zy}(-s), \quad G_{yz}(-s) = G_{zy}(s), \quad s \geq 0, \tag{4.1.8}$$

$$G_z(s) = T^{-1} \sum_{0}^{T-s} z(t)z(t+s)' = G_z(-s)', \qquad s \geq 0, \tag{4.1.9}$$

and put

$$\Gamma_y(s) = \int_{-pi}^{\pi} e^{i\omega s} l\, dF\, l^* + \frac{1}{2\pi} \int e^{i\omega s} k \Sigma k^*\, d\omega \tag{4.1.10}$$

$$\Gamma_{yz}(s) = \int_{-\pi}^{\pi} e^{i\omega s} l\, dF. \tag{4.1.11}$$

We have again abbreviated $l(e^{i\omega})\, dF_z(\omega)$ to $l\, dF_z$, and so on, for simplicity and will often do that. *When* $s = 1$ *we shall use lowercase letters [i.e.,* $g_y(s)$, $\gamma_y(s)$, *and so on].*

Theorem 4.1.1. *If* $y(t)$ *is generated by a stable ARMAX system for which (4.1.1) holds,* $\varepsilon(t)$ *is wholly independent of the* $z(t)$ *sequence and*

$$\frac{1}{T} \sum_{1}^{T-s} \varepsilon(t)\varepsilon(t+s)' \to \delta_{0s}\Sigma, \quad a.s., \quad E\{\varepsilon(s)\varepsilon(t)'\} = \delta_{st}\Sigma, \qquad E\{\varepsilon(t)\} \equiv 0,$$

then

$$G_y(s) \to \Gamma_y(s), \qquad G_{yz}(s) \to \Gamma_{yz}(s), \quad a.s.$$

Proof. We may put

$$y(t) = \sum_0^{t-1} K(j)\varepsilon(t-j) + \sum_1^{t-1} L(j)z(t-j) + f(t).$$

As has been stated in Section 1.2, $K(j)$ and $L(j)$ converge to zero at a geometric rate, that is,

$$\|K(j), L(j)\| \le C z_0^{-j} \qquad \text{for some } z_0 > 1.$$

Equally, $f(t)$ converges to zero at the same rate, a.s. Indeed, from (1.2.23) we see that $f(t) = HF^t x(0)$, using a minimal (prediction error) state-space representation. Thus we may put

$$y(t) = \sum_0^{\infty} K(j)\varepsilon(t-j) + \sum_0^{\infty} L(j)z(t-j) + f(t),$$

where, now, $\varepsilon(t)$, $z(t)$ are defined to be zero for $t \le 0$. Then

$$\begin{aligned} &\frac{1}{T}\sum_1^{T-s}\left(\sum_0^{\infty} K(j)\varepsilon(t-j)\right)\left(\sum_0^{\infty} K(j)\varepsilon(t+s-j)\right)' \\ &\quad = \sum_{j,k=0}^{\infty} K(j)\frac{1}{T}\sum_1^{T-s}\varepsilon(t-j)\varepsilon(t+s-k)'K(k)'. \end{aligned} \tag{4.1.12}$$

Now for fixed, j, k

$$\lim_{T\to\infty}\frac{1}{T}\sum_1^{T-s}\varepsilon(t-j)\varepsilon(t+s-k)' = \Sigma\delta_{j,k-s}, \quad \text{a.s.}$$

Moreover,

$$\|T^{-1}\sum\varepsilon(t-j)\varepsilon(t+s-k)'\| \le \frac{1}{T}\sum_1^{T}\varepsilon(t)\varepsilon(t)' \to \Sigma, \quad \text{a.s.}$$

Thus (4.1.12) is

$$\sum_{j,k=0}^{a} K(j)\frac{1}{T}\sum_1^{T-s}\varepsilon(t-j)\varepsilon(t+s-k)'K(k)' + r_a(T),$$

where $r_a(T)$ converges a.s. to zero at a geometric rate as $a \to \infty$, uniformly in T. Thus (4.1.12) is

$$\sum_{j,k=0}^{a} K(j)\Sigma\delta_{j,k-s}K(k)' + r_a(T) + o(1), \tag{4.1.13}$$

where the term that is $o(1)$ converges to zero almost surely. Since the first term in (4.1.13) is

$$(2\pi)^{-1}\int_{-\pi}^{\pi}\left(\sum_{0}^{a} K(j)e^{i\omega j}\right)\Sigma\left(\sum_{0}^{a} K(j)e^{i\omega j}\right)^{*} e^{i\omega j}\,d\omega,$$

then allowing a to increase indefinitely, we find that (4.1.12) converges to

$$\frac{1}{2\pi}\int_{-\pi}^{\pi} e^{i\omega j}k\Sigma k^{*}\,d\omega.$$

Since

$$\lim_{T\to\infty}\frac{1}{T}\sum_{1}^{T-s} z(t-j)z(t+s-k) = \Gamma_z(s+j-k),$$

we see that

$$\frac{1}{T}\sum_{1}^{T-s}\left(\sum_{0}^{\infty} L(j)z(t-j)\right)\left(\sum_{0}^{\infty} L(j)z(t+s-j)\right)'$$

converges to

$$\int_{-\pi}^{\pi} e^{i\omega j} l\,dF\,l^{*}$$

in precisely the same way. Finally, we need to show that

$$\lim_{T\to\infty}\frac{1}{T}\sum_{1}^{T-s} \varepsilon(t-j)z(t+s-k)' = 0, \quad \text{a.s.}$$

We may as well simplify and consider, treating $z(t)$ as a fixed sequence,

$$a(T) = \frac{1}{T}\sum_{1}^{T} \varepsilon_j(t)z_k(t),$$

as the proof is the same in any case. We use the method of subsequences. The variance is $\sigma_{jj}^2 T^{-2} \sum z_k(t)^2 = O(T^{-1})$, by (4.1.1). Then $a(T^2)$ converges a.s. to zero by Chebychev's inequality and the Borel-Cantelli lemma. Indeed, for any $\varepsilon > 0$,

$$\sum_{T=1}^{\infty} P(|a(T^2)| > \varepsilon) \leq \frac{1}{\varepsilon^2} \sum_{1}^{\infty} E\{a(T^2)^2\} < \infty.$$

On the other hand, writing, for brevity, $\max_N$ for the maximum over $N^2 + 1 \leq T < (N+1)^2$,

$$E\left\{\max_N |a(T) - T^{-1} N^2 a(N^2)|^2\right\} \leq N^{-4} E\left\{\max_N \left(\sum_{N^2+1}^{T} \varepsilon_j(t) z_k(t)\right)^2\right\}$$

and we may use *Menchoff's inequality* (Stout, 1974, p. 18) to bound the right side by

$$cN^{-4}\{\log(2N+1)\}^2 \sum_{N^2+1}^{(N+1)^2} z_k(t)^2,$$

where, as we customarily shall, we use c for a positive constant, not always the same one. Thus

$$\begin{aligned}\sum_{N=1}^{\infty} P\left\{\max_N |a(T) - T^{-1} a(N^2)|^2 > \varepsilon\right\} &\leq c\varepsilon^{-2} \sum_{N=1}^{\infty} (\log N)^2 N^{-4} \sum_{N^2+1}^{(N+1)^2} z_k(t)^2 \\ &\leq c\varepsilon^{-2} \sum_{N=1}^{\infty} (\log N)^2 N^{-2} < \infty\end{aligned}$$

since

$$\begin{aligned}\frac{1}{(N+1)^2} \sum_{N^2+1}^{(N+1)^2} z_k(t)^2 &= \frac{1}{(N+1)^2} \sum_{1}^{(N+1)^2} z_k(t)^2 \\ &\quad - \frac{N^2}{(N+1)^2} \frac{1}{N^2} \sum_{1}^{N^2} z_k(t)^2 \to 0,\end{aligned}$$

by (4.1.1). Since $\max_N\{a(N^2) - T^{-1}N^2 a(N^2)\} \leq \{(2N+1)/(N+1)^2\} a(N)^2 \to 0$, a.s., we have shown that $\max_N |a(T) - a(N)^2| \to 0$, a.s. [i.e., $a(T) \to 0$, a.s.]. Thus the second part of the theorem is established and the proof is completed.

We should mention that, for example, $G_y(s)$ and (say) $(1/T)\sum_1^T y(t)y(t+s)'$ converge to the same limits. Indeed, $T^{-1/2}y(T)$ converges a.s. to zero since $T^{-1/2}z(T)$ does so, from (4.1.1), and so also does $T^{-1/2}\varepsilon(T)$ from the fact that $T^{-1}\sum \varepsilon(t)\varepsilon(t)' \to \Sigma$.

We shall, in the proof of the central limit theorem in Section 4, need the convergence of the quantities $T^{-1}\sum_1^T z(t)$. Thus we require that

$$\lim_{T\to\infty} \frac{1}{T}\sum_1^T z(t) = \mu_z, \quad \text{a.s.} \tag{4.1.14}$$

So far we have introduced conditions only in relation to $z(t)$ save for the unexceptionable requirement of Theorem 4.1.1. However, in Section 4, and also later in the book, more will be needed. It could, of course, be required that the $\varepsilon(t)$ be Gaussian, but this is never needed and much weaker conditions will suffice. Let us assume $\varepsilon(t)$ stationary and introduce the σ-algebras $\mathscr{F}_t$ generated by $\varepsilon(s)$, $s \le t$. Thus we now take $\mathscr{F} = \mathscr{F}_\infty$ for the σ-algebra of measurable sets with respect to which all probability statments refer. As before, we may put $y(t) = u(t) + v(t)$, $u(t) = \sum K(j)\varepsilon(t-j)$. Then as in Section 1.4, we may argue that it is reasonable to impose the condition

$$E\{\varepsilon(t)|\mathscr{F}_{t-1}\} = 0. \tag{4.1.15}$$

As will be seen later, the central limit theorem for the system parameter estimates depends on the same theorem for quantities:

$$T^{-1/2}\sum_t \varepsilon_j(t)\varepsilon_k(t-a), \quad a > 0; \qquad T^{-1/2}\sum_t \varepsilon_j(t)z_k(t-b).$$

Thus if this theorem is to give a covariance matrix for the system parameter estimates that is as for the Gaussian case we need, using σ_{jk} for the typical element of Σ,

$$E\{T^{-1/2}\sum_t \varepsilon_j(t)\varepsilon_k(t-a)T^{-1/2}\sum_t \varepsilon_l(t)\varepsilon_m(t-b)\} \to \sigma_{jl}\sigma_{km}\delta_{ab},$$

with an analogous result for $T^{-1/2}\sum \varepsilon_j(t)z_k(t-b)$. There is no problem with this last quantity since, for $s \ge t$, $b = 0$ for example, using (4.1.15),

$$E\{\varepsilon_j(s)z_k(s)\varepsilon_l(t)z_m(t)\} = E[E\{\varepsilon_j(s)\varepsilon_l(t)|\mathscr{F}_{s-1}\}]z_k(s)z_m(t) = \delta_{st}\sigma_{jl}z_k(t)z_m(t),$$

as for the Gaussian case. Using (4.1.15) we have $E\{\varepsilon_j(s)\varepsilon_k(t-a)\varepsilon_l(t)\varepsilon_m(t-b)\} = 0$, $s > t$, $a > 0$, $b > 0$, but for $s = t$, in order that $T^{-1/2}\sum \varepsilon_j(t)\varepsilon_k(t-a)$, $a > 0$, should have the required convariance property, we need $E\{\varepsilon_j(t)\varepsilon_l(t)|\mathscr{F}_{t-1}\} = \sigma_{jl}$, that is,

$$E\{\varepsilon(t)\varepsilon(t)'|\mathscr{F}_{t-1}\} = \Sigma. \tag{4.1.16}$$

This requirement is hard to check. [Since (4.1.15) is related to nonlinearity or its absence, that condition is, to a certain extent, checkable.] For a central limit theorem (4.1.16) can be replaced by much weaker conditions, although from what has been said, the covariance matrix in the limiting distribution will not then be the classical one. The conditions we require are

$$\lim_{k\to\infty} E\{\varepsilon(t)\varepsilon(t)'|\mathscr{F}_{t-k}\} = \Sigma, \quad \text{a.s.} \tag{4.1.17}$$

$$\begin{aligned}&\lim_{k\to\infty} E\{\varepsilon_a(t)\varepsilon_b(t)\varepsilon_c(t-r)|\mathscr{F}_{t-k}\}\\ &\quad = E\{\varepsilon_a(t)\varepsilon_b(t)\varepsilon_c(t-r)\} = \sigma_{abc}(r), \quad \text{a.s.,}\\ &\qquad a, b, c = 1, 2, \ldots, s; \quad r > 0.\end{aligned} \tag{4.1.18}$$

The latter implies that $\varepsilon(t)$ has finite third moments, of course. Both (4.1.17), (4.1.18) are implied by (4.1.16), but they are much weaker than that, for they amount merely to requirments of (nonlinear) nondeterminism, that is, that apart from their mean values there is nothing from the infinitely far past influencing certain random variables [i.e., $\varepsilon_a(t)\varepsilon_b(t)$, $\varepsilon_a(t)\varepsilon_b(t)\varepsilon_c(t-r)$]. One condition, often used, that (needless to say) ensures that (4.1.17), (4.1.18) hold is that $\mathscr{F}_{-\infty} = \bigcap_t \mathscr{F}_t$ be trivial (i.e., contains only events of probability zero or 1). This is called regularity (see Section 1.4). The limit on the left in (4.1.17), (4.1.18) always exists. For example, put $\xi_k = E\{\varepsilon_a(t)\varepsilon_b(t)|\mathscr{F}_{t-k}\}$ and $\mathscr{G}_k = \mathscr{F}_{t-k}$. Then $\mathscr{G}_k \supset \mathscr{G}_{k+1}$ and $E\{\xi_k|\mathscr{G}_{k+1}\} = \xi_{k+1}$, so that ξ_k is a reversed martingale and (Neveu, 1975, Cor. V-3-12) ξ_k converges almost surely to a limit. What is being asserted in (4.1.17), (4.1.18) is that the limit is a constant, in which case it must be as shown. It also follows from the reference just cited that ξ_k converges in the L_1 norm, so that, in particular,

$$\lim_{k\to\infty} E[|E\{\varepsilon_a(t)\varepsilon_b(t)|\mathscr{F}_{t-k}\} - \sigma_{ab}|] = 0, \qquad a, b = 1, \ldots, s. \tag{4.1.19}$$

This result will be used below. Finally, when (4.1.16) does not hold [but only (4.1.17), (4.1.18)] we will sometimes need a condition to ensure the existence of $E\{\varepsilon_a(t)\varepsilon_b(t-r)\varepsilon_c(t)\varepsilon_d(t-s)\}$, $r, s > 0$. For this purpose we require

$$E\{\varepsilon_j(t)^4\} < \infty, \qquad j = 1, \ldots, s. \tag{4.1.20}$$

One aspect of the replacement of (4.1.16) by (4.1.17), (4.1.18) may be noted. As has been said, the latter are much weaker but the covariance matrix in the limiting distribution is then more complex. However, its complexity depends only on the differences $\sigma_{abcd}(r, s) - \delta_{rs}\sigma_{ab}\sigma_{cd}$; r, $s > 0$, where $\sigma_{abcd}(r, s) = E\{\varepsilon_a(t)\varepsilon_b(t)\varepsilon_c(t - r)\varepsilon_d(t - s)\}$, which is assumed to be finite. It might be reasonable to assume that these quantities will be very small, albeit nonzero. In that case the covariance matrix in the limiting distribution will be correspondingly near to that obtaining when (4.1.16) holds. This seems a preferable justification to the use of that limiting distribution than (4.1.16) itself.

2. CONSISTENCY OF THE MAXIMUM LIKELIHOOD ESTIMATION

In this section strong consistency for the *maximum likelihood estimators* (MLE) of (k, l, Σ) will be established. Although coordinates will be needed for the actual optimization of the likelihood, the main results as well as the proofs given here will be coordinate free, in the sense that they relate only to the transfer functions k, l (and to the innovation variance Σ) and not (directly) to the parameter τ. In general, there is no explicit expression for the MLE and the optimum of the likelihood has to be calculated by numerical procedures. The problem of algorithmic construction of the MLEs is discussed in Chapter 6.

We do not here make any compactness assumptions on the parameter space and in particular do not bound the zeros of $\det b(z)$ away from the unit circle or the eigenvalues of Σ away from zero. This noncompactness of the parameter space causes the main complication in the proof. Proofs under such general conditions were first given for the scalar ARMA case in Hannan (1973), for the vector ARMA case in Dunsmuir and Hannan (1976) (see also Deistler, Dunsmuir, and Hannan, 1978), and for the vector ARMAX case in Hannan, Dunsmuir, and Deistler (1980). The first precise results, for the scalar ARMA case, are probably those of Walker (1962), but his results, like those of Caines (1978), Ljung (1978), and Rissanen and Caines (1979), required compactness and usually also independence of the $\varepsilon(t)$. For further results, see Deistler and Pötscher (1984), Kohn (1979), P-D Tuan (1978), and Pötscher (1987).

Let $U \subset U_A$ denote the set of transfer functions (k, l) corresponding to the model class under consideration. We emphasize again that the main consistency result given here is coordinate free, in the sense that we do not refer to an ARMAX (or state space) parametrization, since the likelihood depends only on (k, l, Σ). Therefore, as an underlying parameter space for our coordinate-free approach, we may take $\Theta \subset U \times \underline{\Sigma}$, where $\underline{\Sigma} = \{\sigma(\Sigma) | \Sigma > 0, \Sigma' = \Sigma\}$ and $\sigma(\Sigma) \in \mathbf{R}^{s(s+1)/2}$ is the vector of on- and above-diagonal elements of Σ. By this general formulation, restrictions on $\sigma(\Sigma)$ and joint restrictions between (k, l) and Σ are also allowed. We recall that sets of transfer functions are always endowed with the pointwise topology. Sets like Θ are endowed with the relative topology in the product space $U_A \times \mathbf{R}^{s(s+1)/2}$. In the process of optimizing the likelihood, certain boundary points will not be excluded. Thus the likelihood function is considered over

$$\hat{\Theta} = \{(k, l, \sigma(\Sigma)) \in \bar{\Theta} \,|\, (k(z), l(z)) \text{ has no pole for } |z| = 1 \text{ and } \Sigma > 0\}$$

rather than over Θ. We also consider the set

$$\Theta^* = \{(k, l, \sigma(\Sigma)) \in \hat{\Theta} \,|\, k(z) \text{ has no zero for } |z| = 1\}.$$

We use θ for the parameter $(k, l, \sigma(\Sigma)) \in \hat{\Theta}$ and $\theta_0 = (k_0, l_0, \sigma(\Sigma_0))$ for the true parameter value.

The *standard assumptions of this chapter* are the standard assumptions of Chapter 1, namely $K(0) = I$, (1.1.5), $\Sigma > 0$, the stability assumption (1.2.3), the miniphase assumption (1.3.15) for b, $Ez(s)\varepsilon(t)' = 0$, the assumption (4.1.1) that $z(t)$ is asymptotically stationary and the persistent exciting condition (4.1.6), and in addition the assumption

(4.2.1) U has finite degree; that is, there exists an ARMAX parameterization ψ: $U \to T$ and a $p \in \mathbf{N}$ such that $v(a, b, d) \leq p$, for all $(a, b, d) \in T$ holds.

Clearly, by Theorem 2.6.5, (4.2.1) is equivalent to saying that there is a $n \in \mathbf{N}$ such that $U \subset \overline{M(n)}$. Thus also $\bar{U} \subset \overline{M(n)}$ and also then $\bar{U}$ is of finite degree. In the following we will use p to denote an upper bound for the $v(a, b, d)$, where (a, b, d) correspond to (a suitably chosen parameterization of) $\bar{U}$. In this section we also assume that

(4.2.2) $$\theta_0 \in \Theta^*$$

and *thus the true system is contained in the model class* (or on its boundary) and $\det k_0(z)$ has no zeros for $|z| = 1$.

In the following, $y(t)$ is used as a symbol for the observed outputs (i.e., the outputs corresponding to the data or to the true θ_0) as well as for the outputs of an ARMAX system corresponding to the (variable) parameter $\theta \in \hat{\Theta}$. The actual meaning of $y(t)$ will become clear from the context. Put $y(t) = u(t) + v(t)$, where

(4.2.3) $$u(t) = k(z)\varepsilon(t)$$

is assumed to be already in stationary form. Then $v(t)$ satisfies $a(z)v(t) = d(z)z(t)$. Starting this system with initial values $e = (v(0)', v(-1)', \ldots, v(-p+1)', z(0), \ldots, z(-r+1))'$ gives a solution of the form

$$v(t) = \sum_{j=1}^{t-1} L(j)z(t-j) + y_e(t), \tag{4.2.4}$$

where $y_e(t)$ depends only on e and decreases geometrically to zero as t increases. The likelihood function depends on θ and on e. *However, for simplicity we shall put* $e = 0$, and thus we shall not optimize the likelihood with respect to e. It is not really difficult to be more general, but the argument is already complex and it is rather unlikely that optimization with respect to e would be performed. All theorems below remain true, however, if optimization is effected with respect to e.

Now put $u_T = (u(1)', \ldots, u(T)')'$, where $u(t)$ is given by (4.2.3) and thus depends on k and Σ, and put $\Gamma_T(\theta) = Eu_T u_T' = \Gamma_T(f_u)$, where we indicate that $\Gamma_T(\theta)$ depends on θ only through the spectral density, $f_u(\omega) = f_u(\omega; \theta)$, of $u(t)$. Note that here $u(t)$ has been used for a process depending on the (varying) parameter θ and is not necessarily connected with the process generating the data. Furthermore, put

$$u(t, \theta) = y(t) - \sum_{j=1}^{t-1} L(j)z(t-j).$$

and $u_T(\theta) = (u(1, \theta)', \ldots, u(T, \theta)')'$. Thus $u(t, \theta) = y(t) - v(t)$, where $v(t)$ is given by (4.2.4) for $e = 0$; $u(t, \theta)$ depends on θ (only via the first $t-1$ coefficients of l) and on the observations, $y(t)$, $z(t)$.

From the expression for the conditional density of $y_T = (y(1)', \ldots, y(T)')'$ given $z_T = (z(1)', \ldots, z(T)')'$, on Gaussian assumptions, we see that $-2T^{-1}$ times the log of the (Gaussian) likelihood is given (omitting a constant) by

$$L_T(\theta) = T^{-1} \log \det \Gamma_T(\theta) + T^{-1} u_T(\theta)' \Gamma_T^{-1}(\theta) u_T(\theta). \tag{4.2.5}$$

We shall often speak of L_T as the likelihood function. Although the likelihood has been written down as if the observations were Gaussian, no Gaussianity assumption on the observations will be made. Note that L_T is finite on $\hat{\Theta}$, since $\Gamma_T(\theta) > 0$, $\theta \in \hat{\Theta}$. Indeed, for an arbitrary $x \in \mathbf{R}^{sT}$, with $x(j)$ as the jth block of s elements, we have

$$x'\Gamma_T(\theta)x = \int_{-\pi}^{\pi} \left(\sum_{j=1}^{T} x(j)e^{i\omega j} \right)^* f_u(\omega) \left(\sum_{j=1}^{T} x(j)e^{i\omega j} \right) d\omega.$$

Now for $x \neq 0$, the expression $\Sigma x(j)e^{i\omega j}$ is null at only finitely many points, and $f_u(\omega)$ is nonsingular with the exception of finitely many points by our standard assumptions. Thus $x'\Gamma_T(\theta)x > 0$ for $x \neq 0$, as has been asserted.

Let $\theta = (k, l, \sigma(\Sigma)) \in \bar{\Theta}$ be such that Σ is singular or k has a pole for $|z| = 1$. Then L_T has the property that [under the additional assumption (4.2.21)], for fixed (and for large enough) T, we have that $\theta_n \to \theta$, $\theta_n \in \hat{\Theta}$ implies that

$$L_T(\theta_n) \to \infty \qquad \text{for } n \to \infty, \quad \text{a.s.} \tag{4.2.6}$$

For the case where Σ (corresponding to θ) is singular, this will become clear from formula (4.2.31) in the proof of Theorem 4.2.1. Thus we here only have to show that (4.2.6) holds for the case where k has a pole for $|z| = 1$. Let $\lambda_1(\Gamma_T(\theta_n))$ and $\lambda_T(\Gamma_T(\theta_n))$ denote the smallest and the largest eigenvalue of $\Gamma_T(\theta_n)$. Then

$$T^{-1} \log \det \Gamma_T(\theta_n) \geq T^{-1} \log \lambda_T(\Gamma_T(\theta_n)) + T^{-1}(Ts - 1) \log \lambda_1(\Gamma_T(\theta_n)). \tag{4.2.7}$$

Let $\theta_n = (k_n, l_n, \Sigma_n)$, $\theta = (k, l, \Sigma)$. Then $k_n \to k$ and $\Sigma_n \to \Sigma$. Thus the Σ_n and the power series coefficients $K_n(j)$ of k_n are bounded in n, the latter for every $j \in \mathbf{N}$. As easily seen, $|\det a(e^{i\lambda})|^2 \leq 2^{ps}$ holds since $\det a(z) \neq 0$ $|z| \leq 1$, and thus, due to (4.2.1), the coefficients $P_{n,j}$ of $P_n(z) = \det a_n(z) k_n(z) = \operatorname{adj} a_n(z) b_n(z) = \sum_{j=0}^{r} P_{nj} z^j$; $r = ps$, are bounded. Here $a_n^{-1} b_n$ is a suitably chosen MFD of k_n satisfying $A(0) = B(0) = I$. Now let $\tilde{P}_n = (Q_n, S_n)$, where, taking $T > r$, for example,

$$Q_n = \begin{pmatrix} \Sigma_n^{1/2} & P_{n1}\Sigma_n^{1/2} & \cdots & P_{nr}\Sigma_n^{1/2} & 0 & \cdots & 0 \\ 0 & \Sigma_n^{1/2} & \cdots & P_{nr-1}\Sigma_n^{1/2} & & & 0 \\ \vdots & \ddots & \ddots & & & & \vdots \\ \vdots & & \ddots & \ddots & & & \vdots \\ \vdots & & & \ddots & \ddots & & \vdots \\ \vdots & & & & \ddots & \ddots & \vdots \\ 0 & & & & & 0 & \Sigma_n^{1/2} \end{pmatrix}$$

$$S_n = \begin{pmatrix} 0 & \cdots & 0 \\ \vdots & & \vdots \\ P_{nr}\Sigma_n^{1/2} & \cdots & 0 \\ \vdots & & \vdots \\ P_{n1}\Sigma_n^{1/2} & \cdots & P_{nr}\Sigma_n^{1/2} \end{pmatrix}.$$

Here Q_n is $sT \times sT$ and S_n is $sT \times sr$. Then $k_n \Sigma_n k_n^* \geq 2^{-ps} P_n \Sigma_n P_n^*$, and therefore

$$\Gamma_T(k_n \Sigma_n k_n^*) \geq \Gamma_T(2^{-ps} P_n \Sigma_n P_n^*) \geq 2^{-ps} Q_n Q_n'.$$

Here we have omitted the argument $\exp(i\omega)$ from k_n and P_n. From these inequalities we see that $\lambda_1(\Gamma_T(\theta_n))$ is, uniformly in n, bounded away from zero, since $\det(Q_n Q_n') = \det \Sigma_n^T$ holds and $\operatorname{tr}(Q_n Q_n')$ [and therefore also $\lambda_T(Q_n Q_n')$] is bounded, uniformly in n. Thus from (4.2.7) it is clear that it remains to show that $\lambda_T(\Gamma_T(\theta_n)) \to \infty$, for $n \to \infty$. For this, it is sufficient to show that there is a $u_1 \in \mathbf{C}^s$, with $u_1 u_1^* = 1$, such that

$$\int_{-\pi}^{\pi} u_1^* f_n u_1 \, d\omega \to \infty \qquad \text{for } n \to \infty$$

holds; here $2\pi f_n = k_n \Sigma_n k_n^*$. Now, by assumption, $\int \operatorname{tr} k\Sigma k^* \, d\omega = \infty$ and thus there is a u_1 such that $\int u_1^* k\Sigma k^* u_1 \, d\omega = \infty$. Note that $k_n \to k$ (in T_{pt}) together with (4.2.1) implies that there is a subsequence, $k_{n(m)}$, say, such that $k_{n(m)}(e^{i\lambda}) \to k(e^{i\lambda})$, λ a.e. holds. This is easily seen using a compactness argument, since, as has been shown before, the coefficients of $\det a_n$ and of $P_n = \operatorname{adj} a_n b_n$ are bounded. Applying Fatou's lemma to such a subsequence, we obtain

$$\liminf\left(\int_{-\pi}^{\pi} u_1^* f_{n(m)} u_1 \, d\omega\right) \geq \int_{-\pi}^{\pi} u_1^* f u_1 \, d\omega = \infty$$

and thus (4.2.6) holds.

The maximum likelihood estimator $\hat{\theta}_T = (\hat{k}_T, \hat{l}_T, \sigma(\hat{\Sigma}_T))$ is obtained by minimizing L_T over $\hat{\Theta}$. It is not apparent that L_T has a minimum in $\hat{\Theta}$ (see Deistler and Pötscher, 1984). However, $L_T(\hat{\theta}_T)$ has a well-defined meaning as $\inf L_T(\theta)$, $\theta \in \hat{\Theta}$. We may regard $\hat{\theta}_T$ (for fixed T) as a sequence with $L_T(\theta_j)$ converging to that infimum. For the term $T^{-1} u_T(\theta)' \Gamma_T(\theta)^{-1} u_T(\theta)$ we sometimes use the symbol $Q_T(\theta)$.

Let us define a function by

$$L(\theta) = \log\det\Sigma + Q(\theta), \tag{4.2.8}$$

where

$$\begin{aligned} Q(\theta) = (2\pi)^{-1} &\int_{-\pi}^{\pi} \operatorname{tr}\{(k\Sigma k^*)^{-1}(k_0 \Sigma_0 k_0^*)\} \, d\omega \\ &+ \int_{-\pi}^{\pi} \operatorname{tr}\{(k\Sigma k^*)^{-1}(l - l_0) \, dF_z(\omega)(l - l_0)^*\}. \end{aligned} \tag{4.2.9}$$

Here again we have omitted the argument $\exp(i\omega)$ from k, l, and so on.

As we shall see (from Lemmas 4.2.2 and 4.2.3), L is the asymptotic form of L_T. L is finite when defined on Θ^*, whereas for $\theta \in \bar{\Theta} - \Theta^*$ the integrals in (4.2.9) might not converge [in this case we may put $L(\theta) = +\infty$].

We now establish three lemmas to prepare the consistency result. The first lemma is concerned with the infimum of the asymptotic likelihood L.

Lemma 4.2.1

(i) $\min_{\theta\in\Theta^*} L(\theta) = \log\det\Sigma_0 + s$ *for* $\theta_0 \in \Theta^*$.

(ii) $L(\theta) \geq \log\det\Sigma_0 + s$; $\theta \in \bar{\Theta}$, *for* $\theta_0 \in \hat{\Theta}$.

(iii) For $\theta \neq \theta_0$, *strict inequality holds in (ii).*

Proof. We first establish (ii) and (iii). By Jensen's inequality we have

$$(4.2.10)\qquad \begin{aligned}&(2\pi)^{-1}\int_{-\pi}^{\pi}\operatorname{tr}\{(k\Sigma k^*)^{-1}(k_0\Sigma_0 k_0^*)\}\,d\omega\\ &\geq \exp\left\{(2\pi)^{-1}\int_{-\pi}^{\pi}\log\operatorname{tr}\{(k\Sigma k^*)^{-1}(k_0\Sigma_0 k_0^*)\}\,d\omega\right\}.\end{aligned}$$

If the left-hand integral diverges, we allocate to it the value $+\infty$, in which case (ii) evidently holds, so that from now on we may assume that the integral converges. The inequality is strict unless the integrand is constant λ a.e., where λ indicates Lebesgue measure on $[-\pi,\pi]$. By the inequality between the arithmetic and the geometric means (which is also an aspect of Jensen's inequality) for the eigenvalues $\lambda_1(A),\ldots,\lambda_s(A)$ of the matrix

$$A = (k\Sigma k^*)^{-1/2^*}(k_0\Sigma_0 k_0^*)(k\Sigma k^*)^{-1/2},$$

we obtain

$$\begin{aligned}\operatorname{tr}(k\Sigma k^*)^{-1}(k_0\Sigma_0 k_0^*) &= \sum_{i=1}^{s}\lambda_i(A) \geq s\left[\prod_{i=1}^{s}\lambda_i(A)\right]^{1/s}\\ &= s[\det(k\Sigma k^*)^{-1}(k_0\Sigma_0 k_0^*)]^{1/s}\end{aligned}$$

and thus, due to (4.2.10),

$$(4.2.11)\qquad \begin{aligned}&\log\det\Sigma + (2\pi)^{-1}\int_{-\pi}^{\pi}\operatorname{tr}\{(k\Sigma k^*)^{-1}(k_0\Sigma_0 k_0^*)\}\,d\omega\\ &\geq \log\det\Sigma + s\exp\left\{(2\pi)^{-1}\int_{-\pi}^{\pi}\frac{1}{s}\log\det\{(k\Sigma k^*)^{-1}(k_0\Sigma_0 k_0^*)\}\,d\omega\right\}\\ &= \log\det\Sigma + s(\det\Sigma_0/\det\Sigma)^{1/s}.\end{aligned}$$

Thereby the last equation can be shown as follows. The function $f =$

$\det\{(kk^*)^{-1}(k_0k_0^*)\}$ is evidently 2π by a spectral density with prediction error variance unity. [Note that if the left-hand side of (4.2.10) is finite, f is integrable.] Thus by (1.3.11) (which is sometimes called Szegö's formula) we have

$$\exp\left\{(2\pi)^{-1}\int_{-\pi}^{\pi}\log\det(kk^*)^{-1}k_0k_0^*\,d\omega\right\} = 1.$$

Now the inequality between the extreme members of (4.2.11) is an equality if and only if $\operatorname{tr}\{(k\Sigma k^*)^{-1}(k_0\Sigma_0 k^*)\}$ is constant λ a.e. and the eigenvalues of $(k\Sigma k^*)^{-1/2^*}(k_0\Sigma_0 k_0^*)(k\Sigma k^*)^{-1/2}$ are equal to each other for almost all ω. This in turn is equivalent to $k\Sigma k^* = ck_0\Sigma_0 k_0^*$, λ a.e., $c > 0$, and since $k(0) = I$ we must have $k = k_0$, $\Sigma = c\Sigma_0$, $c > 0$. Since the function $\log x + s(a/x)^{1/s}$; $a > 0$, is uniquely minimized at $x = a$, the right-hand member of (4.2.11) is greater than or equal to $\log\det\Sigma_0 + s$ and equality can hold only if $\det\Sigma = \det\Sigma_0$, so that $c = 1$. The second term in $Q(\theta)$ is strictly positive unless $l = l_0$ in virtue of the persistent exciting condition (4.1.6). Then (ii) and (iii) are established.

If $\theta_0 \in \Theta^*$, equality holds in (4.2.11) at $\theta = \theta_0$ by what was said before and thus (i) follows from (ii).

Lemma 4.2.2

(i) $T^{-1}\log\det\Gamma_T(\theta) \geq \log\det\Sigma$, $\theta \in \hat{\Theta}$.

(ii) $\lim_{T\to\infty} T^{-1}\log\det\Gamma_T(\theta) = \log\det\Sigma$, $\theta \in \hat{\Theta}$.

Proof. Let (a, b, d) be an ARMAX realization of (k, l) corresponding to θ, where, without restriction of generality, $a(0) = b(0) = I$ holds and let A_T be the lower triangular block Toeplitz matrix corresponding to $a(z)$, that is,

$$A_T = \begin{pmatrix} I_s & 0 & 0 & \cdots & 0 \\ A(1) & I_s & 0 & \cdots & \cdots \\ A(2) & A(1) & I_s & 0 & \cdots \\ \cdots & \cdots & \cdots & \cdots & \cdots \\ A(T-1) & \cdots & \cdots & \cdots & I_s \end{pmatrix}.$$

Similarly define B_T in relation to b. Then

$$A_T u_T = B_T \varepsilon_T + \delta_T, \tag{4.2.12}$$

where $\varepsilon_T = (\varepsilon(1)', \varepsilon(2)', \ldots, \varepsilon(T)')'$ and δ_T is null outside the first $p[=\max v(a, b, d)$; $a^{-1}(b, d) \in \bar{U}]$ blocks and has linear combinations of $\varepsilon(t)$, $u(t)$, $-p < t \leq 0$ in the first p blocks; thus δ_T and ε_T are uncorrelated and

$$A_T\Gamma_T(\theta)A_T' = B_T(I_T \otimes \Sigma)B_T' + \Delta_T, \tag{4.2.13}$$

where $\Delta_T = E\delta_T\delta_T'$ has null elements outside the first p block rows and columns. Since $\det A_T = \det B_T = 1$ and $\Delta_T \geq 0$, (i) follows. If we write

$$B_T = \begin{pmatrix} B_p & 0 \\ C & B_{T-p} \end{pmatrix},$$

then we obtain from (4.2.13) and a well-known lemma for the determinant of partitioned matrices,

$$\begin{aligned}\det \Gamma_T(\theta) = {} & \det[B_p(I_p \otimes \Sigma)B_p' + \Delta_p - B_p(I_p \otimes \Sigma)C'\{B_{T-p}(I_{T-p} \otimes \Sigma)B_{T-p}' \\ & + C(I_p \otimes \Sigma)C'\}^{-1}C(I_p \otimes \Sigma)B_p'] \det[B_{T-p}(I_{T-p} \otimes \Sigma)B_{T-p}' \\ & + C(I_p \otimes \Sigma)C'].\end{aligned}$$

Now (ii) follows since p is fixed.

Since we are maximizing the likelihood over $\hat{\Theta}$, we need some device to avoid problems arising from the fact that the term $(k\Sigma k^*)^{-1}$ occurring in $Q(\theta)$ may have poles on the unit circle. Let c denote the least common denominator of the elements of k satisfying $c(0) = 1$ and let $N = ck$. Further let (a, b, d) be an ARMAX realization of (k, l) which is left coprime, where a satisfies the stability condition, where b satisfies the miniphase condition, where $a(0) = b(0) = I$ and where $v(a, b, d) \leq p$. Clearly, such an (a, b, d) exists for all $(k, l) \in \bar{U}$. For $\eta > 0$ we put

$$\phi_\eta(\omega; \theta) = 2\pi|c(e^{i\omega})|^2\{N(e^{i\omega})\Sigma N(e^{i\omega})^* + \eta I\}^{-1}$$

and

$$\begin{aligned}Q_\eta(\theta) = {} & (2\pi)^{-1}\int_{-\pi}^{\pi} \operatorname{tr}\{\phi_\eta(\omega; \theta)f_u(\omega; \theta_0)\}\, d\omega \\ & + (2\pi)^{-1}\int_{-\pi}^{\pi} \operatorname{tr}\{\phi_\eta(\omega; \theta)(l - l_0)\, dF_z(\omega)(l - l_0)^*\}\end{aligned} \tag{4.2.14}$$

and similarly, we put

$$Q_{T,\eta}(\theta) = T^{-1}u_T(\theta)'\Gamma_T^{-1}(\phi_\eta^{-1})u_T(\theta), \tag{4.2.15}$$

where we have used $\Gamma_T(\phi_\eta^{-1})$ to indicate the covariance matrix corresponding to the spectral density ϕ_η^{-1}.

Let

$$\Theta_{c_1,c_2} = \{\theta \in \bar{\Theta} | \lambda_1(\Sigma) \geq c_1, \lambda_s(\Sigma) \leq c_2\}, \tag{4.2.16}$$

where $c_2 > c_1 > 0$ and $\lambda_1(\Sigma)$ and $\lambda_s(\Sigma)$ denote the smallest and the largest eigenvalue of Σ, respectively, and call Θ_{c_1,c_2,c_3} the set of all elements of Θ_{c_1,c_2} where in addition

$$\begin{aligned}\|(R_1 R_2)\|^2 &= (2\pi)^{-1} \int_{-\pi}^{\pi} \operatorname{tr}\{(R_1, R_2)(R_1, R_2)^*\}\, d\omega \\ &= \sum_{j=0}^{r} \operatorname{tr}\{(R_{i,j}, R_{2,j})(R_{1,j}, R_{2,j})'\} \leq c_3; \qquad 0 < c_3 < \infty\end{aligned}$$

holds, where we have used the notation $R_1(z) = \operatorname{adj}(b)a = \sum_{j=0}^{r} R_{1,j} z^j$; $R_2(z) = \operatorname{adj}(b)d = \sum_{j=0}^{r} R_{2,j} z^j$, $r = (s-1)p + p = sp$.

Note that Θ_{c_1,c_2,c_3} in compact, which can be shown as follows. Take a sequence $(k_n, l_n, \sigma(\Sigma_n)) = \theta_n \in \Theta_{c_1,c_2,c_3}$. Clearly, there is a subsequence $\Sigma_{n(m)}$ converging to a matrix Σ with $\lambda_1(\Sigma) \geq c_1$, $\lambda_s(\Sigma) \leq c_2$. We again write $k_n^{-1} = (\det b_n)^{-1} \operatorname{adj}(b_n) a_n$, where $b_n(0) = a_n(0) = I$ and $\det b_n(z) \neq 0\, |z| < 1$. Then we again can find a subsequence $\det b_{n(m)}$ converging to a polynomial c with $c(0) = 1$, $c(z) \neq 0$, $|z| < 1$, and with $v(c) \leq sp$, since the coefficients of $\det b_n(z)$ are bounded due to the assumptions $\det b_n(0) = 1$, $\det b_n(z) \neq 0$, $|z| < 1$; and $v(\det b_n) \leq sp$. Since $\|\operatorname{adj}(b_n)a_n\| \leq c_3$ and since $v(\operatorname{adj} b_n a_n) \leq (s-1)p + p$, for a suitable subsequence we have (using now n also as the index for the subsequence) $\operatorname{adj}(b_n)a_n \to N$. All subsequences can be chosen to have the same index set. Setting $k^{-1} = c^{-1}N$ we have $k_n^{-1} \to k^{-1}$ (in T_{pt}) and thus $k_n \to k$ (in T_{pt}). With a similar argument for $(\det b_n)^{-1} \operatorname{adj}(b_n) d_n = k_n^{-1} l_n$, we see that l_n converges to a transfer function l for a suitable subsequence.

Combining the results of the previous lemma with those of the next lemma, we obtain the asymptotic behavior of the likelihood function.

Lemma 4.2.3. *If $\theta_0 \in \hat{\Theta}$ and the conditions of Theorem 4.1.1 hold then*

(i) $\lim_{T\to\infty} Q_T(\theta) = Q(\theta)$, *a.s.* $\theta \in \Theta^*$.

(ii) $\lim_{T\to\infty} Q_{T,\eta}(\theta) = Q_\eta(\theta)$, *a.s.*, $\eta > 0$, *uniformly in* $\Theta_{c_1,c_2,c_3} \cap \hat{\Theta}$.

Proof. To prove (i) we write

$$Q_T(\theta) = T^{-1}\{y_T'\Gamma_T^{-1}(\theta)y_T - 2z_T'L_T'\Gamma_T^{-1}(\theta)y_T + z_T'L_T'\Gamma_T^{-1}(\theta)L_T z_T\}, \tag{4.2.17}$$

where

$$L_T = \begin{pmatrix} 0 & \cdots\cdots\cdots\cdots & & 0 \\ L(1) & 0 & & \\ \vdots & & \ddots & \\ L(T-1) & L(T-2) & \cdots & 0 \end{pmatrix}.$$

According to Fejer's theorem (see, e.g., Rudin, 1973), every continuous periodic function $f\colon \mathbf{R} \to \mathbf{C}$ can be uniformly approximated by trigonometric polynomials, and in particular by the Cesaro sum of its finite Fourier series representations. Thus, for $f_u(\omega; \theta)$, $\theta \in \Theta^*$ we can find spectral densities P_1, P_2, whose entries are trigonometric polynomials, such that

$$0 < P_2 \leq f_u^{-1} \leq P_1, \qquad \sup_{\omega \in [-\pi, \pi]} \|P_1 - P_2\| < \varepsilon \tag{4.2.18}$$

for arbitrary $\varepsilon > 0$. Clearly, (4.2.18) implies that

$$\Gamma_T^{-1}(P_2^{-1}) \leq \Gamma_T^{-1}(f_u) \leq \Gamma_T^{-1}(P_1^{-1}). \tag{4.2.19}$$

Now for notational simplicity let us omit the index i and write P for P_i. P^{-1} may be taken as spectral density of an autoregression

$$\sum_{i=0}^{M} C(j)x(t-j) = \xi(t)$$

satisfying the stability condition (1.2.3), where $C(0) = I$, $E\xi(s)\xi(t)' = \delta_{st}\Sigma_P$. For $T > M$ we define the $Ts \times Ts$ matrices

$$C = \begin{pmatrix} I_{M_s} & \vdots & & & & \\ \cdots & & & & 0 & \\ C(M) & \cdots & C(0) & & & \\ 0 & \ddots & & \ddots & & \\ \vdots & \ddots & \ddots & & \ddots & \\ 0 & \cdots & 0 & C(M) & \cdots & C(0) \end{pmatrix}.$$

Then

$$\Gamma_T^{-1}(P^{-1}) = C' \begin{pmatrix} \Gamma_M^{-1}(P^{-1}) & \vdots & & 0 \\ \cdots & & & \\ & & \Sigma_P^{-1} & \\ & 0 & & \ddots & \\ & & & & \Sigma_P^{-1} \end{pmatrix} C$$

and thus

$$\begin{aligned} & T^{-1}y_T'\Gamma_T^{-1}(P^{-1})y_T \\ & \quad = T^{-1}y_M'\Gamma_M^{-1}(P^{-1})y_M \\ & \qquad + T^{-1}\sum_{t=M+1}^{T}\left(\sum_{j=0}^{M}C(j)y(t-j)\right)'\Sigma_P^{-1}\left(\sum_{j=0}^{M}C(j)y(t-j)\right). \end{aligned} \tag{4.2.20}$$

Since M is fixed, the first term on the right-hand side of (4.2.20) goes to zero for $T \to \infty$; the second term can be written as

$$\operatorname{tr}\left(\Sigma_P^{-1/2}\cdot T^{-1}\sum_{t=M+1}^{T}\left(\sum_{j=0}^{M}C(j)y(t-j)\right)\left(\sum_{j=0}^{M}C(j)y(t-j)\right)'\Sigma_P^{-1/2}\right),$$

and therefore by Theorem 4.1.1, this term converges a.s. to

$$\operatorname{tr}\left\{\int_{-\pi}^{\pi} Pk_0\Sigma_0k_0^*\,d\omega + (2\pi)^{-1}\int_{-\pi}^{\pi} Pl_0\,dF_z(\omega)l_0^*\right\}.$$

This together with (4.2.18) and (4.2.19) implies that

$$\begin{aligned} \lim_{T\to\infty} T^{-1}y_T'\Gamma_T^{-1}(\theta)y_T = \operatorname{tr}\Bigg\{&\int_{-\pi}^{\pi} f_u^{-1}k_0\Sigma_0k_0^*\,d\omega \\ &+ (2\pi)^{-1}\int_{-\pi}^{\pi} f_u^{-1}l_0\,dF_z(\omega)l_0^*\Bigg\}. \end{aligned}$$

Using the same kind of argument for the other terms on the right-hand side of (4.2.17) finally gives (i). Now, let us prove (ii). It is straightforward to show that the set $\{(N\Sigma N^* + \eta I)^{-1} | \theta \in \Theta_{c_1,c_2,c_3}\}$ is relatively compact in the space C_∞ of continuous functions $f\colon [-\pi,\pi] \to \mathbf{C}^{s\times s}$ endowed with the supremum norm, $\sup_{\omega\in[-\pi,\pi]}\|f(\omega)\|$; that an analogous result holds for the set $\{|c|^2 | \theta \in \Theta_{c_1,c_2,c_3}\}$. Therefore, every sequence in the set $\{\phi_\eta(\omega;\theta) | \theta \in \Theta_{c_1,c_2,c_3}\}$ contains a subsequence,

$$\phi_{\eta,n} = 2\pi|c_n|^2(N_n\Sigma_nN_n^* + \eta I)^{-1},$$

say, such that $|c_n|^2$ and $(N_n\Sigma_nN_n^* + \eta I)^{-1}$ converge, and thus also $\phi_{\eta,n}$ converges in the supremum norm and therefore $\{\phi_\eta(\omega;\theta) | \theta \in \Theta_{c_1,c_2,c_3}\}$ is relatively compact in C_∞. This, together with Fejer's theorem, implies that there exists a finite number of spectral densities $P_{1,i}$, $P_{2,i}$; $i = 1 \ldots n$, whose entries are trigonometric polynomials, such that for every $\phi_\eta(\omega;\theta)$, $\theta \in \Theta_{c_1,c_2,c_3}$ there is an index i such that

$$0 < P_{2,i} \leq \phi_\eta \leq P_{1,i}, \qquad \sup_{\omega \in [-\pi,\pi]} \|P_{1,i} - P_{2,i}\| < \varepsilon$$

holds, for arbitrary $\varepsilon > 0$. The rest of the proof then follows the same way as the proof of (i).

Theorem 4.2.1. *Let the assumptions $\theta_0 \in \Theta^*$, the standard assumptions including (4.1.1), (4.1.6), (4.2.1) and the other conditions of Theorem 4.1.1 hold as well as*

(4.2.21) *For $s > 1$ the matrix $\Gamma_z = [\Gamma_z(i-j)]_{i,j=1,\ldots,sp}$ has full rank.*

Then the maximum likelihood estimators $\hat{\theta}_T$ (over $\hat{\Theta}$) are strongly consistent, that is,

$$\lim_{T\to\infty} \hat{\theta}_T \to \theta_0, \quad a.s. \tag{4.2.22}$$

Proof. From the definition of the MLE we have $L_T(\hat{\theta}_T) \leq L_T(\theta)$; $\theta \in \Theta^*$, and thus using Lemmas 4.2.2, 4.2.3, and 4.2.1, we obtain

$$\begin{aligned} \limsup_{T\to\infty} L_T(\hat{\theta}_T) &\leq \inf_{\theta\in\Theta^*} \limsup_{T\to\infty} L_T(\theta) \\ &= \inf_{\theta\in\Theta^*} L(\theta) = \log\det\Sigma_0 + s, \quad \text{a.s.} \end{aligned} \tag{4.2.23}$$

and thus [since $Q_T(\hat{\theta}_T) \geq 0$ and by Lemma 4.2.2(i)] $\det\hat{\Sigma}_T$ is bounded (in T) for almost every trajectory.

We next show that the smallest eigenvalue $\lambda_1(\hat{\Sigma}_T)$ of $\hat{\Sigma}_T$ is bounded away from zero by a positive constant a.s., for $T \geq T_0$, where T_0 in general depends on the point ω in sample space. We shall consistently use T_0 in the foregoing sense.

From Lemma 4.2.2(i) we have

$$T^{-1}\log\det\Gamma_T(\theta) \geq \log\det\Sigma \geq s\log\lambda_1(\Sigma); \qquad \theta \in \hat{\Theta}. \tag{4.2.24}$$

Let $k = a^{-1}b$, with (a,b) left coprime $a(0) = b(0) = I$ and $v(a,b) \leq p$, and let $f_u(\omega) = f_u(\omega;\theta)$ be the corresponding spectral density. Note that due to $\det b(z) \neq 0, |z| < 1$, and $b(0) = I$ we have $|\det b(e^{-i\omega})|^2 \leq \gamma$, where for suitably chosen $\gamma > 0$, due to $v(a,b) \leq p$, the relation holds uniformly in $\hat{\Theta}$. We repeat that $R_1(z) = \operatorname{adj}(b(z))a(z) = \sum_0^r R_{1,j}z^j$ and $R_2(z) = \operatorname{adj}(b(z))d(z) = \sum_0^r R_{2,j}z^j$, where $r = sp$. Then we have

$$R_1 f_u R_1^* = (2\pi)^{-1}|\det b|^2\Sigma \leq (2\pi)^{-1}\gamma^2\Sigma; \qquad \theta \in \hat{\Theta}. \tag{4.2.25}$$

Now let A be the $(T-r)s \times Ts$ matrix with $A_{ij} = R_1(r+i-j)$ as the (i,j)th block entry (which is of size $s \times s$), $i \leq j \leq i+r$ and with zeros elsewhere. Then, from (4.2.25),

$$A\Gamma_T(\theta)A' = \Gamma_{T-r}(R_1 f_u R_1^*) \leq \gamma^2(I_{T-r} \otimes \Sigma), \qquad \theta \in \hat{\Theta}. \tag{4.2.26}$$

Now for an arbitrary matrix $B \in \mathbf{R}^{p\times s}$ with $p \leq s$, $BB' \leq I$ implies that $\lambda_s(B'B) \leq 1$. This can be shown as follows. Let $P(\Lambda, 0)Q$ (where P and Q are orthogonal matrices) be the singular-value decomposition of B. Then

$$\begin{aligned} z'B'Bz &= (w_1', w_2')\begin{pmatrix}\Lambda^2 & 0\\ 0 & 0\end{pmatrix}(w_1', w_2')' = w_1'P'BB'Pw_1 \\ &= x'BB'x \leq x'x \leq 1 \qquad \text{for } z'z = 1, \end{aligned}$$

where $(w_1', w_2') = z'Q'$ and $x' = w_1'P'$. Using this result and choosing B as $(I_{T-r} \otimes \Sigma^{-1/2})\, A\Gamma_T(\theta)^{1/2}$ we obtain from (4.2.26) [pre- and postmultiplying it by $(I_{T-r} \otimes \Sigma^{-1/2})$]:

$$\begin{aligned} T^{-1}u_T(\theta)'\Gamma_T^{-1}(\theta)u_T(\theta) &\geq T^{-1}\gamma^2 u_T(\theta)'A'(I_{T-r} \otimes \Sigma^{-1})Au_T(\theta) \\ &= \gamma^{-2}\operatorname{tr}\{\Sigma^{-1}q(\theta)\}, \end{aligned} \tag{4.2.27}$$

where

$$q(\theta) = \sum_{i,j=0}^{r} R_{1,i}\left[T^{-1}\sum_{t=r+1}^{T} u(t-i,\theta)u(t-j,\theta)'\right]R_{1,j}.$$

Let

$$R = (I, R_{1,1}, \ldots, R_{1,r}, -R_{2,1}, \ldots, -R_{2,r})$$

and

$$G = \begin{pmatrix} G_y & G_{yz} \\ G_{zy} & G_z \end{pmatrix},$$

where G is partitioned conformally with the partitions in R into parts involving $R_{1,j}$ and parts involving $R_{2,j}$. The blocks in G are composed of the $G_y(j-k)$, $G_{yz}(j-k)$, $G_z(j-k)$ in the (j,k)th place except that, for example, $G_y(j-k)$ is replaced by

(4.2.28) $$\frac{1}{T}\sum_{r+1}^{T} y(t-j)y(t-k)'.$$

Then

$$q(\theta) = RGR'.$$

Now

$$G = \Gamma + E, \qquad \Gamma = \begin{pmatrix} \Gamma_y(\theta_0) & \Gamma_{yz}(\theta_0) \\ \Gamma_{zy}(\theta_0) & \Gamma_z \end{pmatrix}$$
$$= \begin{pmatrix} \Gamma_u(\theta_0) & 0 \\ 0 & 0 \end{pmatrix} + \begin{pmatrix} \Gamma_v(\theta_0) & \Gamma_{vz}(\theta_0) \\ \Gamma_{zv}(\theta_0) & \Gamma_z \end{pmatrix},$$

where, for example, the (j, k)th block element of $\Gamma_u(\theta_0)$ is $E\{u(j)u(k)'\}$, where $u(t)$ corresponds to θ_0. Since r [due to assumption (4.2.1)] is fixed, $E \to 0$ a.s. by Theorem 4.1.1. Now

(4.2.29)
$$\lambda_1(q(\theta)) = \inf_{x'x=1} x'RGR'x \geq \lambda_1(G) \inf_{x'x=1} x'RR'x$$
$$\geq \left(\lambda_1(\Gamma) - \sup_{x'x=1} x'Ex\right) \inf_{x'x=1} x'RR'x.$$

But $\lambda_1(\Gamma) > 0$ since $\lambda_1(\Gamma_u) > 0$ and $\lambda_1(\Gamma_z) > 0$ the latter by (4.2.21). When $s = 1$ then $\mathrm{tr}\{\sigma^{-2}q(\theta)\} = \sigma^{-2}\mathrm{tr}\{q(\theta)\}$ and, by the same argument as above, but without (4.2.21), $\mathrm{tr}\{q(\theta)\} > 0$ for $T > T_0$. Also, $x'RR'x \geq 1$, $x'x = 1$, since $R_{1,0} = I_s$ and $\sup x'Ex \to 0$, $x'x = 1$. Thus uniformly in $\theta \in \hat{\Theta}$, for $T > T_0$,

(4.2.30) $$\lambda_1(RGR') \geq c > 0, \quad \text{a.s.}$$

Now from (4.2.24), (4.2.27) and (4.2.30) we have, uniformly in $\theta \in \hat{\Theta}$, for $T > T_0$

(4.2.31) $$L_T(\theta) \geq s\log\lambda_1(\Sigma) + \gamma^{-2}\cdot c\cdot\lambda_1^{-1}(\Sigma), \quad \text{a.s.}$$

Now it follows that

(4.2.32) $$\liminf_{T\to\infty} \lambda_1(\hat{\Sigma}_T) > 0, \quad \text{a.s.}$$

Indeed, if (4.2.32) did not hold, there would be a set $S \subset \Omega$ with $P(S) > 0$ such that

$$\liminf_{T\to\infty} \lambda_1(\hat{\Sigma}_T) = 0$$

and along such a sequence $\hat{\Sigma}_T$, from (4.2.31),

$$\limsup_{T\to\infty} L_T(\hat{\theta}_T) = \infty,$$

contrary to (4.2.23). Thus, as $\det \hat{\Sigma}_T$ is bounded [due to (4.2.23)], there are c_1, c_2, and T_0 such that for $T > T_0$,

$$0 < c_1 \leq \lambda_1(\hat{\Sigma}_T) \leq \lambda_s(\hat{\Sigma}_T) \leq c_2 < \infty, \quad \text{a.s.} \tag{4.2.33}$$

Now, returning to (4.2.27) we obtain

$$L_T(\theta) \geq s \log \lambda_1(\Sigma) + \gamma^{-2} \lambda_s(\Sigma)^{-1} \operatorname{tr}(q(\theta)). \tag{4.2.34}$$

Thus

$$L_T(\theta) \geq s \log \lambda_1(\Sigma) + \gamma^{-2} \lambda_s(\Sigma)^{-1} \lambda_1(G) \operatorname{tr}(R'R). \tag{4.2.35}$$

Now,

$$\lambda_1(G) = \lambda_1(\Gamma + E) \geq c > 0, \quad \text{a.s.}$$

for $T > T_0$, as was shown above. By (4.2.33), $\operatorname{tr} \hat{R}_T' \hat{R}_T$, where $\hat{R}_T$ is the matrix R corresponding to $\hat{\theta}_T$, must be bounded a.s. for $T > T_0$ by the same kind of argument as was used in relation to $\lambda_1(\hat{\Sigma}_T)$. Thus, using (4.2.6), we have shown that $\hat{\theta}_T \in \Theta_{c_1, c_2, c_3} \cap \hat{\Theta}$, a.s. for $T > T_0$ and for suitable $c_1 > 0$, $c_2 > 0$, $c_3 > 0$. From $\Gamma_T^{-1}(\theta) \geq \Gamma_T^{-1}(\phi_\eta^{-1})$ and from Lemma 4.2.2(i) we have

$$\liminf_{T\to\infty} L_T(\hat{\theta}_T) \geq \sup_{\eta > 0} \liminf_{T\to\infty} \{\log \det \hat{\Sigma}_T + Q_{T,\eta}(\hat{\theta}_T)\}. \tag{4.2.36}$$

Since $\hat{\theta}_T \in \Theta_{c_1, c_2, c_3} \cap \hat{\Theta}$ for $T > T_0$, by Lemma 4.2.3(ii),

$$\begin{aligned} &\sup_{\eta > 0} \liminf_{T\to\infty} \{\log \det \hat{\Sigma}_T + Q_{T,\eta}(\hat{\theta}_T)\} \\ &\quad = \sup_{\eta > 0} \liminf_{T\to\infty} \{\log \det \hat{\Sigma}_T + Q_\eta(\hat{\theta}_T)\}, \text{ a.s.} \end{aligned} \tag{4.2.37}$$

Now, if $\lim \hat{\theta}_T = \theta_0$, a.s., would not hold, then due to the compactness of Θ_{c_1,c_2,c_3}, there would be a subsequence of $\hat{\theta}_T$, with limit $\theta = (k, l, \sigma(\Sigma)) \in \Theta_{c_1,c_2,c_3}$ a.s., and $\theta \neq \theta_0$. Here, for simplicity, subsequences are denoted by the same indices. It is straightforward to show [using a compactness argument], that $(k_T, l_T) \to (k, l)$ (in T_{pt}) implies the existence of a subsequence such that $(k_T(\omega), l_T(\omega)) \to (k(\omega), l(\omega))$, λ-a.e., and that $c_T \to \bar{c}$, $N_T \to \bar{N}$ and $k = \bar{c}^{-1}\bar{N}$, where c_T is the least common denominator of k_T, satisfying $c_T(0) = 1$ and $N_T = c_T k_T$. [Note that $\bar{c}$ is not necessarily the least common denominator of k]. Therefore, since the c_T are bounded [due to $c_T(z) \neq 0$, $|z| < 1$, $c_T(0) = 1$ and (4.2.1)] and since $(N\Sigma N^* + \eta I)^{-1} \leq \eta^{-1} I$, using the dominated convergence theorem, we obtain for a suitably chosen subsequence

$$\lim_{T\to\infty} \int_{-\pi}^{\pi} \operatorname{tr}\{\phi_\eta(\omega; \hat{\theta}_T) f_u(\omega; \theta_0)\}\, d\omega$$

$$= \int_{-\pi}^{\pi} \operatorname{tr}\{2\pi |\bar{c}(e^{i\omega})|^2 [\bar{N}(e^{i\omega})\Sigma\bar{N}(e^{i\omega})^* + \eta I]^{-1} f_u(\omega; \theta_0)\}\, d\omega.$$

Analogously, since for example,

$$\frac{1}{2\pi}\operatorname{tr}\{\phi_\eta l l^*\} \leq \operatorname{tr}\{(k\Sigma k^*)^{-1} l l^*\} = \operatorname{tr}\{R_2 R_2^* \Sigma^{-1}\}$$

holds, and thus the left hand side in the expression above is bounded in Θ_{c_1,c_2,c_3}, we get, again using the dominated convergence theorem, convergence for the second part of Q_η and thus

$$\lim_{T\to\infty} Q_\eta(\hat{\theta}_T) = Q_\eta(\theta). \tag{4.2.38}$$

Since $\eta \downarrow 0$ implies $\phi_\eta(\omega, \theta) \uparrow f_u(\omega, \theta)^{-1}$ λ-a.e., from the monotone convergence theorem we obtain

$$\sup_{\eta > 0} Q_\eta(\theta) = Q(\theta).$$

Thus, from (4.2.36), (4.2.37), (4.2.38) and the last expression, we get, for a suitably chosen subsequence

$$\liminf_{T\to\infty} L_T(\hat{\theta}_T) \geq \log\det\Sigma + Q(\theta) \geq \log\det\Sigma_0 + s \tag{4.2.39}$$

where the last inequality follows from Lemma 4.2.1. Thus, we have a subsequence $\hat{\theta}_T \to \theta \neq \theta_0$ such that [by (4.2.23) and (4.2.39)]

(4.2.40)
$$L(\theta) = \lim_{T \to \infty} L_T(\hat{\theta}_T) = \log\det\Sigma_0 + s.$$

On the other hand, since $\theta \neq \theta_0$, from Lemma 4.2.1 we have $L(\theta) > L(\theta_0) = \log\det\Sigma_0 + s$ in contradiction to (4.2.40). Thus $\theta = \theta_0$ and the assertion of the theorem has been proved.

REMARK 1. An essential point in the last proof was that L is uniquely minimized over Θ^* at $\theta = \theta_0$. If

(4.2.41)
$$\inf_{\theta \in \Theta^*} L(\theta) = L(\theta_0) = \log\det\Sigma_0 + s \qquad \text{for } \theta_0 \in \hat{\Theta}$$

holds instead of Lemma 4.2.1 (i), Theorem 4.2.1 is true for $\theta_0 \in \hat{\Theta}$. Now, (4.2.41) is not valid for arbitrary parameter space. However, if Θ has the following property:

(4.2.42)
For every $\theta_0 \in \hat{\Theta} - \Theta^*$ there is a sequence $\theta_n = (k_n, l_0, \sigma(\Sigma_0)) \in \Theta^*$ such that $\{k_n\Sigma_0 k_n^*\} \uparrow \{k_0\Sigma_0 k_0^*\}$ λ a.e. holds,

then (4.2.41) is true. This can be seen as follow. The term $\operatorname{tr}\{(k_n\Sigma_0 k_n^*)^{-1}(k_0\Sigma_0 k_0^*)\}$ is nonnegative and by (4.2.42) uniformly bounded in ω and n. By Fatou's lemma, then,

$$\begin{aligned}
&\inf_n (2\pi)^{-1} \int_{-\pi}^{\pi} \operatorname{tr}\{(k_n\Sigma_0 k_n^*)^{-1}(k_0\Sigma_0 k_0^*)\}\, d\omega \\
&\le \limsup_{n\to\infty} (2\pi)^{-1} \int_{-\pi}^{\pi} \operatorname{tr}\{(k_n\Sigma_0 k_n^*)^{-1}(k_0\Sigma_0 k_0^*)\}\, d\omega \\
&\le (2\pi)^{-1} \int_{-\pi}^{\pi} \limsup_{n\to\infty} \operatorname{tr}\{(k_n\Sigma_0 k_n^*)^{-1}(k_0\Sigma_0 k_0^*)\}\, d\omega = s.
\end{aligned}$$

Since the remaining term in $Q(\theta_n)$ is equal to zero, this together with Lemma 4.2.1 gives (4.2.20).

Condition (4.2.42) holds, for example, for $\Theta = M(n) \times \underline{\Sigma}$ and $s = 1$ (i.e., the single-output case). As is easily seen, in this case a sequence k_n satisfying (4.2.42) is obtained by multiplying the zeors of b_0 which are on the unit circle by $c_n \in \mathbf{R}$, $c_n \downarrow 1$.

REMARK 2. If $y(t)$ contains constant and periodic components, this can be taken into account by adding a term $C\xi(t)$ to the right-hand side of (1.1.7), where $\xi(t)' = (1, \cos\omega_1 t, \sin\omega_1 t, \cos\omega_2 t, \ldots, \cos\omega_a t, \sin\omega_a t)$ is a vector of dimension m_1, say (which may be regarded as a component vector of the input vector), where $0 < \omega_1 < \omega_2 < \cdots < \omega_a \leq \pi$ and where C is a (constant) $s \times m_1$ matrix. Of course, this case is practically important. Theorem 4.2.1 can be applied to this situation as follows. The expressions R and G below (4.2.27) are changed (using the old symbols for new quantities) to

$$R = (I, R_{1,1}, \ldots, -R_{2,1}, \ldots, -R_{2,r}, -R_3)$$

and

$$G = \begin{pmatrix} G_y & G_{yz} & G_{y\xi} \\ G_{zy} & G_z & G_{z\xi} \\ G_{\xi y} & G_{\xi z} & G_\xi \end{pmatrix},$$

where the meaning of the new quantities $R_3 \in \mathbf{R}^{s \times m_1}$, $G_\xi \in \mathbf{R}^{m_1 \times m_1}$, and so on, is clear from the proof of Theorem 4.2.1. Then again we have

$$q(\theta) = RGR'.$$

Now, let

$$G = \Gamma + E, \qquad \Gamma = \begin{pmatrix} \Gamma_y(\theta_0) & \Gamma_{yz}(\theta_0) & \Gamma_{y\xi}(\theta_0) \\ \Gamma_{zy}(\theta_0) & \Gamma_z & \Gamma_{z\xi} \\ \Gamma_{\xi y}(\theta_0) & \Gamma_{\xi z} & \Gamma_\xi \end{pmatrix},$$

where

$$\Gamma_\xi = \begin{pmatrix} 1 & 0 \\ 0 & \frac{1}{2} I_{2a} \end{pmatrix} \quad \text{or} \quad \Gamma_\xi = \begin{pmatrix} 1 & 0 & 0 \\ 0 & \frac{1}{2} I_{2a} & 0 \\ 0 & 0 & 1 \end{pmatrix}$$

for $\omega_a < \pi$ and $\omega_a = \pi$, respectively (and where, again, the meaning of $\Gamma_{z\xi}$ is obvious). Then (4.2.30) holds if (4.2.21) is changed to

$$\begin{pmatrix} \Gamma_z & \Gamma_{z\xi} \\ \Gamma_{\xi z} & \Gamma_\xi \end{pmatrix} > 0 \tag{4.2.43}$$

in order to ensure (4.2.30). Now (4.2.43) is equivalent to

$$\begin{pmatrix} \Gamma_z - \Gamma_{z\xi}\Gamma_\xi^{-1}\Gamma_{\xi z} & 0 \\ \Gamma_{\xi z} & \Gamma_\xi \end{pmatrix} > 0$$

and thus to

$$(\Gamma_z - \Gamma_{z\xi}\Gamma_\xi^{-1}\Gamma_{\xi z}) > 0. \tag{4.2.44}$$

Condition (4.2.44), as can easily be seen, has a nice interpretation. It means that for $z(t)$, after correction for mean and periodic effects by regression for $\xi(t)$, a condition like (4.2.21) must hold.

REMARK 3. Note that condition (4.2.21) is implied by the persistent exciting (1.3.22). This can easily be seen by the same kind of argument as was used below (4.2.5) to show that $\Gamma_T(\theta) > 0$ holds.

REMARK 4. Of course in many cases in estimation the main interest is in the ARMAX or state-space parameters. Also, as pointed out already, the actual calculation has to be made in coordinates. This raises the question about the consequences of Theorem 4.2.1 for ARMAX or state-space parameter estimation. We will discuss this now for the case $\Theta = U \times \underline{\Sigma}$, where U is either V_α or U_α or W_α, as defined in Chapter 2. Let ψ: $U \to T$ denote one of the corresponding ARMAX or state-space parameterizations and let T be the corresponding set of free ARMAX or state-space parameters. As described in Remark 1 following Theorem 2.5.3, we have to distinguish between the following three cases:

(i) If $(k_0, l_0, \Sigma_0) \in \Theta$ holds, and thus if α (in V_α or U_α or W_α) is correctly specified, then $(\hat{k}_T, \hat{l}_T, \hat{\Sigma}_T) \to (k_0, l_0, \Sigma_0)$ implies that the MLEs $\hat{\tau}_T = \psi(\hat{k}_T, \hat{l}_T)$ for $\tau_0 = \psi(k_0, l_0)$ are defined from a certain T_0 onward (since U is open in $\bar{U}$) and moreover that, due to the continuity of ψ, $\hat{\tau}_T$ *is strongly consistent for* τ_0, that is,

$$\lim_{T \to \infty} \hat{\tau}_T = \tau_0 \quad \text{a.s.}$$

(ii) For $(\hat{k}_T, \hat{l}_T) \in \pi(\bar{T}) - U$, every element $\hat{\tau}_T$ of the equivalence class $\pi^{-1}(\hat{k}_T, \hat{l}_T) \cap \bar{T}$ is a maximum likelihood estimator. If $(k_0, l_0) \in \pi(\bar{T}) - U$ holds and if suitable prior bounds on the norms of the elements of $\bar{T}$ have been imposed, then $\hat{\theta}_T \to \theta_0$ implies that $\hat{\tau}_T$ converges to the true equivalence class $\pi^{-1}(k_0, l_0) \cap \bar{T}$. Thus in this case of misspecification of α, we still have *convergence to the true equivalence class*, but not necessarily to a special point within this class.

(iii) In the case $(k_0, l_0) \in \bar{U} - \pi(\bar{T})$, the MLEs $\hat{\tau}_T$, if they exist [i.e., if $(\hat{k}_T, l_T) \in \pi(\bar{T})$ holds], will have the property $\|\hat{\tau}_T\| \to \infty$ and thus they will *not converge*.

It should be pointed out that even if $(k_0, l_0) \in U$, but if (k_0, l_0) is "near" to a point in $\bar{U} - U$, then, in finite samples, problems similar to those described in (ii) or (iii) may arise. The reason for this is that the mapping ψ, although being continuous, may be very distorting at such points, in the sense that a "small" variation of transfer functions causes a "large" variation of τ-parameters.

3. THE CENTRAL LIMIT THEOREM AND THE LAW OF THE ITERATED LOGARITHM

Throughout this book we use the notation o_P, O_P, o, O. By $X_T = o(f_T)$, for example, we mean that $|X_T|/f_T \to 0$, a.s. and by $X_T = O(f_T)$ that $\limsup|X_T|/f_T \leq c < \infty$, *a.s. In the case of o_P, O_P the meaning is the same with a.s. replaced by "in probability." Thus, by $|X_T| = O_P(f_T)$, we mean that there is a c, $0 < c < \infty$, so that $P\{|X_T|/f_T > c\} \to 0$.*

We shall now be concerned mainly with the case where the true system is a rational transfer function system of known order. Of course in Chapter 5 we consider the estimation of order and obtain strongly consistent estimates of order, so that the results of this section will be relevant when the order is not known. We treat only the case where $\det\{k(z)\} \neq 0$, $|z| \leq 1$. It would be possible to consider some case where $\det(k)$ has a zero on $|z| = 1$, although the limiting distribution would not then be normal, but we have not done that.

We first consider the case where the parameter set is $M(n) \times \underline{\Sigma}$, $\underline{\Sigma} = \{\sigma(\Sigma)|\Sigma > 0\}$, but the other relevant cases will then be respond to the same treatment. Of course, each $(k, l) \in M(n)$ is now assumed to be analytic for $|z| \leq 1$ and satisfy $\det(k) \neq 0$, $|z| \leq 1$. See the Remark 2 below Theorem 2.5.3 and Remark 1 below Theorem 2.6.5. In this section and later sections we also use θ to indicate a vector of the form $(\tau', \sigma(\Sigma)')'$, i.e., the coordinatised form of $(k, l, \sigma(\Sigma))$. However, θ_0 will be allowed to satisfy constraints defined by twice continuously differentiable functions $\psi_j(\theta)$, $j = 1, \ldots, c$, that is,

$$\psi_j(\theta_0) = 0, \qquad j = 1, \ldots, c. \tag{4.3.1}$$

To avoid redundant complexity it is further assumed that

$$\operatorname{rank} \partial_\theta \psi(\theta_0) = c. \tag{4.3.2}$$

Here $\partial_\theta \psi(\cdot)$ is the gradient matrix of the vector $\psi(\cdot)$ composed of the $\psi_j(\cdot)$ evaluated in some coordinate system. The matrix $\partial_\theta \psi(\theta_0)$ would take different values in different coordinate systems, but its rank would be independent of the system. *We consistently use a notation such as $\partial_\theta \psi(\theta_0)$ to mean $\partial_\theta \psi(\theta)$ evaulated at θ_0.* The space Θ in Section 4.2 is now just the part of $M(n) \times \underline{\Sigma}$ satisfying the constraints. Of course, now $\theta_0 \in \Theta^*$ because of the condition on det(k). However, we do not here consider cases where, say, the McMillan degree for θ_0 is lower than n, so that always $\theta_0 \in M(n) \times \underline{\Sigma}$.

We know that $\hat{\theta}_T \to \theta_0$, a.s., from Theorem 4.2.1 and thus $\hat{\theta}_T$ eventually enters some neighborhood of θ_0 in $M(n) \times \underline{\Sigma}$. Thus from some T on we may express $\hat{\theta}_T$ in a fixed system of coordinates and in future we shall allow θ, $\hat{\theta}_T$, θ_0 to stand for these quantities when expressed as vectors in this system of coordinates. Write $\theta' = (\tau', \sigma')$ for the partition of θ into a component, τ, in $M(n)$ and a component, σ, in $\underline{\Sigma}$.

A common situation of importance is that where each $\psi_j(\theta)$ is a function either of τ alone or of σ alone. We shall then, for brevity, say that the constraints do not "mix parameters" (i.e., do not mix system parameters with covariances). Although prior constaints on Σ are not common, they can arise, as we shall see in later examples, in connection with constraints on τ.

We shall consider the central limit theorem (CLT) mainly for $\hat{\tau}_T$, although $\hat{\sigma}_T$ will also be dealt with. The simplest case is the ARMA case, where constraints do not mix parameters. Then, in addition to the ergodicity of the $\varepsilon(t)$ and (4.3.1), (4.3.2) we minimally need, for the CLT for $\hat{\tau}_T$,

Conditions A: (4.1.15), (4.1.20).

However, under this condition the covariance matrix in the limiting distribution is more complicated because $\sigma_{abcd}(r, s)$; r, $s > 0$ (see the end of Section 4.1) is not $\delta_{rs}\sigma_{ab}\sigma_{cd}$, and thus is not that obtaining when the $\varepsilon(t)$ are Gaussian. However, that is so under the following

Conditions A′: (4.1.15), (4.1.16).

When constraints *do* mix parameters, the distribution of $\hat{\tau}_T$ depends on

$$\dot{\Sigma} = \frac{1}{T}\sum_{1}^{T} \varepsilon(t)\varepsilon(t)'$$

and now, even in the ARMA case, we need

Conditions AM: (4.1.15), (4.1.16), (4.1.20) and

$$\lim_{k\to\infty} E\{\varepsilon_a(t)\varepsilon_b(t)\varepsilon_c(t)|\mathscr{F}_{t-k}\} = \sigma_{abc}(0), \qquad 1 \le a, b, c \le s. \tag{4.3.3}$$

However, again the covariance matrix in the limiting distribution will be complicated unless these are strengthened to

Conditions AM′: (4.15), (4.1.16), (4.1.20) and

$$E\{\varepsilon_a(t)\varepsilon_b(t)\varepsilon_c(t)|\mathscr{F}_{t-1}\} = \sigma_{abc}(0), \qquad 1 \le a, b, c \le s. \tag{4.3.3$'$}$$

In the ARMAX case, when constraints do not mix parameters, near-to-minimal conditions seem to be, in addition to those of Theorem 4.2.1 not mentioned below and (4.3.1) (4.3.2),

Conditions AX: (4.1.1), (4.1.14), (4.1.15), (4.1.17), (4.1.18), (4.1.20).

For a simple covariance matrix we need, however,

Conditions AX′: (4.1.1), (4.1.14), (4.1.15), (4.1.16).

When constraints do mix parameters, we need

Conditions AMX: (4.1.1), (4.1.14), (4.1.15), (4.1.16), (4.1.20). (4.3.3).

However, again for a simple covariance structure in the limiting distribution, this has to be strengthened to

Conditions AMX′: (4.1.1), (4.1.14), (4.1.15), (4.1.16), (4.1.20), (4.3.3)′.

Since (4.1.16), (4.3.3)′ are not easy to justify, except on Gaussian assumptions, the best justification for the standard procedure, which acts as if the $\varepsilon(t)$ are Gaussian, seems to be, when constraints do not mix parameters, Conditions A or AX combined with the implicit assumption that $\sigma_{abc}(r)$, $\sigma_{abcd}(r, s)$, $r, s > 0$ (see Section 4.1), are near enough to 0 and $\delta_{rs}\sigma_{ab}\sigma_{cd}$ for use of the Gaussian asymptotic covariance matrix to be reasonable.

Consider

$$M_T(\theta, \lambda) = L_T(\theta) + \lambda'\psi(\theta),$$

where λ is a vector of Lagrange multipliers. Then, $\hat{\theta}_T$, $\hat{\lambda}_T$ being the optimizing values,

$$\partial_\lambda M_T(\hat{\theta}_T, \hat{\lambda}_T) = \psi(\hat{\theta}_T)' = 0. \tag{4.3.4}$$

We adopt this formulation because it forms a basis for a perspicuous presentation of the central limit theorem. In particular, $\hat{\lambda}_T$, as is well known, provides a test statistic for the validity of the constraints. Of course, as has been said, we know that $\hat{\theta}_T \to \theta_0$, from Theorem 4.2.1. Then also $\partial_\theta M_T(\theta, \lambda) = \partial_\theta L_T(\theta) + \lambda'\partial_\theta\psi_T(\theta)$. Now it is not difficult to show (we give details in a moment) that $\partial_\theta L_T(\hat{\theta}_T) \to \partial_\theta L(\theta_0)$, which is zero, since $L(\theta)$ has a minimum at θ_0, from Lemma 4.2.1,(ii). Thus, from (4.3.2), Lemma 4.3.1 follows.

Lemma 4.3.1. *Under the conditions of Theorem 4.2.1, $\hat{\lambda}_T \to 0$, a.s..*

The proof that $\partial_\theta L_T(\hat{\theta}_T) \to \partial_\theta L(\theta_0) = 0$ is essentially the same as for Theorems 4.1.1 and 4.2.1. Consider, for example, $T^{-1}\partial_\theta \log\det\Gamma_T(\theta)$. From the proof of Lemma 4.2.2 this is $T^{-1}\partial_\theta \log\det\Sigma + T^{-1}\partial_\theta \log\det C_T(\theta)$, where $C_T(\theta)$ is described in the proof of Lemma 4.2.2, and we may now use the matrix fraction description and covariates described in Theorem 2.6.4, for example. The first term clearly converges to zero uniformly in θ, while $C_T(\theta)$ is continuously differentiable in its elements and is of fixed dimension and again converges to zero uniformly in θ in the closure of some nondegenerate neighborhood of θ_0. As far as $\partial_\theta Q_T(\theta)$ is concerned, the proof that this converges to $\partial_\theta Q(\theta)$ uniformly in some neighborhood of θ_0 again proceeds as in the proof of Theorem 4.2.1 (i.e., Lemma 4.2.3), the proof now being easier because we may choose the neighborhood so that Σ has its smallest eigenvalue bounded and, of course, so that the elements of a, b, d in a canonical matrix fraction description have uniformly bounded coefficients and $\det a$, $\det b$ have zeros bounded away from $|z| = 1$. Thus $\partial_\theta M_T(\hat{\theta}_T, \hat{\lambda}_T) \to \partial_\theta L(\theta_0) + \hat{\lambda}_T' \partial_\theta \psi(\theta_0) = \hat{\lambda}_T' \partial_\theta \psi(\theta_0)$ and Lemma 4.3.1 follows, as pointed out above, from (4.3.2). We now state the main theorem of this section. To state the theorem succinctly, we introduce some notation. By I_0 we mean $\frac{1}{2}\partial_\tau^2 Q(\theta_0)$ [see (4.2.9)], that is, $\partial_\tau^2 Q(\theta_0)$ has $\partial^2 Q(\theta)/\partial_{\tau_i}\partial_{\tau_j}$ evaluated at θ_0, to the (i,j)th place. By S_0 we mean the covariance matrix in this limiting distribution of $T^{1/2}\partial_\sigma L_T(\theta_0)$ [see (4.3.24), (4.3.26) below]. Thus we write the elements of $T^{1/2}\partial_\sigma L_T(\theta_0)$ by stacking the derivatives in a vector, taking them in dictionary order first according to column number and then row number and taking the derivatives corresponding to the points of the covariance matrix on or below the diagonal. This would usually be written as $\operatorname{vech}(T^{1/2}\partial_\sigma L_T(\theta_0))$.

Theorem 4.3.1. *Let $\theta_0 \in M(n) \times \Sigma$ and be constrained so that (4.3.1) and (4.3.2) hold where the $\psi_j(\theta)$ are twice continuously differentiable in some neighborhood of θ_0. Let the conditions of Theorem 4.2.1 hold and let $\varepsilon(t)$ be ergodic. Then the CLT holds for the following cases under the indicated conditions. (See below for the definition of $\lambda^{(1)}$.)*

ARMA:	*Constraints do not mix parameters,*	$T^{1/2}(\hat{\tau}_T - \tau_0), \hat{\lambda}_T^{(1)}$;	*A or A'*
		$T^{1/2}(\hat{\theta}_T - \theta_0), \hat{\lambda}_T$;	*AM or AM'*
ARMA:	*Constraints mix parameters,*	$T^{1/2}(\hat{\theta}_T - \theta_0), \hat{\lambda}_T$;	*AM or AM'*
ARMAX:	*Constraints do not mix parameters,*	$T^{1/2}(\hat{\tau}_T - \tau_0), \hat{\lambda}_T^{(1)}$;	*AX or AX'*
		$T^{1/2}(\hat{\theta}_T - \theta_0), \hat{\lambda}_T$;	*AMX or AMX'*
ARMAX:	*Constraints mix parameters,*	$T^{1/2}(\hat{\theta}_T - \theta_0), \hat{\lambda}_T$;	*AMX or AMX'.*

Under the primed conditions the covariance matrices in the limiting distributions are

$$\begin{bmatrix} 2I_0 & 0 & \partial_\tau \psi(\theta_0)' \\ 0 & \partial_\sigma^2 L(\theta_0) & \partial_\sigma \psi(\theta_0)' \\ \partial_\tau \psi(\theta_0) & \partial_\sigma \psi(\theta_0) & 0 \end{bmatrix}^{-1} \begin{bmatrix} 4I_0 & 0 & 0 \\ 0 & S_0 & 0 \\ 0 & 0 & 0 \end{bmatrix} \times \begin{bmatrix} 2I_0 & 0 & \partial_\tau \psi(\theta_0)' \\ 0 & \partial_\sigma^2 L(\theta_0) & \partial_\sigma \psi(\theta_0)' \\ \partial_\tau \psi(\theta_0) & \partial_\sigma \psi(\theta_0) & 0 \end{bmatrix}^{-1}. \tag{4.3.5}$$

REMARK. It is evident that the same theorem holds for cases where $\mathrm{M}(n)$ *is, for example, replaced by* $M(p,q,r)$, V_α, U_α. Indeed, there are many other examples in Sections 2.7 and 2.5.

We now discuss some examples. We observe first that when constraints do not mix parameters we many partition λ' as $(\lambda^{(1)'}, \lambda^{(2)'})$, according as the constraints apply, respectively, to τ or σ, and correspondingly partition $\psi(\theta)'$ as $(\psi^{(1)}(\tau)', \psi^{(2)}(\sigma)')$. Now

$$\begin{bmatrix} \partial_\tau \psi(\theta_0)' \\ \partial_\sigma \psi(\theta_0)' \\ 0 \end{bmatrix} = \begin{bmatrix} \partial_\tau \psi^{(1)}(\tau_0)' & 0 \\ 0 & \partial_\sigma \psi^{(2)}(\sigma_0)' \\ 0 & 0 \end{bmatrix}$$

and (4.3.5) simplifies. Indeed, now $T^{1/2}(\hat{\tau} - \tau_0)$, $T^{1/2}\hat{\lambda}^{(1)}$ have an asymptotic distribution depending only on that of $T^{1/2}\partial_\tau L_T(\theta_0)$ (i.e., on I_0) while $T^{1/2}(\hat{\sigma} - \sigma_0)$, $T^{1/2}\hat{\lambda}^{(2)}$ depends only on the asymptotic distribution of $T^{1/2}\partial_\sigma L_T(\theta_0)$ (i.e, on S_0).

Example (i)

Let $\Theta = M(p,q,r) \times \underline{\Sigma}$. Arrange the elements of τ in dictionary order, first according to row index, $i = 1, \ldots, s$, then according to whether the element is an element of a coefficient matrix in $a(z)$, $b(z)$ or $d(z)$. We indicated this choice by an index $j = 1, 2, 3$. The third index in the dictionary ordering is to be the column index k, which will run from 1 to s for $j = 1, 2$ and from 1 to m for $j = 3$. Finally in the dictionary ordering we take u, respresenting the lag, running from 1 to p, q, or r according as $j = 1, 2, 3$. We adopt this dictionary ordering because it results in the most simply written down formulas in general. We must evaluate

(4.3.6)

$$I_0 = \left[\int_{-\pi}^{\pi} \operatorname{tr}\{(k_0\Sigma_0 k_0^*)^{-1}(\partial_{\tau_a} k_0)\Sigma_0(\partial_{\tau_b} k_0^*)\, d\omega/2\pi + (\partial_{\tau_a} l_0)\, dF_z(\omega)(\partial_{\tau_b} l_0^*)\}\right],$$

where the (a, b)th element is shown. We use the notation $A \otimes B$ for the tensor or Kronecker product of two matrices so that $A \otimes B = [a_{ij}B]$, where a typical block is shown. We write e_i for the vector with unity in the ith place and zeros elsewhere and

$$c_p(\omega) = (e^{iu\omega})_{u=1,\ldots,p}$$

for the p-rowed vector with $\exp(iu\omega)$ in the uth place. We use 0_m for the $m \times m$

matrix of zeros. Then

$$I_0 = \int_{-\pi}^{\pi} (b_0 \Sigma_0 b_0^*)'^{-1} \otimes \left\{ \begin{bmatrix} -k_0 \otimes c_p(\omega) \\ I_s \otimes c_q(\omega) \\ 0_m \otimes c_r(\omega) \end{bmatrix} \right.$$

$$\times \Sigma_0[-k_0^* \otimes c_p(\omega)^* \vdots I_s \otimes c_q(\omega)^* \vdots 0_m \otimes c_r(\omega)^*]\, d\omega/2\pi \tag{4.3.7}$$

$$\left. + \begin{bmatrix} -l_0 \otimes c_p(\omega) \\ 0_s \otimes c_q(\omega) \\ I_m \otimes c_r(\omega) \end{bmatrix} dF_z(\omega)[-l_0^* \otimes c_p(\omega)^* \vdots 0_s \otimes c_q(\omega)^* \vdots I_m \otimes c_r(\omega)^*] \right\}.$$

For example, taking $\tau_a = a_{ik}(u)$, $\tau_b = a_{i'k'}(v)$ and using e_j for the vector with unity in the jth place and zeros elsewhere, the integrand in (4.3.6) is

$$\mathrm{tr}\{(k_0 \Sigma_0 k_0^*)^{-1}(a_0^{-1} e_i e_k' e^{iu\omega} a_0^{-1} b_0 \Sigma_0 b_0^* a_0^{*-1} e^{-iv\omega} e_{k'} e_{i'}' a_0^{*-1}\, d\omega/2\pi$$
$$+ a_0^{-1} e_i e_k' e^{iu\omega} a_0^{-1} d_a\, dF_z(\omega) d_0^* a_0^{*-1} e^{-iv\omega} e_{k'} e_{i'}' a_0^{*-1})\},$$

which gives the contribution

$$(b_0 \Sigma_0 b_0^*)^{-1} \otimes \{k_0 \Sigma_0 k_0^* + l_0\, dF_z l_0^*\} \otimes c_p(\omega) c_p(\omega)^*.$$

In case $s = 1$, then $\Sigma_0 = \sigma_0^2$, a_0, b_0, k_0 are scalars, while l_0 is an r-column, row vector. We omit below-diagonal elements in the Hermitian matrix, I_0, and also the argument ω or $e^{i\omega}$ in c_p, k_0, l_0, dF, and so on. Then, for $s = 1$, showing only on- and above-diagonal elements in the Hermitian matrix in this integrand,

$$I_0 = \int_{-\pi}^{\pi} \begin{bmatrix} (|a_0|^{-2}(d\omega/2\pi) + l_0\, dF_z l_0^*) c_p c_p^* & \cdot & -(a_0 \bar{b}_0)^{-1} c_p c_q^*\, d\omega/2\pi & \cdot & -(\sigma_0^2 |b_0|^2)^{-1} l_0\, dF_z \otimes c_p c_r^* \\ \cdots & \cdot & \cdot\cdot & \cdot & \cdots \\ & \cdot & |b_0|^{-2} c_q c_q^*\, d\omega/2\pi & \cdot & 0 \\ \cdots & \cdot & \cdot\cdot & \cdot & (\sigma_0^2 |b_0|^2)^{-1}\, dF_z \otimes c_r c_r^* \end{bmatrix}.$$

In particular for $m = 0$ we obtain

$$I_0 = \frac{1}{2\pi} \int_{-\pi}^{\pi} \begin{bmatrix} |a_0|^{-2} c_p c_p^* & -(a_0 \bar{b}_0)^{-1} c_p c_q^* \\ \cdots & |b_0|^{-2} c_q c_q^* \end{bmatrix} d\omega. \tag{4.3.8}$$

Example (ii)

Next consider the V_α for α indexing some set of Kronecker indices n_j, $j = 1, \ldots, s$. There are two complications compared to the situation arising in example (i). The first arises from the fact that the maximum lag is now row dependent and the second from the fact that $A(0) = B(0)$ and is not necessarily the unit matrix. The first difficulty is accommodated by exhibiting the elements of I_0 for each pair of lags. Thus let (i, j, k, u), (i', j', k', v) be the indexes of the two elements τ_a, τ_b so that we are considering, in the first case, the ith row in the jth type of matrix ($a(z)$, $b(z)$, or $d(z)$), the kth column, and the uth lag. Then we shall exhibit a matrix $I_0(u, v)$ for fixed u, v for all values of (i, j, k), (i', j', k'). Some of these rows [or columns in the case of elements in $A(0) = B(0)$] may not correspond to any freely varying parameters so that when I_0 is composed from the elements of the various $I_0(u, v)$ some elements of the latter will be neglected. As far as the second difficulty is concerned, this occurs only when u (or v) $= 0$. Then for τ_a corresponding to a freely varying element of a_0 or b_0, we have

$$\partial_{\tau_a} k_0 = -a_0^{-1} e_i e_k' \{a_0^{-1} b_0 - I_s\}.$$

When choosing elements of $I_0(u, v)$ for inclusion in I_0, we choose elements corresponding to freely varying elements of $A(0)$, or $B(0)$, but not both, needless to say.

Then, now writing $O_{m,s}$ for an $m \times s$ matrix of zeros,

$$I_0(u, v) = \int_{-\pi}^{\pi} e^{i(u-v)w} (b_0 \Sigma_0 b_0^*)'^{-1} \otimes \left\{ \begin{bmatrix} -(k_0 - \delta_{0,u} I_s) \\ I_s \\ O_{m,s} \end{bmatrix} \right.$$

$$(4.3.9) \qquad \times \Sigma_0 [-(k_0 - \delta_{0,v} I_s)^* \vdots I_s \vdots O_{m,s}'] \, d\omega / 2\pi$$

$$\left. + \begin{bmatrix} -l_0 \\ O_{m,s}' \\ I_m \end{bmatrix} dF_z(\omega) [-l_0^* \vdots O_{m,s} \vdots I_m] \right\}.$$

Thus if $s = 2$, $m = 0$, $n_1 = 2$, $n_2 = 1$ (so that the $V_\alpha = U_\alpha =$ "generic neighborhood"), the element of I_0 corresponding to $a_{21}(0)$ and $b_{11}(1)$ is

$$-\frac{1}{2\pi} \int_{-\pi}^{\pi} e_2' (b_0 \Sigma_0 b_0^*)'^{-1} e_1 e_1' \{(k_0 - I_s) \Sigma_0\} e_1 e^{-i\omega} \, d\omega.$$

Example (iii)

If $\Theta = U_\alpha \times \Sigma$ and α does not correspond to the generic neighborhood, then, as explained in Chapter 2, the elements of the coefficient matrices of $b(z)$

that are not 0 or 1 are not all freely varying. In this case it will be easier to parameterize via the representations

$$k(z) = H\{z^{-1}I_n - F\}^{-1}K + I_s, \qquad l(z) = H\{z^{-1}I_n - F\}^{-1}L.$$

We shall not go into details. The formulas will be more complicated than those in (4.3.6) and (4.3.9).

Example (iv)

Consider a scalar autoregressive process observed subject to noise. Thus

$$\sum_0^p \alpha(j)x(t-j) = \xi(t), \qquad y(t) = x(t) + \eta(t)$$

$$E\{\xi(s)\xi(t)\} = \delta_{st}\sigma_\xi^2, \qquad E\{\eta(s)\eta(t)\} = \delta_{st}\sigma_\eta^2, \qquad E\{\xi(s)\eta(t)\} \equiv 0.$$

We assume that $a(z) = \sum \alpha(j)z^j \neq 0$, $|z| \leq 1$. Now

$$\sum_0^p \alpha(j)y(t-j) = \sum_0^p \beta(j)\varepsilon(t-j), \qquad E\{\varepsilon(s)\varepsilon(t)\} = \delta_{st}\sigma^2,$$

where the $\varepsilon(t)$ are the innovations and

$$f_y(\omega) = \frac{\sigma^2}{2\pi}\frac{|b(e^{i\omega})|^2}{|a(e^{i\omega})|^2} = \frac{\sigma_\xi^2}{2\pi}\frac{1}{|a(e^{i\omega})|^2} + \frac{\sigma_\eta^2}{2\pi}.$$

Here, of course, $b(z) = \sum \beta(j)z^j$. Then

$$\sigma_\eta^2 \sum_0^{p-1} \alpha(j)\alpha(j+l) + \sigma_\xi^2\delta_{0,l} = \sigma^2 \sum_0^{p-1} \beta(j)\beta(j+l), \qquad l = 0, 1, \ldots, p.$$

Thus, following Dunsmuir (1976),

$$\sigma_\eta^2\sum\alpha(j)^2 + \sigma_\xi^2 = \sigma^2\sum\beta(j)^2, \qquad \sigma_\eta^2\alpha(p) = \sigma^2\beta(p) \tag{4.3.10}$$

$$\psi_l(\tau) = \alpha(p)\sum_0^{p-1}\beta(j)\beta(j+l) - \beta(p)\sum_0^{p-1}\alpha(j)\alpha(j+l) = 0; \qquad l = 1, \ldots, p-1. \tag{4.3.11}$$

We have assumed that $\alpha(p) \neq 0$ and thus, certainly, $\beta(p) \neq 0$, from the second part of (4.3.10). Then (4.3.11) follows. We may regard (4.3.11) as the $(p-1)$ constraints on the system parameters τ. The equations (4.3.10) serve to express σ_η^2, σ_ξ^2 in terms of τ and σ^2. *The constraints do not mix parameters and the* *central limit theorem holds for τ under Conditions A (assuming ergodicity). Thus we may find the limiting distribution from (4.3.5), (4.3.8), and (4.3.11).*

Example (v)

The matrix (4.3.5) gives, putting $b = \partial_\tau \psi^{(1)}(\tau_0)'$, for the part involving $\hat{\tau}$, $\hat{\lambda}^{(1)}$, when constraints do not mix parameters

$$\begin{bmatrix} I_0^{-1} - I_0^{-1}b\{b'I_0^{-1}b\}^{-1}b'I_0^{-1} & \cdot & 0 \\ \cdots & \cdot & \cdots \\ 0 & \cdot & 4(b'I_0^{-1}b)^{-1} \end{bmatrix}. \tag{4.3.12}$$

Thus $T^{1/2}\hat{\lambda}^{(1)}$ is asymptotically normally distributed with the bottom right-hand element as covariance matrix. As is well known, $T^{1/2}\hat{\lambda}^{(1)}$ may be used via this distribution to test the constraints (or some subset of them) by means of a test that is asymptotically equivalent to the likelihood ratio test.

The matrix (4.3.5) in its entirety may be evaluated similarly.

Example (vi)

In case $n = 1$, $s = 1$, $m = 0$, then from example (i),

$$I_0 = \begin{bmatrix} \dfrac{1}{1-\alpha^2} & \dfrac{-1}{1-\alpha\beta} \\ \dfrac{-1}{1-\alpha\beta} & \dfrac{1}{1-\beta^2} \end{bmatrix}. \tag{4.3.13}$$

However, one may *not* proceed as in example (v) to use that procedure to test $\alpha(1) = \beta(1)$, as is evident from the fact that (4.3.13) is then singular. This is because the point $\alpha(1) = \beta(1)$ does not lie in $M(n)$. *We study this situation further in Chapter 5.*

We shall say that the law of the iterated logarithm holds for, say, $(\hat{\tau} - \tau_0)$ if it holds for each element, that is, if the sequence comprised by that element of $\{T/(2\log\log T)\}^{1/2}(\hat{\tau} - \tau_0)$ for every $T \geq 3$ has its limit points constituting precisely the interval $[-a^{1/2}, a^{1/2}]$, where a is the variance of the element in its limiting distribution. We then have the following.

Theorem 4.3.2. *Let $\theta_0 \in M(n) \times \Sigma$ for the ARMA case and the conditions of Theorem 4.2.1 hold. Let θ_0 be constrained so that (4.3.1) and (4.3.2) hold where the $\psi_j(\theta)$ are twice continuously differentiable in some neighborhood of θ_0. If constraints do not mix parameters, then $(\hat{\tau} - \tau_0)$, $\hat{\lambda}_T^{(1)}$ obey the law of the iterated logarithm if* Conditions A *hold. If* Conditions AM *hold, the same is true of $(\hat{\theta} - \theta_0)$, $\hat{\lambda}_T$.*

REMARK. *Of course, this theorem extends to other parameter spaces, for example, with $M(n)$ replaced by $M(p, q, r)$, V_α, or U_α.*

For the ARMAX case we have used some strengthening of the conditions on $z(t)$, although it is probable that can be avoided. (This is because of the need for a.s. proofs.) We then have the following.

Theorem 4.3.3. *Let all of the conditions of Theorem 4.3.2 hold but for the ARMAX case and let $z(t)$ be generated by a stationary process independent of that generating $\varepsilon(t)$, and let Conditions A and (respectively) AM be strengthened to AX and (respectively) AMX. Then the conclusions of Theorem 4.3.2 hold.*

REMARK. Again Theorem 4.3.3 generalizes in the same way as indicated in the remark below Theorem 4.3.2.

We now go on to prove these theorems.

Put $\partial^2 M_T(\theta, \lambda)$ for the Hessian of second derivatives with respect to the elements of θ, λ. Then because of (4.3.4),

$$\partial M_T(\theta_0, 0) = \begin{pmatrix} \partial_\theta L_T(\theta_0) \\ 0 \end{pmatrix} = -\partial^2 M_T(\bar{\theta}, \bar{\lambda}) \begin{pmatrix} \hat{\theta}_T - \theta_0 \\ \hat{\lambda}_T \end{pmatrix}. \tag{4.3.14}$$

Here the Hessian is being evaluated, *row by row*, at values of θ, λ intermediate between $\hat{\theta}_T$, $\hat{\lambda}_T$ and θ_0, 0. Thus $\bar{\theta}$, $\bar{\lambda}$ stands for this assembly of values, which are row dependent. Thus $\bar{\theta}$, $\bar{\lambda} \to \theta_0$, 0 as $T \to \infty$ (i.e., the value along each row so converges). We must thus establish the CLT for $T^{1/2} \partial M_T(\theta_0, 0)$ and find the limiting value of $\partial^2 M_T(\bar{\theta}, \bar{\lambda})$. The proof that the latter converges to $\partial^2 M(\theta_0, 0)$. where $M(\theta, \lambda) = L(\theta) + \lambda' \psi(\theta)$, is not essentially different from that involved in the proof of Theorem 4.2.1 (but is easier for the same reason as was given in the argument after Lemma 4.3.1). Thus, partitioning $(\theta', \lambda')'$ as $(\tau', \sigma', \lambda')$,

$$\partial^2 M_T(\bar{\theta}, \bar{\lambda}) \to \begin{bmatrix} \partial_\tau^2 Q(\theta_0) & 0 & \{\partial_\tau \psi(\theta_0)\}' \\ 0 & \partial_\sigma^2 L(\theta_0) & \{\partial_\sigma \psi(\theta_0)\}' \\ \partial_\tau \psi(\theta_0) & \partial_\sigma \psi(\theta_0) & 0 \end{bmatrix}, \quad \text{a.s.} \tag{4.3.15}$$

The matrix in (4.3.15) is partitioned accordingly. All of these evaluations are simple, except possibly for $\partial_{\sigma,\tau}^2 M(\theta_0, 0) = \partial_{\sigma,\tau}^2 Q(\theta_0)$, which has to be shown to be null. Taking, for example, elements with which differentiation is to be carried out to be σ_{ij}, τ_u, we have to consider only the first term in (4.2.9), since evidently the required derivative of the second, at $\theta = \theta_0$, will be null. The derivative with respect to σ_{ij} of the first term is

$$-(2 - \delta_{ij})(2\pi)^{-1} \int_{-\pi}^{\pi} \operatorname{tr}\{(k\Sigma k^*)^{-1} k(E_{ij} + E_{ji})k^*(k\Sigma k^*)^{-1}(k_0 \Sigma_0 k_0^*)\}\, d\omega,$$

where E_{ij} has unity in row i, column j, and zeros elsewhere. Now the differentiation with respect to τ_u produces a sum of six terms according as to which factor, k or k^*, is differentiated. The result is the same for each. Taking the first k, for example, we obtain, after putting $\theta = \theta_0$,

$$(2 - \delta_{ij})(2\pi)^{-1} \int_{-\pi}^{\pi} \operatorname{tr}\{\Sigma_0^{-1} k_0^{-1}(\partial_{\tau_u} k_0)(E_{ij} + E_{ji})\}\, d\omega,$$

where $\partial_{\tau_u} k_0$ is the derivative of k evaluated at θ_0. This is null since $k(z)$ has $k(0) = I$, so that $\partial_{\tau_u} k_0$ contains only positive powers of z.

We turn now to $T^{1/2}\partial_\theta L_T(\theta_0)$. Consider $T^{1/2}\partial_\theta\{T^{-1}\log\det\Gamma_T(\theta_0)\} = T^{-1/2}\partial_\theta \log\det\Gamma_T(\theta_0)$. Use the development of Lemma 4.2.2. Then again it is evident that

$$T^{-1/2}\partial_\theta \log\det\Gamma_T(\theta_0) = T^{1/2}\partial_\theta \log\det\Sigma_0 + O(T^{-1/2}). \tag{4.3.16}$$

Of course by $\partial_\theta \log\det\Sigma_0$ we mean $\partial_\theta \log\det\Sigma$ evaluated at θ_0. Next use the expansion of $T^{1/2}Q_T(\theta)$ from the first line of the proof of Lemma 4.2.3. Consider first $T^{-1/2}y_T'\partial_\theta\{\Gamma_T(\theta)^{-1}\}y_T$. In (4.2.13) put $K_T = A_T^{-1}B_T$, $C_T = K_T(I_T \otimes \Sigma)K_T'$, $D_T = A_T^{-1}\Delta_T(A_T')^{-1}$. Then $\Gamma_T(\theta) = C_T + D_T$ and

$$\begin{aligned} T^{-1/2}y_T'\partial_\theta\{\Gamma_T(\theta_0)^{-1}\}y_T &= T^{-1/2}y_T'\partial_\theta C_T^{-1}(\theta_0)y_T \\ &\quad - T^{-1/2}y_T'\partial_\theta\{\Gamma_T^{-1}(\theta_0)D_T(\theta_0)C_T^{-1}(\theta_0)\}y_T, \end{aligned}$$

where now the dependence of D_T and C_T on θ has been exhibited. (See the proof of Lemma 4.2.3.) The matrices $\Gamma_T^{-1}(\theta_0)$, $\partial_\theta\Gamma_T(\theta_0)$, $D_T(\theta_0)$, $\partial_\theta D_T(\theta_0)$, $C_T(\theta_0)$, and $\partial_\theta C_T(\theta_0)$ all have elements decreasing geometrically in magnitude as $|j - k|$ increases, j, k being the row and column indices. For example, $\Gamma_T(\theta) = C_T(\theta) + A_T(\theta)^{-1}\Delta_T(\theta)A_T'(\theta)$ and $A_T(\theta_0)^{-1}$, $C_T(\theta_0)$, $\partial_\theta A_T(\theta_0)$, $\partial_\theta C_T(\theta_0)$ have this property. Indeed, $A_T(\theta_0)^{-1}$ has elements that are coefficients in the expansion of $a_0(z)^{-1}$ down its diagonals. (In this connection it may be observed that lower triangular block Toeplitz matrices, with blocks all of the same size, have inverses, if they exist, and products that are again lower triangular block Toeplitz and that the generating functions of the inverses and products are the inverses and products of the generating functions. This is easily established. Of course, if all blocks depend on a parameter vector, θ, then the derivative of the matrix is block Toeplitz with generating function obtained by differentiation.) For $\Gamma_T^{-1}(\theta_0)$ the result follows from the representation of $\Gamma_T^{-1}(\theta_0)$ as $C_T(\theta_0)^{-1}A_T(\theta_0)\{I + \Delta_T B_T'(\theta_0)^{-1}(I_T \otimes \Sigma_0^{-1})B_T(\theta_0)^{-1}\}^{-1}A_T(\theta_0)^{-1}$. The expression within braces is of the form

$$\begin{bmatrix} I_{ms} + \Delta_T X & \Delta_T Y \\ 0 & I_{(T-m)s} \end{bmatrix},$$

and since $\Gamma_T(\theta_0)$ has it smallest eigenvalue bounded away from zero, the same is true of $I_{ms} + \Delta_T X$. It is easily seen that $\Delta_T Y$ has elements whose modulus decreases to zero geometrically as $|j-k|$ increases. (Here we use the fact that Δ_T has null elements outside the first ms rows and columns.) Thus $T^{-1/2} y_T' \partial_\theta \{\Gamma_T^{-1}(\theta_0) D_T(\theta_0) C_T^{-1}(\theta_0)\} y_T$ is composed of a fixed number of expressions of the form $T^{-1/2} y_T' E_T F_T G_T y_T$, where E_T, G_T have elements decreasing, in modulus, geometrically with $|j-k|$, while F_T is null outside the first ms rows and columns. [Of course, F_T is Δ_T or $\partial_\theta \Delta_T(\theta_0)$.] Thus this expression is $O_P(T^{-1/2})$. Indeed, it is $T^{-1/2}$ by products of expressions that are bounded in norm by expressions of the form $\sum_j \|M(j)\| \, \|y(t-j)\|$, where $\|M(j)\|$ decreases geometrically, and these expressions occur only for a fixed finite number of values of t. [For later use we mention that the overall quantity is $o(1)$ and indeed is $o(T^{-a})$, $a < \frac{1}{2}$.]

The same argument can be applied to the other terms in $T^{1/2} \partial_\theta Q_T(\theta_0)$ [see (4.2.17)], so that

$$T^{1/2} \partial_\theta Q_T(\theta_0) = T^{-1/2} \partial_\theta [\{K_T(\theta_0)^{-1} y_T - K_T(\theta_0)^{-1} L_T(\theta_0) z_T\}' \\ \times (I_T \otimes \Sigma_0)\{K_T(\theta_0) y_T - K_T(\theta_0)^{-1} L_T(\theta_0) z_T\}] + o_P(1).$$

Indeed, the error term is $O_P(T^{-1/2})$ [and $O(T^{-a})$, $a < \frac{1}{2}$, a.s.] Thus

(4.3.17)

$$T^{1/2} \partial_\theta Q_T(\theta_0) = T^{1/2} \partial_\theta \frac{1}{2\pi} \int_{-\pi}^{\pi} \operatorname{tr}\{(k_\tau \Sigma k_\tau^*)^{-1} w_\tau(\omega) w_\tau(\omega)^*\} \, d\omega|_{\theta=\theta_0} + o_P(1),$$

wherein we now use a more explicit notation to indicate the differentiation followed by putting $\theta = \theta_0$, and

$$w_\tau(\omega) = T^{-1/2} \sum_1^T \{y(t) - l_\tau z(t)\} e^{it\omega},$$

$$l_\tau z(t) = \sum_1^\infty L_\tau(j) z(t-j), \qquad z(t) = 0, \quad t \le 0.$$

Again, the $o_P(1)$ term is actually $O(T^{-a})$, $a < \frac{1}{2}$. Thus we must consider the first term on the right side of (4.3.17). We may rewrite $w_\tau(\omega)$ as

$$w_\tau(\omega) = T^{-1/2} \sum_1^T \{k_0 \varepsilon(t) - (l_\tau - l_0) z(t)\} e^{it\omega}, \qquad z(t) = 0, \quad t \le 0.$$

Now when constraints do not mix parameters, we need consider, for the asymptotic distribution of $T^{1/2}(\hat{\tau}_T - \tau_0)$,

$$T^{1/2}\partial_\tau \frac{1}{2\pi}\int_{-\pi}^{\pi} \operatorname{tr}\{(k_\tau \Sigma_0 k_\tau^*)^{-1} w_\tau(\omega) w_\tau(\omega)^*\}\, d\omega \bigg|_0 .$$

In this expression "squared terms" [i.e., terms that are elements of $\varepsilon(t)\varepsilon(t)'$ for some t] can occur only through

$$T^{1/2}\partial_\tau \frac{1}{2\pi}\int_{-\pi}^{\pi} \operatorname{tr}\{(k_\tau \Sigma_0 k_\tau^*)^{-1} I_u(\omega)\}\, d\omega \bigg|_0 , \tag{4.3.18}$$

$$I_u(\omega) = w_u(\omega) w_u(\omega)^*, \qquad w_u(\omega) = T^{-1/2}\sum_1^T u(t) e^{it\omega}, \qquad u(t) = k_0 \varepsilon(t).$$

Lemma 4.3.2. *The contribution to (4.3.18) from the squared terms converges in probability to zero if* $E\{\varepsilon(t)\varepsilon(t)'\} < \infty$.

Put $\phi(\omega) = \partial_j (k_0 \Sigma_0 k_0^*)^{-1}$, where ∂_j indicates differentiation with respect to the jth element of τ and again the notation indicates that τ is put equal to τ_0 after differentiation. We need to show that the squared terms may be eliminated from

$$T^{1/2}\frac{1}{2\pi}\int_{-\pi}^{\pi} \operatorname{tr}\{\phi(\omega) I_u(\omega)\}\, d\omega. \tag{4.3.19}$$

For any symmetric S,

$$\frac{1}{2\pi}\int_{-\pi}^{\pi} \operatorname{tr}\{\phi(\omega) k_0 S k_0^*\}\, d\omega = -\frac{1}{\pi}\int_{-\pi}^{\pi} \operatorname{tr}\{S \Sigma_0^{-1} k_0^{-1} (\partial_j k_0)\}\, d\omega.$$

However, this is null since k_0^{-1} is analytic for $z \geq 1$ and $\partial_j k_0$ has null constant term. Thus (4.3.19) is equal to

$$T^{1/2}\frac{1}{2\pi}\int_{-\pi}^{\pi} \operatorname{tr}[\phi(\omega)\{I_u(\omega) - k_0 \dot{\Sigma} k_0^*\}]\, d\omega, \qquad \dot{\Sigma} = \frac{1}{T}\sum_1^T \varepsilon(t)\varepsilon(t)'. \tag{4.3.20}$$

The contribution from the squared terms to (4.3.20) [and hence to (4.3.19)] is

$$T^{-1/2}\frac{1}{2\pi}\int \operatorname{tr}\left[\phi(\omega)\left\{\sum_{j,k=0}^{\infty}\sum K_0(j)\left(\sum_{jk} - \sum_T\right) K_0(k)^* e^{i(j-k)\omega}\right\}\right] d\omega,$$

where $\sum_T$ and $\sum_{jk}$ are sums of the $\varepsilon(t)\varepsilon(t)'$, the former for $1 \leq t \leq T$ and the latter for $1 - k \leq t \leq T - j, j \geq k$; $1 - j \geq t \leq T - k, k \geq j$. The origin of the sum $\sum_T$ is obvious and $\sum_{jk}$ is obtained from the fact that

$$I_u(\omega) = \frac{1}{T}\sum_{j,k=0}^{\infty}\sum K_0(j)\left(\sum_{s,t=1}^{T}\sum \varepsilon(s-j)\varepsilon(t-k)'\right)K_0(k)^* e^{i(s-t)\omega}$$

and the squared terms occur for $s - t = j - k$, $1 \le s, t \le T$. Now $\sum_{jk} - \sum_T = \sum'_{jk} - \sum''_{jk}$, where $\sum'_{jk}$ is over t such that $t + 1 \le j, k \le T + t$, $T \ge 0$ while $\sum''_{jk}$ is over $T - t + 1 \le \max(j, k)$, $1 \le t \le T$. Now

$$\begin{aligned}
&\sum_{j,k=0}^{\infty}\sum K_0(j)\sum_{jk}{}' K_0(k)^* e^{i(j-k)\omega} \\
&\quad = \sum_{t\ge 0}\left\{\sum_{j,k=t+1}^{\infty}\sum K_0(j)\varepsilon(t)\varepsilon(t)'K_0(k)^*(e^{i(j-k)\omega})\right\} \ge 0,
\end{aligned}$$

so that

$$\begin{aligned}
&T^{-1/2}\frac{1}{2\pi}\int \operatorname{tr}\left[\phi(\omega)\left\{\sum_{j,k=t+1}^{\infty}\sum K_0(j)\sum_{jk}{}' K_0(k)^* e^{i(j-k)\omega}\right\}\right]d\omega \\
&\quad \le CT^{-1/2}\frac{1}{2\pi}\int \operatorname{tr}\left\{\sum_{j,k}\sum K_0(j)\sum_{jk}{}' K_0(k)^* e^{i(j-k)\omega}\right\}d\omega,
\end{aligned}$$

where C bounds the eigenvalues of $\phi(\omega)$, in modulus. The expectation of this positive quantity is

$$\begin{aligned}
&CT^{-1/2}\sum_{t=0}^{\infty}\left\{\sum_{t+1}^{T+t} K_0(j)\Sigma_0 K_0(j)^*\right\} \\
&\quad = CT^{-1/2}\sum_{j=0}^{\infty}\min(j, T)k_0(j)\Sigma_0 K_0(j)^* = O(T^{-1/2}).
\end{aligned}$$

The contribution from $\sum''_{jk}$ may be broken up into three terms,

$$\begin{aligned}
&T^{-1/2}\sum_{0}^{T-1}\frac{1}{2\pi}\int \operatorname{tr}\left[\phi(\omega)\left\{k_0\varepsilon(T-t)\varepsilon(T-t)'\sum_{t+1}^{\infty}K_0(k)^* e^{-ik\omega}\right.\right. \\
&\quad + \sum_{t+1}^{\infty}K_0(j)\varepsilon(T-t)\varepsilon(T-t)'k_0^* e^{ij\omega} \qquad (4.3.21) \\
&\quad \left.\left. + \sum_{t+1}^{\infty}\sum K_0(j)\varepsilon(T-t)\varepsilon(T-t)'K_0(k)^* e^{i(j-k)\omega}\right\}\right]d\omega.
\end{aligned}$$

The last term in braces is again nonnegative definite and its contribution may be shown to converge in probability to zero in the same way as for that for $\sum'_{jk}$. If $\phi(\omega)$ were a trigonometric polynomial, it would be sufficient to show that the first two terms converge in probability to zero if $\phi(\omega)$ is replaced by $A\exp(ik\omega)$. However, then, the first term is

$$T^{-1/2}\sum_{1}^{T} \operatorname{tr}\{\textstyle\sum''' AK_0(j)\varepsilon(t)\varepsilon(t)'K_0(j+k)^*\},$$

where the inner sum $\sum'''$, is for $T-t+k+1 \le j < \infty$. Using Schwarz's inequality we may consider

$$T^{-1/2}\sum_{1}^{T}\textstyle\sum''' \varepsilon(t)'K_0(j+k)^*K_0(j+k)\varepsilon(t) \tag{4.3.22}$$

and a similar expression in which $K_0(j+k)$ is replaced $AK_0(j)$. However, (4.3.22) is positive with expectation

$$T^{-1/2}\sum_{1}^{T}\textstyle\sum''' \operatorname{tr}\{K_0(j+k)\Sigma_0K_0(j+k)^*\} = O(T^{-1/2}),$$

the last relation following from a reordering of the double sum, as before.

Finally, we may approximate to $\phi(\omega)$ by a trigonometric polynomial $\phi_N(\omega)$ with an error, $\|\phi(\omega)-\phi_N(\omega)\| \le e$, where e may be chosen arbitrarily small uniformly in ω. We may bound the expectation of the modulus of the first term in (4.3.21) by

$$Ce\sum_{0}^{\infty}[\|K_0(j)\|^2\{\min(j,T)\}^{1/2}],$$

where C is an appropriate constant. Indeed,

$$\begin{aligned} &E\left|\operatorname{tr}\left[\{\phi(\omega)-\phi_N(\omega)\}k_0\varepsilon(T-t)\varepsilon(T-t)'\sum_{t+1}^{\infty}K_0(k)^*e^{-ik\omega}\right]\right| \\ &\quad\le Ce\|\Sigma_0\|\left\|\sum_{t+1}^{\infty}K_0(k)^*e^{-ik\omega}k_0 + k_0^*\sum_{t+1}^{\infty}K_0(j)e^{ij\omega}\right\|. \end{aligned}$$

Hence the lemma is proved.

REMARK. In the proof of the lemma we used only the finiteness of $E\{\varepsilon(t)\varepsilon(t)'\}$ and the fact that

$$\sum_{0}^{\infty} j^{1/2} \|K_0(j)\|^2 < \infty, \tag{4.3.23}$$

so that the result does not depend on the fact $\sum K_0(j)z^j$ converges at a geometric rate. Note also that the critical step is that in (4.3.20) and this depends on $\int \operatorname{tr}\{\phi k_0 S k_0^*\}\, d\omega = 0$, which depends on the fact that we are differentiating $k_\tau \Sigma k_\tau^*$ with respest to τ. The result does not hold for differentiation with respect to σ. *Thus the lemma is relevant only when the limiting distribution of* $T^{1/2}(\hat{\tau}_T - \tau_0)$ *depends only on* $T^{1/2}\partial_\tau L_T(\theta_0)$ *(i.e., only when constraints do not mix parameters).*

Thus consider now

$$\partial_\tau T^{1/2} \frac{1}{2\pi} \int_{-\pi}^{\pi} \operatorname{tr}\{(k_\tau \Sigma_0 k_0^*)^{-1} \tilde{I}_\tau(\omega)\}\, d\omega|_0, \qquad I_\tau(\omega) = w_\tau(\omega) w_\tau(\omega)^*,$$

and $\tilde{I}_\tau(\omega)$ is $I_\tau(\omega)$ with all "squared terms" in the $\varepsilon(t)$ removed. This has a component corresponding to differentiation with respest to τ_j, which is

$$\begin{aligned} &T^{1/2} \frac{1}{2\pi} \int_{-\pi}^{\pi} \operatorname{tr}\{\phi(\omega) \tilde{I}_u(\omega)\}\, d\omega \\ &\quad - 2T^{1/2} \frac{1}{2\pi} \int_{-\pi}^{\pi} \{(k_0 \Sigma_0 k_0^*)^{-1} w_u(\omega) w_v(\omega)^*\}\, d\omega. \end{aligned} \tag{4.3.24}$$

Here

$$w_v(\omega) = T^{-1/2} \sum_{1}^{T} (\partial_{\tau_j} l_0) z(t) e^{it\omega}, \qquad z(t) = 0, \quad t \le 0,$$

and $\partial_{\tau_j} l_0 = \partial_{\tau_j} \sum L_\tau(j) z^{-j}|_0$ is operating as a lag operation on $z(t)$. Although $\phi(\omega)$ and $w_v(\omega)$ depend on j, we have suppressed that in the notation for convenience. Consider the first term in (4.3.24). The function $\phi(\omega)$ has coefficient matrices, $\Phi(j)$ say, decreasing to zero at a geometric rate with $|j|$. Indeed, ϕ is $\partial_{\tau_j}(k\Sigma_0 k^*)^{-1}$ evaulated at θ_0 and $\partial_{\tau_j}(k\Sigma_0 k^*)^{-1} = -k^{*-1}\partial_{\tau_j} k^* k^{*-1} \Sigma_0^{-1} k^{-1} - k^{*-1}\Sigma_0^{-1} k^{-1} \partial_{\tau_j} k k^{-1}$. Moreover, $\partial_{\tau_j} k = \sum_1^\infty \partial_{\tau_j} K(u) z^{-u}$ and $\partial_{\tau_j} K(u) = \partial_{\tau_j} H F^{u-1} K$ and these decrease geometrically to zero due to the fact that F has all eigenvalues of less than unit modulus. Since k_0^{-1} has coefficients decreasing geometrically to zero because of the assumption concerning the zeros of $\det k_0$, the $\Phi(j)$ decrease geometrically to zero. Thus

$$T^{1/2} \frac{1}{2\pi} \int_{-\pi}^{\pi} \operatorname{tr}\{\phi(\omega) \tilde{I}_u(\omega)\}\, d\omega = T^{1/2} \sum_{-\infty}^{\infty} \operatorname{tr}\{\Phi(j) \tilde{G}_u(j)\},$$

where $\tilde{G}_u(j)$ is $G_u(j)$ (see Section 4.2) with the squared terms removed. However,

$$TE[\{\operatorname{tr}\Phi(j)\tilde{G}_u(j)\}^2] \le c^2\rho^{2j}TE[\operatorname{tr}\{\tilde{G}_u(j)\tilde{G}_u(j)^*\}], \qquad \|\Phi(j)\| \le c\rho^j$$

and

$$\begin{aligned} &TE[\operatorname{tr}\{\tilde{G}_u(j)\tilde{G}_u(j)^*\}] \\ &= T^{-1}E\Big[\operatorname{tr}\sum\sum\sum\sum K_0(a)\sum_s \varepsilon(s-a)\varepsilon(s+j-n)'K_0(b)'K_0(c) \\ &\qquad \times \sum_t \varepsilon(t+j-c)\varepsilon(t-d)'K_0(d)'\}\Big] \end{aligned} \tag{4.3.25}$$

and the outer sum, over a, b, c, d, is such as to exclude $s-a=s+j-b$ or $t-d=t+j-c$. Now because of (4.1.15) for given a, b, c, d, for example with $a<b-j, d<c-j$, we need consider only $s-a=t-d$ and there are at most T such terms. If (4.1.16) or (4.1.20) holds (see the note below), each has finite expectation and because of the geometric rate of decrease of the $K_0(a)$, then $TE[\operatorname{tr}\{\Phi(j)\tilde{G}_u(j)\}^2] \le c\rho^{2j}$. Now we may use what is sometimes called Bernstein's lemma (see Hannan, 1970, p. 242, for example).

Lemma 4.3.3. *If x_T is a sequence of random vectors and for every $\varepsilon>0, \zeta>0, \eta>0$, there exist sequences $y_T(\varepsilon)$, $z_T(\varepsilon)$ so that $x_T=y_T(\varepsilon)+z_T(\varepsilon)$ and $y_T(\varepsilon)$ has a distribution converging to the multivariate normal distribution with zero mean and covariance matrix $\Sigma(\varepsilon)$, while*

$$\lim_{\varepsilon\to 0}\Sigma(\varepsilon)=\Sigma, \qquad P\{z_T(\varepsilon)'z_T(\varepsilon)>\zeta\}<\eta,$$

the distribution of x_T converges to the multivariate normal distribution with zero mean and covariance matrix Σ.

We may use this with ε determined by a truncation point in the series

$$T^{1/2}\sum_{-\infty}^{\infty}\operatorname{tr}\{\Phi(j)\tilde{G}_u(j)\}$$

(i.e., truncate the series at $|j|=m$ and put $\varepsilon=m^{-1}$). Then the conditions of the theorem are met with x_T the first term in (4.3.24) and $y_T(\varepsilon)$ the truncated expression. Thus it is sufficient to consider only $\tilde{G}_u(j), j=0,\ldots,m$. The same argument goes through for the second term in (4.3.24), and hence to establish a central limit theorem we need consider, jointly, only $T^{1/2}\tilde{G}_u(j)$, $j = 0, \pm 1, \ldots, \pm m$; $T^{1/2}G_{uz}(j), j = 0, \pm 1, \ldots, \pm m$.

We note the following. In the argument above, $I_u(\omega)$ was replaced by $\tilde{I}_u(\omega)$ only so as to eliminate squared terms and avoid (4.1.16), which is needed if $\dot{\Sigma}$ is to obey the CLT. If (4.1.16) holds and the constraints do not mix parameters, then (4.1.20) is not needed because the expectations not zero in (4.3.25) involve, for example,

$$K_0(a)\varepsilon(t-d)\varepsilon(t+a+j-b-a)'K_0(b)'K_0(c)\varepsilon(t+j-c)\varepsilon(t-d)'K_0(d)'$$

and $t-d > t-d+(a+j-b)$, since $a < b-j$, $t-d > t+j-c$, since $d < c-j$. Thus using (4.1.16), without (4.1.20) the expectation is finite. If constraints do mix parameters, the argument below (4.3.25) goes through unchanged if (4.1.20) holds because the nature of the outer sum for $\tilde{I}_u$ [noted below (4.3.25)] played no part in the proof. In this case, where constraints do mix parameters, we shall need (4.1.16), however, for the proof relating to $G_u(j)$, as distinct from $\tilde{G}_u(j)$.

We may use Lemma 4.3.3 once again to replace $T^{1/2}\tilde{G}_u(j)$, $T^{1/2}G_{uz}(j)$ by

$$\begin{aligned} &T^{-1/2}\sum_1^T \varepsilon(t)\varepsilon(t-j)', \quad j = 1, \ldots, m; \\ &T^{-1/2}\sum_1^T \varepsilon(t)z(t+j)', \quad j = 0, \pm 1, \ldots, \pm m. \end{aligned} \tag{4.3.26}$$

If constraints do mix parameters, then $I_u(\omega)$ must be considered and

$$T^{-1/2}\sum_1^T \varepsilon(t)\varepsilon(t)' \tag{4.3.27}$$

must be adjoined to (4.3.26).

Lemma 4.3.4. *Assuming the $\varepsilon(t)$ to be ergodic, the set of random variables in (4.3.26) is jointly asymptotically normal under Conditions AX or AX′. If the set (4.3.27) is added, the same is true under Conditions AMX or AMX′. If only the quantities $T^{-1/2}\sum \varepsilon(t)\varepsilon(t-j)'$, $j = 1, 2, \ldots, m$ are considered, A or A′ suffice.*

We shall use the results in Hall and Heyde (1980), Theorem 3.2. For an alternative proof using Macleish (1977) see Hannan, Dunsmuir, and Deistler (1980). It is sufficient to establish the lemma for

$$T^{1/2}\sum_1^T \left\{ b'\varepsilon(t)\sum_1^m b(j)'\varepsilon(t-j) + c'\varepsilon(t)\sum_{-m}^m c(j)'z(t+j) \right\} \tag{4.3.28}$$

with arbitrary but fixed vectors b, $b(j)$, c, $c(j)$. To avoid a complex notation we consider

$$(4.3.29) \qquad T^{1/2} \sum_{1}^{T} \{\varepsilon_i(t)\varepsilon_j(t-1) + c\varepsilon_k(t)z_l(t)\}.$$

The proof is no different for (4.3.28), merely more notationally complex. Put $X(t)$ for the summand in (4.3.29) and $X_T(t) = T^{-1/2}X(t)$, $t = 1, \ldots, T$. [This takes the place of X_{ni}, $i = 1, \ldots, n$ in Hall and Heyde (1980), Th. 3.2).] It is sufficient to show that

(i) $\max_{1 \le t \le T} X_T(t)^2 \underset{p}{\to} 0$,
(ii) $\sum_{t=1}^{T} X_T(t)^2 \underset{p}{\to} \eta^2$,
(iii) $\sup_T E\{\max_{1 \le t \le T} X_T(t)^2\} \le c < \infty$,

where the notation indicates convergence in probability. However,

$$E\left\{\sum_{t=1}^{T} X_T(t)^2\right\} = \frac{1}{T}E\left\{\sum_{1}^{T} X(t)^2\right\}$$
$$= E\{\varepsilon_i(t)^2\varepsilon_j(t-1)^2\} + c^2\sigma_{kk}\frac{1}{T}\sum_{1}^{T} z_l(t)^2$$
$$+ 2c\frac{1}{T}\sum_{1}^{T} E\{\varepsilon_i(t)\varepsilon_k(t)\varepsilon_j(t-1)\}z_l(t)$$
$$\to E\{\varepsilon_i(t)^2\varepsilon_j(t-1)^2\} + c^2\sigma_{kk}\Gamma_{z,ll}(0) + 2c_1\mu_l,$$

where $\Gamma_{z,ll}(0)$ is the indicated element of $\Gamma_z(0)$ and $c_1 = cE\{\varepsilon_i(t)\varepsilon_k(t)\varepsilon_j(t-1)\}$. Call this last expression $\eta^2 + 2c_1\mu_l$. Now (i) holds because

$$\max_{1 \le t \le T} X_T(t)^2 \le \max_{1 \le t \le T} X_t(t)^2$$

$$E\{X_t(t)^2\} = E\left\{\sum_{1}^{t} X_t(j)^2 - \sum_{1}^{t-1} X_t(j)^2\right\}$$
$$= E\left\{\sum_{1}^{t} X_t(j)^2 - \sum_{1}^{t-1} X_{t-1}(j)^2\right\} + \frac{1}{t}E\left\{\sum_{1}^{t-1} X_{t-1}(j)^2\right\} \to 0.$$

Moreover,

$$E\left\{\max_{1 \le t \le T} X_T(t)^2\right\} \le \sum_{1}^{T} E\{X_T(t)^2\} \to \eta^2 + 2c_1\mu_l,$$

so that (iii) also holds. Thus we need to establish only (ii). Thus since $T^{-1}\sum \varepsilon_i(t)^2\varepsilon_j(t-1)^2 \to E\{\varepsilon_i(t)^2\varepsilon_j(t-1)^2\}$, by ergodicity, we need only to establish

$$\frac{1}{T}\sum_1^T \varepsilon_i(t)\varepsilon_j(t-1)\varepsilon_k(t)z_l(t) \xrightarrow{p} 0,$$

$$\frac{1}{T}\sum_1^T \varepsilon_k(t)^2 z_l(t)^2 \xrightarrow{p} \sigma_{kk}\Gamma_{z,ll}(0).$$

Consider the second of these, for example. It is sufficient to show that

$$\frac{1}{T}\sum_1^T z_l(t)^2\{\sigma_{kk} - \varepsilon_k(t)^2\} \xrightarrow{p} 0.$$

Now

$$\begin{aligned} E&\left[\left|\frac{1}{T}\sum_1^T z_l(t)^2\{\sigma_{kk} - E\{\varepsilon_k(t)^2|\mathscr{F}_{t-r}\}\}\right|\right] \\ &\leq \frac{1}{T}\sum_1^T z_l(t)^2 E[|\sigma_{kk} - E\{\varepsilon_k(t)^2|\mathscr{F}_{t-r}\}|] \\ &= \frac{1}{T}\sum_1^T z_l(t)^2 E[|\sigma_{kk} - E\{\varepsilon_k(0)^2|\mathscr{F}_{-r}\}|] \\ &\leq \varepsilon_r\frac{1}{T}\sum_1^T z_l(t)^2, \end{aligned}$$

where ε_r may be taken as small as is desired by taking r large. This follows from (4.1.19), which is a consequence of (4.1.17). By Lemma 4.3.3 (with the limiting normal distribution degenerate at the origin) it is sufficient to prove that

$$\frac{1}{T}\sum_1^T z_l(t)^2[\varepsilon_k(t)^2 - E\{\varepsilon_k(t)^2|\mathscr{F}_{t-r}\}] \xrightarrow[p]{} 0$$

that is,

$$\frac{1}{T}\sum_1^T z_l(t)^2 \sum_{j=1}^r u(t;t-j) \xrightarrow[p]{} 0$$

$$u(t;t-j) = E\{\varepsilon_k(t)^2|\mathscr{F}_{t-j+1}\} - E\{\varepsilon_k(t)^2|\mathscr{F}_{t-j}\}$$

[see (1.4.9) for an analogous decomposition] and thus it is sufficient to prove that

$$\frac{1}{T}\sum_1^T z_l(t)^2 u(t; t-j) \underset{p}{\rightarrow} 0. \tag{4.3.30}$$

Now we use the "three series" criterion in Hall and Heyde (1980, p. 29). Note that the $u(t; t-j)$ are martingale differences with respect to the σ algebra sequence $\mathscr{F}_{t-j}$. Now put

$$X_T(t) = z_l(t)^2 u(t, t-j) I\{z_l(t)^2 |u(t, t-j)| \leq T\},$$

where $I\{\cdot\}$ is the indicator function. Thus (4.3.30) holds if

(i) $\sum_1^T P\{z_l(t)^2|u(t, t-j)| > T\} \rightarrow 0.$
(ii) $T^{-1}\sum_1^T E\{X_T(t)|\mathscr{F}_{t-j-1}\} \overset{p}{\rightarrow} 0.$
(iii) $T^{-2}\sum_1^T E[X_T(t)^2 - \{E(X_T(t)|\mathscr{F}_{t-j-1}\}^2] \rightarrow 0.$

For the first of these, using $dP(x)$ for the differential elements of probability for $|u(t, t-j)|$, we have

$$\sum_1^T \int_{T/z_l(t)^2}^{\infty} \leq \frac{1}{T}\sum z_l(t)^2 \int_{T/z_l(t)^2} x\, dP \rightarrow 0,$$

the latter holding because $T^{-1}\sum z_l(t)^2$ converges to $\Gamma_{z,ll}(0)$ while

$$\max_{1\leq t\leq T} z_l(t)^2/T \rightarrow 0. \tag{4.3.31}$$

$$T^{-1}\sum_1^T E\{X_T(t)|\mathscr{F}_{t-j-1}\} = \frac{1}{T}\sum_1^T z_l(t)^2 \int_{A(t)} x\, dP_j(x),$$

where $dP_j(x)$ in the differential element of probability for $u(t, t-j)$ given $\mathscr{F}_{t-j-1}$ and $A(t) = \{x||x| \leq T/z_l(t)^2\}$, which also depends on T. Of course, $\int x\, dP_j(x) = 0$. Thus the expectation of the modulus of the expression on the right is dominated by

$$\frac{1}{T}\sum_1^T z_l(t)^2 \int_{T/z_l(t)^2}^{\infty} x\, dP,$$

which converges to zero because of (4.3.31) and

$$\int_0^{\infty} x\, dP < \infty.$$

Next

$$
\begin{aligned}
T^{-2} \sum_1^T z_l^4(t) \int_{A(T)} x^2 \, dP(x)
&\le T^{-1} \sum_1^T z_l^2(t) \int_a^{T/z_l^2(t)} x \, dP + T^{-2} \sum_1^T z_l^4(t) \int_0^a x^2 \, dP \\
&\le T^{-1} \sum_1^T z_l^2(t) \int_a^{\infty} x \, dP + T^{-2} \sum_1^T z_l^4(t) \int_0^a x^2 \, dP.
\end{aligned}
$$

Choose a so that

$$
\int_a^{\infty} x \, dP \le \varepsilon, \qquad \int_0^a x^2 \, dP = C;
$$

then since

$$
T^{-2} \sum_1^T z_l^4(t) \le \left\{ \max_{1 \le t \le T} z_l(t)^2 / T \right\} \frac{1}{T} \sum_1^T z_l^2(t) \to 0,
$$

we see that

$$
\limsup_{T \to \infty} T^{-2} \sum_1^T E\{X_T(t)^2\} \le \varepsilon
$$

and since ε is arbitrary,

$$
T^{-2} \sum_1^T E\{X_T(t)^2\} \to 0.
$$

Finally,

$$
T^{-2} \sum_1^T z_l^4(t) E\left\{ \int_{A(t)} x \, dP_j(x) \right\}^2 \le T^{-2} \sum_1^T z_l^4(t) \int_{A(t)} x^2 \, dP_j(x).
$$

This has the expectation

$$
T^{-2} \sum_1^T z_l^4(t) \int_{A(t)} x^2 \, dP(x) \to 0.
$$

Thus the three series criterion is satisfied and hence condition (ii) of Theorem

3.2 of Hall and Heyde (1980), and hence also the first part of Lemma 4.3.4, is true. If (4.3.27) is adjoined, the proof is the same, adding (4.3.3) or (4.3.3)′ and (4.1.16) [in which case (4.1.17) and (4.1.18) are not needed]. The last part of the lemma follows immediately from the fact that the summands $\varepsilon_a(t)\varepsilon_b(t-j)$, $j > 0$, are stationary, ergodic, square-integrable martingale differences for which the CLT holds. (This is a special case of Theorem 3.2 of Hall and Heyde, 1980, due initially to Billingsley and to Ibragimov.)

Thus we have established the asymptotic normality of $T^{1/2}\partial_\tau L_T(\theta_0)$ *under Conditions AX or AX′ if constraints do not mix parameters and under AMX or AMX′ otherwise. This assumes also the remainder of Theorem 4.2.1 and (4.3.1), (4.3.2). For the ARMA case only A or A′ are needed if constraints do not mix parameters or AM, AM′ otherwise.* Next consider [see (4.3.16)]

$$T^{1/2}\partial_\sigma L_T(\theta_0) = T^{1/2}\{\partial_\sigma(\log\det\Sigma_0) + \partial_\sigma Q_T(\theta_0) + o(1).$$

Now

$$\partial_{\sigma_{ij}}\log\det\Sigma_0 = (2-\delta_{ij})e_i'\Sigma_0^{-1}e_j$$

when δ_{ij} is Kronecker's delta and the vector e_i has unity in the ith place and zeros elsewhere. Also,

$$\begin{aligned} T^{1/2}\partial_{\sigma_{ij}}Q_T(\theta_0) &= -\partial_{\sigma_{ij}}T^{1/2}\frac{1}{2\pi}\int \operatorname{tr}\{\Sigma^{-1}k_0^{-1}w_u w_u k_0^{*-1}\}\,d\omega|_0 + o(1) \\ &= -\partial_{\sigma_{ij}}T^{1/2}\operatorname{tr}\{\Sigma^{-1}\dot{\Sigma}\}|_0 + o(1) \\ &= T^{1/2}(2-\delta_{ij})e_i'\Sigma_0^{-1}\dot{\Sigma}\Sigma_0^{-1}e_j. \end{aligned}$$

Thus

$$T^{1/2}\partial_{\sigma_{ij}}L_T(\theta_0) = T^{1/2}(2-\sigma_{ij})e_i'\Sigma_0^{-1}(\dot{\Sigma}-\Sigma_0)\Sigma_0^{-1}e_j + o(1). \tag{4.3.32}$$

However, under (4.1.16), (4.1.20), $\varepsilon(t)\varepsilon(t)' - \Sigma_0$ is a matrix of square-integrable, ergodic martingale differences and again the central limit theorem holds. (This was used in the last statement in the proof of Lemma 4.3.4.) Of course, $T^{1/2}\partial_\tau L_T(\theta_0)$ and $T^{1/2}\partial_\sigma L_T(\theta_0)$ are jointly asymptotically normal by the proof of Lemma 4.3.4.

It remains to elucidate further the covariance matrix in the limiting distribution for $T^{1/2}\partial_\tau L_T(\theta_0)$. We do this only under the primed conditions A′, AM′, AX′, or AMX′, so that simple formulas are obtained. Of course, when *constraints do not mix parameters* the CLT holds under the weaker conditions, but the covariance matrix may be too complicated to use (except in special circumstances) so that we must rely on the principle stated in this section in italics below the list of Conditions A and so on. In the evaluation of this covariance matrix, we must consider the asymptotic distribution of the quantities

$$T^{1/2}\partial_\tau \frac{1}{2\pi}\int_{-\pi}^{\pi} \operatorname{tr}\{(k_\tau \Sigma_0 k_\tau^*)^{-1} I_u(\omega)\}\, d\omega|_0$$

$$- 2T^{1/2}\partial_\tau \frac{1}{2\pi}\int_{-\pi}^{\pi} \operatorname{tr}\{(k_0 \Sigma_0 k_0^*)^{-1} w_u(\omega) w_v(\omega)\}\, d\omega|_0,$$

where $w_v(\omega) = T^{-1/2}\sum l_\tau z(t)\exp(it\omega)$. Call ζ_j a typical element of this and call ξ_j, η_j the first and second terms, so that $\zeta_j = \xi_j - \eta_j$. Put $\zeta_{m,j}$, $\xi_{m,j}$, $\eta_{m,j}$ for the expressions obtained by replacing $\partial_\tau(k_\tau\Sigma_0 k_\tau^*)^{-1}|_0$, $(k_0\Sigma_0 k_0^*)^{-1}$ and $\partial_\tau l_\tau|_0$ by their expansions to only m terms in their Fourier series. Put $\bar{\zeta}_{m,j}$, $\bar{\xi}_{m,j}$, $\bar{\eta}_{m,j}$ for the remainders (e.g., $\zeta_j = \zeta_{m,j} + \bar{\zeta}_{m,j}$). Put $\xi_{m,j} = \xi'_{m,j} + \xi''_{m,j}$, where in $\xi'_{m,j}$ all "squared terms" have been removed. Then $\zeta_j = (\xi'_{m,j} - \eta_{m,j}) + \xi''_{m,j} + \bar{\zeta}_{m,j}$. From (Bernstein's) Lemma 4.4.3 it is sufficient to obtain the limiting covariance matrix of $\zeta_{m,j}$ and allow m to increase. We know that the limiting distribution of $\zeta_{m,j}$ is the same as that of $\zeta'_{m,j} = \xi'_{m,j} - \eta_{m,j}$. The limiting distribution of this is precisely the same as if the $\varepsilon(t)$ had been Gaussian, under (4.1.16) since all covariances are the same as if the $\varepsilon(t)$ had been Gaussian. But then the limiting distribution of $\zeta'_{m,j}$ is the same as that of $\zeta_{m,j}$ computed as if the $\varepsilon(t)$ had been Gaussian. Thus we may evaluate the covariance matrix of the ζ_j in their limiting distribution as if the $\varepsilon(t)$ had been Gaussian, under (4.1.16). But then, using the Gaussian assumption,

$$TE\{\partial_{\tau_j} L_T(\theta_0)\partial_{\tau_k} L_T(\theta_0)\} = 2E\{\partial^2_{\tau_j,\tau_k} L_T(\theta_0)\}, \tag{4.3.33}$$

the factor $-2T^{-1}$ coming from the fact that $L_T(\theta) = -2/T$ (log likelihood). Now by the same type of arguments as were used in establishing Lemmas 4.2.2(ii) and 4.2.3(i), we see that the right side of (4.3.33) coverges to

$$2\partial^2_{\tau_i\tau_j}\{\log\det\Sigma + Q(\theta_0)\} = 2\partial^2_{\tau_i\tau_j} Q(\theta_0).$$

The proof is simpler than in the lemmas cited because θ_0 is fixed. The only essential difference is the use of the expectation operator to evaluate serial covariances instead of ergodic behavior.

As observed earlier under (4.1.16), (4.1.20), the limiting distribution of $\partial_{\sigma_{ij}} L_T(\theta_0)$ is such that this is also asymptotically normal, jointly with the elements of $\partial_\tau L_T(\theta_0)$ and independently of the latter. The distribution of $\partial_\sigma L_T(\theta_0)$ is obtained from (4.3.32). Put $\dot{\sigma}_{ij} = e_i'\dot{\Sigma}e_j$. Then

$$T^{1/2}(\dot{\sigma}_{i,j} - \sigma_{0,ij}) = T^{-1/2} \sum_{1}^{T} \{\varepsilon_i(t)\varepsilon_j(t) - \sigma_{0,ij}\}$$

and the covariance in the limiting distribution between these quantities for subscripts i, j and k, l becomes

$$\sigma_{0,ik}\sigma_{0,jl} + \sigma_{0,il}\sigma_{0,jk} + \kappa_{ijkl}, \tag{4.3.34}$$

where κ_{ijkl} is the fourth cumulant between $\varepsilon_i(t)$, $\varepsilon_j(t)$, $\varepsilon_k(t)$, $\varepsilon_l(t)$. We shall not, in detail, write down the covariance matrix of the elements $T^{1/2}\partial_\tau L_T(\theta_0)$ in their limiting distribution. Writing the elements of the vector in some order, for example, stacking the parts of columns below or on the diagonal, working from left to right [which we write vech $(T^{1/2}\partial_\sigma L_T(\theta_0))$], the corresponding covariance matrix can easily be obtained from (4.3.32) and (4.3.34). *Put S_0 for the covariance matrix in this limiting distribution.* Thus we need to describe

$$\partial_\tau^2 Q(\theta_0) = \partial_\tau^2 \frac{1}{2\pi} \int_{-\pi}^{\pi} \operatorname{tr}\{(k_\tau \Sigma_0 k_\tau^*)^{-1} k_0 \Sigma_0 k_0^*\}\, d\omega|_0$$
$$+ 2\left[\int_{-\pi}^{\pi} \operatorname{tr}\{(k_0\Sigma_0 k_0^*)^{-1} \partial_{\tau_j} l_0\, dF\, \partial_{\tau_k} l_0^*\}\right]_{j,k=1,\ldots,2ns},$$

where by the last we indicate the (j, k)th entry in the matrix. This simplifies further to

$$2\left[\int_{-\pi}^{\pi} \operatorname{tr}\left\{(k_0\Sigma_0 k_0^*)^{-1}(\partial_{\tau_j} k_0)\Sigma_0(\partial_{\tau_k} k_0^*)\frac{d\omega}{2\pi} + \partial_{\tau_j} l_0\, dF\, \partial_{\tau_k} l_0^*)\right\}\right], \tag{4.3.35}$$

where the (j, k)th entry in the matrix is shown. This is because $\partial_{\tau_j} k_0$ has no term in $\exp(ij\omega)$ for $j = 0$, so that if k_r (or k_τ^*) is differentiated twice, there is a contribution canceling with $k_0\Sigma_0 k_0^*$ and the overall contribution vanishes. Of course, $\partial_{\tau_j} k_0$ or $\partial_{\tau_j} l_0$ refer to differentiation with respect to τ_j followed by evaluation at τ_0. We have evaluated (4.3.35) more explicitly in the examples studied below the statement of Theorem 4.3.1. Let us write (4.3.35) as $2I_0$ so that I_0 is the matrix shown in (4.3.35). We return to (4.3.14), (4.3.15) and recall the partition of λ' as $(\lambda^{(1)'}, \lambda^{(2)'})$, where $\lambda^{(1)}$ corresponds to the constraints, $\psi^{(1)}(\tau_0)$, involving only τ and $\lambda^{(2)}(\sigma_0)$ involving only σ. We then have (see the formula immediately following the remark after Theorem 4.3.1)

$$\begin{pmatrix} \partial_\tau \hat{Q}_T(\theta_0) \\ 0 \end{pmatrix} = -\begin{bmatrix} \partial_\tau^2 Q(\theta_0) & \partial_\tau \psi^{(1)}(\tau_0)' \\ \partial_\tau \psi^{(1)}(\tau_0) & 0 \end{bmatrix} \begin{pmatrix} \hat{\tau} - \tau_0 \\ \hat{\lambda}_\tau^{(1)} \end{pmatrix} + 0_p(T^{-1/2}).$$

Thus we then obtain the limiting covariance matrix of $T^{1/2}(\hat{\tau}-\tau_0)$, $T^{1/2}\hat{\lambda}_T^{(1)}$ as

$$\text{(4.3.36)}\quad \begin{bmatrix} \partial_\tau^2 Q(\theta_0) & \partial_\tau \psi^{(1)}(\tau_0)' \\ \partial_\tau \psi^{(1)}(\tau_0) & 0 \end{bmatrix}^{-1} \begin{bmatrix} T_0 & 0 \\ 0 & 0 \end{bmatrix} \begin{bmatrix} \partial_\tau^2 Q(\theta_0) & \partial_\tau \psi^{(1)}(\tau_0)' \\ \partial_\tau \psi^{(1)}(\tau_0) & 0 \end{bmatrix}^{-1},$$

where T_0 is the covariance matrix of $T^{1/2}\partial_\tau \hat{Q}_T(\theta_0)$ in its limiting distribution. Of course, $\partial_\tau^2 Q(\theta_0)$ is given by (4.3.35) and when (4.1.1), (4.1.15), and (4.1.16) hold, then $T_0 = 2\partial_\tau^2 Q(\theta_0)$, so that (4.3.36) becomes

$$\text{(4.3.37)}\quad \begin{bmatrix} 2I_0 & \partial_\tau \psi^{(1)}(\tau_0)' \\ \partial_\tau \psi^{(1)}(\tau_0) & 0 \end{bmatrix}^{-1} \begin{bmatrix} 4I_0 & 0 \\ 0 & 0 \end{bmatrix} \begin{bmatrix} 2I_0 & \partial_\tau \psi^{(1)}(\tau_0)' \\ \partial_\tau \psi^{(1)}(\tau_0) & \end{bmatrix}^{-1}.$$

Of course, when there are no constraints, this becomes I_0^{-1}. The matrix I_0 is the limit of $-T^{-1}E\{\partial_\tau^2 \log \text{likelihood}\}$, when there are no constraints and this is essentially T^{-1} by the information matrix in the Gaussian case. Then when Conditions AM′, AMX′ hold, the limiting distribution, whether or not parameters mix, of $T^{1/2}(\hat{\tau}-\tau_0)$, $T^{1/2}(\hat{\sigma}-\sigma_0)$, $\hat{\lambda}_T$ have the covariance matrix

$$\text{(4.3.38)}\quad \begin{bmatrix} 2I_0 & 0 & \partial_\tau \psi(\theta_0)' \\ 0 & \partial_\sigma^2 L(\theta_0) & \partial_\sigma \psi(\theta_0)' \\ \partial_\tau \psi(\theta_0) & \partial_\sigma \psi(\theta_0) & 0 \end{bmatrix}^{-1} \begin{bmatrix} 4I_0 & 0 & 0 \\ 0 & S_0 & 0 \\ 0 & 0 & 0 \end{bmatrix} \times \begin{bmatrix} 2I_0 & 0 & \partial_\tau \psi(\theta_0)' \\ 0 & \partial_\sigma^2 L(\theta_0) & \partial_\sigma \psi(ho)' \\ \partial_\tau \psi(\theta_0) & \partial_\sigma \psi(\theta_0) & 0 \end{bmatrix}^{-1},$$

where S_0 is described above and we have

$$\begin{aligned} \partial^2_{\sigma ij,\sigma_{kl}} L(\theta_0) &= 2\{\sigma_0^{ik}\sigma_0^{lj} + \sigma_0^{il}\sigma_0^{kj}\}, && i=j,\quad k=l \\ &= 2\sigma_0^{ik}\sigma_0^{li}, && i=j,\quad k=l \\ &= (\sigma_0^{ik})^2, && i=j,\quad k=l, \end{aligned}$$

where $[\sigma_0^{ij}] = \Sigma_0^{-1}$. This completes the proof of Theorem 4.3.1.

We now turn to the proof of the law of the iterated logarithm. Although this is a strong convergence result, it belongs in this section rather than in Section 2 since it demands the introduction of coordinates and also conditions of the nature of those used for the central limit theorem, to which it is closely related. *First we study the ARMA case* and use a modification of a technique due to Bouaziz (1983, 1984; see also Azencott and Dacunha-Castelle, 1984).

This technique could have been used in the proof of Theorem 4.3.1, but most of the proof would still be needed and that proof seems to give a slightly more general result, in the sense of the remark containing (4.3.23). Examining the proof down to Lemma 4.3.2, we see that $T^{1/2}\partial_\theta L_T(\theta_0)$ may be reduced to (4.3.17) in the sense that the omitted terms converge a.s. to zero. We call the observed sequence $u(t)$ so that later, when the ARMAX case is considered, we shall put, as usual, $y(t) = u(t) + v(t)$. Now let us be a little more general for a moment and consider $k_0(e^{i\omega})$, $\phi(\omega)k_0(e^{i\omega})$, $\phi(\omega) = \partial_j(k_0\Sigma_0 k_0^*)^{-1}$, to satisfy

$$\text{(4.3.39)} \qquad \sum_{u=1}^{\infty} u^2 \|K_0(u)\|^2 < \infty, \qquad \sum_{u=1}^{\infty} u \|H(u)\|^2 < \infty,$$

where $H(u)$ is the uth Fourier coefficient of $\phi(\omega)k_0(e^{i\omega})$. Of course, (4.3.35) holds in the ARMA case.

Then we shall establish the following.

Lemma 4.3.5. *Under (4.1.15), (4.1.20), (4.3.39) and ergodicity for the $\varepsilon(t)$, the result in Lemma 4.3.2 holds a.s.*

The proof of the lemma is unaltered down to the consideration of

$$CT^{-1/2}\frac{1}{2\pi}\int_{-\pi}^{\pi} \operatorname{tr}\left\{\sum_{u,v}\sum K_0(u) \sum_{uv}' K_0(v)^* e^{i(u-v)\omega}\right\} d\omega$$

[see above (4.3.21)]. This is

$$CT^{-1/2}\sum_k K_0(u)\sum_{uu}' K_0(u)^* \le CT^{-1/2}\sum_u uK_0(u)K_0(u)^*$$
$$= O(T^{-1/2}).$$

Indeed,

$$\sum_{uu}' \varepsilon(t)\varepsilon(t)' = \sum_{\max(u-T,0)}^{u-1} \varepsilon(t)\varepsilon(t)' = O(u),$$

by ergodicity. Next examine (4.3.21). The last term is

$$T^{-1/2}\sum_1^{\infty} \operatorname{tr}\left\{K_0(u) \sum_{\max(T-u+1,1)}^{T} \varepsilon(t)\varepsilon(t)' K_0(u)^*\right\}.$$

However, because of (4.1.20) $\|\varepsilon(t)\varepsilon(t)'\| = O(t^{1/2})$, so that we obtain a bound for the norm of the last expression, that is,

$$o(1)\left[T^{-1/2}\left\{\sum_1^T \|K_0(u)\|^2 uT^{1/2}\right\} + cT^{-1/2}\left\{\sum_{T+1}^{\infty} \|K_0(u)\|^2 T\right\}\right],$$

which converges to zero because of (4.3.39). Finally, we may examine only one of the remaining terms in (4.3.21) since they are of the same form. The first term is

$$T^{-1/2}\sum_0^{T-1} \operatorname{tr}\left\{\sum_{t+1}^{\infty} H(u)\varepsilon(T-t)\varepsilon(T-t)'K_0(u)^*\right\}$$

and this converges a.s. to zero, using (4.3.39), by almost the same argument as was first used. Thus the lemma is established.

Now consider, as above (4.3.24),

$$\frac{1}{2\pi}\int_{-\pi}^{\pi} \operatorname{tr}\{\phi(\omega)\tilde{I}_u(\omega)\}\, d\omega.$$

As in Bouaziz (1983, 1984; see also Hannan, 1970, Chap. IV)

$$\sum_1^T u(t)e^{it\omega} = \sum_0^{\infty} K_0(j)e^{ij\omega}\sum_1^T e(t)e^{it\omega} + Y_0 - Y_T$$

$$Y_s = \sum_{t\geq s+1}\left\{\sum_{j\leq s} K_0(t-j)\varepsilon(j)\right\}e^{it\omega}, \qquad s = 0, 1, \ldots.$$

We leave the reader to check this equality. We may rewrite Y_s as

$$Y_s = e^{is\omega}\sum_{k=0}^{\infty} \varepsilon(s-k)\left(\sum_{u=1}^{\infty} K_0(u+k)e^{iu\omega}\right),$$

so that it is $e^{is\omega}$ by a (complex-valued) stationary process. Now consider

$$T^{1/2}\frac{1}{2\pi}\int_{-\pi}^{\pi} \operatorname{tr}[\phi(\omega)\{k_0(e^{i\omega})w_\varepsilon(\omega) + T^{-1/2}(Y_0 - Y_T)\}$$

$$\times \{k_0(e^{i\omega})w_\varepsilon(\omega) + T^{-1/2}(Y_0 - Y_T)\}]\, d\omega, \tag{4.3.40}$$

$$w_\varepsilon(\omega) = T^{-1/2}\sum_1^T \varepsilon(t)e^{it\omega},$$

which is the essential component in $T^{1/2}\partial_\tau \hat{Q}_T(\theta_0)$. In (4.3.40) we may eliminate the squared terms under the conditions of Lemma 4.3.5. Now in

$$T^{-1/2}\frac{1}{2\pi}\int_{-\pi}^{\pi}\operatorname{tr}\{\phi(\omega)Y_sY_s^*\}\,d\omega, \tag{4.3.41}$$

we may also, of course, eliminate the squared terms. Since $\phi(\omega)$ is uniformly bounded, we may consider

$$T^{-1/2}\sum_{u\geq 1}\sum_{j,k\geq 0}\sum \varepsilon(s-j)'K_0(u+j)K_0(u+k)\varepsilon(s-k), \qquad s=0,1,\ldots,$$

which is $T^{-1/2}$ by a strictly stationary process whose mean square is bounded by

$$\left(\sum_{u\geq 1}\left[\sum_{j\neq k}\|K_0(u+j)\|^2\|K_0(u+k)\|^2E\{\|\varepsilon(j)\|^2\|\varepsilon(k)\|^2\}\right]^{1/2}\right)^2$$
$$\leq C\left\{\sum_{u\geq 1}\sum_{j\geq 0}\|K_0(u+j)\|^2\right\}^2 = C\left\{\sum_1^\infty j\|K_0(j)\|^2\right\} < \infty.$$

Thus (4.3.41) is $T^{-1/2}o(s^{1/2})$ and this is $o(1)$, $s\leq T$. Thus the contributions to (4.3.40) from Y_0, Y_T may be ignored provided that, putting $Q_T^{-1} = (T/\log\log T)^{1/2}$,

$$Q_T^{-1}\frac{1}{2\pi}\int_{-\pi}^{\pi}\operatorname{tr}[\phi(\omega)k_0(e^{i\omega})w_\varepsilon(\omega)w_\varepsilon(\omega)^*k_0(e^{i\omega})^*]\,d\omega$$

is a.s. bounded. However, again eliminating squared terms, this is

$$(T\log\log T)^{-1/2}\sum_{-\infty}^{\infty}\operatorname{tr}\left\{F(j)\sum_j\varepsilon(t)\varepsilon(t-j)'\right\},$$

where the $F(j)$ are the matrices of Fourier coefficients of $k_0^*\phi k_0$. The sum $\sum_j$ is over $1\leq t, t-j\leq T, t\neq t-j$. Rearranging this, we obtain

$$(T\log\log T)^{-1/2}\sum_1^T\varepsilon(t)\left\{\sum_{0<|j|<t}F(j)'\varepsilon(t-|j|)\right\}'. \tag{4.3.42}$$

Consider the martingale

$$\sum_1^T\varepsilon(t)\left\{\sum_{|j|\geq t}F(j)'\varepsilon(t-|j|)\right\}',$$

which has mean square

$$
\begin{aligned}
&\sum_{1}^{T} \sum_{|j|,|k| \geq k} E\{\varepsilon(t-|j|)' F(j) \varepsilon(t) \varepsilon(t)' F(k)' \varepsilon(t-|k|)\} \\
(4.3.43) \qquad &\leq c \sum_{1}^{T} \sum_{|j|,|k| \geq t} \|F(j)\| \, \|F(k)\| \leq s \sum_{-\infty}^{\infty} \|F(j)\| \, \|F(k)\| \min(|j|,|k|),
\end{aligned}
$$

which is finite if

$$
\sum_{-\infty}^{\infty} |j|^{1/2} \|F(j)\| < \infty,
$$

and this is certainly true since the $F(j)$ decrease geometrically. Thus (see, e.g., Neveu, 1975, p. 149) we may replace (4.3.42) by

$$
(4.3.44) \qquad (T \log\log T)^{-1/2} \sum_{1}^{T} \varepsilon(t) \left\{ \sum_{0<|j|<\infty} F(j)' \varepsilon(t-|j|) \right\}'.
$$

The summands over t are now stationary, square-integrable martingale differences and thus the law of the iterated logarithm obtains (Hall and Heyde, 1980, Chap. 4). Thus the sequence (4.3.44) has, a.s., its limit points comprising the interval $[-(2a)^{1/2}, (2a)^{1/2}]$, where a is the variance of the summands in (4.3.44). When (4.1.16) holds,

$$
a = \operatorname{tr}\{F(j)\Sigma_0 F(j)'\Sigma(0)\},
$$

but, in general, when (4.1.16) does not hold, this constant is too complicated to be usefully evaluated. When (4.1.16), (4.1.20) hold together with (4.1.14), then

$$
\frac{1}{2\pi} \int_{-\pi}^{\pi} \operatorname{tr}\{\phi(\omega) I_u(\omega)\} \, d\omega
$$

obeys the law of the iterated logarithm even if $\phi(\omega) = \partial_\theta (k_0 \Sigma_0 k_0^*)^{-1}$ and the constraints do mix parameters. The proof is essentially the same as that first given. Returning to (4.3.14) the analogous results obviously apply, in the ARMA case, to $\hat{\theta}_T - \theta_0$, $\hat{\lambda}_T$. Thus $Q_T^{-1}(\hat{\theta}_T - \theta_0)$, $Q_T^{-1}\hat{\lambda}_T$ are sequences whose limit points are precisely comprised by an interval $[-(2a)^{1/2}, (2a)^{1/2}]$, but for which we have not always evaluated $0 < a < \infty$. This completes the proof of Theorem 4.3.2.

We go on to consider the ARMAX case and [see (4.3.17)] must consider

$$
Q_T^{-1} \frac{1}{2\pi} \int_{-\pi}^{\pi} \operatorname{tr}\{\phi(\omega) I_u(\omega) - (k_0 \Sigma_0 k_0^*)^{-1} 2T^{-1/2} \sum_{1}^{T} (\partial_j l_\tau) z(t) e^{it\omega} w_u(\omega)^*\}|_{\theta=\theta_0} \, d\omega,
$$

no terms involving $z(s)$ with $z(t)$ occurring, as before, because one factor, $(l_\tau - l_0)$, in $\partial_j\{w_\tau(\omega)w_\tau(\omega)^*\}$, at $\theta = \theta_0$, will not be differentiated and vanish. Thus we must additionally consider the behavior of

$$Q_T^{-1}\frac{1}{2\pi}\int_{-\pi}^{\pi} \operatorname{tr}\left[(k_0\Sigma_0 k_0^*)^{-1}\left\{T^{-1/2}\sum_1^T (\psi z(t))e^{it\omega}\right\} w_u(\omega)^*\right] d\omega$$

$$\psi(z) = \partial_j l_\tau(z).$$

Of course, ψ is a lag operator operating on $z(t)$. Now as above [see the argument following (4.3.40)] this may be replaced by

$$Q_T^{-1}\frac{1}{2\pi}\int_{-\pi}^{\pi} \operatorname{tr}\{(k_0\Sigma_0 k_0^*)^{-1}\psi(e^{i\omega})w_z(\omega)w_u(\omega)^*\}\, d\omega,$$

provided that $z(t)$ is part of a stationary sequence, independent of the $\varepsilon(t)$ sequence, with finite variances, and $E\{\varepsilon(t)\varepsilon(t)'\} < \infty$. Indeed, then the argument above hardly changes. Now, as before, this is

$$(4.3.45) \qquad (T\log\log T)^{-1/2}\sum_{t=1}^T \varepsilon(t)\sum_{|j|\le t}\tilde{F}(j)'z(t-|j|) + o(1), \quad \text{a.s.},$$

where the $\tilde{F}(j)$ are the Fourier coefficients of $(k_0\Sigma_0 k_0^*)^{-1}\psi(e^{i\omega})$. Now we may replace (4.3.45) by

$$(4.3.46) \qquad (T\log\log T)^{-1/2}\sum_{t=1}^T \varepsilon(t)\left\{\sum_{-\infty}^{\infty}\tilde{F}(j)'z(t-j)\right\}'$$

since, for example, the martingale

$$\sum_{t=1}^T \dot{\varepsilon}(t)\left\{\sum_t^{\infty}\tilde{F}(j)'z(t-j)\right\}'$$

has mean square

$$\sum_1^T\sum_{j,k\ge t}\sum E\{z(t-j)'\tilde{F}(j)\varepsilon(t)\varepsilon(t)'\tilde{F}(k)'z(t-k)\}$$
$$\le C\sum_1^T\sum_{j,k\ge t}\sum \|\tilde{F}(j)\|\,\|\tilde{F}(k)\| < \infty.$$

Now (4.3.46) again is $(T\log\log T)^{-1/2}$ by a martingale with stationary martingale differences. The differences will not necessarily be ergodic. However, the law of the iterated logarithm will still hold (see again Hall and Heyde, 1980, Chap. 4), the constant in the statement of the law now being a random variable defined by

$$a = \lim_{T\to\infty} \frac{1}{T}\sum_{1}^{T}\left[a(t)\left\{\sum_{-\infty}^{\infty} \tilde{F}(j)'z(t-j)\right\}'\right]^2,$$

where a will be the expectation of the summands on the right side given the σ-algebra of invariant events. Thus Theorem 4.3.3 is established.

CHAPTER 5

Estimating the Order of a Linear System

1. INTRODUCTION

In Chapter 4 a central limit theorem was obtained for the case where the true system order is known. In this chapter methods are discussed for estimating the order of the system consistently. The theory then attains some theoretical completeness, since the central limit theorem will apply to the system parameter estimates, corresponding to the order estimated by the consistent procedure.

However, it is again emphasized that there may be no true rational transfer function system and hence no true system order. This more realistic situation will be discussed again in Chapter 7. Here it is mentioned so that results relating to consistency in order estimation will not mislead the reader into the complete rejection of methods that are inconsistent, when it is assumed that there is a true rational transfer function system.

A characteristic of order estimation is that to determine the true order, orders higher than the true order must examined (i.e., cases $n > n_0$, $\alpha > \alpha_0$, or $p > p_0$, $q > q_0$, $r > r_0$). Then we are in the situation described in Theorem 2.5.3 (see the remarks below the theorem and the example) and the asymptotic theory becomes more difficult since the coordinate vector, $\hat{\tau}$, will not converge in any reasonable sense, as will be shown later. (However, see Section 6 of this chapter for a method that avoids this problem.) It follows from Theorem 4.2.1 that the transfer functions, $\hat{k}$, $\hat{l}$, will converge to k_0, l_0 even when, for example, $n > n_0$, but for order determination asymptotics much more than that is needed, since very accurate evaluations are required. Indeed, consider $BIC(n)$, which is one of the criteria functions introduced in the next section. This is

$$BIC(n) = \log \det \hat{\Sigma}_n + n(2s + m) \log T/T.$$

Here $\hat{\Sigma}_n$ is the maximum likelihood estimate of Σ from a model of order n and $\hat{n}$ is chosen to minimize $BIC(n)$ over suitable range of values of n. If the asymptotic behavior of $\hat{n}$ is to be determined, it will be necessary to account for random variation in $\hat{\Sigma}_n$ up to an order of magnitude $\log T/T$. At first sight this seems impossible since $\hat{\Sigma}_n$ could not be accurate, in the a.s. sense, above the order of accuracy given by the law of the iterated logarithm [i.e., $(\log\log T/T)^{1/2}$]. However, the terms of that order are independent of n and it will be shown that the remaining terms involve squared errors that are $O(\log\log T/T)$. However, these evaluations are difficult and involve the consideration of a number of autocovariances [e.g., of the $\varepsilon(t)$] up to lags increasing as $\log T$ with T. For this reason, in Section 3 some new strong theorems, of the nature of the law of the iterated logarithm, are derived. In Section 4 the case $s = 1$ is discussed, partly because of its importance and partly because the discussion, then being simpler, will serve to introduce the discussion of the general case in Section 5. In Section 6 a separate discussion of the autoregressive case will be given together with some associated cases.

2. ESTIMATION CRITERIA

A more detailed discussion of the basis for the criteria that we here introduce is given in Chapter 7 because the main justifications have been in connection with situations where it is not assumed that there is a true rational transfer function system. The criteria to be minimized are of the form, when the $M(n)$ are being considered,

$$\chi(n) = \log\det\hat{\Sigma}_n + d(n)C_T/T, \qquad d(n) \le H_T. \tag{5.2.1}$$

Here $d(n) = n(2s + m)$ is the dimension of the vector, τ, of system parameters. For that matter we might consider the echelon forms indexed by α, with $d(\alpha)$ given in Theorem 2.5.3. In general, (5.2.1) can be used with $d(n)$ replaced by the number of system parameters used in forming $\hat{\Sigma}$. If $M(p, q, r)$ is being considered, we call (5.2.1) $\chi(p, q, r)$ and similarly $\chi(\alpha)$ in the case of the echelon forms.

The constants C_T and H_T have to be prescribed. Part of the purpose of the analysis here (and in Chapter 7) is to help in a specification of C_T. It will always be assumed that

$$C_T > 0, \qquad C_T/T \to 0. \tag{5.2.2}$$

In practice, no upper bound, H_T, might be prescribed since it might be expected that the minimum of (5.2.1) will be found for quite small values of n (or α or p, q, r). However some upper bound will be needed in the proof of theorems, and this suggests that an uncritical minimization of $\chi(\cdot)$ over too large a range of values of its argument could lead to trouble. This is indicated also by the following considerations. In fitting an autoregression to $y(t)$ for

$s = 1$ by the standard Toeplitz (Yule-Walker) procedure, what is minimized is (see Section 6.2)

(5.2.3)

$$\frac{1}{2\pi}\int_{-\pi}^{\pi} I(\omega)\left|\sum_{0}^{h}\phi(j)e^{ij\omega}\right|^2 d\omega, \qquad \phi(0) = 1, \qquad I(\omega) = \frac{1}{T}\left|\sum_{1}^{T} y(t)e^{it\omega}\right|^2.$$

By a famous theorem due to Szegö (see Hannan, 1970, Chap. III, and Theorem 1.3.2 herein) the minimum value will converge, as $h \to \infty$, to

$$\exp\left[\frac{1}{2\pi}\int_{-\pi}^{\pi} \log I(\omega)\, d\omega\right]$$

and as $T \to \infty$, under fairly general circumstances it is shown in An (1982) that this converges to $\sigma^2 \exp(-\gamma)$, where $\gamma > 0$ is Euler's constant. Thus if (5.2.1) in this circumstance was examined for very large values of h, then, noting (5.2.2), it seems that a ridiculously large value might be chosen. Thus some reasonable limit, H_T, on the dimension of the system being fitted is needed and again it is part of the purpose of the analysis here to suggest a reasonable bound.

Special cases of (5.2.1) of importance are

$$AIC(n) = \log\det\hat{\Sigma}_n + 2d(n)/T, \qquad BIC(n) = \log\det\hat{\Sigma}_n + d(n)\log T/T.$$

As for $\chi(n)$ we shall write $AIC(p,q,r)$, $BIC(p,q,r)$ when $M(n)$ is replaced by $M(p,q,r)$. These formulas were introduced in Akaike (1969, 1977), and $BIC(n)$ was justified on another basis in Rissanen (1983); see also Schwarz, (1978). These discussions are dealt with in Chapter 7. Here we merely point out that the second term in (5.2.1) is a penalty term for fitting parameters, whose inaccurate determination adds to the uncertainty of the method even though an increase in their number allows a better fit to the data, as measured by $\hat{\Sigma}_n$.

Many other procedures have also been suggested for order determination. One alternative, developed mainly in relation to autoregressions, is the criterion due to Parzen (1974),

$$\begin{aligned} CAT(h) &= \frac{1}{T}\sum_{j=1}^{h}\frac{1}{\tilde{\sigma}_j^2} - \frac{1}{\tilde{\sigma}_h^2}, \qquad j = 1, 2, \ldots, H_T, \qquad \tilde{\sigma}_j^2 = \left(1 - \frac{j}{T}\right)^{-1}\hat{\sigma}_j^2, \\ &= -(1 + T^{-1}), \qquad h = 0, \end{aligned}$$

where $\hat{\sigma}_j^2$ is the minimum of the expression (5.2.3). It has been suggested by Tong (1979) that the term for $j = 0$ should be included in the sum, which is then taken to define the criterion for all h. Tong also points out that

$$CAT(h) - CAT(h-1) = \frac{1}{\tilde{\sigma}_{h-1}^2}\{AIC(h) - AIC(h-1)\} + \frac{1}{T}O\left\{\frac{\tilde{\sigma}_h^2 - \tilde{\sigma}_{h-1}^2}{\tilde{\sigma}_{h-1}^2}\right\}.$$

Thus, ignoring the last term, $CAT(h)$ and $AIC(h)$ rise and fall together and thus would be minimized at the same value. For this reason we shall not deal further with the CAT criterion. For a discussion of its properties in relation to autoregressions, see Bhansali (1986).

Another criterion of the same kind is (Mallows, 1973) C_h, which, for an autoregression once more, would be

$$C_h = T\left\{\frac{\hat{\sigma}_h^2}{\hat{\sigma}_H^2}\left(1 - \frac{H}{T}\right) - 1 + \frac{2h}{T}\right\}, \qquad h = 0, 1, \ldots, H.$$

Thus $T^{-1}C_h$, which we may as well consider, satisfies

$$T^{-1}(C_h - C_{h-1}) = \frac{\hat{\sigma}_h^2 - \sigma_{h-1}^2}{\hat{\sigma}_H^2}\left(1 - \frac{H}{T}\right) + \frac{2}{T},$$

while

$$AIC(h) - AIC(h-1) \approx \frac{\hat{\sigma}_h^2 - \hat{\sigma}_{h-1}^2}{\hat{\sigma}_{h-1}^2} + \frac{2}{T}.$$

Again the two are closely related. However, our considerations are mainly for ARMA and ARMAX models and for these, presumably, a generalization of C_h will involve fitting a high-order model to obtain the analogue of $\hat{\sigma}_H^2$. We do not discuss C_h further, although it may have virtues.

A different kind of procedure altogether was suggested in Gray, Kelley, and McIntire (1978), for ARMA processes. Their idea is to extend the criterion of partial autocorrelation so as to form an array, $\hat{\rho}(i,j)$ let us say, which will estimate an array, $\rho(i,j)$, which will show a certain pattern of zeros if the data are generated by an ARMA process. Because the technique depends to a certain degree on pattern recognition, which is subjective, such methods would be difficult to study analytically. There is some evidence, also, that they do not perform well (see de Gooijer et al., 1985). A related method, which could be studied analytically, was suggested by Tiao and Tsay (1983) and Tsay and Tiao (1984), particularly in connection with the estimation of the autoregressive parameters for ARMA models. Their methods are related to those presented in Section 6.4.

3. UNIFORM CONVERGENCE OF AUTOCOVARIANCES AND AUTOCORRELATIONS

The first strong theorem relating to autocovariances, involving convergence uniform in the lag, seems to be that in Hannan (1974). It is assumed there that the data are generated by a stationary, purely nondeterministic process, that is, one for which

$$E\{y(t)|\mathscr{F}_{-\infty}\} = 0 \tag{5.3.1}$$

and for which the variances are finite. It is shown that, putting $G(t) = 0, t \geq T$,

$$\lim_{T\to\infty} \sup_{t\geq 0} \|G(t) - \Gamma(t)\| = 0, \quad \text{a.s.} \tag{5.3.2}$$

Of course, a process that is purely nondeterministic in the sense of (5.3.1) is also linearly purely nondeterministic (see Ibragimov and Linnik, 1971). Thus $y(t)$ has an absolutely continuous spectrum and $\Gamma(t)$ converges to zero as $t \to \infty$. We shall establish (5.3.2) in Chapter 7 (see Theorem 7.4.1), where the case where the spectrum is not rational is discussed, but we use the result in this chapter. However, (5.3.2) is not strong enough for our purposes and a more accurate prescription concerning the rate of decrease of $\|G(t) - \Gamma(t)\|$, albeit for $t \leq H_T = o(T)$, is needed. The first results of this type are in An, Chen, and Hannan (1982), although results involving independence assumptions were given in Hannan (1980a, 1981). In Hannan and Kavalieris (1983, 1984) these results were further improved by avoiding (4.1.16). The proofs in some of these papers were faulty in part, and that will be corrected below.

The notation of Section 4.3 will continue to be used, namely,

$$y(t) = u(t) + v(t), \qquad u(t) = \sum_0^\infty K(j)\varepsilon(t-j), \qquad v(t) = \sum_1^\infty L(j)z(t-j).$$

In this section it is assumed that $z(t)$ is generated by a stationary, ergodic process with zero mean and finite fourth moment. As a consequence, (4.1.1) and (4.1.14) both hold. The reasons for the stronger assumptions concerning $z(t)$ are the same as those that caused us to make these assumptions in connection with Theorem 4.3.3. *It is assumed that the process generating $z(t)$ is independent of that generating $\varepsilon(t)$.* The zero mean assumption is not necessary and is made only for simplicity. We put

$$Q_T = (\log\log T/T)^{1/2}, \qquad \phi(T) = TQ_T = (T\log\log T)^{1/2}.$$

By $O(\cdot)$, $o(\cdot)$ it is always meant that the indicated order relations hold a.s. (see Section 4.3).

Theorem 5.3.1. *Let $y(t)$ be generated by an ARMAX system with $z(t)$ stationary and ergodic with finite fourth moment, and with $\varepsilon(t)$ satisfying (4.1.15), (4.1.17) and with finite fourth moment. Then for $H_T \leq (\log T)^a$, $a < \infty$,*

$$\max_{1 \leq t \leq H_T} \left\| \frac{1}{T} \sum_{t+1}^{T} \varepsilon(s)u(s-t)' \right\| = O(Q_T),$$

(5.3.3)

$$\max_{0 \leq |t| \leq H_T} \left\| \frac{1}{T} \sum_{1 \leq s, s-t \leq T} \varepsilon(s)v(s-t)' \right\| = O(Q_T).$$

This theorem is of the form of the law of the iterated logarithm save that the constants needed accurately to prescribe the right sides have not been evaluated. More important, the results are uniform in t up to H_T. The case $t = 0$ is not included in the first part of (5.3.3), *but if (4.1.16) holds, we may alter the first component in (5.3.3) to read*

$$\max_{0 \leq t \leq H_T} \left\| \frac{1}{T} \sum_{t+1}^{T} \varepsilon(u)u(s-t)' - \delta_{ot}\Sigma \right\| = O(Q_T).$$

Indeed, by (4.1.15) and (4.1.16),

$$E\{\varepsilon(t)u(t)' | \mathscr{F}_{t-1}\} = E\{\varepsilon(t)\varepsilon(t)' | \mathscr{F}_{t-1}\} = \Sigma,$$

and thus $\varepsilon(t)u(t)' - \Sigma$ is composed of stationary, square-integrable, ergodic martingale differences. Thus the law of the iterated logarithm holds for the mean of these (Stout, 1970). However, in the applications in Sections 4 and 5, (4.1.16) can be avoided.

The proof of Theorem 5.3.1 is long and involved. It has therefore been placed at the end of this section, and first some consequences are deduced from it, in the ARMA case. Often, proofs will be given for the scalar case since the vector case proof is only notationally more complex. We do this now and consider the sample autocovariances, $g(t)$, of $u(t)$ [see (4.1.7)–(4.1.11) and following those formulas] and their "true"values $\gamma(t)$. Then, using a lowercase $\kappa(j)$ for $K(j)$ in this scalar situation,

$$g(t) - \gamma(t) = T^{-1} \sum_{0}^{\infty}\sum \kappa(j)\kappa(k) \sum_{s=1}^{T-t} \{\varepsilon(s-j)\varepsilon(s+t-k) - \sigma^2 \delta_{j,k-t}\}$$
$$- \sigma^2 \frac{t}{T} \sum_{0}^{\infty} \kappa(j)\kappa(j+t).$$

The last term is $t\gamma(t)/T \leq cH_T/T = o(Q_T)$, $t \leq H_T$. Call $S_{T-t}(j, k, t)$ the innermost sum in the second term. Then, using "max" for a maximum over $0 \le t \le H_T$, and assuming (4.1.15), (4.1.16),

$$P\left\{\max T^{-1} \sum_{c \log T}^{\infty} \sum_{k=0}^{\infty} \kappa(j)\kappa(k) S_{T-t}(j,k,t) \ge \varepsilon Q_T\right\}$$

$$\le (\varepsilon Q_T)^{-2} \sum_{t=0}^{T} E\left[\left\{T^{-1} \sum_{c \log T}^{\infty} \sum_{0}^{\infty} \kappa(j)\kappa(j) S_{T-t}(j,k,t)\right\}^2\right] \tag{5.3.4}$$

$$\le cT^{-1} H_T (\varepsilon Q_T)^{-2} \left\{\sum_{c \log T}^{\infty} |\kappa(j)|\right\}^2 \left\{\sum_{0}^{\infty} |\kappa(j)|\right\}^2,$$

using the triangle inequality and $E\{S_{T-t}(j,k,t)^2\} \le cT$, which follows immediately from (4.1.15), (4.1.16). Since the $\kappa(j)$ converge to zero at a geometric rate, (5.3.4) may be bounded by cT^{-a}, $a > 1$, taking c sufficiently large. By the Borel-Cantelli lemma, then,

$$\max T^{-1} \sum_{c \log T}^{\infty} \sum_{0}^{\infty} \kappa(j)\kappa(k) S_{T-t}(j,k,t) = o(Q_T).$$

The same results hold with the status of j, k reversed and with $j, k \ge c \log T$. Thus we need consider only

$$T^{-1} \sum_{j,k=0}^{c \log T}\sum \kappa(j)\kappa(k) S_{T-t}(j,k,t)$$

and from Theorem 5.3.1 this is $O(Q_T)$ uniformly in $t \le H_T$ [taking $u(t) = \varepsilon(t)$ in that theorem, i.e., $\kappa(j) = 0$, $j \ne 0$]. The proof for the vector case being the same, the following theorem results.

Theorem 5.3.2. *Let $\varepsilon(t)$ be ergodic, satisfy (4.1.15), (4.1.16), and have finite fourth moments, and let $y(t)$ be an ARMA process with innovations $\varepsilon(t)$. Then*

$$\max_{0 \le t \le H_T} \|G(t) - \Gamma(t)\| = O(Q_T), \qquad H_T \le (\log T)^a, \qquad a < \infty.$$

If (4.1.16) is not assumed but only (4.1.17), then

$$G(t) - \Gamma(t) = O(Q_T) + \sum_{0}^{\infty} K(j) \left[\frac{1}{T} \sum_{1}^{T-t} \{\varepsilon(s-j)\varepsilon(s-j)' - \Sigma\}\right] K(j+t)',$$

by the same argument, since only for the last term on the right is (4.1.16) involved. Moreover,

$$\max \sum_0^\infty K(j)\left[\frac{1}{T}\sum_1^{T-t} \varepsilon(s-j)\varepsilon(s-j)' - \frac{1}{T}\sum_1^T \varepsilon(s)\varepsilon(s)'\right]K(j+t)' \tag{5.3.5}$$
$$= o(T^{-1/2}),$$

as will now be shown. First the expression in brackets is

$$\frac{1}{T}\sum_{-j+1}^{m(j,t)} \varepsilon(s)\varepsilon(s)' - \frac{1}{T}\sum_{M(j,t)}^{T} \varepsilon(s)\varepsilon(s)',$$

where $m(j,t) = \min(0, T-t-j)$, $M(j,t) = \max(1, T-t-j+1)$. Consider the second term, which is the more troublesome. Because of later uses of the result, we relax the requirement on the $K(j)$ to be

$$\sum_0^\infty j\|K(j)\|^2 < \infty, \qquad \limsup_{j\to\infty} j\|K(j)\| < \infty. \tag{5.3.6}$$

[See the remark after Lemma 4.3.2 and (4.3.39) for analogous conditions.] Of course, (5.3.6) is satisfied in the ARMA case. Now $\Gamma(t) = o(t^{-1/2})$, as is easily seen. Putting $G(t) = 0$, $t \geq T$, then $\|G(t) - \Gamma(t)\| = o(T^{-1/2})$, $t \geq T$. Thus considering the second term in (5.3.5), we obtain a contribution dominated in norm, uniformly in $t < T$, by

$$o(1)\frac{1}{T}\sum_0^\infty \|K(j)\|\,\|K(j+t)\|\,\{T^{1/2}(j+t)\}$$

because $\varepsilon(s)\varepsilon(s)' = o(s^{1/2})$, by the finiteness of the fourth moment. Because of (5.3.6) this is $o(T^{-1/2})$. Also,

$$\frac{1}{T}\sum_0^\infty \|K(j)\|\,\|K(j+t)\|\left\|\sum_{j=1}^{m(j,t)} \varepsilon(s)\varepsilon(s)'\right\|$$
$$\leq c\frac{1}{T}\left\{\sum_0^{T-t} \|K(j)\|\,\|K(j+t)\|j + \sum_{T-t+1}^\infty \|K(j)\|\,\|K(j+t)\|j^{1/2}(T-t)\right\}$$

since the sum over s is $O(j)$ for $m(j,t) = 0$, by ergodicity, and $o\{j^{1/2}(T-t)\}$ otherwise, by the fourth moment condition. Again this last expression is $o(T^{-1/2})$ by (5.3.6). Thus

$$G(t) - \Gamma(t) = O(Q_T) + \sum_0^\infty K(j)\left\{\frac{1}{T}\sum_1^T \varepsilon(s)\varepsilon(s)' - \Sigma\right\}K(j+t)'$$

or, putting

$$\dot{\Gamma}(t) = \sum_0^\infty K(j)\dot{\Sigma}K(j+t)', \qquad \dot{\Sigma} = \frac{1}{T}\sum_1^T \varepsilon(s)\varepsilon(s)',$$

$$G(t) - \dot{\Gamma}(t) = O(Q_T). \tag{5.3.7}$$

Under the weaker condition

$$\sum_0^\infty j\|K(j)\|^2 < \infty, \tag{5.3.6$'$}$$

a reconsideration of the proof above shows that

$$G(t) - \Gamma(t) = O(Q_T) + \sum_0^\infty K(j)\left\{\frac{1}{T}\sum_1^{T-t} \varepsilon(s)\varepsilon(s)' - \Sigma\right\}K(j+t)', \tag{5.3.7$'$}$$

but this is somewhat more difficult to handle.

We may use (5.3.7) for $s = 1$. Put $r(t) = g(t)/g(0)$, $\rho(t) = \gamma(t)/\gamma(0)$. Then

$$r(t) - \rho(t) = \{g(t)\gamma(0) - g(0)\gamma(t)\}/\{g(0)\gamma(0)\}$$

and

$$\begin{aligned}\{g(t)\gamma(0) - g(0)\gamma(t)\} &= \{g(t) - \gamma(t)\}\gamma(0) - \{g(0) - \gamma(0)\}\gamma(t)\\ &= O(Q_T) + \left\{\sum_0^\infty \kappa(j)\kappa(j+t)\gamma(0) - \sum_0^\infty \kappa(j)^2\gamma(t)\right\}\\ &\quad\times\left\{\frac{1}{T}\sum_1^T \varepsilon(s)^2 - \sigma^2\right\} = O(Q_T)\end{aligned}$$

since $\gamma(t) = \sigma^2\sum\kappa(j)\kappa(j+t)$. Under (5.3.6)$'$ we obtain

$$O(Q_T) + \gamma(0)\gamma(t)\left\{T^{-1}\sum_{T-t+1}^{T} \varepsilon(s)^2/\sigma^2\right\} = O(Q_T) + o\{t\gamma(t)T^{-1/2}\},$$

so that, for $s = 1$, $t\gamma(t)$ bounded in t would be a sufficient condition in addition to (5.3.6)$'$ [i.e., (5.3.6)]. Since (5.3.6) holds in the ARMA, case, we have

Corollary 5.3.3. *Let $y(t)$ be ARMA for $s = 1$, (4.1.15), (4.1.17) hold and the fourth moment of $\varepsilon(t)$ be finite. Then for $H_T \le (\log T)^a$, $a < \infty$,*

$$\max_{0 \le t \le H_T} |r(t) - \rho(t)| = O(Q_T).$$

For later purposes we need the following

Theorem 5.3.4. *Under the conditions of Theorem 5.3.1,*

$$\max_{|t| \le H_T} \left\| \frac{1}{T} \sum_1^T u(s)v(s-t)' \right\| = O(Q_T), \qquad H_T \le (\log T)^a, \qquad a < \infty.$$

Proof

$$\frac{1}{T} \sum_1^{T-t} u(s)v(s+t)' = \sum_0^\infty K(a) \frac{1}{T} \sum_1^{T-t} \varepsilon(s-a)v(s-t)'.$$

However, as in the proof of Theorem 5.3.2, we need consider only terms for $a < c \log T$, for sufficiently large c. Now using Theorem 5.3.1, the result follows, for the change in the limits of summation introduces only the term

$$\frac{1}{T} \sum_{-a+1}^{0} \varepsilon(s)v(s+a-t)' = o(T^{-1} \log T).$$

Finally, in this section, it will be convenient to point out that the first part of Theorem 5.3.1 will hold when $u(t)$ is much more general than there considered.

Theorem 5.3.5. *Let $\varepsilon(t)$ satisfy (4.1.15), (4.1.16) and be ergodic with finite fourth moment and*

$$u(t) = \sum_0^\infty K(j)\varepsilon(t-j), \qquad \sum_0^\infty \|K(j)\|^2 < \infty.$$

Then

$$\max_{1 \le t \le H_T} \left\| \frac{1}{T} \sum_{t+1}^T \{\varepsilon(s)u(s-t)'\} \right\| = O(Q_T), \qquad H_T \le (\log T)^a, \qquad a < \infty.$$

The condition (4.1.16) is now used in place of (4.1.17). It is probable that the latter will suffice. The proof will be given after the proof of Theorem 5.3.1. The corresponding theorem for the autocovariances of $u(t)$ now does not hold because the argument around (5.3.4), allowing truncation of the sums at $c \log T$, would not be valid. Although $t < 0$ has not been considered in the theorem, it is clear that, for example,

$$\frac{1}{T}\sum_{1}^{T-t}\{\varepsilon(s)u(s+t)'-\Sigma K(t)'\}=O(Q_T),$$

and indeed that, suitably modified, the theorem would hold for $|t| \leq H_T$. However, we use below only the result in the stated theorem. We now go on to establish the main theorem. Some readers may prefer to go on to Section 4.

Proof of Theorem 5.3.1. If the result holds with the summands replaced by $\varepsilon_a(s)u_b(s-t)$, $\varepsilon_a(s)v_b(s-t)$, it holds in general, so that we may as well consider $s = m = 1$. Now

$$\frac{1}{T}\sum_{t+1}^{\infty}\varepsilon(s)u(s-t)=\sum_{0}^{\infty}\kappa(j)\frac{1}{T}\sum_{t+1}^{T}\varepsilon(s)\varepsilon(s-t-j) \tag{5.3.8}$$

and

$$\left|\kappa(j)\frac{1}{T}\sum_{t+1}^{T}\varepsilon(s)\varepsilon(s-t-j)\right|\leq c\rho^{j}, \quad \text{a.s.,} \qquad 0<\rho<1,$$

since the $\kappa(j)$ decrease geometrically, and

$$\left|\frac{1}{T}\sum_{t+1}^{T}\varepsilon(s)\varepsilon(s-t-j)\right|\leq\left\{\frac{1}{T}\sum_{1}^{T}\varepsilon(s)^2\right\}^{1/2}\left\{\frac{1}{T}\sum_{1}^{T}\varepsilon(s-j)^2\right\}^{1/2}\leq cj^{1/4}$$

by ergodicity, since

$$\frac{1}{T}\sum_{-j+1}^{T-j}\varepsilon(s)^2\leq c+\frac{1}{T}\sum_{-j+1}^{\min(0,T-j)}\varepsilon(s)^2\leq c(1+j^{1/2}),$$

by the finiteness of the fourth moment. Thus only the first $c\log T$ terms need to be considered on the right in (5.3.8). The same kind of argument applies to $T^{-1}\sum\varepsilon(s)v(s-t)$ and reduces consideration to that of

(5.3.9)

$$\frac{1}{T}\sum_{t+1}^{T}\varepsilon(s)\varepsilon(s-t), \quad t>0; \qquad \frac{1}{T}\sum_{t+1}^{T}\varepsilon(s)z(s-t), \qquad \frac{1}{T}\sum_{1}^{T-t}\varepsilon(s)z(s+t).$$

Consider the first part of (5.3.9). First we truncate $\varepsilon(t)$, replacing that by $d(t)=\varepsilon(t)I\{|\varepsilon(t)|\leq\lambda(|t|)^{1/2}\}$. Here $I\{\cdot\}$ is the indicator function. It is required that

$$\lambda(t) \uparrow \infty, \qquad \eta(t) = \lambda(t)Q_t \downarrow 0. \tag{5.3.10}$$

It is necessary to show that

(5.3.11)

$$\phi(T)^{-1}\left(\sum_1^{T-t} \varepsilon(s)\varepsilon(s+t) - \sum_1^{T-t} [d(s)d(s+t) - E\{d(s)d(s+t)|\mathscr{F}_{s+t-1}\}]\right)$$

converges a.s. to zero, uniformly in $t \le H_T$. As will be shown, this follows from

$$\sum_1^{\infty} E\{e(t)^2\}/\phi(t) < \infty, \qquad e(t) = \varepsilon(t) - d(t). \tag{5.3.12}$$

We first show (5.3.11) assuming (5.3.12) and then discuss (5.3.12). Observe that

$$\lambda(s+t)^{1/2} \le \lambda(s)^{1/2} + \lambda(t)^{1/2}$$

since this is true of $t^{1/2}$ and $\eta(t)/(\log \log t)^{1/2} \downarrow 0$. Now

$$\phi(T)^{-1} \sum_1^{T-t} e(s)e(s+t) \le \phi(T)^{-1} \sum_1^{T} e(s)^2$$

and this will converge to zero if $\sum e(s)^2/\phi(s) < \infty$, by Kronecker's lemma, and a.s. this will hold if (5.3.12) holds. We shall call this "this argument using Kronecker's lemma" for brevity. Similarly,

$$\left|\phi(T)^{-1} \sum_1^{T-t} e(s)d(s+t)\right| \le \phi(T)^{-1} \sum_1^{T-t} |e(s)|\lambda(s+t)^{1/2}$$

$$\le \phi(T)^{-1} \sum_1^{T} |e(s)|\{\lambda(s)^{1/2} + \lambda(t)^{1/2}\}$$

and the part involving $\lambda(s)^{1/2}$ will converge to zero by the argument involving Kronecker's lemma since $E\{|e(s)|\lambda(s)^{1/2}\phi(s)^{-1}\} \le E\{e(s)^2\}/\phi(s)$. Also, $\phi(T)^{-1}|e(s)|\lambda(t)^{1/2} \le \phi(T)^{-1}\lambda(T)^{1/2}|e(s)|$, $s \le T$ and $\sum \phi(s)^{-1}\lambda(s)^{1/2}|e(s)|$ converges, as has just been pointed out. Finally,

$$\phi(T)^{-1} \max_t \sum_1^{T} |e(s+t)d(s)| \le \phi(T)^{-1} \max_t \sum_1^{T} |e(s+t)|\lambda(s+t)^{1/2}$$

$$\le \phi(T)^{-1} \sum_1^{2T} |e(s)|\lambda(s)^{1/2} \to 0.$$

Thus

$$\max_t \phi(T)^{-1} \sum_1^{T-t} |e(s)d(s+t) + d(s)e(s+t) + e(s)e(s+t)| \to 0, \qquad \text{a.s.}$$

The proof has also shown that

$$\max_t \phi(T)^{-1} \sum_1^{T-t} E\{|e(s)d(s+t) + d(s)e(s+t) + e(s)e(s+t)|\} \to 0.$$

Now (5.3.11) is

$$\phi(T)^{-1} \sum_1^{T-t} [e(s)d(s+t) + d(s)e(s+t) + e(s)e(s+t) - E\{e(s)d(s+t) + d(s)e(s+t) + e(s)e(s+t)|\mathscr{F}_{s+t-1}\}]$$

and

$$E|E\{e(s)d(s+t) + d(s)e(s+t) + e(s)e(s+t)|\mathscr{F}_{s+t-1}\}| \leq E\{|e(s)d(s+t) + d(s)e(s+t) + e(s)e(s+t)|\}.$$

Thus (5.3.11) converges a.s. to zero uniformly in $t \leq H_T$ and the problem is reduced to the consideration of

$$\phi(T)^{-1} \sum_1^{T-t} [d(s)d(s+t) - E\{d(s)d(s+t)|\mathscr{F}_{s+t-1}\}].$$

The condition (5.3.12) has been shown to hold, for $\lambda(t)$ satisfying (5.3.10), in Hartman and Wintner (1941). Their proof relates to a sequence of identically distributed random variables with finite mean square and in our context this sequence is constituted by $\varepsilon(t)^2$. We do not reproduce their proof. In fact, a careful examination of the proof shows that we can take $\lambda(t) = cQ_t^{-1}$, for some $c > 0$, basically because we are not trying to evaluate the $O(Q_t)$ term very precisely. If $\lambda(t) = cQ_t^{-1}$, then (5.3.12) is easily established because

$$\sum_1^{\infty} E\{e(t)^2\}/\phi(t) = \sum_1^{\infty} \frac{1}{\phi(t)} \int_{\lambda(t)}^{\infty} x\, dF(x),$$

where $F(x)$ is the distribution function of $e(t)^2$. The last expression is

$$\sum_1^\infty \frac{t}{\phi(t)} \int_{\lambda(t)}^{\lambda(t+1)} x\, dF(x) \le \sum_1^\infty c \int_{\lambda(t)}^{\lambda(t+1)} x^2\, dF(x)$$
$$= c \int_{\lambda(1)}^{\infty} x^2\, dF(x) < \infty.$$

[For $t < 3$, $\log\log t$ is not defined. Then $\lambda(t)$, $\phi(t)$ may be allotted some arbitrary value.] However, we shall use the Hartman-Wintner result since the fact that $\eta(t)$ converges to zero makes some parts of the argument a little simpler. Put

$$s_t^2(T) = \sum_{t+1}^{T} (E\{d^2(s)d^2(s-t)|\mathscr{F}_{s-1}\} - [E\{d(s)d(s-t)|\mathscr{F}_{s-1}\}]^2).$$

This is the "increasing process" (Neveu, 1975) associated with the square-integrable martingale $S_t(T)$, where

$$S_t(T) = \sum_{t+1}^{T} Y_t(u), \qquad Y_t(u) = d(u)d(u-t) - E\{d(u)d(u-t)|\mathscr{F}_{u-1}\}.$$

Thus

$$s_t^2(T) = \sum_{t+1}^{T} E\{Y_t(u)^2|\mathscr{F}_{u-1}\}.$$

It will now be shown that, uniformly in $t \le H_T$,

$$s_t^2(T) = TE\{\varepsilon(0)^2\varepsilon(t)^2\}\{1 + o(1)\}. \tag{5.3.13}$$

First

$$\max_t T^{-1} \sum_1^{T-t} [E\{d(s)d(s+t)|\mathscr{F}_{s+t-1}\}]^2$$
$$= \max_t T^{-1} \sum_1^{T-t} d(s)^2 [E\{e(s+t)|\mathscr{F}_{s+t-1}\}]^2$$
$$\le \max_t T^{-1} \sum_1^{T-t} \lambda(s+t) [E\{e(s+t)|\mathscr{F}_{s+t-1}\}]^2$$
$$\le \max_t cT^{-1} \sum_1^{T-t} Q_{s+t}^{-1} [E\{e(s+t)|\mathscr{F}_{s+t-1}\}]^2,$$

and this converges to zero uniformly in t by the argument involving Kronecker's lemma. Thus

$$s_t^2(T) = \sum_1^{T-t} E\{d^2(s)d^2(s+t)|\mathscr{F}_{s+t-1}\} + o(T).$$

Also,

$$T^{-1}\sum_1^{T-t} e^2(s)E\{d^2(s+t)|\mathscr{F}_{s+t-1}\} \le T^{-1}\sum_1^{T-t} e^2(s)\lambda(s+t)$$
$$\le T^{-1}\sum_1^{T} e^2(s)Q_s^{-1} + (TQ_T)^{-1}\sum_1^{T}\varepsilon^2(s),$$

and each of these two terms converges to zero, the first by the argument involving Kronecker's lemma. Thus it is necessary to consider

$$\sum_1^{T-t}\varepsilon^2(s)[E\{\varepsilon^2(s+t)|\mathscr{F}_{s+t-1}\} - E\{e^2(s+t)|\mathscr{F}_{s+t-1}\}].$$

However,

$$\sum_1^{T-t} d^2(s)E\{e^2(s+t)|\mathscr{F}_{s+t-1}\} \le \sum_1^{T-t}\lambda(s)E\{e^2(s+t)|\mathscr{F}_{s+t-1}\}$$
$$\le \sum_1^{T}\lambda(s)E\{e^2(s)|\mathscr{F}_{s-1}\} = o(T)$$

by the same argument, so that we are reduced to

$$\sum_1^{T-t}\varepsilon^2(s)E\{\varepsilon^2(s+t)|\mathscr{F}_{s+t-1}\} - \sum_1^{T-t} e^2(s)E\{e^2(s+t)|\mathscr{F}_{s+t-1}\}.$$

However, if $e_b(s) = \varepsilon(s) - \varepsilon(s)I\{|\varepsilon(s)| \le b\}$, then $|e_b(s)| \ge e(s)$, at least for $\lambda(s)^{1/2} \ge b$ (i.e., for all but finitely many values). Moreover, by ergodicity,

$$\frac{1}{T}\sum_1^{T} e_b(s)^4 \to E\{e_b(0)^4\},$$

which may be made arbitrarily small by taking b large. Thus

$$\sum_1^{T-t} e(s)^4 = o(T)$$

and the same argument applies to show

$$\sum_{1}^{T-t} [E\{e^2(s+t)|\mathscr{F}_{s+t-1}\}]^2 \leq \sum_{1}^{T} E\{e^4(s)|\mathscr{F}_{s-1}\} = o(T).$$

Thus, uniformly in $t \leq H_T$,

$$s_t(T)^2 = \sum_{1}^{T-t} \varepsilon^2(s)E\{\varepsilon^2(s+t)|\mathscr{F}_{s+t-1}\} + o(T).$$

Now use (5.3.2) taking the two components of the vector $y(t)$, there referred to as $\varepsilon^2(s) - \sigma^2$, $E\{\varepsilon^2(s)|\mathscr{F}_{s-1}\} - \sigma^2$. These are jointly stationary, have finite variance, and are purely nondeterministic since, for example,

$$E[E\{\varepsilon^2(s)|\mathscr{F}_{s-1}\} - \sigma^2|\mathscr{F}_{-\infty}] = E\{\varepsilon^2(s)|\mathscr{F}_{-\infty}\} - \sigma^2 = 0,$$

by (4.1.17). Thus, uniformly in t, (5.3.13) holds. Now

$$\lim_{t\to\infty} E\{\varepsilon(0)^2\varepsilon(t)^2\} = \sigma^4. \tag{5.3.14}$$

Indeed,

$$E\{\varepsilon(0)^2\varepsilon(t)^2\} - \sigma^4 = E[\{\varepsilon(0)^2 - \sigma^2\}\varepsilon(-t)^2].$$

Put $\varepsilon_a(-t)^2 = \varepsilon(-t)^2 I\{\varepsilon(-t)^2 \leq a\}$. Then

$$E[\{\varepsilon(0)^2 - \sigma^2\}\{\varepsilon(-t)^2 - \varepsilon_a(-t)^2\}] = c_t, \qquad |c_t| \leq c_\infty,$$

where c_∞ may be chosen arbitrarily small by taking a large, as follows from Schwarz's inequality. Thus

$$E\{\varepsilon(0)^2\varepsilon(t)^2\} - \sigma^4 = E[E\{\varepsilon^2(0) - \sigma^2|\mathscr{F}_{-t}\}\varepsilon_a(-t)^2] + c_t.$$

However, $\xi(t) = E\{\varepsilon^2(0)|\mathscr{F}_{-t}\}$ is a reversed martingale relative to the decreasing sequence of σ-algebras $\mathscr{G}_t = \mathscr{F}_{-t}$ [i.e., $E\{\xi(t)|\mathscr{G}_{t+1}\} = \xi(t+1)$]. Also, $\xi(t)$ is square integrable and hence converges to $E\{\xi(t)|\mathscr{G}_\infty\} = E\{\varepsilon(0)^2|\mathscr{F}_{-\infty}\} = \sigma^2$, a.s. and in the L_1 norm (Neveu, 1975, p. 118). Thus

$$|E\{\varepsilon(0)^2\varepsilon(t)^2\} - \sigma^4| \leq aE[|E\{\varepsilon(0)^2|\mathscr{F}_{-t}\} - \sigma^2|] + |c_t| \to |c_t|.$$

Since c_∞ is arbitrarily small, (5.3.14) follows.

Now, using i.o. to stand for "infinitely often," consider

(5.3.15)

$$P\left\{\max_t S_t(T) > b\phi(T), \text{ i.o. in } T\right\}, \qquad P\left\{\max_t \{-S_t(T)\} > b\phi(T), \text{ i.o. in } T\right\}.$$

If both of these are shown to converge to zero, the first part of the theorem will be proved and since the proof is the same for each, we take the first. Introduce the random time T_k, which is the smallest T such that

$$\max_{t \le H_{T+1}} s_t^2(T+1) \ge p^{2k}, \tag{5.3.16}$$

where $p > 1$ and for convenience $p > e$. Put $H^{(k)}$ for H_T at $T = T_k$. Now from (5.3.13), (5.3.14),

$$T_k \le c_0 p^{2k}\{1 + o_k(1)\},$$

where c_0^{-1} is a lower bound for $E\{\varepsilon(0)^2\varepsilon(t)^2\}$ and $o_k(1) \to 0$ as $k \to \infty$. Thus $H^{(k)} \le c_1 k^a\{1 + o_k(1)\}$, where $c_1 = (2\log p)^a > 0$. It is evident that T_k is a stopping time for the martingales $S_t(T)$, that is, the event $T_k = s$ is measurable $\mathscr{F}_s$ for each s. [It was for this reason that $s_t^2(T+1)$ occurs in (5.3.16) and not $s_t^2(T)$, because $s_t^2(T+1)$ is measurable $\mathscr{F}_T$.) Put $S_t^{(k)}(T) = S_t(T)$, $T \le T_k$, $S_t^{(k)}(T) = S_t(T_k)$, $T > T_k$. Then $S_t^{(k)}(T)$ is again a martingale relative to the $\mathscr{F}_T$ σ-algebras (Stout, 1974, p. 38). Now it is sufficient to show that

$$P\left\{\sup_T \max_{t \le H^{(k)}} S_t^{(k)}(T) > b\phi(T_{k-1} + 1), \text{ i.o in } k\right\} = 0 \tag{5.3.17}$$

since if the events involved in the first part of (5.3.15) happen, they must happen in infinitely many of the intervals (T_{k-1}, T_k) and $H_T \le H^{(k)}$ in this interval, while $\phi(T) \ge \phi(T_{k-1} + 1)$. Now replace $Y_t(u)$ in (5.3.17) by

(5.3.18)

$$Y_t(u)' = Y_t(u) I\left[\lambda(u)^2 \le \left\{\left(\max_{t \le H_u} s_t(u)^2\right) \Big/ \left(\log\log \max_{t \le H_u} s_t(u)^2\right)\right\} \eta c_1\right]$$

and call $S_t^{(k)}(T)'$ the corresponding expression replacing $S_t^{(k)}(T)$. The event under the indicator function symbol fails to happen only finitely often in u, $t \le H_u$, by (5.3.13), (5.3.14) for appropriate values of $c_1 < \infty$, $\eta > 0$, and thus the problem has not been essentially altered. Also, we may replace $T_{k-1} + 1$ by cp^{2k} and $H^{(k)}$ by $P_k = c_1 k^a$, choosing $c_1 < \infty$ appropriately. This follows because the events $T_{k-1} + 1 > c_1 p^{2k}$, $H^{(k)} > P_k$ happen only finitely often in k, a.s. Indeed, if

$$\limsup_{k\to\infty} \max_{t\le P_k} S_t^{(k)}(T)'/\phi(c_1 p^{2k}) \le b, \quad \text{a.s.,} \tag{5.3.19}$$

then also

$$\limsup_{k\to\infty} \max_{t\le H^{(k)}} S_t^{(k)}(T)/\phi(T_{k+1}+1) \le b, \quad \text{a.s.}$$

since the two sequences differ for only finitely many values of k, a.s. It is still true that $E\{Y_t(u)'|\mathscr{F}_{u-1}\} = 0$ since the event under the indicator function symbol in (5.3.18) is in $\mathscr{F}_{u-1}$. Thus consider

$$\begin{aligned}&\sum_1^\infty P\left\{\sup_T \max_{t\le P_k} S_t^{(k)}(T)' > b\phi(c_1 p^{2k})\right\}\\ &\quad\le \sum_1^\infty \sum_1^{P_k} P\left\{\sup_T S_t^{(k)}(T)' > b\phi(c_1 p^{2k})\right\}\end{aligned} \tag{5.3.20}$$

because $P\{\max X_j > c\} \le \sum P(X_j > c)$ for any X_j. If (5.3.20) is finite, then (5.3.19) holds, by the Borel-Cantelli lemma. Now we follow Stout (1974, p. 301) precisely but, somewhat redundantly, write out the details to make reference to that work easier. [It is for this reason also that p^{2k} was used in (5.3.16), rather, say, than 2^k.]

$$P\left\{\sup_T S_t^{(k)}(T)' > b\phi(c_1 p^{2k})\right\} = P\left\{\sup_T S_t^{(k)}(T)' > b'\lambda_k p^{2k}\right\},$$
$$\lambda_k = (\log\log c_1 p^{2k})^{1/2}/p^k, \qquad b' = c_1 b.$$

This probability equals

$$\begin{aligned}&P\left[\sup_T \exp\{\lambda_k S_t^{(k)}(T)'\} > \exp(b'\lambda_k^2 p^{2k})\right]\\ &\quad\le P\left[\sup_T V_t^{(k)}(T) > \exp\{b'\lambda_k^2 p^{2k} - \tfrac{1}{2}\lambda_k^2(1+\mu_k\lambda_k/2)s_k^2(T_k)\}\right],\end{aligned} \tag{5.3.21}$$

where

$$V_t^{(k)}(T) = \exp\{\lambda_k S_t^{(k)}(T)' - \tfrac{1}{2}\lambda_k^2(1+\mu_k\lambda_k/2)s_t^{(k)}(T)^2\}.$$

Here

$$s_t^{(k)}(T)^2 = \sum_{t+1}^{\min(T, T_k)} (E\{Y_t'(u)^2|\mathscr{F}_{u-1}\} - [E\{Y_t(u)'|\mathscr{F}_{u-1}\}]^2),$$

while u_k bounds $Y_t(u)'$, $u \le T_k$. We see that $s_t^{(k)}(T)^2 \le s_t(T_k)^2$, a.s. and

$$\mu_k \le \eta c_1 p^{2k}/(\log\log p^{2k})$$

by (5.3.18) and the fact that $s_t^2(T_k) \le p^{2k}$, by the choice of T_k. Thus $\lambda_k \mu_k < 1$ by an appropriate choice of η. Moreover, as has been said,

$$\max_t s_t^2(T_k) \le p^{2k}$$

by (5.3.16), so that in place of (5.3.21) we may consider

$$P\left\{\sup_T V_t^{(k)}(T) > \exp\{b'\lambda_k^2 p^{2k} - \tfrac{1}{2}\lambda_k^2(1 + \mu_k\lambda_k/2)p^{2k}\}.\right.$$

Now by Stout (1974, Th. 5.4.1, p. 209), $V_t^{(k)}(T)$ is, for each k, t a positive super martingale and by Corollary 5.4.1, on the same page of this reference, the last probability is bounded by

(5.3.22) $\quad \exp - \{b'\lambda_k^2 p^{2k} - 1/2\lambda_k^2(1 + \lambda_k\mu_k/2)p^{2k}\} \le \exp - \{b'' \log\log c_1 p^{2k}\},$

where $b'' = b' - 1/2(1 + \lambda_k\mu_k/2) \ge b' - 1$. Thus (5.3.20) is bounded by

$$c\sum_1^\infty k^a k^{-b''} < \infty,$$

since $P_k \le c_1 k^a$, the inequality following by taking b'' (i.e., b', i.e., b) sufficiently large. This completes the proof of the first half of the theorem.

The proof of the second part of the theorem is almost the same. Consider, for example,

$$\frac{1}{T}\sum_{t+1}^{T} \varepsilon(s)z(s+t).$$

Since $z(t)$ is stationary with finite fourth moment, the truncation argument is, as before, putting $d_z(t) = z(t)I\{|z(t)| \le \lambda(t)^{1/2}\}$, $d_z(t) + e_z(t) = z(t)$, and, as before, obtaining

(5.3.12)$'$ $\quad \sum_1^\infty E\{e_z(t)^2\}/\phi(t) < \infty,$

using $\lambda(t)$ [i.e., $\eta(t)$] suitable for both $z(t)$ and $\varepsilon(t)$. The remainder of the proof of is almost the same as that for the first part of the theorem except for the proof that, uniformly in t,

(5.3.13)′
$$\lim_{T\to\infty} T^{-1}\sum_{t+1}^{T}(E\{d^2(s)|\mathscr{F}_{s-1}\}d_z^2(s-t)-[E\{d(s)|\mathscr{F}_{s-1}\}]^2 d_z(s-t)^2)$$
$$=\sigma^2 E\{z(0)^2\}.$$

Now

$$T^{-1}\sum_{t+1}^{T}E\{e(s)^2|\mathscr{F}_{s-1}\}d_z^2(s-t)\le cT^{-1}\sum_{t+1}^{T}E\{e(s)^2|\mathscr{F}_{s-1}\}\lambda(s-t)$$
$$< cT^{-1}\sum_{1}^{\infty}E\{e(s)^2|\mathscr{F}_{s-1}\}Q_s^{-1}\to 0,$$

since $\lambda(s)=Q_s^{-1}\eta(s)$, $\eta(s)\downarrow 0$, and Q_s/Q_{s-t} is uniformly bounded for $t\le H_T$. Thus we need to consider

$$T^{-1}\sum_{t+1}^{T}[E\{\varepsilon^2(s)|\mathscr{F}_{s-1}\}-\sigma^2]d_z^2(s-t) \tag{5.3.23}$$

since

$$\lim_{T\to\infty}\frac{1}{T}\sum_{1}^{T}d_z^2(s-t)=\lim_{T\to\infty}\left\{\frac{1}{T}\sum_{1}^{T-t}z^2(s)-\frac{1}{T}\sum_{1}^{T-t}e_z^2(s)\right\}$$
$$=E\{z(0)^4\}$$

by ergodicity and (5.3.12)′. Now, putting [see Section 1.4, at and followng (1.4.8)]

$$u(s,s-k)=E\{\varepsilon^2(s)|\mathscr{F}_{s-k}\}-E\{\varepsilon^2(s)|\mathscr{F}_{s-k-1}\},$$

then

$$E\{\varepsilon^2(s)|\mathscr{F}_{s-1}\}=\sum_{k=1}^{\infty}u(s,s-k)+\sigma^2,\qquad \sum_{1}^{\infty}E\{u(s,s-k)^2\}<\infty,$$

because of (4.1.17). The $u(s, s-k)$ are martingale differences relative to $\mathscr{F}_{s-k}$ for each fixed k. Thus (5.3.23) becomes

$$\sum_{k=1}^{\infty} \frac{1}{T} \sum_{t+1}^{T} u(s, s-k) d_z^2(s-t)$$

and

$$\begin{aligned}
&\frac{1}{T} \sum_{t+1}^{T} \left\{ \sum_{k=a}^{\infty} u(s, s-k) \right\} d_z^2(s-t) \\
&\leq \left[\frac{1}{T} \sum_{t+1}^{T} \left\{ \sum_{k=a}^{\infty} u(s, s-k) \right\}^2 \frac{1}{T} \sum_{t+1}^{T} d_z^4(s-t) \right]^{1/2} \\
&\leq \left[\frac{1}{T} \sum_{1}^{T} \left\{ \sum_{a}^{\infty} u(s, s-k) \right\}^2 \frac{1}{T} \sum_{1}^{T} z^4(s) \right]^{1/2} \\
&= \left[E\{z(0)^4\} \sum_{k=a}^{\infty} E\{u(s, s-k)^2\} \right]^{1/2} \{1 + o(1)\}
\end{aligned}$$

uniformly in t, by ergodicity. This may be made arbitrarily small by taking a large. Thus it is necessary only to show that for any fixed k, uniformly in t,

$$\lim_{T \to \infty} \frac{1}{T} \sum_{t+1}^{T} u(s, s-k) d_z^2(s-t) = 0, \quad \text{a.s.} \tag{5.3.24}$$

to establish (5.3.13)′ and hence the theorem. Now

$$\begin{aligned}
&P\left\{ \max_{|t| \leq H_T} \left| \frac{1}{T} \sum_{t+1}^{T} u(s, s-k) d_z^2(s-t) \right| \geq \delta \right\} \\
&\leq H_T \sum_{t+1}^{T} d_z^4(s-t) E\{u(s, s-k)^2\} / \delta^2 T^2 \leq c H_T / \delta^2 T \\
&\leq c (\log T)^a / T.
\end{aligned}$$

Thus along the subsequence $T = n^2$, then (5.3.24) holds by the Borel-Cantelli lemma, since the series $n^{-2}(\log n^2)^a$ converges. Moreover,

$$\begin{aligned}
&\max_{n^2 \leq T < (n+1)^2} \max_{|t| \leq H_T} \left| \frac{1}{T} \sum_{t+1}^{T} u(s, s-k) d_z^2(s-t) - \frac{1}{n^2} \sum_{t+1}^{n^2} u(s, s-k) d_z^2(s-t) \right| \\
&\leq \max_{T} \max_{|t|} \left| \frac{1}{T} \sum_{n^2+1}^{T} u(s, s-k) d_z^2(s-t) \right| \qquad (5.3.25) \\
&\quad + \max_{T} \max_{|t|} \left| \left(\frac{1}{n^2} - \frac{1}{T} \right) \sum_{t+1}^{n^2} u(s, s-k) d_z^2(s-t) \right|
\end{aligned}$$

and the second term converges to zero by what has been proved since $n^{-2} - T^{-1} \leq (2n+1)/n^4$. Moreover, using Doob's inequality (Hall and Heyde, 1980, p. 15) for the martingale,

$$\sum_{n=1}^{\infty} P\left\{\max_T \max_{|t|} \left|\frac{1}{T}\sum_{n^2+1}^{T} u(s, s-k)d_z^2(s-t)\right| > \delta\right\}$$
$$\leq \sum_{n=1}^{\infty} \sum{}' \frac{4}{n^4} \sum_{n^2+1}^{(n+1)^2} E\{u(s, s-k)^2\}d_z^4(s-t)/\delta^2,$$

where $\sum'$ is a sum over $-H_T \leq t \leq H_T$, for $T = (n+1)^2$. Thus the following bound is obtained:

$$\sum_1^{\infty}\sum_t \frac{c}{n^4} \sum_{n^2+1}^{(n+1)^2} z^4(s-t) \leq c \sum_1^{\infty} z^4(t)/t^3 < \infty, \quad \text{a.s.}$$

(The last inequality is fairly gross and we leave the reader to check it.) Thus (5.3.25) converges to zero and Theorem 5.3.1 is established.

The proof of Theorem 5.3.5 is almost the same. We observe that taking $s = 1$, for example, and using (4.1.15), (4.1.16), yields

$$E\{u(s)^4\} = \sigma^4\left(\sum_0^{\infty} \kappa(j)^2\right)^2 + \kappa_4 \sum_0^{\infty} \kappa(j)^4 < \infty,$$

where κ_4 is the fourth cumulant for $\varepsilon(t)$. A slightly more elaborate calculation shows the fourth moments to be finite also for $s > 1$. Because of (4.1.16), then [taking $u(s)$ as $z(s)$ in the second part of the proof] (5.3.23) is zero and the proof goes through as before.

4. ASYMPTOTIC PROPERTIES OF ORDER ESTIMATES FOR $s = 1$

Now $s = 1$ but $m \geq 0$, so that

$$\sum_0^{p} \alpha(j)y(t-j) = \sum_1^{r} \Delta(j)z(t-j) + \sum_0^{q} \beta(j)\varepsilon(t-j),$$

where $\Delta(j)$ is $1 \times m$. Consider the maximization of the Gaussian likelihood or, rather, the minimization of $L_T(\theta)$ [see (4.2.5)]

$$L_T(\theta) = T^{-1}\log\det\{\Gamma_T(\theta)\} + T^{-1}u_T'\Gamma_T^{-1}(\theta)u_T, \qquad (5.4.1)$$
$$u_T' = (u(1), u(2), \ldots, u(T)),$$

Thus

$$u(t) = y(t) - v(t), \qquad v(t) = \sum_{1}^{\infty} L(j)z(t-j), \qquad z(t) = 0, \quad t \leq 0,$$

where the $L(j)$ are also $1 \times m$. Let $\hat{\sigma}^2_{p,q,r}$ be the maximum likelihood estimate of σ^2. Then p, q, r are chosen to minimize

$$\log \hat{\sigma}^2_{p,q,r} + (p + q + rm)C_T/T, \qquad p \leq P, \quad q \leq Q, \quad r \leq R. \tag{5.4.2}$$

In fact, P, Q, R may be allowed to increase with T, but that will be discussed later; for the moment it is assumed that $p_0 \leq P$, $q_0 \leq Q$, $r_0 \leq R$. It has not been found possible to prove a theorem about $\hat{p}$, $\hat{q}$, $\hat{r}$ that minimize (5.4.2) without another condition on θ, or rather on

$$\tau' = (\alpha(1), \alpha(2), \ldots, \alpha(p), \beta(1), \ldots, \beta(q), \Delta(1), \ldots, \Delta(r)). \tag{5.4.3}$$

Thus it is assumed, for δ known a priori,

(5.4.4) $a_0(z)$, $b_0(z)$, $d_0(z)$ are coprime and $a_0(z) \neq 0$, $|z| \leq 1$, $b_0(z) \neq 0$, $|z| < 1 + \delta$, $\delta > 0$.

It is the last part of (5.4.4) that is the extra condition. It seems probable that a theorem can be proved with $\delta = 0$, but the proof would be complex for the following reason. In optimizing (5.4.1) we shall search a region for which (5.4.4) holds for $a(z)$, $b(z)$, $d(z)$. If $\delta = 0$ were allowed, then, as we shall see, once $p > p_0$, $q > q_0$, $r > r_0$ there will be zeros of $\hat{a}(z)$, $\hat{b}(z)$ that will approach the unit circle, at values nearly common to the two polynomials, and as $T \to \infty$ the behavior of these zeros will be erratic since they may vary around the circle. Heuristically, it can be seen why this might happen from the following. (In remarks following the proof of Theorem 5.4.2, below these heuristics will be supported.) Consider $p_0 = q_0 = 0$, $r_0 = r = 0$, $p = q = 1$. Then, as has been seen in Chapter 4, the optimization of the likelihood is essentially equivalent to the optimization of

$$Q(\alpha, \beta) = \frac{1}{2\pi} \int_{-\pi}^{\pi} I(\omega) \frac{|1 + \alpha e^{i\omega}|^2}{|1 + \beta e^{i\omega}|^2} \, d\omega.$$

From Chapter 4 it is known that $\hat{\alpha}$, $\hat{\beta}$ optimizing this must satisfy $|\hat{\alpha} - \hat{\beta}| \to 0$, since $\hat{k} \to 1$, uniformly on compact subsets of the open unit disk. If $|\beta|$ stays bounded away from 1, then

$$|(1 + \alpha e^{i\omega})/(1 + \beta e^{i\omega})|^2 - 1 = (\alpha - \beta)\{\alpha + \beta + 2\cos\omega\}/\{1 + \beta^2 + 2\beta\cos\omega\}$$

converges to zero uniformly in ω as $(\alpha - \beta) \to 0$. However, say sequences α_T, β_T are chosen as

$$\alpha_T = -1 + \varepsilon_T, \qquad \beta_T = -1 + \eta_T, \qquad \varepsilon_T, \eta_T \to 0, \qquad \varepsilon_T/\eta_T \to 0.$$

Then

$$|(1 + \alpha_T e^{i\omega})/(1 + \beta_T e^{i\omega})|^2 - 1 \to 0$$

uniformly in $|\omega| \geq \varepsilon > 0$, but at $\omega = 0$ the expression is $\varepsilon_T^2/\eta_T^2 \to 0$. Thus we shall tend to reduce the value of $Q(\alpha, \beta)$, given that $(\alpha - \beta)$ is near zero, by taking α, β both near -1 (or $+1$) since this may produce a "notch" in $|[1 + \alpha \exp(i\omega)]/[1 + \beta \exp(i\omega)]|^2$ at $\omega = 0$ (or π) that takes this function down from a value near unity at all other ω to a value near zero. Thus it can be expected that $\hat{\alpha}$, $\hat{\beta}$ will, in the case considered, fluctuate as $T \to \infty$ from values near $(-1, -1)$ to values near $(1, 1)$. What actual values will arise will be difficult to determine because $I(\omega)$ near $0, \pi$ will be wildly fluctuating, with many relative maxima near these points. Here we have neglected the fact that $y(t)$ will be mean corrected. Presumably, when this is done, $\hat{\alpha}, \hat{\beta} \to -1$. However, for $p - p_0 > 1$, $q - q_0 > 1$ there is a much greater range of possibilities, in any case.

If $\delta = 0$ were allowed, it would be necessary accurately to assess the effects of this notch. [Of course, for $\delta > 0$ the notch cannot appear since $(\hat{\alpha} - \hat{\beta}) \to 0$ then implies that $\hat{\alpha}$ is bounded away from ± 1.] The form of (5.4.2) shows that this assessment would need to be to an order $o(C_T/T)$ [e.g., $o(\log T/T)$ for BIC]. The task will not be attempted here. However, in Chapter 6 an algorithm of a special type for the optimization of (5.4.1) will be considered. Such an explicit formula is often easier (in some ways) to handle than the estimates implicitly defined by optimizing (5.4.1) and $\delta > 0$ will then not be required. However, there are, then, other conditions, on C_T, for example.

These considerations also show the need for the introduction of new coordinates (in the τ-space), which, in the case just studied, would be $\chi = (\alpha - \beta)/2$, $\psi = (\alpha + \beta)/2$. Then we know $\hat{\chi} \to 0$, a.s., but ψ will have to be treated as freely varying and an expansion about $\psi = \hat{\psi}$ will not be useful. This way of proceeding was first suggested in Hannan (1980a), as also were the coordinates now introduced (see also Deistler and Hannan, 1981). New coordinates are needed only for $\rho = \min(p - p_0, q - q_0, r - r_0) > 0$. Consider the equivalence class in $\mathbf{R}^{p+q+rm}$ corresponding to (k_0, l_0). [For the moment ignore the restrictions (5.4.4).] Then k_0, l_0 will be the image of some points in $\mathbf{R}^{p+q+rm}$. These points are evidently those for which $a = ha_0, b = hb_0, d = hd_0$, where h is of degree $\leq \rho$ and has $h(0) = 1$. As $h - 1$ (i.e., its ρ coefficients) varies the vector, τ, of parameters varies over an affine subspace of $\mathbf{R}^{p+q+rm}$ of dimension ρ. Indeed, this subspace will consist of vectors of the form

$$\psi = A_0 \tilde{h} + \tau_0,$$

where $\tilde{h}$ is the vector of the ρ coefficients of $h - 1$ and A_0 depends only on a_0, b_0, d_0. We shall use ψ for a representative vector in this ρ-dimensional space. In the space orthogonal to the ψ space we may institute coordinates, χ, of dimension $N = (p + q + rm) - \rho$. Then $\chi = 0$ implies that the image of the point (χ, ψ) in $\mathbf{R}^{p+q+rm}$ is k_0, l_0 and since it is known that $(\hat{k}, \hat{l}) \to (k_0, l_0)$, then $\hat{\chi} \to 0$, a.s., under the conditions of Theorem 4.2.1. It is therefore $\hat{\psi}$ whose behavior has to be carefully handled. Of course, $\tilde{h}$ (i.e., h) cannot vary arbitrarily since $b = hb_0$ must satisfy the last part of (5.4.4) [i.e., h must itself satisfy the last part of (5.4.4)]. *Nevertheless, ψ varies over a compact subset of $\mathbf{R}^\rho$ since evidently $\tilde{h}$ does so.*

The following theorem is essentially due to Hannan (1980a) but is here improved using An, Chen, and Hannan (1982).

Theorem 5.4.1. *Let $y(t)$ be generated by an ARMAX process with $s = 1$, as in Theorem 4.2.1, and let (5.4.4) be known to hold. Let $\hat{p}$, $\hat{q}$, $\hat{r}$ be obtained by minimizing (5.4.2) subject to the known restriction $p_0 \le P$, $q_0 \le Q$, $r_0 \le R$. Let (4.1.15), (4.1.17) hold and $\varepsilon(t)$ have finite fourth moment. Let $z(t)$ be stationary and ergodic with finite fourth moment and (4.1.6) hold. Then there are c_0, c_1, $0 < c_0 \le c_1 < \infty$, such that the following are true:*

(a) If $\liminf C_T/(2 \log\log T) > c_1$, then $(\hat{p}, \hat{q}, \hat{r}) \to (p_0, q_0, r_0)$, a.s. When (4.1.16) holds we may take $c_1 = 1$.

(b) If $\limsup C_T/(2 \log\log T) < c_0$, then a.s. convergence of $(\hat{p}, \hat{q}, \hat{r})$ to (p_0, q_0, r_0) fails. When (4.1.16) holds, we may take $c_0 = 1$.

(c) If $C_T \uparrow \infty$, then $(\hat{p}, \hat{q}, \hat{r}) \to (p_0, q_0, r_0)$ in probability.

(d) If $\limsup C_T < \infty$, then convergence in probability of $(\hat{p}, \hat{q}, \hat{r})$ to (p_0, q_0, r_0) fails unless $P = p_0$, $Q = q_0$, $R = r_0$.

The fact that strong consistency holds for $C_T/\log\log T$ sufficiently large, or when (4.1.16) holds when that is eventually greater than 2, is not of itself a reason for using $C_T = c \log\log T$. For one thing $\log\log T$ increases too slowly with T for the asymptotic theory to have much meaning. Some other remarks relating to this theorem are made at the conclusion of the proof. Before going on to the proof, we state another result.

Theorem 5.4.2. *Let the conditions of Theorem 5.4.1 hold and also (4.1.16). Assume that $p_0 < P$, $q_0 < Q$, $r_0 < R$ and*

$$\limsup_{T \to \infty} C_T < \infty.$$

Then

$$\lim_{\delta\to 0}\lim_{T\to\infty} P\{\hat{p} > p_0, \quad \hat{q} > q_0, \quad \hat{r} > r_0\} = 1.$$

This result says that in a certain sense not only is AIC inconsistent but it is also certain to over estimate the orders. The theorem will be complemented by some precise evaluations for the ARMA case when $p = p_0$, $q > q_0$ or $p > p_0$, $q = q_0$, in Theorem 5.6.1 below. *Again, however, we stress that Theorems 5.4.1 and 5.4.2 do not constitute a condemnation of AIC, which is designed for a situation where there is no true rational transfer function model.* The result in Theorem 5.4.2 probably holds without (4.1.16). The result is very asymptotic, as the proof shows, and may be relevant only for very large T. The theorem is derived from a more precise result relating to a special case that is of independent interest. Consider the model

$$\begin{gathered} y(t) = x(t) + \eta(t), \qquad x(t) = \alpha x(t-1) + \xi(t), \qquad |\alpha| < 1. \\ E\{\xi(t)\} = E\{\eta(t)\} = 0, \qquad E\{\xi(s)\xi(t)\} = \delta_{st}\sigma_\xi^2, \\ E\{\eta(s)\eta(t)\} = \delta_{st}\sigma_\eta^2, \qquad E\{\xi(s)\eta(t)\} \equiv 0. \end{gathered} \tag{5.4.5}$$

Then (5.4.5) expresses $y(t)$ as a (1, 1) ARMA process (Hannan, 1970, p. 171)

(5.4.6)

$$\begin{gathered} y(t) - \alpha y(t-1) = \varepsilon(t) + \beta\varepsilon(t-1), \qquad |\beta| < 1, \qquad E\{\varepsilon(s)\varepsilon(t)\} = \delta_{st}\sigma^2 \\ \alpha\beta \geq 0, \qquad |\beta| \leq |\alpha|. \end{gathered}$$

Equality holds in the last two cases only for $\alpha = 0$. If $E\{\xi(s)\eta(t)\} = \delta_{st}\sigma_{\xi\eta}$ is allowed, the condition on β becomes just $|\beta| \leq 1$. The likelihood ratio test for $\alpha = 0$ is based on the statistic

$$\lambda_T = -2(\log \text{likelihood ratio})$$

and is equivalent to testing $(\alpha, \beta) \neq (0, 0)$ against $\alpha = \beta = 0$, in (5.4.6).

Theorem 5.4.3. *Let (5.4.6) hold with $\varepsilon(t)$ ergodic and satisfying (4.1.15), (4.1.16) and have finite fourth moment. Let the likelihood be optimized subject to (5.4.4). Then when $\alpha = \beta = 0$, the distribution of λ_T converges to that of*

$$\max_{-v\leq \nu \leq v} \phi(\nu)^2, \qquad v = \log\{(2+\delta)/\delta\},$$

where $\phi(v)$ is a stationary Gaussian process on the real line with zero mean and absolutely continuous spectrum having density $(\cosh \omega\pi)^{-1}$, $-\infty < \omega < \infty$. *As $\delta \to 0$, $v \to \infty$, and (Berman, 1971),*

$$\lim_{v\to\infty} P\left[(2\log 2v)^{1/2}\left\{\max_{-v\le v\le v} |\phi(v)| - (2\log 2v)^{1/2}\right\} \le x - \log 8\pi\right]$$
$$= \exp(-2e^{-x}).$$

Theorem 5.4.2 will be proved using Theorem 5.4.3. The result is of interest because the likelihood ratio statistic has a very unusual distribution. If σ_η^2 in (5.4.5) is small relative to σ^2, then β will be near to zero and λ_T near to $Tr(1)^2$, but the significance point will be of order $2\log[\log\{(2+\delta)/\delta\}]$ as $\delta \to 0$. However, even for $\delta = 0.01$, this is only 4.7. The proof of Theorem 5.4.3 shows that $(\hat{\alpha}, \hat{\beta})$ will be near a point $(1, 1)$ or $(-1, -1)$, and supports the heuristic argument given below (5.4.4). The proof, as well as these arguments, shows how difficult the optimization will become if δ is too small. This will be discussed again at the end of the proof. Of course, in practice there will be no true rational transfer function system and the problem may not be as great as theory suggests. The reader may consult Veres (1987) for a generalization of Theorem 5.4.3, which is due to Hannan (1982), to the case

$$(p, q, r) = (p_0 + 1, q_0 + 1, r_0 + 1)$$

for any p_0, q_0, r_0.

Now the proofs of the theorems will be given.

Proof of Theorem 5.4.1. A technical point is dealt with first. It will be assumed when proving (a) and (c) of the theorem that for $p \ge p_0$, $q \ge q_0$, $r \ge r_0$, $\partial_\chi L_T(\hat{\tau}) = 0$, where $\partial_\chi L_T(\hat{\tau})$ is the row vector of derivatives with respect to the elements of χ, evaluated at the optimising point, $\hat{\tau}$. This could fail to be true only because $\hat{\tau}$ is at a point for which $\hat{b}$ [i.e., $b(z)$ at $\hat{\tau}$] has a zero of modulus $1 + \delta$. Indeed, $\hat{\tau}$ eventually cannot be at a point where $\hat{\alpha}(z)$ has a zero on $|z| = 1$ since $\hat{k} \to k_0$ [see the proof below in relation to the proof that $\partial \log P_T(\tau) = O(1)$]. At a point τ where $a(z)$, $b(z)$, $d(z)$ are not prime, but $\hat{b}$ has no zero on $|z| = 1 + \delta$, then if $\partial_\chi L_T(\tau) \ne 0$ a slight variation in χ, leaving ψ fixed, will move τ to a point satisfying (5.4.4) and lowering $L_T(\tau)$ so that $\partial_\chi L_T(\hat{\tau}) \ne 0$ cannot arise for this reason. Consider the region, $\mathscr{S}_\varepsilon$ composed of all points (χ, ψ) for which $(0, \psi)$ corresponds to a, b, d satisfying (5.4.4) for δ replaced by $\varepsilon > \delta$ (i.e., for which h has a zero of modulus no greater than $1 + \varepsilon$) and for which $\|\chi\| \le d$, $d > 0$. Then $\mathscr{S}_\varepsilon$ may contain points outside the region, $\mathbf{R}_\delta$ defined by (5.4.4), for $\chi \ne 0$ and ψ such that $(0, \psi)$ is near the boundary of (5.4.4). However, $\mathscr{S}_\varepsilon \cup \mathbf{R}_\delta$ for ε, d sufficiently small, certainly lies in some $\mathbf{R}_{\delta'}$ for

δ' slightly smaller than δ. Moreover if optimization is over such $\mathscr{S}_\varepsilon \cup \mathbf{R}_\delta$, then the optimum point, $\hat{\tau}$, must now, for τ large enough, be at a point for which $\partial_\chi L_T(\hat{\tau}) = 0$, for the same reason as before, namely that $\hat{\tau}$ must be, for large T, at a point near the affine subspace $\psi = 0$ and if $\partial_\chi L_T(\hat{\tau}) \neq 0$, then a small variation in χ will move $\hat{\tau}$ to a point for lower $L_T(\tau)$. If it is shown that p_0, q_0, r_0 is asymptotically preferred when optimization is over $\mathscr{S}_\varepsilon \cup \mathbf{R}_\delta$, the same must be true for $\mathbf{R}_\delta$ since we shall show that $L_T(\tau) = \log \hat{\sigma}^2_{p,q,r} + O(T^{-1})$, uniformly in τ, so that moving to the smaller region $\mathbf{R}_\delta$, can increase (5.2.1) by at most $O(T^{-1})$ and the proof of (a), (c) of the theorem shows that quantities of this order are negligible. For parts (b), (d) of the theorem (which may be called the necessity parts) we may consider a region $\mathscr{S}_\varepsilon \cup \mathbf{R}_{\delta'}$, for δ' slightly larger than δ and d sufficiently small, that lies within $\mathbf{R}_\delta$. Now if $(p, q, r) \neq (p_0, q_0, r_0)$ is asymptotically preferred to p_0, q_0, r_0, then by the same kind of argument as before, it will be asymptotically preferred when $\mathbf{R}_\delta$ is considered. *Thus in future we act as if* $\partial_\chi L_T(\hat{\tau}) = 0$, *without further comment. We also retain* $(1 + \delta)$ *as the bound on the zero of* $b(z)$ *to avoid a more complicated notation.*

Now examining (5.4.1) it is seen that $\Gamma_T(\theta) = \sigma^2 P_T(\tau)$, where $P_T(\tau)$ depends only on $k(z)$ and not actually on τ. Thus

$$\hat{\sigma}^2_{p,q,r}(\tau) = T^{-1} u'_T P_T^{-1}(\tau) u_T$$

and τ is determined by minimizing

$$L_T(\tau) = T^{-1} \log \det P_T(\tau) + \log \hat{\sigma}^2_{p,q,r}(\tau).$$

Of course u_T depends on τ through $v(t)$. It is known from the proofs in Section 4.2, which would now be much easier because of (5.4.4), that $\hat{\sigma}^2_{p,q,r}(\tau)$ converges to

$$\frac{\sigma_0^2}{2\pi} \int_{-\pi}^{\pi} |k_0|^2 |k_\tau|^{-2} \, d\omega + \int_{-\pi}^{\pi} |k_\tau|^{-2} (l_\tau - l_0) \, dF_z(\omega) (l_\tau - l_0)^* \tag{5.4.7}$$

and the convergence is uniform in τ. (*In the remainder of this chapter the limits of integration,* $-\pi$, π *will be omitted with no loss of clarity.*) The expression (5.4.7) is strictly greater than σ_0^2 unless $k_\tau = k_0$ and $l_\tau = l_0$. Thus if $p < p_0$ or $q < q_0$ or $r < r_0$, then (5.4.7) is bounded away from, and above, σ_0^2 since otherwise there would be a sequence of τ values such that along this sequence $k_\tau \to k_0$, $l_\tau \to l_0$ and this is impossible since $p < p_0$, or $q < q_0$ or $r < r_0$. Thus

$$\begin{aligned} &\log \hat{\sigma}^2_{p,q,r} - \log \hat{\sigma}^2_{p_0,q_0,r_0} + \{p - p_0 + q - q_0 + (r - r_0)m\} C_T/T \\ &\quad = \log(\hat{\sigma}^2_{p,q,r}/\sigma_0^2) + o(1) + \{p - p_0 + q - q_0 + (r - r_0)m\} C_T/T \end{aligned}$$

since we know that $\hat{\sigma}^2_{p_0, q_0, r_0} \to \sigma_0^2$, a.s. Since we have shown that the first term is bounded away from, and above, zero for T large enough and $C_T/T \to 0$, it follows that for sufficiently large T, Theorem 5.4.1 cannot be minimized for $p < p_0$ or $q < q_0$ or $r < r_0$ and henceforth only $p \geq p_0$, $q \geq q_0$, $r \geq r_0$ needs to be examined.

Next it is shown that $\partial \log \det P_T(\hat{\tau})$ is composed of elements that are $O(1)$. The notation of Section 4.3 is used, so that $\partial_\tau \log \det P_T(\tau)$ is the row vector of partial derivatives with respect to the elements of τ and $\partial_\tau^2 f(\tau)$ is a Hessian matrix of second derivatives. Now

$$\begin{aligned}\partial \log \det P_T(\tau)/\partial \tau_j &= tr\{P_T(\tau)^{-1} \partial P_T(\tau)/\partial \tau_j\} \\ &= tr[P_T(\tau)^{-1} \partial \{A_T^{-1}(B_T B_T' + \Delta_T)(A_T')^{-1}\}/\partial \tau_j]\end{aligned}$$

using (4.2.13). Consider $\tau_j = \alpha(k)$, for example. The derivative then is

$$\begin{aligned}&-tr[P_T(\tau)^{-1}\{A_T^{-1}\partial A_T/\partial\alpha(k) P_T(\tau) + P_T(\tau)\partial A_T'/\partial\alpha(k)(A_T')^{-1} \\ &\quad - A_T^{-1}\partial\Delta_T/\partial\alpha(k)(A_T')^{-1}\}] = tr[P_T(\tau)^{-1}A_T^{-1}\partial\Delta_T/\partial\alpha(k)(A_T')^{-1}],\end{aligned}$$

since $tr\{A_T^{-1}\partial A_T/\partial\alpha(k)\} = 0$. For T large, $\hat{a}(z)$ and $\hat{b}(z)$ have all zeros bounded away from $|z| = 1$ since $\hat{k} \to k_0$ by Theorem 4.2.1 and (5.4.4) holds. Indeed, $\hat{k} \to k_0$ uniformly on compact subjects of the open unit disk, so that if $\hat{a} = \Pi(1 - z_{j,T}z)$ has $z_{j,T}$ with

$$\limsup_{T \to \infty} \max_j |z_{j,T}| = 1,$$

then for some c independent of ε, $c < \infty$, and $\varepsilon > 0$ arbitrary,

$$\limsup_{T \to \infty} \inf_{|z| \leq 1 - \varepsilon} \hat{a} \leq c\varepsilon$$

and

$$\limsup_{T \to \infty} \sup_{|z| \leq 1 - \varepsilon} \hat{k}(z) \geq c\varepsilon^{-1},$$

whereas

$$\sup_{z \leq 1} k_0(z) \leq c_1 < \infty.$$

Thus it may be assumed that A_T, B_T, Δ_T have elements that are uniformly bounded and A_T^{-1}, B_T^{-1} are Toeplitz, lower triangular, with elements decreasing at a geometric rate to zero, as the diagonal on which the element lies moves away from the main diagonal. Also, $P_T(\hat{\tau})^{-1}$ has its largest eigenvalue bounded, since this is the maximum of

$$\left\{\frac{1}{2\pi}\int\left|\sum x_j e^{ij\omega}\right|^2 |\hat{k}(e^{i\omega})|^2\, d\omega\right\}^{-1}, \qquad x'x = 1$$
$$\leq \left\{c_1 \frac{1}{2\pi}\int\left|\sum x_j e^{ij\omega}\right|^2 d\omega\right\}^{-1} = c_1^{-1},$$

where c_1 is a lower bound for $\hat{k}$, which is strictly positive because of (5.4.4). Thus

$$tr\{P_T(\hat{\tau})^{-1} A_T^{-1}\partial\Delta_T/\partial\alpha(k)(A_T')^{-1}\} \leq c_1^{-1} tr\{A_T^{-1}\partial\Delta_T/\partial\alpha(k)(A_T')^{-1}\} \leq c < \infty.$$

The same is true of a derivative with respect to $\beta(k)$. Thus

$$T^{-1}\partial_\tau \log\det P_T(\hat{\tau}) = O(T^{-1}), \quad \text{a.s.} \tag{5.4.8}$$

Now

$$\hat{\sigma}^2_{p,q,r}(\hat{\chi},\hat{\psi}) = \hat{\sigma}^2 + \partial_\chi \hat{\sigma}^2_{p,q,r}(o,\hat{\psi})\hat{\chi} + \tfrac{1}{2}\hat{\chi}'\partial_\chi^2\hat{\sigma}^2_{p,q,r}(\bar{\chi},\hat{\psi})\hat{\chi}, \tag{5.4.9}$$

where $\hat{\sigma}^2 = \hat{\sigma}^2_{p,q,r}(o,\hat{\psi}) = \hat{\sigma}^2_{p_0,q_0,r_0}(k_0,l_0)$, since $\chi = 0$ implies that $k = k_0, l = l_0$. Here also $\partial_\chi^2\hat{\sigma}^2_{p,q,r}(\bar{\chi},\hat{\psi})$ means that the Hessian has been evaluated at values $\bar{\chi}$, differing perhaps at different places in the Hessian but with $\|\bar{\chi}\| \leq \|\hat{\chi}\| \to 0$. Also,

$$\hat{\sigma}^2 - \hat{\sigma}^2_{p_0,q_0,r_0} = c + O(T^{-1}), \qquad c \geq 0$$

since $\hat{\sigma}^2_{p_0,q_0,r_0}$ is at a minimized value for $L_T(\tau)$, whereas $\hat{\sigma}^2$ is at k_0, l_0. The $O(T^{-1})$ term comes from the $T^{-1}\log\det P_T(\tau)$ terms. Thus

$$\log\hat{\sigma}^2_{p,q,r} - \log\hat{\sigma}^2_{p_0,q_0,r_0} = \log\{1 + (\hat{\sigma}^2_{p,q,r} - \hat{\sigma}^2_{p_0,q_0,r_0})/\hat{\sigma}^2_{p_0,q_0,r_0}\}$$
$$\geq \{(\hat{\sigma}^2_{p,q,r} - \hat{\sigma}^2)/\sigma_0^2\}\{1 + o(1)\} + O(T^{-1})$$

since $\hat{\sigma}^2_{p_0,q_0,r_0} \to \sigma_0^2$. Thus for parts (a), (c) of the theorem we must consider

$$(\hat{\sigma}^2_{p,q,r} - \hat{\sigma}^2)/\sigma_0^2 + \{p - p_0 + q - q_0 + (r - r_0)m\}C_T/T \tag{5.4.10}$$

and must show that this is eventually positive (a.s. or in probability) when $\{p - p_0 + q - q_0 + (r - r_0)m\} > 0$. For simplicity in the following, put $\sigma_0^2 = 1$, as may be done since the first term in (5.4.10) is scale free. Now

$$\partial_\chi \hat{\sigma}^2_{p,q,r}(o, \hat{\psi}) = -\hat{\chi}' \partial^2_\chi \hat{\sigma}^2_{p,q,r}(\tilde{\chi}', \hat{\psi}) + O(T^{-1}), \tag{5.4.11}$$

because

$$0 = \partial_\chi L_T(\hat{\tau}) = \{\partial_\chi \hat{\sigma}^2_{p,q,r}(o, \hat{\psi}) + \hat{\chi}' \partial^2_\chi \hat{\sigma}^2_{p,q,r}(\tilde{\chi}', \hat{\psi})\}/\hat{\sigma}^2_{p,q,r}(\hat{\chi}, \hat{\psi}) + O(T^{-1}),$$

the last following from (5.4.8). Here again $\|\tilde{\chi}'\| \leq \|\hat{\chi}\| \to 0$. The result is uniform in $\hat{\psi}$. Thus from (5.4.9), (5.4.11), we obtain

$$\hat{\sigma}^2_{p,q,r} - \hat{\sigma}^2 = -\hat{b}'\hat{B}^{-1}\hat{b}\{1 + o(1)\} + \hat{b}'\hat{B}^{-1}O(T^{-1}) + O(T^{-2}),$$

where

$$\hat{b}' = \tfrac{1}{2}\partial_\chi \hat{\sigma}^2_{p,q,r}(o, \hat{\psi}), \qquad \hat{B} = \tfrac{1}{2}\partial^2_\chi \hat{\sigma}^2(o, \hat{\psi}).$$

It will be shown that $\hat{b} = O(Q_T)$, so that the next-to-last term is $o(T^{-1})$ and is negligible in all that follows. The fact that $\bar{\chi}, \tilde{\chi}' \to 0$ has been used together with the fact that $\{\partial^2_\chi \hat{\sigma}^2_{p,q,r}(\chi, \psi)\}^{-1}$ is continuous in χ near $\chi = 0$, uniformly in ψ, as will be shown. To show part (a) of the theorem under (4.1.16) it is necessary to show that

$$\limsup_{T\to\infty} \hat{b}'\hat{B}^{-1}\hat{b}/(2Q_T^2) = 1, \quad \text{a.s.} \tag{5.4.12}$$

since then (5.4.10) will satisfy

$$\begin{aligned} &\liminf_{T\to\infty} T\{\hat{\sigma}^2_{p,q,r} - \hat{\sigma}^2\} + \{p - p_0 + q - q_0 + (r - r_0)m\}C_T \\ &\qquad \geq -1 + \{p - p_0 + q - q_0 + (r - r_0)m\}c, \qquad c > 1 \end{aligned}$$

and eventually (5.4.10) will be positive. For part (a) without (4.1.16) it is, similarly, necessary to show that

$$\limsup_{T\to\infty} \hat{b}'\hat{B}^{-1}\hat{b}/(2Q_T^2) \leq c_1. \tag{5.4.13}$$

The results (5.4.12), (5.4.13) will be established as follows. First let $\hat{\psi} = \psi$, some fixed value in the compact set over which $\hat{\psi}$ varies. Then $\hat{B}(\psi)$ will be shown to converge to a matrix continuous in ψ and with smallest eigenvalue bounded away from zero, uniformly in ψ. Then $\hat{b}(\psi)$ will be shown to obey the law of the iterated logarithm (LIL) for each fixed ψ. Finally, $\hat{b}(\psi)/Q_T$ will be shown to be continuous in ψ, uniformly in T [i.e., to be equicontinuous in ψ (with respect to T)]. This is the essential part of the proof and it uses

Theorem 5.3.1. Then if $f_T(\psi)$ is the quantity whose lim sup is being taken in (5.4.12) or (5.4.13), we reason as follows. Choose ψ_j, $j = 1, \ldots, M$, so that, for each ψ, there is a j_0 such that $|\psi - \psi_{j_0}| \le \varepsilon$. This is possible for any $\varepsilon > 0$ since the ψ space is compact. Let $\hat{j}$ correspond to $\hat{\psi}$. Then

$$\begin{aligned}\limsup_{T\to\infty}\{f_T(\hat{\psi})\} &= \limsup_{T\to\infty}[f_T(\psi_{\hat{j}}) + \{f_T(\hat{\psi}) - f_T(\psi_{\hat{j}})\}]\\ &\le \limsup_{T\to\infty} f_T(\psi_{\hat{j}}) + \eta \le c + \eta,\end{aligned}$$

where c is an upper bound for lim sup $f_T(\psi_j)$, $j = 1, \ldots, M$ and η is arbitrarily small for M large. It will be shown that $c \le c_1 < \infty$ for all $M < \infty$ and, when (4.1.16) holds, c may be taken as unity for all $M < \infty$, so that (5.4.12) holds.

In (4.3.17) it is, effectively, shown that, uniformly in ψ, remembering that $\sigma_0^2 = 1$,

$$T^{1/2}\partial_\chi \hat{\sigma}^2_{p,q,r}(o,\psi) = T^{1/2}\partial_\chi \frac{1}{2\pi}\int |k_{\chi=0}|^2 |w_{\chi=0}|^2\, d\omega + o(1),$$

where the notation indicates that χ is put at 0 after differentiation. Here also

$$w_\tau(\omega) = T^{-1/2}\sum_1^T \{y(t) - l_\tau z(t)\}e^{it\omega},$$

$$l_\tau z(t) = \sum_1^\infty L_\tau(j)z(t-j), \qquad z(t) = 0, \quad t \le 0.$$

The fact that the formulas hold uniformly in ψ follows from the last part of (5.4.4), which ensures that the geometric rate of decrease of all expansions involved in the proof of (4.3.17) is uniform in χ, ψ, since we need consider only χ near 0, so that $\hat{a}(z)$ has zeros bounded away from the unit circle. Thus

$$\text{(5.4.14)} \qquad \partial_\chi \hat{\sigma}^2_{p,q,r}(o,\psi) = \partial_\chi \frac{1}{2\pi}\int |k_{\chi=0}|^2 |w_{\chi=0}|^2\, d\omega + o(T^{-1/2}).$$

Also, see (4.3.6), uniformly in ψ,

$$\partial_\chi^2 \hat{\sigma}^2_{p,q,r}(o,\psi) \to \partial_\chi^2 \sigma^2_{p,q,r}(o,\psi)$$

$$\sigma^2_{p,q,r}(\chi,\psi) = \sigma^{-2}Q(\tau) = \frac{1}{2\pi}\int |k_\tau|^{-2} f_0(\omega)\, d\omega + \int |k_\tau|^{-2}(l_\tau - l_0)\, dF_z(l_\tau - l_0)^*.$$

For any vector x of dim χ components

$$\begin{aligned}\tfrac{1}{2}x'\partial_\chi^2\hat{\sigma}_{p,q,r}^2(o,\psi)x &= \frac{1}{2\pi}\int\frac{1}{|a_0b_0h|^2}[\|a_xb_0-b_xa_0\|^2\,d\omega \\ &\quad+(a_xd_0'-d_x'a_0)^*\,dF_z(\omega)(a_xd_0'-d_x'a_0)]+o(1).\end{aligned}\tag{5.4.15}$$

Here h corresponds to ψ (see above the statement of the theorem) and a_xb_x are a, b at $(\chi = x, \psi)$. We omit the details of the derivation of (5.4.15), which are elementary but tedious. One thing to be observed in the derivation is that a term of the form $\{(\partial_\chi b)'(\partial_\chi a)\bar{b}\bar{a}/|b|^4\}f_0(\omega)$, after evaluation at $\chi = 0$, vanishes on integration since it contains only positive powers of $e^{i\omega}$. Also, $x'(\partial_\chi ab_0 - \partial_\chi ba_0)$ is, at $\chi = 0$, $(a_x - aha_0)b_0 - (b_x - hb_0)a_0 = a_xb_0 - b_xa_0$.

Now the vector of coefficients in $a_xb_0 - b_xa_0$ is of the form C_0x, since it is linear in x and zero at $x = 0$, for any ψ. Similarly, the vector of coefficients in $a_xd_0' - d_x'a_0$, arranged, say, in dictionary order first according to lag, are of the form D_0x. Also, $x'C_0'C_0x + x'D_0'D_0x$ can be null only for $x = 0$ since otherwise $a_xb_0 = b_xa_0$, $a_xd_0 = d_xa_0$ for $x \neq 0$, which is impossible. Hence $x'C_0'C_0x + x'D_0'D_0x \geq c > 0$ for $x'x = 1$, because the fixed matrix

$$[C_0' \quad D_0']\begin{bmatrix}C_0\\ D_0\end{bmatrix}$$

is nonsingular. Since $|a_0b_0h|^2 \leq c < \infty$, then (5.4.15) is not less than $c > 0$ for $x'x = 1$, using (4.1.6). Indeed, we know that, because of (4.1.6),

$$\int(l_x - l_0)\,dF_z(l_x - l_0)^* > 0$$

unless $l_x = l_0$ and hence also

$$\int(a_xd_0' - d_x'a_0)^*\,dF_z(a_xd_0' - d_x'a_0) > 0$$

unless $a_xd_0' = d_x'a_0$, because $|a_xa_0|^{-2}$ is bounded away from zero. Thus (5.4.15) is not less than $c\{x'C_0'C_0x + x'D_0'D_0x\}$ for some $c > 0$ and the smallest eigenvalue of $\partial_\chi^2\hat{\sigma}_{p,q,r}^2(o,\psi)$ is bounded away from zero, since this is a positive continuous function of ψ on a compact set.

Now

$$\partial_\chi\hat{\sigma}_{p,q,r}^2(o,\psi) = \frac{1}{2\pi}\int\partial_\chi|k_{\chi=0}|^{-2}|w_0(\omega)|^2 + 2|k_0|^{-2}\{\overline{w_0(\omega)}\partial_\chi w_{0v}(\omega)\}\,d\omega$$

$$w_0(\omega) = T^{-1/2}\sum_1^T u_0(t)e^{it\omega}, \qquad u_0(t) = \sum_0^\infty \kappa_0(j)\varepsilon(t-j),$$

$$\partial_\chi w_{0v}(\omega) = T^{-1/2} \sum_1^T \zeta_0(t) e^{it\omega}, \qquad \zeta_0(t) = \partial_\chi \cdot l_{\chi=0} z(t).$$

Thus carrying out the differentiations

(5.4.16)

$$\tfrac{1}{2}\partial_\chi \hat{\sigma}^2_{p,q,r}(o,\psi) + o(T^{-1/2}) = -\frac{1}{2\pi}\int |k_0|^{-2}[(a_0 b_0 h)^{-1}\partial_{\chi=0}(ab_0 - ba_0)|w_0(\omega)|^2$$
$$+ \overline{w_0(\omega)}\, T^{-1/2}\sum_1^T h^{-1} a_0^{-2}\{\partial_{\chi=0}(ad_0' - d'a_0\}' z(t) e^{it\omega}]\, d\omega.$$

Now we may replace $|w_0(\omega)|^2$ by $|k_0|^2 |w_\varepsilon(\omega)|^2$,

$$w_\varepsilon(\omega) = T^{-1/2}\sum_1^T \varepsilon(t) e^{it\omega},$$

and thus the term under the integral sign in (5.4.16) by

$$\begin{aligned}
&(a_0 b_0 h)^{-1}\partial_{\chi=0}(ab_0 - ba_0)|w_\varepsilon(\omega)|^2 \\
&\quad + |k_0|^{-2}\bar{k}_0 h^{-1} a_0^{-2}\overline{w_\varepsilon(\omega)}\{\partial_{\chi=0}(ad_0' - d'a_0)'\} w_z(\omega) \\
&= (ab_0 h)^{-1}\{|w_\varepsilon(\omega)|^2 \partial_{\chi=0}(ab_0 - ba_0) \\
&\quad + \overline{w_\varepsilon(\omega)}\{\partial_{\chi=0}(ad_0' - d'a_0)\}' w_z(\omega)\},
\end{aligned} \tag{5.4.17}$$

where

$$w_z(\omega) = \frac{1}{\sqrt{T}}\sum_1^T z(t) e^{it\omega}.$$

That this replacement of $w(\omega)$ by $k_0 w_\varepsilon(\omega)$ introduces a negligible error [i.e., $o(T^{-1/2})$] follows from the proof below Lemma 4.3.5, although here the proof would be simpler since the existence of a fourth moment is required, this being needed for the truncation argument in the proof of Theorem 5.3.1. Thus finally, from (5.4.16), (5.4.17), we have

$$\begin{aligned}
&\tfrac{1}{2}\partial_\chi \hat{\sigma}^2_{p,q,r}(o,\psi) + o(T^{-1/2}) \\
&= -\frac{1}{2\pi}\int \frac{1}{a_0 b_0 h}[\partial_{\chi=0}(ab_0 - ba_0)|w_\varepsilon(\omega)|^2 \\
&\quad + \{\partial_{\chi=0}(ad_0' - d'a_0)\}' w_z(\omega)\overline{w_\varepsilon(\omega)}]\, d\omega.
\end{aligned} \tag{5.4.18}$$

Call the right side of (5.4.18), at $\psi = \hat{\psi}$, $\hat{b}_0'$ and put $\partial_\chi^2 \hat{\sigma}_{p,q,r}^2(o, \hat{\psi}) = \hat{B}_0$. Also, write $b_0(\psi)$ for (5.4.18), evaluated at arbitrary ψ. Thus it is $\hat{b}_0' \hat{B}_0^{-1} \hat{b}_0$ that must be considered in (5.4.13) because we shall show that [in relation to (a), for example] $\hat{b}_0$ is $O(Q_T)$ so that the error in this replacement is $o(Q_T^2)$ and thus negligible. Put $B_0(\psi)$ for $\partial_\chi^2 \sigma_{p,q,r}^2(o, \psi)$.

Rewrite $b_0'(\psi)$, first, as a vector of $N = (p + q + rm) - \rho$ components, of which the jth is

(5.4.19)
$$\sum_{k=1}^{T-1} \{\mu_{jk}(\psi) c_\varepsilon(k) + \nu_{jk}(\psi)' c_{\varepsilon z}(k)\}$$
$$c_\varepsilon(k) = \frac{1}{T} \sum_{k+1}^{T} \varepsilon(s)\varepsilon(s-k), \qquad c_{\varepsilon z}(k) = \frac{1}{T} \sum_{1}^{T} \varepsilon(s) z(s-k).$$

Here $\mu_{jk}(\psi)$ is the coefficient of z^k in $(a_0 b_0 h)^{-1} \partial_{\chi=0}(a b_0 - b a_0)$ and in the kth column of that row vector while $\nu_{jk}(\psi)$ is the jth column of coefficients of z^k in $(a_0 b_0 h)^{-1} \{\partial_{\chi=0}(a d_0' - d' a_0)\}'$.

We may rewrite (5.4.19) as

(5.4.20)
$$\frac{1}{T} \sum_{1}^{T} \varepsilon(t) \xi_j(t, \psi) + O(T^{-1})$$
$$\xi_j(t, \psi) = \sum_{1}^{\infty} \{\mu_{jk}(\psi)\varepsilon(t-k) + \nu_{jk}(\psi)' z(t-k)\}$$

because the error is, for example, from the first term in (5.4.19),

$$-\sum_{k=1}^{T} \mu_{jk}(\psi) \frac{1}{T} \sum_{t=1}^{k} \varepsilon(t)\varepsilon(t-k) - \sum_{T}^{\infty} \mu_{jk}(\psi) \frac{1}{T} \sum_{1}^{T} \varepsilon(t)\varepsilon(t-k),$$

and this is dominated by

$$\frac{c}{T} \sum_{1}^{\infty} |\mu_{jk}(\psi)| \min(k, T) = O(T^{-1}),$$

using Schwarz's inequality and ergodicity to bound the series over t. Thus we are reduced to the consideration of the first term in (5.4.20). However,

$$\sum_{1}^{T} \varepsilon(t) \xi_j(t, \psi)$$

is a martingale with stationary ergodic square-integrable differences and thus obeys the law of the iterated logarithm. Moreover, when (4.1.16) holds the vector with components $T^{-1/2}\sum \varepsilon(t)\xi_j(t, \psi)$ has covariance matrix converging to $B_0(\psi)$. This follows from the evaluations

$$E\left\{T^{-1/2}\sum_1^T \varepsilon(t)\varepsilon(t-j)T^{-1/2}\sum_1^T \varepsilon(t)\varepsilon(t-k)\right\} = \delta_{jk}, \qquad j, k \geq 1$$

$$E\left\{T^{-1/2}\sum_1^T \varepsilon(t)\varepsilon(t-j)T^{-1/2}\sum_1^T \varepsilon(t)z(t-k)\right\} = 0, \qquad j \geq 1$$

$$E\left\{T^{-1/2}\sum_1^T \varepsilon(t)z(t-j)T^{-1/2}\sum_1^T \varepsilon(t)z(t-k)'\right\}$$

$$= \frac{1}{T}\sum_1^T z(t-j)z(t-k)' \to \Gamma_z(j-k).$$

In proving that $B_0(\psi)$ is the covariance matrix the uniform geometric rate of convergence of $\mu_{jk}(\psi)$, $v_{jk}(\psi)$ is used so that only finitely many of the expressions $T^{-1}\sum z(t-j)z(t-k)'$ have to be evaluated.

Thus when (4.1.16) holds, remembering the definition of N above (5.4.19),

$$-b_0(\psi)'B_0(\psi)b_0(\psi) = -T^{-1}\sum_1^N \zeta_j(T, \psi)^2,$$

where $\zeta_j(T, \psi)$ are the elements of $T^{1/2}B_0(\psi)^{-1/2}b_0(\psi)$ and are again martingales but now have differences that are asymptotically orthogonal for different j values and have unit mean square. Thus under (4.1.16),

$$\limsup_{T\to\infty} 1/2Q_T^2 b_0(\psi)'B_0(\psi)^{-1}b_0(\psi)$$

$$= \limsup_{T\to\infty}\sum_1^N \zeta_j(T, \psi)^2/(2\log\log T) = 1, \quad \text{a.s.}$$

Indeed, the result holds for each individual j. Let $\phi(u)$, $u = 1, \ldots, M$, be M unit vectors, with coefficients $\phi_j(u)$, $j = 1, \ldots, N$, chosen so that any vector, x, makes an angle $\theta(u_x)$ with some $\phi(u_x)$, with cosine at least $1 - \varepsilon$. Call $\phi(\tilde{u})$ the vector with such an angle to the vector with components $\zeta_j(T, \psi)/(2\log\log T)^{1/2}$. Then

$$\limsup_{T\to\infty}\sum_1^T \zeta_j(T, \psi)^2/(2\log\log T)$$

(5.4.21) $$= \limsup_{T\to\infty}\left\{\sum_j \phi_j(\tilde{u})\zeta_j(T, \psi)/(2\log\log T)^{1/2}\right\}^2 /\cos^2\theta(\tilde{u})$$

$$\leq \limsup_{T\to\infty} \max_u \left\{\sum_j \phi_j(u)\zeta_j(T,\psi)/(2\log\log T)^{1/2}\right\}^2 /(1-\varepsilon)^2$$
$$= (1-\varepsilon)^{-2}$$

since the maximum is over finitely many terms each of which has limit superior unity. Since ε is arbitrary this establishes (5.4.12) for fixed ψ. Without (4.1.16), then (5.4.13) follows easily from the fact that

$$\limsup_{T\to\infty} \sum_1^T \varepsilon(t)\xi_j(t,\psi)/(2Q_T) \leq c_0 < \infty,$$

so that c_1 in (5.4.13) may be taken as c_0 by the greatest eigenvalue of $B_0(\psi)^{-1}$, maximizing these eigenvalues also over ψ.

What remains for the proof of (a) is the proof of the equicontinuity [see below (5.4.13)]. In other words the equicontinuity must be shown of

$$(T\log\log T)^{-1/2}\sum_1^T \varepsilon(t)\xi_j(t,\psi)$$
$$= (T\log\log T)^{-1/2}\left\{\sum_1^{T-1} \mu_{jk}(\psi)\sum_1^T \varepsilon(t)\varepsilon(t-k) + \nu_{jk}(\psi)'\sum_1^T \varepsilon(t)z(t-k)\right\}.$$

Since $\mu_{jk}(\psi)$, $\nu_{jk}(\psi)$ decrease to zero at a geometric rate, uniformly in ψ, it is necessary to consider only $c\log T$ values of k, for c large, since the remainder will make a contribution that is $o(1)$ uniformly in ψ. Now use Theorem 5.3.1 to obtain the bound

$$\Bigg|\sum_1^{c\log T} \{\mu_{jk}(\psi_1) - \mu_{jk}(\psi_2)\}\sum_1^T \varepsilon(t)\varepsilon(t-k)/(T\log\log T)^{1/2}$$
$$+ \sum_1^{c\log T} \{\nu_{jk}(\psi_1)' - \nu_{jk}(\psi_2)'\}\sum_1^T \varepsilon(t)z(t-k)(T\log\log T)^{-1/2}\Bigg|$$
$$\leq c\sum_1^{c\log T}\{|\mu_{jk}(\psi_1) - \mu_{jk}(\psi_2)| + \|\nu_{jk}(\psi_1) - \nu_{jk}(\psi_2)\|\} \leq c\varepsilon$$

if $\|\psi_1 - \psi_2\|$ is sufficiently small, since $\mu_{jk}(\psi)$, $\nu_{jk}(\psi)$ are continuous in ψ uniformly in k and decrease to zero, uniformly in ψ, at a geometric rate. This establishes equicontinuity and completes the proof of (a).

For (b) it is sufficient to show that, say for $p = p_0$, $q = q_0 + 1$, $r = r_0$, $(\hat\sigma^2_{p,q,r} - \hat\sigma^2_{p_0,q_0,r_0})/(2Q_T^2) < -c_0$, infinitely often, for some $0 < c_0 < \infty$. When (4.1.16) holds we must show that c_0 can be taken arbitrarily near, although less than, unity. Put $\tau' = (\tau_1', \beta(q_0+1))$ and take $\hat{\psi}$ as $\hat{\tau}_1$, $\chi = \beta(q_0+1)$ in previous formulas. Now $\hat{\psi} \to \tau_0$, $\hat{\chi} \to 0$. Let $\hat{\tau}$ be the estimate of τ at p_0, q_0, r_0. Examining (5.4.18) we see that, h now being unity, $(a_0 b_0)^{-1}\partial_{\chi=0}(ab_0 - ba_0)$ has components $a_0^{-1}\exp(ij\omega)$, $j = 1, \ldots, p_0$, corresponding to differentiation with respect to elements of $\alpha(j)$, and $-b_0^{-1}\exp(ij\omega)$, $j = 1, \ldots, q_0 + 1$ corresponding to differentiation with respect to the $\beta(j)$ with similar expressions for $(a_0 b_0)^{-1}\partial_{\chi=0}(ad_0' - d'a_0)$ as $(a_0 b_0)^{-1} d_0 \exp(ij\omega)$, $j = 1, \ldots, p_0$ and $-b_0^{-1}\exp(ij\omega)$, $j = 1, \ldots, r_0$, as the jth column for each row of d. Thus $\hat{\psi}$, and in the same way $\hat{\tau}$, is obtained as a vector of regression coefficients, to $O(T^{-1/2})$, of $\varepsilon(t)$ on $b_0^{-1}\{a_0^{-1}b_0\varepsilon(t-j) + a_0^{-1}d_0 z(t-j)\} = b_0^{-1}y(t-j)$, $j = 1, \ldots, p_0$; $b_0^{-1}\varepsilon(t-k)$, $k = 1, \ldots, q_0+1$; $b_0^{-1}z(t-j)$, $j = 1, \ldots, r_0$. The same is true for $\hat{\tau}$ save that $k = 1, \ldots, q_0$ only. Thus, to order $o(T^{-1/2})$, $\hat{\sigma}^2$ and $\hat{\sigma}^2_{p_0,q_0,r_0}$ are the same since the coefficient in the regression of $\varepsilon(t)$ on any of the variables involved is $O(Q_T)$. Thus $\hat{\sigma}^2_{p,q,r} - \hat{\sigma}^2_{p_0,q_0,r_0} = -b_0' B_0^{-1} b_0\{1 + o(1)\} + o(T^{-1/2})$, as before, but now b_0, B_0 are evaluated at $\psi = \tau_0$, $\chi = 0$ and there is no need for an equicontinuity argument. Precisely as before, the law of the iterated logarithm holds and (b) follows.

For (c) it is sufficient to show that

$$\max_{\psi} \left| \sum_k \mu_{jk}(\psi)c_\varepsilon(k) + \nu_{jk}(\psi)' c_{\varepsilon z}(k) \right| T^{1/2}/C_T^{1/2} \tag{5.4.22}$$

converges in probability to zero, if $C_T \uparrow \infty$, for then, for $p - p_0 + q - q_0 + (r - r_0)m > 0$,

$$\lim_{T\to\infty} P[\log \hat{\sigma}^2_{p,q,r} - \log \hat{\sigma}^2_{p_0,q_0,r_0} + \{p - p_0 + q - q_0 + (r - r_0)m\}C_T/T > 0] = 1.$$

Now (5.4.22) is dominated by

$$\sum_k \rho^k \{|c_\varepsilon(k)| + \|c_{\varepsilon z}(k)\|\} T^{1/2}/C_T^{1/2}, \qquad 0 < \rho < 1.$$

However,

$$E\left[\left\{T^{-1/2}\sum_{k+1}^{T} \varepsilon(s)\varepsilon(s-k)\right\}^2\right] = T^{-1}\sum_{k+1}^{T} E\{\varepsilon(s)^2\varepsilon(s-k)^2\} \le c < \infty$$

and similarly for the components of $c_{\varepsilon z}(k)$. Thus (5.4.21) has root mean square dominated by

$$c \sum_k \rho^k / C_T^{1/2} \to 0$$

and (c) follows.

The result (d) is almost obvious. It is enough to consider, for example, $q = q_0 + 1$, $p = p_0$, $r = r_0$ and to show that when C_T is uniformly bounded, this p, q, r will be preferred to p_0, q_0, r_0 with positive probability, as $T \to \infty$. However, it has been shown that $T\{\log \hat{\sigma}^2_{p,q,r} - \log \hat{\sigma}^2_{p_0,q_0,r_0}\}$ is, to order $o(1)$, the likelihood ratio statistic for testing p, q, r again p_0, q_0, r_0. From the results of Chapter 4, under (4.1.16) this is asymptotically distributed as chi-square with one degree of freedom, this following from the asymptotic normality there established. Thus under (4.1.16),

$$\limsup_{T \to \infty} P[T\{\log \hat{\sigma}^2_{p,q,r} - \log \hat{\sigma}^2_{p_0,q_0,r_0}\} < -C_T] \geq P\{\chi_1^2 < -a\} > 0,$$

where a is the limit superior of C_T. When (4.1.16) does not hold, then $T\{\log \hat{\sigma}^2_{p,q,r} - \log \hat{\sigma}^2_{p_0,q_0,r_0}\}$ is not asymptotically chi-square with one degree of freedom, since $B(\tau_0)$ is no longer the covariance matrix of $T^{1/2} b(\tau_0)$. However,

$$Tb(\tau_0)' B(\tau_0)^{-1} b(\tau_0) = T\{B_1^{-1/2} b(\tau_0)\}' B_1^{1/2} B(\tau_0)^{-1} B_1^{1/2} \{B_1^{-1/2} b(\tau_0)\},$$

where B_1 is the covariance matrix of $Tb(\tau_0)$, which will involve fourth moments, such as $E\{\varepsilon(t)^2 \varepsilon(0)^2\}$. However, $B_1^{1/2} B(\tau_0)^{-1} B_1^{1/2} \geq cI_{p+q+rm}$ for some $c > 0$ and thus once more the inconsistency of the method follows.

REMARK 1. P, Q, R have been taken as fixed, but *the theorem holds, except for the parts involving the use of (4.1.16), when P, Q, R are set at the integer part of* $c \log T$. The fact that (4.1.16) cannot be used to characterize the constants involved in the theorem precisely is due to the fact that $\hat{b}$ now may contain up to $c \log T$ components, so that the argument involving (5.4.21) is no longer valid. However, the remaining arguments can be amended. We omit the details.

REMARK 2. Since the theorem holds for $\delta > 0$, fixed but arbitrary, it is almost obvious that it holds when δ depends on T and converges to zero as $T \to \infty$. It seems probable that $\delta_T = c/\log T$ would suffice, but this has not been checked and will not be further discussed here.

We now go on to prove Theorems 5.4.2 and 5.4.3. It is sufficient in the former to consider $P = p_0 + 1$, $Q = q_0 + 1$, $R = r_0 + 1$. Now from the developments in the proof of Theorem 5.4.1,

$$T\{\log \hat{\sigma}^2_{p,q,r} - \log \hat{\sigma}^2_{p_0,q_0,r_0}\} + (2+m)C_T$$
$$= [T\{\hat{\sigma}^2_{p,q,r} - \hat{\sigma}^2\} + T\{\hat{\sigma}^2 - \hat{\sigma}^2_{p_0,q_0,r_0}\}]\{1 + o(1)\} + o(1) + (2+m)C_T.$$

Now $T\{\hat{\sigma}^2 - \hat{\sigma}^2_{p_0,q_0,r_0}\}$ is asymptotically distributed as chi-square with $p_0 + q_0 + r_0 m$ degrees of freedom, by the results in Chapter 4, since (4.1.16) now holds. This follows from standard likelihood ratio theory and Theorem 4.3.1 (remembering that $\sigma_0^2 = 1$) since $T\log\{\hat{\sigma}^2/\hat{\sigma}^2_{p_0,q_0,r_0}\}$ is essentially the likelihood ratio for testing $\tau = \tau_0$ [see above (5.4.7) for the formula for $L_T(\tau)$]. Thus it is sufficient to show that

$$\lim_{\delta\to 0}\lim_{T\to\infty} P\{-T\{\hat{\sigma}^2_{p,q,r} - \hat{\sigma}^2\} > c\} = 1$$

for any $c > 0$, to prove Theorem 5.4.2. From the proof of Theorem 5.4.1 we know that

$$-T\{\hat{\sigma}^2_{p,q,r} - \hat{\sigma}^2\} = T\hat{b}_0'\hat{B}_0^{-1}\hat{b}_0\{1 + o(1)\} + o(1).$$

Now, again taking $\sigma_0^2 = 1$, ignoring a $o(1)$ term,

$$B_0(\psi) = \frac{1}{2\pi}\int |a_0 b_0 h|^{-2}[\partial_{\chi=0}(ab_0 - ba_0)^* \partial_{\chi=0}(ab_0 - ba_0)\,d\omega$$
$$+ \{\partial_{\chi=0}(ad_0' - d'a_0)\}^* \,dF_z(\omega)\{\partial_{\chi=0}(ad_0' - da_0')\}]$$
$$\geq c\int |h|^{-2}[|\partial_{\chi=0}(ab_0 - ba_0)|^2\,d\omega$$
$$+ \{\partial_{\chi=0}(ad_0' - d'a_0)]^* \,dF_z(\omega)\{\partial_{\chi=0}(ad_0' - d'a_0)\}]$$
$$\geq cI_N(2\pi)^{-1}\int |h|^{-2}\,d\omega,$$
$$N = (p_0 + q_0 + r_0 m) + m + 1.$$

Put $h = 1 + \gamma z$. (Thus $\tilde{h}$ is one-dimensional with γ as its component; see above Theorem 5.4.1. It is notationally simpler to call $\tilde{h}$ just γ.) Then $|\gamma| \leq (1+\delta)^{-1}$, so that $B_0(\psi) \leq cI_N(1-\gamma^2)^{-1}$, where c is independent of γ and depends only on k_0, l_0. Indeed, the formula for $B_0(\psi)$ follows immediately from $(2\pi)^{-1}\int(1-\gamma^2)|h|^{-2}\,d\omega = 1$. Thus for c independent of δ,

$$T\hat{b}_0'\hat{B}_0^{-1}\hat{b}_0 \geq c(1-\hat{\gamma}^2)T\hat{b}_0'\hat{b}_0 \geq c(1-\hat{\gamma}^2)T|x'\hat{b}_0^2|/x'x.$$

Thus it is sufficient to show that for some $x \neq 0$ and $c < \infty$,

$$\lim_{\delta\to 0}\lim_{T\to\infty} P\{(1-\hat{\gamma}^2)T|x'\hat{b}_0|^2 \geq c\} = 1.$$

Now [see following (5.4.15) for C_0, D_0]

$$\partial_{\chi=0}(ab_0 - ba_0)x = \phi_a' C_0 x = a_x b_0 - b_x a_0,$$
$$\partial_{\chi=0}(ad_0' - d' a_0)x = \phi_d' D_0 x = a_x d_0' - d_x' a_0.$$

Here ϕ_a has $\exp(ij\omega)$ in the jth place, $j = 1,\ldots,p_0 + q_0 + 1$ and $\phi_d = I_m \otimes \{\exp(ij\omega), j = 1,\ldots,p_0 + r_0 + 1\}$. Of course, a_x is just the $a(z)$ transfer function with coefficients other than unity for the constant term, determined by x, as a particular value for χ, and similarly for b_x, d_x. Now choose x so that $a_x b_0 - b_x a_0 = a_0 b_0 z$, $a_x d_0 - d_x a_0 = 0$. Indeed, this can be done by choosing $a_x = h_x a_0$, $d_x = h_x d_0$, $b_x = h_x b_0 - b_0 z$. This is possible since we can choose h_x with $|\gamma_x| < (1+\delta)^{-1}$, $|-1+\gamma_x| < (1+\delta)^{-1}$, at least for δ small enough. Then

$$x'\hat{b}_0 = -\frac{1}{2\pi}\int e^{i\omega}\hat{h}^{-1}|w_\varepsilon(\omega)|^2\,d\omega, \qquad \hat{h} = 1 + \hat{\gamma}z$$

and to prove the theorem it is necessary to establish that for all $c < \infty$,

$$\lim_{\delta\to 0}\lim_{T\to\infty} P\{\hat{\zeta}_T^2 \geq c\} = 1,$$
$$\hat{\zeta}_T^2 = T(1-\hat{\gamma}^2)\left\{(2\pi)^{-1}\int e^{i\omega}\hat{h}^{-1}|w_\varepsilon(\omega)|^2\,d\omega\right\}.$$

This is clearly the form of the problem for $p_0 = q_0 = 0$, $p = q = 1$, $r = r_0 = 0$. Moreover, the proof of Theorem 5.4.1 shows that this is, for those p, q, r; p_0, q_0, r_0,

$$\lambda_T = T\log(\hat{\sigma}_{1,1}^2/s^2) + o(1), \qquad s^2 = T^{-1}\sum_1^T y(t)^2.$$

Thus to prove both Theorems 5.4.2 and 5.4.3 it must be shown that $\hat{\zeta}_T^2$ has the distribution described in Theorem 5.4.3, since $s^2 \to 1$ and therefore $\lambda_T = T(\hat{\sigma}_{1,1}^2 - 1)\{1 + o(1)\} + o(1)$. Now [see below (5.4.19)], putting $c_\delta = (1+\delta)^{-1}$, and writing

$$\hat{\zeta}_T^2 = \max_{-c_\delta\leq\gamma\leq c_\delta}\left\{(1-\gamma^2)^{1/2}T^{-1/2}\sum_1^T \varepsilon(t)\sum_0^\infty \gamma^k\varepsilon(t-k-1)\right\}^2$$
$$\times\{1 + o(1)\} + o(1),$$

and we must consider the expression within the braces. Let $\gamma_1, \gamma_2, \ldots, \gamma_M$ be any M values of γ in its range. Then calling $\xi_T(\gamma)$ the expression within braces, these $\xi_T(\gamma_j)$ are jointly asymptotically normal with zero mean and covariance matrix

$$(5.4.23)\quad (1-\gamma_i^2)^{1/2}(1-\gamma_j^2)^{1/2}\sum_0^\infty (\gamma_i\gamma_j)^k = \{(1-\gamma_i^2)(1-\gamma_j^2)\}^{1/2}/(1-\gamma_i\gamma_j).$$

Indeed, the $\xi_T(\gamma)$ are $(1-\gamma^2)^{1/2}T^{-1/2}$ by a martingale with stationary, ergodic square-integrable differences. Now it is asserted that $\xi_T(\gamma)$ converges weakly to a Gaussian random process on $[-c_\delta, c_\delta]$ with zero mean and covariance function given by (5.4.23); that is, considering $\xi_T(\gamma)$ as a continuous function on $[-c_\delta, c_\delta]$, this random function induces a probability law, $P_T(\cdot)$, on the space of continuous functions on $[-c_\delta, c_\delta]$, this space being topologized with the topology of uniform convergence of such functions (i.e., the uniform topology), so that for any Borel set, A, in this space of functions,

$$(5.4.24)\qquad P_T(A) = P\{\xi_T(\gamma) \in A\}.$$

Then it is to be shown that $P_T(A) \to P(A)$, where $P(A)$ is induced by a Gaussian random process on $[-c_\delta, c_\delta]$, with (5.4.23) as its covariance function, and zero mean. To show this we use Theorem 8.2 in Billingsley (1968) (see also Theorem 8.1). These theorems refer to continuous functions on $[0, 1]$ but easily translate into the present context. Two conditions are required. (Herein $x(\gamma)$ is a continuous function on $[-c_\delta, c_\delta]$.)

(i) For each $\eta > 0$ there is an α, so that, for some $T_0 < \infty$,

$$P_T\{x\colon |x(-c_\delta)| > \alpha\} \le \eta, \qquad T \ge T_0$$

(ii) For each $\varepsilon > 0$, $\eta > 0$ there is a β, $0 < \beta < 1$, and a T_1 such that

$$P_T\left\{x\colon \sup_{|\gamma_1-\gamma_2|<\beta} |x(\gamma_1) - x(\gamma_2)| > \varepsilon\right\} \le \eta, \qquad T \ge T_1.$$

(In Billingsley's theorems T_0 is taken as unity, but clearly we can take $T_0 > 1$.) The probability in (i) can be written as [see (5.4.24)] $P\{\xi_T(-c_\delta) > \alpha\}$. Thus it is necessary first to check that there is a $T_0 < \infty$ so that for some α, $P\{\xi_T(-c_\delta) > \alpha\} \le \eta$, $T \ge T_0$, for any $\eta > 0$. However, $\xi_T(-c_\delta)$ converges in distribution to a standard normal variate so that (i) is immediate. Then (ii) becomes

$$(5.4.25)\qquad P\left\{\sup_{|\gamma_1-\gamma_2|<\beta} |\xi_T(\gamma_1) - \xi_T(\gamma_2)| \ge \varepsilon\right\} < \eta, \qquad T > T_1,$$

for every ε, $\eta > 0$ and some $T_1 < \infty$, $0 < \beta < 1$. However,

$$|(1-\gamma_1^2)^{-1/2}\xi_T(\gamma_1) - (1-\gamma_2^2)^{-1/2}\xi_T(\gamma_2)| \leq \sum_{k=1}^{T-1} |\gamma_1^k - \gamma_2^k| \left| T^{-1/2} \sum_{k+1}^{T} \varepsilon(t)\varepsilon(t-k) \right| + o(1)$$

[see (5.4.19)], where the $o(1)$ term is of that order uniformly in γ. Thus at least for $|\beta|$ small this is less than or equal to

$$c \sum_{1}^{T-1} k|\gamma_1 - \gamma_2|^k |T^{-1/2} \sum_{k+1}^{T} \varepsilon(t)\varepsilon(t-k)| + o(1).$$

Thus

$$\begin{aligned}
&\sup_{|\gamma_1-\gamma_2|<\beta} |\xi_T(\gamma_1) - \xi_T(\gamma_2)| \\
&\leq \sup_{|\gamma_1-\gamma_2|<\beta} \left[\left\{ |(1-\gamma_1^2)^{1/2} - (1-\gamma_2^2)^{1/2}| \sup_{\gamma} |\xi_T(\gamma)(1-\gamma^2)^{-1/2}| \right.\right. \\
&\quad \left.\left. + c \sup_{\gamma} (1-\gamma^2)^{1/2} \sum_{1}^{T-1} k|\gamma_1 - \gamma_2|^k \left| T^{-1/2} \sum_{k+1}^{T} \varepsilon(t)\varepsilon(t-k) \right| \right\} \right] + o(1) \\
&\leq \sup_{|\gamma_1-\gamma_2|<\beta} \left[|(1-\gamma_1^2)^{1/2} - (1-\gamma_2^2)^{1/2}| \right. \\
&\quad \times \sum_{1}^{T-1} (1+\delta)^{-k} \left| T^{-1/2} \sum_{k+1}^{T} \varepsilon(t)\varepsilon(t-k) \right| + c \sum_{1}^{T-1} k|\gamma_1 \\
&\quad \left. - \gamma_2|^k \left| T^{-1/2} \sum_{k+1}^{T} \varepsilon(t)\varepsilon(t-k) \right| \right] + o(1).
\end{aligned}$$

However,

$$\left[E\left\{ \left| T^{-1/2} \sum_{k+1}^{T} \varepsilon(t)\varepsilon(t-k) \right| \right\} \right]^2 \leq T^{-1} E\left[\left\{ \sum_{k+1}^{T} \varepsilon(t)\varepsilon(t-k) \right\}^2 \right] \leq 1.$$

Since the term in brackets is an increasing function of β and has expectation that is arbitrarily small for β small, then (5.4.25) holds and hence $\xi_T(\gamma)$ converges weakly to $\xi(\gamma)$, which is a Gaussian process on $[-c_\delta, c_\delta]$ with covariance function (5.4.23). Now $\sup \xi_T(\gamma)^2$ converges in distribution to $\sup \xi(\gamma)^2$, since the supremum of a continuous function is a continuous functional of the function in the uniform norm (see Billingsley, 1968, Th. 5.1). Now if any monotone transformation $\gamma \to v$ is made then, putting $\phi(v) = \xi(\gamma(v))$,

$\sup \phi(v)^2$ will have the same distribution as $\sup \xi(\gamma)^2$. Choose

$$(5.4.26) \qquad v = \log \frac{1+\gamma}{1-\gamma}, \qquad \gamma = \tanh \tfrac{1}{2} v, \qquad \upsilon = \log\{(2+\delta)/\delta\}.$$

Then

$$(1-\gamma_j^2)^{1/2}(1-\gamma_k^2)^{1/2}/(1-\gamma_j\gamma_k) = \{\cosh \tfrac{1}{2}(v_j - v_k)\}^{-1}, \qquad v_j = \log \frac{1+\gamma_j}{1-\gamma_j}.$$

[This is easily established using $\cosh^2 x - \sinh^2 x = 1$, $\tanh x = \sinh x/\cosh x$, $\cosh(x-y) = \cosh x \cosh y - \sinh x \sinh y$.] Thus $\phi(v) = \xi(\tanh \frac{1}{2} v)$ is a Gaussian stationary process on $[-\upsilon, \upsilon]$ and since

$$\int_{-\infty}^{\infty} (\cosh \pi\omega)^{-1} e^{iv\omega}\, d\omega = (\cosh \tfrac{1}{2} v)^{-1},$$

the spectral density of $\phi(v)$ is $(\cosh \pi\omega)^{-1}$. Finally,

$$\int_{-\infty}^{\infty} \omega^2 (\cosh \pi\omega)^{-1}\, d\omega = \tfrac{1}{4}.$$

(Both of the last two integrals may be evaluated by integrating with respect to $s = \sigma + i\omega$ around the contour bounded by $s = i\omega$, $s = 1 + i\omega$ and horizontal lines crossing the imaginary axis at $\pm iA$ and allowing A to increase.)

Now the final result of Theorem 5.4.3 follows immediately from the reference stated in the theorem and Theorem 5.4.2 follows from this. This completes the proof of all theorems of this section.

Remark. The proof shows how $\hat{\psi}$ (i.e., $\hat{\gamma}$) behaves. [In case $p = q = 1$, $p_0 = q_0 = 0$, then γ is just the common value for $\alpha(1)$, $\beta(1)$ and thus γ, as it varies, traces out the line along which $\psi = \frac{1}{2}\{\alpha(1) + \beta(1)\}$ varies.] The transformation (5.4.26) spreads out the interval $[-c_\delta, c_\delta]$ and the regions near $\pm c_\delta$ are spread out most for δ small. Thus there will be many local maxima of $\xi(\tanh \frac{1}{2} v)$ for v varying over such large intervals, and when the transformation is reversed, these will be compressed into small intervals near $\pm c_\delta$. The largest value of $\xi(\gamma)^2$ can be expected to be there also since most local maxima are there and $\xi(\tanh \frac{1}{2} v)$ is stationary in v. Thus for T large, $\xi_T(\gamma)$ can be expected to behave similarly and there will be many local maxima of the likelihood near $\pm c_\delta$ for δ small [i.e., near the line $\chi = 0$ and at the corners, $\alpha(1) = \beta(1) = \pm c_\delta$]. This is a more precise explanation of the phenomenon discussed in another way below (5.4.4).

5. ORDER ESTIMATION IN THE MULTIVARIATE OUTPUT CASE

As we have seen in Chapter 2, there are many systems for representing the ARMAX structures once $s > 1$. Examples are $M(n)$, V_α, $M(p, q, r)$. Our theorems below will cover all of these cases, but we shall deal with cases other than $M(n)$ by remarks after the statement of the theorem. The order, $\hat{n}$, will be chosen by minimizing

$$\begin{aligned} &\log \det \hat{\Sigma}_n + d(n)C_T/T, \qquad d(n) = n(2s + m), \\ &n \le N, \qquad C_T > 0, \qquad C_T/T \to 0. \end{aligned} \tag{5.5.1}$$

However, additional restrictions are required, of the kind of (5.4.4). Thus consider a coordinate neighborhood U_α with $(k_0, l_0) \in U_\alpha$ and the associated mapping $\psi_\alpha^{(2)}$: $U_\alpha \to \mathbf{R}^{d(n_0)}$, where n_0 is the true order (McMillan degree). Then we assume, for known δ, ρ, that there is α so that

$$\det(k_0) \ne 0, \quad |z| < 1 + \delta, \qquad \delta > 0, \qquad \|\phi_\alpha(k_0, l_0)\| \le \rho < \infty. \tag{5.5.2}$$

Correspondingly, in searching coordinate neighborhoods we consider only τ such that $\|\tau\| \le \rho$. Such restrictions complicate a search procedure, but it might in any case be unwise to institute a procedure without some prior constraints such as (5.5.2). In Chapter 6 a specific algorithm will be given for identifying a system and for that algorithm (5.5.2) will not be needed.

Theorem 5.5.1. *If $y(t)$ is generated by an ARMAX system satisfying (4.1.15), (4.1.17) with $\varepsilon(t)$ having finite fourth moment and if $z(t)$ is generated by a stationary process with finite fourth moment with (4.1.6) holding and (5.5.2) holds, the following are true. There are $0 < c_0 \le c_1 < \infty$ such that: (a) If*

$$\liminf_{T \to \infty} C_T/(2 \log \log T) > c_1,$$

then $\hat{n} \to n_0$, a.s. If (4.1.16) holds, we may take $c_1 = 1$. (b) If

$$\limsup_{T \to \infty} C_T/(2 \log \log T) < c_0,$$

the conclusion under (a) fails, and this is so with $c_0 = 1$ if (4.1.16) holds. (c) If $C_T \uparrow \infty$, $\operatorname{plim} \hat{n} = n_0$. (d) If

$$\limsup_{T \to \infty} C_T < \infty,$$

the conclusion under (c) fails.

This theorem is the obvious generalization of Theorem 5.4.1. Theorem 5.4.2 also generalizes as follows. (For earlier forms of both theorems, see Hannan, 1981, in which, however, there are some errors.)

Theorem 5.5.2. *If the conditions of Theorem 5.5.1 hold and also (4.1.16), when* $\limsup C_T < \infty$,

$$\lim_{\delta \to 0} \lim_{T \to \infty} P(\hat{n} > n_0) = 1.$$

REMARK. *The theorem fully generalizes to the estimation of the true Kronecker indices* α_0. For $M(p,q,r)$ the situation is complicated by the fact that (k_0, l_0) could belong to more than one $M(p,q,r)$. Indeed, it would be possible to have that happening with different values of $d(p,q,r) = (p+q)s^2 + rsm$. However, it is evident that (5.5.1) would, under (a) say, choose $(\hat{p}, \hat{q}, \hat{r})$ converging to a set of (p,q,r) values for $d(p,q,r)$ minimal among those for which $(k_0, l_0) \in M(p,q,r)$. Thus what will follow under (a) or (c) is the convergence of $(\hat{p}, \hat{q}, \hat{r})$ to this minimal set. The criterion cannot discriminate between members of the set. Of course, uniqueness of choice could be achieved by choosing, say, q as small as possible, then p, then r, among members of the set.

Proof. Much of the proof is the same as that of Theorem 5.4.1 and can be dealt with briefly. Thus again it may be shown that

$$T^{-1} \log\det \Gamma_T(\theta) = \log\det \Sigma + O(T^{-1}),$$

where the last term is of the indicated order uniformly in θ. (Recall that θ stands for k, l, Σ.) Also, $T^{-1} u_T' \Gamma_T^{-1} u_T$ converges to

$$\frac{1}{2\pi} \int tr[(k\Sigma k^*)^{-1}\{(k_0 \Sigma_0 k_0^*)\, d\omega + 2\pi(l - l_0)\, dF_z(\omega)(l - l_0)^*\}].$$

Thus $L_T(\theta) = \log\det\Sigma + tr(\Sigma^{-1}A) + o(1)$, where

$$A = \frac{1}{2\pi} \int \{k^{-1}k_0\Sigma_0(k^{-1}k_0)^*\, d\omega + 2\pi k^{-1}(l - l_0)\, dF_z(\omega)(l - l_0)^*(k^*)^{-1}\}$$

and is strictly greater than Σ_0 unless $l = l_0, k = k_0$. Thus $L_T(\theta) \geq \log\det A + s + o(1)$ and for $n > n_0$ this is strictly greater than $\log\det\Sigma_0 + s$ for all T sufficiently large. Since for $n = n_0$ we know that $\hat{k} \to k_0, \hat{l} \to l_0, \hat{\Sigma} \to \Sigma_0$, so that $L_T(\hat{\theta}) = \log\det\Sigma_0 + s + o(1)$ at $n = n_0$, then for T sufficiently large we cannot have $\hat{n} < n_0$ and it is necessary to consider only the case $n \geq n_0$. Moreover, let α_0 be the unique α such that $(k_0, l_0) \in V_\alpha$. Then, certainly, $(k_0, l_0) \in U_{\alpha_0}$. Of course, (k_0, l_0) will also be in other U_α. Nevertheless, if $\tau_0 \in \overline{T_\alpha^{(2)}}$, $\pi(\tau_0) = (k_0, l_0)$, then because of Theorem 2.6.5, (iii), (vi) α is "proper" in relation to α_0 (i.e., $\alpha \geq \alpha_0$ in the sense that $n_i \geq n_{0i}$, $i = 1, \ldots, s$; see Theorem 2.6.5). *Thus only U_α need be considered for proper α since otherwise* (k, l) cannot approach (k_0, l_0) and hence $L_T(\theta)$, $\theta \in U_\alpha$, will be strictly greater than $\log\det\Sigma_0 + s$. Thus α is proper from now on.

In any U_α introduce new coordinates, χ_α, ψ_α as before. Thus ψ_α is taken to coordinate the $(n - n_0)s$-dimensional subspace introduced in Theorem 2.6.5, so that

$$\psi_\alpha = A_0\tilde{h} + \tau_0,$$

and where $\tilde{h}$ is the vector of $(n - n_0)s$ freely varying coefficients in the transfer function h satisfying $a = ha_0$, $b = hb_0$, $d = hd_0$, where a_0, b_0, d_0 are given by the uniquely defined matrix fraction description for an element of U_{α_0} (see Theorem 2.6.1). In the space orthogonal to the ψ_α space, we introduce coordinates χ_α so that χ_α varies over a space of dimension $N = n(2s + m) - (n - n_0)s = (n + n_0)s + nm$. Again the region of optimization defined by (5.5.2) must be adjusted. Again, because $(\hat{k}, \hat{l}) \to (k_0, l_0)$, the only boundary points that $\hat{\tau}$ can approach are those due to the condition $\det(k_0) \neq 0$, $|z| < 1 + \delta$ and again the region $\mathbf{R}_{\delta,p}$, defined by (5.5.2) can be extended or reduced, just as at the beginning of the proof of Theorem 5.4.1, so that $\partial_\chi L_T(\hat{\tau}) = 0$, for T sufficiently large. Thus as in Section 4, we arrive at the expression

$$\log\det\hat{\Sigma}_n(\hat{\tau}) - \log\det\hat{\Sigma} = -b_0(\hat{\psi}_\alpha)' B_0(\hat{\psi}_\alpha)^{-1} b_0(\hat{\psi}_\alpha)\{1 + o(1)\} + o(T^{-1}),$$

where the order terms are uniform in ψ_α and $\hat{\Sigma} = \Sigma_{n_0}(o, \hat{\psi}_\alpha)$. Here also [see (4.3.9)]

$$b_0(\psi)' = \tfrac{1}{2}\partial_{\chi=0}\log\det\left[\frac{1}{2\pi}\int\{k_\tau^{-1}w_\tau(\omega)w_\tau(\omega)^*(k_\tau^*)^{-1}\}\,d\omega\right]$$

$$w_\tau(\omega) = T^{-1/2}\sum_1^T\{y(t) - l_\tau z(t)\}e^{it\omega},$$

$$l_\tau z(t) = \sum_1^\infty L_\tau(j)z(t-j), \qquad z(t) = 0, \qquad t \leq 0.$$

Again it is here meant that after differentiation with respect to χ, we put $\chi = 0$. Carrying out the differentiation and replacing $k_0^{-1}w_0(\omega)$ by $w_\varepsilon(\omega)$, $w_0(\omega) = T^{-1/2}\sum u_0(t)\exp(it\omega)$, we obtain, as in Section 4,

$$b_0(\psi_\alpha)' = \left(-\frac{1}{2\pi}\int tr[\Sigma_0^{-1}\{b_0^{-1}h^{-1}\{\partial_u(b - ak_0)w_\varepsilon(\omega)w_\varepsilon(\omega)^* + \partial_u(d - al_0)w_z(\omega)w_\varepsilon(\omega)^*\}]\,d\omega\right)_{u=1,\ldots,N}. \tag{5.5.3}$$

Here the outside parentheses indicate a row vector with the typical element exhibited and *temporarily, ∂_u indicates differentiation with respect to the* uth *element of χ_α, followed by putting $\chi_\alpha = 0$*. The derivation of (5.5.3) is direct. In the same way,

$$\begin{aligned} B_0(\psi_\alpha) &= \tfrac{1}{2}\partial_\chi^2 \log\det\Sigma(o,\psi_\alpha) \\ &= \tfrac{1}{2}\partial_\chi^2 \log\det\left[\frac{1}{2\pi}\int\{k_\tau^{-1}f_0(\omega)(k_\tau^*)^{-1}\,d\omega + k_\tau^{-1}(l_\tau - l_0)dF_z(\omega)(l_\tau - l_0)^*(k_\tau^*)^{-1}\}\right] \\ &= \left[tr\left(\Sigma_0^{-1}\frac{1}{2\pi}\int b_0^{-1}h^{-1}\{\partial_u(b - ak_0)\Sigma_0\partial_v(b - ak_0)^* + \partial_u(d - al_0)\,dF_z(\omega)\partial_v(d - al_0)^*\}(b_0^*h_0^*)^{-1}\right)\right], \end{aligned} \tag{5.5.4}$$

where the (u, v)th element is shown. Almost exactly as in Section 4, it may be shown that $B_0(\psi_\alpha)$ has its largest eigenvalue bounded away from zero uniformly in ψ_α. The typical element $b_u(\psi_\alpha)$ of $b_0(\psi_\alpha)$ may be represented as

$$T^{-1}\sum_1^T \xi_u(t)'\varepsilon(t) + O(T^{-1}), \qquad \xi_u(t) = \sum_1^\infty \{M_{uk}\varepsilon(t-k) + N_{uk}z(t-k)\}, \tag{5.5.5}$$

where the M_{uk}, N_{uk} are, respectively, the Fourier coefficient matrices of

$$\Sigma_0^{-1}b_0h^{-1}\partial_u(b - ak_0), \qquad \Sigma_0^{-1}b_0^{-1}h^{-1}\partial_u(d - al_0).$$

The remainder of the proof of Theorem 5.5.1 is almost the same as that of Theorem 5.4.1, since the summands in (5.5.5) are stationary, square-integrable, ergodic martingale differences and the equicontinuity argument is almost as before, using Theorem 5.3.1.

The proof of Theorem 5.5.2 proceeds, as in Section 5.4, by reducing the problem to that of Theorem 5.4.3, but the reduction is now more complex. Take $n = n_0 + 1$ and h to be of a special form. Of course, if Theorem 5.5.2 is shown to be true when optimization is over only some special values for ψ, the theorem is established. The special form can be most easily derived if we consider

$$k_0(z^{-1}) = \tilde{k}_0(z) = \tilde{a}_0^{-1}\tilde{b}_0, \qquad l_0(z^{-1}) = \tilde{a}_0^{-1}\tilde{d}_0,$$

where $\tilde{a}_0$, $\tilde{b}_0$, $\tilde{d}_0$ are in echelon form. In Section 2.5 we put $\tilde{k}(z) = k(z^{-1}) - I$ but here find it more convenient to use the definition just given. Consider $\tilde{h}_j$ defined as follows. It has zero arguments everywhere except in the jth row and on the main diagonal. It has units on that diagonal except in the jth row, where it has $(z - \gamma)$, $|\gamma| \leq (1 + \delta)^{-1}$. In the jth row it has zero elements in the ith column if $n_{0,ji} < n_{0i} - 1$, where $n_{0,ji}$ is the degree of $\tilde{a}_{0,ji}$, this being the (j, i)th element of $\tilde{a}_0$. If $n_{0,ji} = n_{0i} - 1$ (the degree cannot be higher), the element in the (j, i)th place in $\tilde{h}_j$ is $-\tilde{\alpha}_{0,ji}(n_{0i} - 1)$, where this is (-1) by the coefficient of the highest power of z in $\tilde{a}_{0,ji}$. Then $h_j = Z_1\tilde{h}_j(z^{-1})Z_0^{-1}$, where Z_1 is diagonal with $z^{n_{0i}}$ in the main diagonal except for the jth row, where it has $z^{n_{0j}+1}$. Z_0 is diagonal with $z^{n_{0i}}$ in the main diagonal. Now $\tilde{h}_j\tilde{a}_0$, $\tilde{h}_j\tilde{b}_0$, $\tilde{h}_j\tilde{d}_0$ are in echelon form (but not, of course, coprime) corresponding to the Kronecker indices n_{0i}, $i \neq j$, $n_{0j} + 1$, $i = j$. Thus $h_j a_0$, $h_j b_0$, $h_j d_0$ are in the corresponding form in relation to the representation $k_0 = a_0^{-1}b_0$, $l_0 = a_0^{-1}d_0$. We write $b_j(\gamma)$, $B_j(\gamma)$ for $b_0(\psi_\alpha)$, $B_0(\psi_\alpha)$ [see (5.5.3), (5.5.4)] when α corresponds to n_{0i}, $i \neq j$; $n_{0j} + 1$, $i = j$ and ψ_α is such that h_j is of the particular form just chosen. Now if it is shown that for any $c < \infty$,

$$\lim_{\delta\to 0}\lim_{T\to\infty} P\left\{\max_\gamma \max_j Tb_j(\gamma)'B_j(\gamma)^{-1}b_j(\gamma) > c\right\} = 1, \tag{5.5.6}$$

the theorem will be proved since optimization will have been over (part of) the possible V_α, $\alpha > \alpha_0$, $|\alpha| = n_0 + 1$ [i.e., the proper V_α in $M(n_0 + 1)$]. Now as in Section 5.4, the largest eigenvalue of $(1 - \gamma^2)B_j(\gamma)$ is bounded uniformly in γ because it is bounded by

$$c\frac{1}{2\pi}\int tr\{(h_j h_j^*)^{-1}\}\,d\omega = c\frac{1}{2\pi}\int tr\{\tilde{h}_j\tilde{h}_j^*\}\,d\omega \leq c(1 - \gamma^2)^{-1}.$$

Thus (5.5.6) is implied by

$$\lim_{\delta\to 0}\lim_{T\to\infty} P\left\{\max_\gamma \max_j T(1 - \gamma^2)|x_j'b_j(\gamma)|^2 > c\right\} = 1,$$

where $x_j'x_j \leq c_1 < \infty$. This may be replaced by

$$\lim_{\delta\to 0}\lim_{T\to\infty} P\left\{\max_{\gamma} T(1-\gamma^2)\left|\sum_j x_j' b_j(\gamma)\right|^2 > cs^2\right\} = 1.$$

Now the x_j may be chosen conveniently. If the uth element of x_j is x_{ju}, then

$$\sum_u x_{ju}\partial_u(b - ak_0) = b_{(j)} - a_{(j)}l_0, \qquad \sum x_{ju}\partial_u(d - al_0) = d_{(j)} - a_{(j)}l_0,$$

where $a_{(j)}$, $b_{(j)}$, $d_{(j)}$ are evaluated at $\chi_\alpha = x_j$, and for any ψ_α since $b - ak_0$, $d - al_0$ are independent of ψ_α because they are linear in χ_α, ψ_α, and zero at $\chi_\alpha = 0$. We use $b_{(j)}$, and so on, to distinguish this from b_j defined earlier. Choose x_j so that

$$a_{(j)} = g_j a_0, \qquad d_{(j)} = g_j d_0, \qquad b_{(j)} = g_j b_0 + z b_{0j1} E_{j1}.$$

Here E_{j1} has zero everywhere save for unity in the $(j, 1)$th place, while b_{0j1} is the element in the $(j, 1)$th place in b_0. The matrix g_j has zero everywhere save in the main diagonal where it has units except in the jth place, where it has $(1 - \beta z)$. Then $a_{(j)}$, $b_{(j)}$, $d_{(j)}$ are in an appropriate form for V_α with $\alpha = (n_{01}, \ldots, n_{0j-1}, n_{0j} + 1, \ldots, n_{0s})$. Also, $d_{(j)} - a_{(j)}l_0 = 0$, $b_{(j)} - a_{(j)}l_0 = z b_{0j1} E_{j1}$. Then

$$\begin{aligned}\sum x_j' b_j &= -\frac{1}{2\pi}\int tr\left[\Sigma_0^{-1}\sum_j b_0^{-1} h_j^{-1} E_{j1} b_{0j1} e^{i\omega} w_\varepsilon(\omega) w_\varepsilon(\omega)^*\right] d\omega \\ &= -\frac{1}{2\pi}\int tr\left[\frac{e^{i\omega}}{1-\gamma e^{i\omega}} b_0^{-1}[b_{01}, 0, \ldots, 0] w_\varepsilon(\omega) w_\varepsilon(\omega)^* \Sigma_0^{-1}\right] d\omega,\end{aligned}$$

where b_{01} is the first column of b_0. This last expression is

$$\begin{aligned}&-\frac{1}{2\pi}\int tr\left[\frac{e^{i\omega}}{1-\gamma e^{i\omega}} E_{11} w_\varepsilon(\omega) w_\varepsilon(\omega)^* \Sigma_0^{-1}\right] d\omega \\ &= -\frac{1}{2\pi}\int \frac{e^{i\omega}}{1-\gamma e^{i\omega}} w_1(\omega)\omega)\overline{w_\xi(\omega)}\, d\omega,\end{aligned} \tag{5.5.7}$$

where $w_1(\omega)$ is the first element of $w_\varepsilon(\omega)$ and $w_\xi(\omega)$ is the first element of $\Sigma_0^{-1} w_\varepsilon(\omega)$. Now (5.5.7) by $T(1 - \gamma^2)$ is of precisely the same form as ξ_T in the proof of Theorem 5.4.3, and thus the proofs of the theorems of this section are completed.

6. SOME FURTHER CONSIDERATIONS

This section is divided into subsections since the subjects are not closely related.

(i) Consider the class of models for $s = 1$, $m = 0$, $q = 0$, that is, autoregressions

$$\sum_0^h \phi(j)y(t-j) = \varepsilon(t), \qquad \phi(0) = 1.$$

For distinction from another case [see above (5.6.4)] this will sometimes be called a "forward" regression.

Now p has been changed to h because it is necessary to distinguish the two in Chapter 6. The results and proofs in Chapter 4 and this chapter show that in estimating h and the $\phi(j)$ it is asymptotically equivalent to consider, using AIC,

$$\text{(5.6.1)} \quad \log\left\{\frac{1}{2\pi}\int |w(\omega)|^2 \left|\sum_0^h \hat{\phi}_h(j)e^{ij\omega}\right|^2 d\omega\right\} + 2h/T, \qquad \hat{\phi}_h(0) = 1,$$

where the $\hat{\phi}(j)$ minimize the first term. This first term involves a "Toeplitz" calculation, since the Yule-Walker equations for its minimization are of the form

$$\text{(5.6.2)} \quad \sum_0^h \hat{\phi}_h(j)g(j-k) = \delta_{0k}\hat{\sigma}_h^2, \qquad k = 0, 1, \ldots, h, \qquad \hat{\phi}_h(0) = 1,$$

where $\hat{\sigma}_h^2$ is the minimized value of the integral. The matrix of the system of equations is nonsingular since

$$\sum_1^h\sum x_j x_k g(j-k) = \frac{1}{2\pi}\int |w(\omega)|^2 \left|\sum x_j e^{ij\omega}\right|^2 d\omega,$$

which is evidently nonzero unless $x_j = 0$, $j = 1, \ldots, h$. The Toeplitz nature of the calculation refers to the fact that this matrix has constant elements down any diagonal. Another characterization is useful, namely that (5.6.2) can be thought of as the equations of estimation of a regression, of $y(t)$ on $y(t-j)$, $j = 1, \ldots, h$ with $y(t)$ defined as zero outside $1 \leq t \leq T$. Indeed, consider the array

$$
\begin{array}{c|ccccc}
y(1) & 0 & 0 & \cdots & 0 \\
y(2) & y(1) & 0 & \cdots & 0 \\
\vdots & \vdots & \vdots & \cdots & \vdots \\
y(t) & y(t-1) & y(t-2) & \cdots & y(t-h) \\
\vdots & \vdots & \vdots & \cdots & \vdots \\
y(T) & y(T-1) & y(T-2) & \cdots & y(T-h) \\
0 & y(T) & y(T-1) & \cdots & y(T-h+1) \\
0 & 0 & y(T) & \cdots & y(T-h+2) \\
\vdots & \vdots & \vdots & \cdots & \vdots \\
0 & 0 & 0 & \cdots & y(T).
\end{array}
\tag{5.6.3}
$$

Then (5.6.2) is equivalent to determining the perpendicular projection, of the vector constituted by the first column, on the space spanned by the remaining h. The introduction of the zeros realizes a fiction, and although its effects disappear asymptotically, its effects must be bad. This is discussed further in Chapters 6 and 7 (see Sections 6.4 and 7.5). The simplifying nature of the Toeplitz form, for calculation, can be seen by considering the matrix that would occur if only rows $h + 1$ to T of the array were used. Then a typical element will be

$$\frac{1}{T}\sum_{h+1}^{T} y(t-j)y(t-k),$$

and this will depend on j, k and not only on $(j - k)$.

The equations (5.6.2) can be solved by a recursion on h due to Levinson (see the appendix to Wiener, 1949). This recursion was rediscovered in a statistical context by Durbin (1960). (However, see also Bartlett, 1955, pp. 264, 265.) The simple idea is to make the hth column (to the right of the line) orthogonal to the previous $h - 1$ by regressing on them. The equations for this reverse or "backward" regression are obtained by minimizing

$$
\begin{aligned}
&\frac{1}{2\pi}\int |w(\omega)|^2 \left|\sum_0^{h-1} a(j)e^{i(h-j)\omega}\right|^2 d\omega, \qquad a(0) = 1 \\
&= \frac{1}{2\pi}\int |w(\omega)|^2 \left|\sum_0^{h-1} a(j)e^{ij\omega}\right|^2 d\omega,
\end{aligned}
\tag{5.6.4}
$$

which shows that the optimal $a(j)$ are just the $\hat{\phi}_{h-1}(j)$. Moreover,

(5.6.5)
$$\begin{aligned}\hat{\sigma}_h^2 &= \hat{\sigma}_{h-1}^2 - \hat{\phi}_h(h)^2 \frac{1}{T} \sum_t \left\{ \sum_0^{h-1} \hat{\phi}_{h-1}(j) y(t-h+j) \right\}^2 \\ &= \hat{\sigma}_{h-1}^2 - \hat{\phi}_h(h)^2 \hat{\sigma}_{h-1}^2 = \{1 - \hat{\phi}_h^2(h)\} \hat{\sigma}_{h-1}^2,\end{aligned}$$

since the second factor in the last term in (5.6.5) is just the minimum of (5.6.4), which is $\hat{\sigma}_{h-1}^2$. Thus Levinson's recursion is

(5.6.6)
$$\begin{aligned}&\hat{\phi}_h(j) = \hat{\phi}_{h-1}(j) + \hat{\phi}_h(h)\hat{\phi}_{h-1}(h-j), \qquad \hat{\phi}_h(0) = 1, \qquad j = 1, \ldots, h-1 \\ &\hat{\phi}_h(h) = \sum_0^{h-1} \hat{\phi}_{h-1}(j) g(h-j)/\hat{\sigma}_{h-1}^2 \\ &\hat{\sigma}_h^2 = \hat{\sigma}_{h-1}^2\{1 - \hat{\phi}_h(h)^2\}, \qquad \hat{\sigma}_0^2 = g(0).\end{aligned}$$

The second line follows from the fact that $\hat{\phi}_h(h)$ is the coefficient of (Toeplitz) regression of $y(t)$ on $\sum \hat{\phi}_{h-1}(j)y(t-h+j)$.

Now if h_0 is the true order of the autoregression, then

(5.6.7)
$$\begin{aligned}&\log \hat{\sigma}_h^2 - \log \hat{\sigma}_{h_0}^2 + 2(h-h_0)/T, \qquad h > h_0 \\ &= \frac{1}{T} \sum_{h_0+1}^{h} [T \log\{1 - \hat{\phi}_j(j)^2\} + 2].\end{aligned}$$

Just as in Section 4, AIC cannot, asymptotically, be minimized at $h < h_0$. Thus it is (5.6.7) that must be considered. Assume (4.1.15), (4.1.17) and finite fourth moments for the $\varepsilon(t)$. Then it will be shown that the quantities $T^{1/2}\hat{\phi}_j(j)$, $j = h_0 + 1, \ldots, H$, are jointly asymptotically normal with unit covariance matrix and zero mean vector. Indeed,

$$-T^{1/2}\hat{\phi}_h(h) = -T^{1/2} \sum_0^{h-1}\sum \hat{\phi}_{h-1}(j)\hat{\phi}_{h-1}(k) g(j+k-h)/\hat{\sigma}_{h-1}^2.$$

The denominator converges a.s. to σ_0^2, $h > h_0$, by the consistency proof in Chapter 4. The numerator is

(5.6.8)
$$\begin{aligned}&T^{1/2} \sum_0^{h-1}\sum \phi_0(j)\phi_0(k) g(j+k-h) \\ &\quad - T^{1/2} \sum_0^{h-1}\sum \{\hat{\phi}_{h-1}(j) - \phi_0(j)\}\{\hat{\phi}_{h-1}(k) - \phi_0(k)\} g(j+k-h).\end{aligned}$$

This follows from the fact that

$$\sum_0^{h-1}\sum \hat{\phi}_{h-1}(j)\phi_0(k)g(j+k-h)$$
$$= \sum_0^{h-1} \hat{\phi}_{h-1}(j)g(j-h)$$
$$= \sum_0^{h-1}\sum \hat{\phi}_{h-1}(j)\hat{\phi}_{h-1}(k)g(j+k-h)$$
$$= \sum_0^{h-1}\sum \phi_0(j)\hat{\phi}_{h-1}(k)g(j+k-h),$$

because of the equations (5.6.2) for $k = 1, \ldots, h$ (used in the last lines at $h-1$). Now

$$T^{1/2}\sum_0^{h-1}\sum \phi_0(j)\phi_0(k)g(j+k-h)$$
$$= T^{-1/2}\sum_t \left\{\sum_j \phi_0(j)y(t-j)\right\}\left\{\sum_j \phi_0(k)y(t-h+k)\right\}$$
$$= T^{-1/2}\sum_{t=1}^{T} \varepsilon(t)\eta(t-h) + o(1), \qquad \eta(t) = \sum_0^{h_0} \phi_0(k)y(t+k).$$

The remainder is as indicated since, for example,

$$T^{-1/2}\sum_t \left\{\sum_t^{h_0} \phi_0(j)y(t-j)\right\}\left\{\sum \phi_0(k)y(t-h+k)\right\}$$
$$= o(1)T^{-1/2}\sum_1^{h_0} t^{1/2} = o(1).$$

By Corollary 5.3.2, or the proof of the law of the iterated logarithm in Section 4.3, $\hat{\phi}_{h-1}(j) - \phi_0(j) = O(Q_T)$, so that the last term in (5.6.8) is $O(Q_T^2 T^{1/2}) = o(1)$. Thus

$$T^{1/2}\hat{\phi}_j(j) = \sigma_0^{-2}\left\{T^{-1/2}\sum_1^T \varepsilon(t)\eta(t-j)\right\}\{1+o(1)\} + o(1), \qquad j > h_0. \tag{5.6.9}$$

The $\eta(t)$ sequence is stationary and serially uncorrelated with variance σ_0^2 and $\eta(t)$ is measurable $\mathscr{F}_{t+h_0}$. Thus

$$\sum_1^T \varepsilon(t)\eta(t-j)/\sigma_0^2, \qquad j = h_0+1, \ldots, H$$

is a set of martingales with stationary ergodic square-integrable differences having unit variance and, using (4.1.16),

$$E\{\varepsilon(t)\eta(t-j)\varepsilon(s)\eta(s-k)\} = 0, \qquad j \neq k.$$

Thus (5.6.7) is

$$T^{-1}\left[\sum_{h_0+1}^{h} \{T\hat{\phi}_j(j)^2 - 2\} + o_p(1)\right], \qquad h_0 < h \leq H,$$

where $o_p(1)$ converges to zero in probability, since $\hat{\phi}_j(j)^2 = O(T^{-1})$ in probability. Thus if h, $h_0 < h \leq H$, minimizes AIC for $h \leq H$, then

$$\sum_{n}^{h} \{T\hat{\phi}_j(j)^2 - 2\} + o_p(1) \leq 0, \qquad h_0 < n \leq h,$$

$$\sum_{h+1}^{n} \{T\hat{\phi}_j(j)^2 - 2\} + o_p(1) \geq 0, \qquad h < n \leq H.$$

Let z_j be a sequence of independent identically distributed random variables with the distribution of $\chi_1^2 - 2$, where χ_1^2 is distributed as chi-square with one degree of freedom. Then

$$\begin{aligned}(5.6.10) \quad & \lim_{T\to\infty} P\{\hat{h} = h\} \\ & = P\{s_1 \leq 0, s_2 \leq 0, \ldots, s_{h-h_0} \leq 0\} P\{s_1 \geq 0, s_2 \geq 0, \ldots, s_{H-h} \geq 0\},\end{aligned}$$

where

$$s_h = \sum_{1}^{h} z_j.$$

These probabilities are easily evaluated (Spitzer, 1964) from the theory of random walk, but we shall not write them down here but shall call the right side of (5.6.10) $\pi(h - h_0, H - h)$. These probabilities were evaluated by Shibata (1976), to whom these developments are due, for h, h_0 and $H = 10$. As $H \to \infty$ for any fixed h_0, Quinn (personal communication) shows that $P\{\hat{h} = h_0\} \to 0.71$, to two significant figures.

Consider next the ARMA, scalar, case where $P = p_0$ or $Q = q_0$. (Actually, only $P = 0$ or $Q = 0$ seem of any practical interest.) Then from (5.4.16) and (5.4.18) we find, after a small amount of manipulation, that for $q \geq q_0$ in the first case and $p \geq p_0$ in the second,

$$\hat{b}_0 = -\left(\frac{1}{2\pi}\int \frac{1}{a_0 b_0}|w_\varepsilon(\omega)|^2 e^{ij\omega}\right)_{j=1,\ldots,m},$$

where $m = p + q_0$ or $q + p_0$, according as $q_0 = Q$ or $p_0 = P$ is the case studied. Thus we are in precisely the same situation, asymptotically, as if an autoregression of order m has been fitted when the true order is $p_0 + q_0$. Thus the following theorem holds.

Theorem 5.6.1. *Let $y(t)$ be ARMA, (4.1.15), (4.1.16) hold and the fourth moment be finite. Then if $\hat{p}$, $\hat{q}$ are determined by AIC,*

$$\lim_{T\to\infty} P\{\hat{p} = p_0, \hat{q} = q_0\} = \pi(q - q_0, Q - q), \qquad q \geq q_0, \quad p_0 = P$$

$$= \pi(p - p_0, P - p), \qquad p \geq p_0, \quad q_0 = Q.$$

In any case

$$\lim_{T\to\infty} P\{\hat{p} < p_0 \text{ or } \hat{q} < q_0\} = 0.$$

The last part of the theorem was proved as part of Theorem 5.4.1. This theorem complements Theorem 5.4.2 for the case $C_T = 2$, $m = 0$, $s = 1$ because, now, the probability of over estimation is precisely evaluated, for all cases, as $T \to \infty$, leaving aside the interpretation of the double limit in Theorem 5.4.2.

For the ARMAX case [unless $r_0 = R_0$ or the $z(t)$ are special] such a neat result would be impossible, but there is a complete generalization to the vector autoregressive case. However, we discuss only the autoregressive computations here. As in (5.6.1) the initial criterion, for an asymptotic theory, can be

$$\log\det\left[\int\left\{\sum_0^h \Phi_h(j)e^{ij\omega}\right\} w(\omega)w(\omega)^* \left\{\sum_0^h \Phi_h(j)e^{ij\omega}\right\}^* d\omega\right]$$
$$+ 2s^2hC_T/T, \qquad \Phi_h(0) = I_s,$$

and again this leads to the (block) Toeplitz equations ("forward" equations),

$$\text{(5.6.11)} \qquad \sum_0^h \hat{\Phi}_h(j)G(j-k) = \delta_{0k}\hat{\Sigma}_h, \qquad k = 0, 1, \ldots, h.$$

Again these equations have a nonsingular matrix and can be viewed as regression relations for vectors in an indefinitely large vector space but with component vectors, $y(t)$, put equal to zero for t outside $1 \leq t \leq T$ [see the array (5.6.3)]. Again the regression can be recursively calculated by replacing

$y(t-h)$ by

$$\sum_0^{h-1} \tilde{\Phi}_{h-1}(j)y(t-h+j), \qquad \tilde{\Phi}_{h-1}(0) = I_s,$$

where the $\tilde{\Phi}_{h-1}(j)$ minimize

$$tr\left[\frac{1}{2\pi}\int\left\{\sum_0^{h-1}\Phi(j)e^{i(h-j)\omega}\right\}w(\omega)w(\omega)^*\left\{\sum_0^{h-1}\Phi(j)e^{i(h-j)\omega}\right\}^* d\omega\right],$$
$$\Phi(0) = I_s.$$

However, now it is not true that $\tilde{\Phi}_{h-1}(j) = \hat{\Phi}_{h-1}(j)$, because the minimization is equivalent to solving

$$\text{(5.6.12)} \qquad \sum_0^{h-1} \tilde{\Phi}_{h-1}(j)G(k-j) = \delta_{0k}\tilde{\Sigma}_{h-1}, \qquad k = 0, 1, \ldots, h-1$$

and $G(k-j) = G(j-k)'$. However the "backward" equations (5.6.12) can be solved recursively also, by replacing $y(t-1)$ by

$$\sum_0^{h-2} \hat{\Phi}_{h-2}(j)y(t-1-j),$$

which is Toeplitz orthogonal to $y(t-2), \ldots, y(t-h+1)$. In this way the first line of (5.6.13) below is obtained. Also,

$$\begin{aligned}\hat{\Phi}_h(h) &= -\left\{\sum_0^{h-1} \hat{\Phi}_{h-1}(j)G(h-j)'\right\}\tilde{\Sigma}_{h-1}^{-1}\\ &= -\left\{\sum_0^{h-1} G(h-j)'\tilde{\Phi}_{h-1}(j)'\right\}\tilde{\Sigma}_{h-1}^{-1}\end{aligned}$$

since it is the matrix of regression coefficients in the Toeplitz regression of $y(t)$ on $-\sum\tilde{\Phi}_{h-1}(j)y(t-h+j)$, giving the second formula, and also of $\sum\hat{\Phi}_{h-1}(j)y(t-j)$ on $-\sum\tilde{\Phi}_{h-1}(j)y(t-h+j)$, giving the first formula, because $\sum\Phi_{h-1}(j)y(t-j)$ is Toeplitz orthogonal to $y(t-h+j)$, $j \le h-1$. Similarly,

$$\begin{aligned}\tilde{\Phi}_h(h) &= -\left\{\sum_0^{h-1} G(h-j)\hat{\Phi}_{h-1}(j)'\right\}\hat{\Sigma}_{h-1}^{-1}\\ &= -\left\{\sum_0^{h-1} \tilde{\Phi}_{h-1}(j)'G(h-j)\right\}\hat{\Sigma}_{h-1}^{-1}.\end{aligned}$$

Putting

$$\Delta_h = \sum_0^h \hat{\Phi}_h(j)G(j-h-1),$$

the second line of (5.6.13) follows. Finally,

$$\hat{\Sigma}_h = \hat{\Sigma}_{h-1} - \hat{\Phi}_h(h)\tilde{\Sigma}_{h-1}\hat{\Phi}_h(h)' = \{I_s - \hat{\Phi}_h(h)\tilde{\Phi}_h(h)\}\hat{\Sigma}_{h-1}$$
$$\tilde{\Sigma}_h = \tilde{\Sigma}_{h-1} - \tilde{\Phi}_h(h)\hat{\Sigma}_{h-1}\tilde{\Phi}_h(h)' = \{I_s - \tilde{\Phi}_h(h)\hat{\Phi}_h(h)\}\tilde{\Sigma}_{h-1},$$

using the second line of (5.6.13) for the last expression in each case. Thus the following recursion (the Levinson-Whittle algorithm), due to Whittle (1963), has been obtained:

(5.6.13)

$$\hat{\Phi}_h(j) = \hat{\Phi}_{h-1}(j) + \hat{\Phi}_h(h)\tilde{\Phi}_{h-1}(h-j), \qquad \tilde{\Phi}_h(j) = \tilde{\Phi}_{h-1}(j) + \tilde{\Phi}_h(h)\hat{\Phi}_{h-1}(h-j),$$
$$j = 1, \ldots, h-1, \qquad \hat{\Phi}_h(0) = \tilde{\Phi}_h(0) = I_s,$$
$$\hat{\Phi}_h(h) = -\Delta_{h-1}\tilde{\Sigma}_{h-1}^{-1}, \qquad \tilde{\Phi}_h(h) = -\Delta_{h-1}'\hat{\Sigma}_{h-1}^{-1},$$
$$\Delta_h = \sum_0^h \hat{\Phi}_h(j)G(j-h-1),$$
$$\hat{\Sigma}_h = \{I_s - \hat{\Phi}_h(h)\tilde{\Phi}_h(h)\}\hat{\Sigma}_{h-1}, \qquad \tilde{\Sigma}_h = \{I_s - \tilde{\Phi}_h(h)\hat{\Phi}_h(h)\}\tilde{\Sigma}_{h-1},$$
$$\hat{\Sigma}_0 = \tilde{\Sigma}_0 = G(0).$$

We now present a proof that

$$\det\left\{\sum_0^h \hat{\Phi}_h(j)z^j\right\} \neq 0, \qquad |z| \leq 1$$

which is essentially that in Whittle (1963). Let us begin from any stationary process, $x(t)$, having a covariance sequence such that

$$[\Gamma(j-k)]_{j,k=1,\ldots,h} > 0. \tag{5.6.14}$$

Then the equations

$$\sum_0^h \Phi_h(j)\Gamma(j-k) = 0, \qquad k = 1, \ldots, h, \qquad \Phi_h(0) = I_s$$

and uniquely solvable and

$$x(t+1|t) = -\sum_1^h \Phi_h(j)x(t+1-j)$$

is evidently the best linear predictor of $x(t+1)$ from h previous values. Then

(5.6.15)

$$\begin{pmatrix} x(t+1) \\ x(t+2) \\ \vdots \\ x(t+h) \end{pmatrix} = \begin{bmatrix} 0 & I_s & \cdots & 0 \\ 0 & 0 & \cdots & 0 \\ \vdots & \vdots & \vdots & \vdots \\ 0 & 0 & \cdots & I_s \\ -\Phi_h(h) & -\Phi_h(h-1) & \cdots & -\Phi_h(1) \end{bmatrix} \begin{pmatrix} x(t) \\ x(t+1) \\ \vdots \\ x(t+h-1) \end{pmatrix} + \begin{pmatrix} 0 \\ 0 \\ \vdots \\ \varepsilon_h(t+h) \end{pmatrix}$$

$$\varepsilon_h(t+h) = x(t+h) - x(t+h|t+h-1).$$

Then if ξ is a left eigenvector of the matrix on the right side for eigenvalue λ and P is the covariance matrix of the vector on the left,

$$\xi' P \xi = \lambda^2 \xi' P \xi + E\{|\xi_h' \varepsilon_h(t+h)|^2\},$$

where ξ_h is the hth block of ξ, since $\varepsilon_h(t+h)$ is certainly orthogonal to $x(t+h-1), \ldots, x(t)$. But then $\lambda^2 < 1$ or $\xi_h' \varepsilon_h(t+h) = 0$, a.s., if $\xi_h' x(t)$ is perfectly predictable from h previous values ($\xi_h \neq 0$, as otherwise it is easily seen that $\xi = 0$.) However, it was shown in Theorem 1.7.2 that the eigenvalues of the matrix in (5.6.15) are the zeros of $\det\{\sum \Phi_h(j) z^{m-j}\}$.

Since $w(\omega)w(\omega)^*$ can always serve as the spectrum of a stationary process, the following theorem holds.

Theorem 5.6.2. *It is always true that*

$$\det\left(\sum \hat{\Phi}_h(j) z^j\right) \neq 0, \qquad |z| < 1$$

and the condition can be strengthened to $|z| \leq 1$ *if*

(5.6.16) $$[G(j-k)]_{j,k=1,\ldots,h+1} > 0.$$

If (5.6.14) holds for y(t) and that is ergodic, (5.6.16) holds for sufficiently large T. When $s = 1$ *these conditions are equivalent to* $-1 < \hat{\phi}_h(h) < 1$, $-1 \leq \hat{\phi}_h(h) \leq 1$, *respectively.*

It may be pointed out that the equations (5.6.11) imply that

$$(5.6.17)\quad \hat{\Sigma}_h \leq \int \left(\sum_0^h A(j)e^{ij\omega}\right) I(\omega) \left(\sum_0^h A(j)e^{ij\omega}\right)^* d\omega, \qquad A(0) = I_s,$$

as can be seen by decomposing $a(e^{i\omega}) = \sum A(j)e^{ij\omega}$ as $a_h + (a - a_h)$ and observing that, because of (5.6.11),

$$\int a_h I(\omega)(a - a_h)^* \, d\omega = 0.$$

Thus $\hat{\Sigma}_h$ actually minimizes the expression on the right side of (5.6.17), not merely its trace. (Of course, here and elsewhere the $\leq$ sign in (5.6.17) relates to the usual partial ordering of symmetric matrices.)

(ii) An important modification of the procedures of this chapter has been suggested by Pötscher (1983), on the basis of a modification of procedures due to Poskitt and Tremayne (1980, 1981, 1982). It is natural to think of determining the order by a series of significance tests. A consideration of Theorem 5.4.3 shows how troublesome this will be if p, q is compared to $(p-1)$, $(q-1)$. Pötscher's idea is to consider a sequence, for the scalar ARMA case, such as (0, 0), (0, 1), (1, 1), (2, 1), (2, 2), (2, 3), (3, 3), (4, 3), ..., and in any case he considers only sequences for which $p + q$ increases by unity at each step. The second idea, suggested also by Poskitt and Tremayne, is to choose P, Q so that $p_0 \leq P$, $q_0 \leq Q$ and carry out a Lagrange multiplier test of (p, q) against P, Q (see Cox and Hinkley, 1974, for Lagrange multiplier tests). The virtue in this kind of test, compared to a likelihood ratio test, is that the likelihood is not maximized at P, Q, but instead the matrices and vectors necessary for the test are calculated on the basis of the maximum likelihood estimates at p, q. Pötscher fairly easily shows that his test procedure, to be defined more fully below, will not asymptotically choose $p < p_0$ or $q < q_0$. Thus the first critical case is p_0, q or p, q_0, and here the advantage of his procedure appears, for then he is not in the nonidentifiable position of Theorem 5.4.3 and thus can avoid the last part of condition (5.4.4). His procedure is to declare $\hat{p}$, $\hat{q}$ the true values if his test statistic fails to be significant, *for the first time*, when these values are considered and shows that this will be asymptotically true at p_0, q, $q \geq q_0$ or p, q_0, $p \geq p_0$. The procedure, briefly, is as follows. The criterion considered is

$$\sigma_\theta^2 = \frac{1}{T}\sum_1^T \varepsilon_\theta(t)^2 = \frac{1}{T}\sum_1^T b_\theta^{-1} a_\theta y(t),$$

adopting a Toeplitz convention once more so that we put $y(t) = 0, t \leq 0$. Then the vector of derivatives with respect to $\alpha(1), \ldots, \alpha(P), \beta(1), \ldots, \beta(Q)$ is

$$(\eta_\theta(t-1), \ldots, \eta_\theta(t-P), -\xi_\theta(t-1), \ldots, -\xi_\theta(t-Q))' = v_\theta(t)',$$

let us say, where

$$b_\theta \eta_\theta(t) = y(t), \qquad b_\theta \xi_\theta(t) = \varepsilon_\theta(t), \qquad b_\theta \varepsilon_\theta(t) = a_\theta y(t),$$
$$\varepsilon_\theta(t) = y(t) = \xi_\theta(t) = \eta_\theta(t) = 0, \qquad t \leq 0.$$

Now $v_\theta(t)$ is evaluated at $a_\theta = \hat{a}_p$, $b_\theta = \hat{b}_q$, where these are the maximum likelihood estimates at p, q. Put $\hat{v}_{p,q}(t)$ for $v_\theta(t)$ when this is done, $\hat{\sigma}^2_{p,q}$ for σ^2_θ and $\hat{\varepsilon}_{p,q}(t)$ for $\varepsilon_\theta(t)$. Put

$$\hat{A}_{p,q} = \frac{1}{T} \sum_1^T \hat{v}_{p,q}(t) \hat{v}_{p,q}(t)'.$$

Then the test statistic is

$$(T\hat{\sigma}^2_{p;q})^{-1} \left\{ \sum_1^T \hat{v}_{p,q}(t) \hat{\varepsilon}_{p,q}(t) \right\}' \hat{A}^+_{p,q} \left\{ \sum_1^T \hat{v}_{p,q}(t) \hat{\varepsilon}_{p,q}(t) \right\}, \tag{5.6.18}$$

where $\hat{A}^+_{p,q}$ is a generalized inverse of the symmetric matrix $\hat{A}_{p,q}$ (obtained, for example, by diagonalizing that matrix by an orthogonal transformation, taking the reciprocals of the nonzero diagonal elements, and reversing the orthogonal transformation).

Pötscher then shows, assuming that (5.4.4) for $r = 0$, $\delta = 0$, (4.1.15), (4.1.16) and the finiteness of the fourth moments, that results fully analogous to (a), (c) of Theorem 5.4.1 hold. Thus choose significance points $c_{p,q}(T)$ for the tests based on (5.6.18) and adopt the procedure described for choosing $\hat{p}$, $\hat{q}$. If $p_0 = q_0$ and the sequence (0, 0), (0, 1), (1, 1), (2, 1) and so on, described above, is used while $c_{p,q}(T) \uparrow \infty$ as $T \to \infty$, for each p, q, then $\hat{p}$, $\hat{q}$ converge to p_0, q_0 in probability, while if $c_{p,q}(T)/(\log \log T) \uparrow \infty$ as $T \to \infty$, then $(\hat{p}, \hat{q}) \to (p_0, q_0)$ a.s. Of course, it is not necessary to take $p_0 = q_0$ since once $\hat{p}$, $\hat{q}$ are determined, these will be p, q_0, or p_0, q, $p \geq p_0$, $q \geq q_0$ in any case and then examination of smaller values than $\hat{p}$, $\hat{q}$ will choose consistent estimates.

Although the procedure is presented for the ARMA scalar case, it probably could be generalized to the scalar ARMAX case and also to the vector case so as to determine the Kronecker indices (by increasing only one of these, by unity, at each step). The problem with the procedure is the choice of P, Q. Of course, P, Q have to be prescribed, in principle, for the procedures using AIC, BIC, but as explained at the end of the proof of Theorem 5.4.1, these can be allowed to increase with T and moreover (and this is the important point) *the effectiveness of the procedures of Theorem 5.4.1 will not be much affected by too large a prescription for P, Q.* However, the effectiveness of Pötscher's procedure may be radically affected by too large a prescription and the procedure may run a severe risk of underestimation of p_0, q_0. By analogy consider fitting a trend to data by orthonormalizing the functions $1, t, t^2, \ldots$ over $t = 1, 2, \ldots, T$, by the Gram-Schmidt method. Call $x_0(t), x_1(t), \ldots,$ the orthonormalized functions and assume that $y(t) = \mu + \varepsilon(t)$, where the $\varepsilon(t)$ are normal and independent with zero mean and unit variance and these last two constants are known. Then the procedure analogous to Pötscher's for determining the degree of the polynomial trend is to consider

$$\sum_{j=p}^{P} \left\{ \sum_{1}^{T} y(t) x_j(t) \right\}^2, \qquad p = 0, 1, 2, 3, \ldots .$$

At $p > 0$ this is distributed as chi-square with $P - p + 1$ degrees of freedom, and at $p = 0$ it is

$$\left\{ \mu + \frac{1}{\sqrt{T}} \sum_{1}^{T} \varepsilon(t) \right\}^2 + \chi_P^2,$$

where the χ_P^2 term in independent of the first term. Clearly, a test based on this quantity becomes less and less sensitive to μ as P is increased.

The fact that the choice of P, Q does not much affect the methods treated in Theorem 5.4.1 is shown by the fact that usually P, Q would not in practice be prescribed, but then another problem emerges since the $\hat{p}$, $\hat{q}$ at which the criterion is minimized is not of the nature of a stopping time (i.e., one can never be sure that an absolute minimum has been reached). Since Pötscher's procedure uses a "first entrance" time (rather than a "last exit" time), no such problem then ensues.

CHAPTER 6

Calculation of the Estimates

1. INTRODUCTION

This book is concerned with the statistical theory of linear systems, not with the applications of that theory. Certainly, no attempt will be made definitively to cover the algorithmic problems associated with the use of the theory. Nevertheless, the theory would be incomplete without reference to algorithms, because much of it has been developed in relation to them and because, as indicated at the beginning of Section 5.5, specific algorithms are easier to deal with than the general maximum likelihood method. However, we shall not deal with the algorithms, as they might be realized as computer programs. For example, the apparatus of the Kalman filter may be used to calculate the Gaussian likelihood, but in writing a program to give effect to that, all kinds of problems arise that we do not discuss. Among these would be the need to take into account the large number of 0, 1 elements in the F matrix so that the number of operations is reduced to a minimum. Again, there are many ways of realizing the Kalman filter so as to reduce problems of numerical instability, but we do not discuss that.

In Section 2, after discussing alternative likelihood forms and the Box-Jenkins (1970) ideas, the calculation of the likelihood via the Kalman filter is discussed. There methods for obtaining initial estimates, and an important method due to Walker (1962), are also dealt with. In Section 3 we deal with a method due to Akaike (1976). Akaike seems to have been the first to realize that, especially for $s > 1$, it will be computationally too burdensome to calculate, say, $\hat{\Sigma}_n$ for each n that might be considered, and that not only an initial estimate of the system parameters, given n is needed, but also an initial estimate of n itself, so that the full likelihood calculation is done only at that estimated n. What seems really to be needed, however, is some iterative procedure that estimates n at each iteration, along with other parameters, and does this via a recursive calculation on n at each iteration, so that the calculation is not too computationally onerous. In Section 5 a procedure of this kind is discussed and an extensive theory concerning the procedure is presented in Section 6. In Section 7 the vector case is discussed. In Section 4 some modified algorithms calculating autoregressions are discussed which relate to the procedures of Sections 5, 6, and 7. It would be foolish of us to intimate that a definitive solution of the problem of calculating estimates is being presented here, for there is no doubt that much ingenuity will yet be used on this problem.

2. ALTERNATIVE LIKELIHOOD FORMS AND THE KALMAN FILTER

From the theoretical developments in Chapters 4 and 5, it is evident that to the first order of magnitude, in minimizing

$$L_T(\theta) = T^{-1}\log\det\Gamma_T(\theta) + T^{-1}u_T'\Gamma_T(\theta)^{-1}u_T,$$

it will not matter if $T^{-1}\log\det\Gamma_T(\theta)$ is replaced by $\log\det\Sigma$. This was first pointed out, it seems, by Whittle (1951), without any very precise justification, in a work that was very influential. Whittle pointed out that in the scalar case with no inputs, but in a more general context than our rational transfer function case, the transformation

$$\text{(6.2.1)}\qquad u(t) \to \frac{1}{\sqrt{T}}\sum_1^T u(t)e^{it\omega_j}, \qquad \omega_j = 2\pi j/T, \qquad -\pi < \omega_j \leq \pi$$

approximately diagonalizes $\Gamma_T(\theta)$, the diagonal elements being $2\pi f(\omega_j)$. (For much more detail, see Grenander and Szegö, 1958.) Using this approximation, and observing that the transformation (6.2.1) is unitary, then omitting the limits of integration, $-\pi$, π as we do throughout this chapter,

$$\text{(6.2.2)}\qquad T^{-1}\log\det\Gamma_T(\theta) \approx T^{-1}\sum\log 2\pi f(\omega_j) \approx \frac{1}{2\pi}\int \log 2\pi f(\omega)\,d\omega$$

and by a famous formula due to Szegö (see, e.g., Hannan, 1970, p. 137), this last is $\log\sigma^2$. In the vector case the same argument may be used, with (6.2.1) making $\Gamma_T(\theta)$ nearly block diagonal and $\log\det\Sigma$ being the final approximation. Again this approximate diagonalization of $\Gamma_T(\theta)$ suggests that

$$T^{-1}u_T'\Gamma_T(\theta)^{-1}u_T \approx \frac{1}{T}\sum w_u(\omega_j)^*\{2\pi f(\omega_j)\}^{-1}w_u(\omega_j),$$

$$\text{(6.2.3)}\qquad w_u(\omega) = \frac{1}{\sqrt{T}}\sum_1^T u(t)e^{it\omega},$$

$$\approx \frac{1}{2\pi}\int w_u(\omega)^*\{2\pi f(\omega)\}^{-1}w_u(\omega)\,d\omega.$$

Thus we might use one or other of these last forms instead of $L_T(\theta)$. The expression

$$\tilde{L}_T(\theta) = \log \sigma^2 + \frac{1}{T}\sum_j |w_u(\omega_j)|^2/\{2\pi f(\omega_j)\} \tag{6.2.4}$$

has come to be called the Whittle likelihood and the same terminology would be applied to its vector form using (6.2.2), (6.2.3). Of course, the formulas (6.2.2), (6.2.3) have appeared repeatedly in Chapters 4 and 5, where they have been shown to provide valid approximations in those contexts. Of course, also, (6.2.4) looks attractive compared to $L_T(\theta)$, especially in the scalar case where σ^2 may be separated as a factor in the second term so that the function to be minimized is just

$$\frac{1}{T}\sum_j |w_u(\omega_j)|^2 |k(\omega_j)|^{-2}.$$

These forms have been extensively studied, therefore, in early treatment (see Walker, 1964; Hannan, 1969b, 1970; and Nicholls, 1976, for example). However, there are problems with the use of the formulas in small samples, as is well known. To begin with the modification of the determinantal term means that (6.2.4) is not a likelihood function since to make $\exp\{-T\tilde{L}_T(\theta)\}$ a probability density, one must multiply by $\exp[-\sum \log\{2\pi f(\omega_j)/\sigma^2\}]$ [i.e., replace σ^2 by $T^{-1}\sum \log 2\pi f(\omega_j)$ in $\tilde{L}_T(\theta)$]. However, this may not adequately deal with the problem because when the zeros and poles of $k(z)$ are well away from the unit circle and T is not small, this replacement will have little effect, whereas when there are zeros or poles near the circle, the replacement may be inappropriate. The situation may be illustrated by the simplest case where $n = s = 1$, $m = 0$, $\alpha(1) = \alpha$, $\beta(1) = \beta$. Then

$$\frac{1}{T}\log\det\Gamma_T(\theta) = \log\sigma^2 + \frac{1}{T}\log\left\{1 + \frac{(\alpha-\beta)^2}{1-\alpha^2}\cdot\frac{1-\beta^{2T}}{1-\beta^2}\right\}.$$

We omit the derivation. This becomes infinite at $\alpha = \pm 1$. At $\beta = \pm 1$ it is $\log\sigma^2 + T^{-1}\log\{1 + T(1-\alpha)/(1+\alpha)\}$, which makes access to $\beta = \pm 1$ a little more difficult because of the $\log T$ occurring. This tends to make the minimization procedure more stable since $\alpha, \beta = \pm 1$ are the troublesome points. On the other hand, for this simple case,

$$T^{-1}\sum \log 2\pi f(\omega_j) = \log\sigma^2 + 2T^{-1}\log\{(1 - |\alpha|^T)/(1 - |\beta|^T)\},$$

which behaves very differently near $\alpha, \beta = \pm 1$. Of course, the argument that we should work with a likelihood, and indeed the whole argument about the determinant, is heuristic.

A second objection to the Whittle likelihood can be seen from the formula, for T odd,

$$u(t) = T^{-1/2}w(0) + 2T^{-1}\sum_j \{\xi_j \cos t\omega_j + \eta_j \sin t\omega_j\}, \qquad \xi_j + i\eta_j = w(\omega_j),$$

where the sum is over $j = 1,\ldots,[1/2(T-1)]$. [If T is even, add $T^{-1/2}w(\pi) \times \cos t\pi$.] Since the Whittle likelihood treats the ξ_j, η_j as independent identically distributed random variables, independent of T, it acts as if $u(t)$ is periodic with period T. For that matter consider the autoregressive case, where $b \equiv 1$. Then the function to be minimized is

$$\frac{1}{T}\sum |w_u(\omega_j)|^2 \left|\sum_0^p \alpha(k)e^{ik\omega j}\right|^2$$

and the estimation equations are

$$\sum_0^p \hat{\alpha}(k)\{g_u(k-j) + g_u(T-k+j)\} = \delta_{oj}\hat{\sigma}^2,$$

$$j = 0, 1, \ldots, p, \qquad g_u(j) = 0, \qquad j \geq T.$$

The term $g_u(T-k+j)$ that occurs is another manifestation of this circularity. Of course, this is $o(T^{-3/4})$ since

$$\frac{1}{T}\sum_1^{T-(T-j)} u(s)u(s+T-j) \leq \frac{1}{T}\left\{\sum_1^j u(s)^2 \sum_1^j u(s+T-j)^2\right\}^{1/2} = o(T^{-3/4}).$$

Nevertheless, the effect may be sizable for T small.

The circularity aspect can be avoided by considering, say in the scalar case, for a (p,q) ARMA process, in place of the Whittle likelihood, the last part of (6.2.3), i.e.

$$\log\sigma^2 + \frac{1}{2\pi}\int |w(\omega)|^2|a_p|^2/|b_q|^2\,d\omega = \log\sigma^2 + g(0)\pi(0) + 2\sum_1^{T-1} g(j)\pi(j),$$

$$a_p = \sum_0^p \alpha(j)e^{ij\omega}, \qquad b_q = \sum_0^q \beta(j)e^{ij\omega},$$

$$\pi(j) = \frac{1}{2\pi}\int e^{ij\omega}|a_p|^2/|b_q|^2\,d\omega, \qquad g(j) = \frac{1}{T}\sum_1^{T-j} u(s)u(s+j).$$

This is now more complex to handle computationally because of the sum over $T-1$ values of j but Godolphin (1984) suggests that this sum be truncated at $m \ll T$, and the resulting expression be maximized, which is not so difficult a task. Nevertheless, one feels it may be best to try to avoid the subjective element in the choice of m, and the use of an expression which is not truly a likelihood. After all in the case of an autoregression (i.e., $b_q \equiv 1$), this likelihood leads to the use of the Yule-Walker equations, which are solved recursively by the Levinson-Whittle algorithm (see Section 5.6). As already mentioned, these suffer from the Toeplitz assumption, which is discussed further in Section 4 below, and which is known to lead to substantial biases when p/T is not sufficiently small.

One of the first attempts to provide a rather complete inferential procedure for rational transfer function models was that put forward, for $s = 1$, in Box and Jenkins (1970), and consequently that book was influential in making these models popular. Box and Jenkins replaced (6.2.4) by

$$\text{(6.2.5)} \qquad \log \sigma^2 + \frac{1}{\sigma^2}\frac{1}{T}\sum \varepsilon_\theta(t)^2, \qquad b_\theta \varepsilon_\theta(t) = a_\theta u(t).$$

[Here and below, in this chapter we customarily use $u(t)$ in the ARMA case, to distinguish that from the ARMAX case, where $y(t)$ is used and which is the basic case considered.] This second equation may be solved for $\varepsilon_\theta(t)$, given a_θ, b_θ, if $\varepsilon_\theta(t)$, $u(t)$ are put at zero for $t \le 0$. However, once an estimate, $\hat{\theta}$, is available, these may be used to "backforecast" $u(t)$ for $t \le 0$. The use of (6.2.5) avoids the circularity assumptions in (6.2.4) and indeed if $u(t)$, $\varepsilon(t)$ were known and treated as constants for $t \le 0$, then (6.2.5) is the true likelihood.

Nevertheless, it is felt that a better procedure will involve the full Gaussian likelihood. Of course, the data will not be Gaussian, so that the best procedure would use the true likelihood, but this would involve theory and calculations of an entirely different order of magnitude. Robust methods relate to this, needless to say. One way to calculate the full Gaussian likelihood is to use the Kalman filter to effect a Gram-Schmidt orthogonalization of the $u_\theta(t)$ sequence. Thus, from Theorem 3.1, we obtain, using a particular coordinate neighborhood or echelon form,

$$\varepsilon_\theta(t) = y(t) - H_\theta \hat{x}_\theta(t|t-1), \quad \hat{x}_\theta(t+1|t) = F_\theta \hat{x}_\theta(t|t-1) + L_\theta z(t) + K_\theta(t)\varepsilon_\theta(t)$$

$$E\{e_\theta(s)e_\theta(t)'\} = \delta_{st}\{H_\theta P_\theta(t) H_\theta' + \Sigma_\theta\} = \Sigma(t), \qquad e_\theta(1) = u_\theta(1).$$

Of course, $\Sigma(t)$ is also a function of θ. Then (see Section 3.3)

$$L_T(\theta) = T^{-1}\sum_1^T \log\det(\Sigma(t)) + T^{-1}\sum_1^T e_\theta(t)'\Sigma(t)^{-1}e_\theta(t).$$

This could be optimized by some descent algorithm, but it will be better numerically to calculate derivatives. This requires an additional pass through a Kalman filter type of calculation for each system parameter. Indeed, examining the formulas in Theorem 3.1 it is seen that $\partial_\tau \hat{x}(t|t-1)$, $\partial_\tau K_\theta(t)$ are needed, assuming that H_θ independent of θ, and for this purpose it is necessary to calculate

$$\begin{aligned}\partial_u \hat{x}(t+1|t) &= \{F_\theta - HK_\theta(t)\}\partial_u \hat{x}(t|t-1) + \partial_u F_\theta \hat{x}(t|t-1)\\ &\quad + \partial_u L_\theta z(t) + \partial_u \mathrm{K}_\theta(t) e_\theta(t)\\ \partial_u P_\theta(t+1) &= \partial_u F_\theta P_\theta(t) F_\theta' + F_\theta \partial_u P_\theta(t) F_\theta' + F_\theta P_\theta(t) \partial_u F_\theta'\\ &\quad + \partial_u(K_\theta(t)\Sigma_\theta) - \partial_u\{K_\theta(t)\Sigma_\theta(t)K_\theta(t)\}'.\end{aligned}$$

These equations are of the same form as for the Kalman filter. Of course, more work is involved since matrix operations are needed using these derivatives to calculate the gradient vector of the likelihood. Thus even for s, n not large, this will be a costly computation, even for the individual value of n. Moreover, an initial value of θ will be needed.

The most difficult part of the problem of obtaining an initial θ is that of obtaining an initial $b(z)$. If this can be avoided, the problem is simpler since initial values for $a(z)$, $d(z)$ are more easily found, as will be shown below. The problem of estimating $b(z)$ may involve spectral factorization since if $\hat{a}$, $\hat{d}$ are available, we may form $\hat{a}\hat{u}(t) = \hat{a}\{y(t) - \hat{v}(t)\}$ and hence an estimate of the covariances of the $b\varepsilon(t)$, namely,

$$\hat{\Gamma}_{b\varepsilon}(t) = \frac{1}{T}\sum \{\hat{a}\hat{u}(s)\}\{\hat{a}\hat{u}(s+t)\}'.$$

From this we calculate

$$\frac{1}{2\pi}\hat{b}\hat{\Sigma}\hat{b}^* = \frac{1}{2\pi}\sum_t \hat{\Gamma}_{b\varepsilon}(t)e^{-it\omega}, \tag{6.2.6}$$

the summation being over limits determined by the Kronecker indices, if the echelon form is being used. Thus the right side of (6.2.6) has to be factored to obtain $\hat{b}$, $\hat{\Sigma}$. This is no simple task, although in the scalar case recursive algorithms have been provided by Bauer (1955) and Wilson (1969) [see remark (ii) in Section 4].

It may also be true that $\hat{b}$, $\hat{\Sigma}$ are never needed. For example, if the end purpose of the calculation is the estimation of $f(\omega)$, these are not needed. For these kinds of reasons an alternative parameterization and an associated alternative form for the Kalman filter were suggested by Solo (1984, 1986) for the ARMA case. In the general ARMAX case let us call $\mathscr{H}_K$ the Hankel matrix constructed from the $K(j)$ and put $\mathscr{G} = \mathscr{H}_K(I_\infty \otimes \Sigma)\mathscr{R}$, where $\mathscr{R}$ has $K(j-k)'$, $j, k = 1, 2, \ldots$, as the (j, k)th block, $K(j) = 0$, $j < 0$. Then $\mathscr{G}$ has $\Gamma(j + k - 1)'$ as the (j, k)th block. Let Γ be composed from the same rows of the first column of blocks of $\mathscr{G}$ as K is composed of the first column of blocks of $\mathscr{H}_K$. Now H, F, K, L, Σ uniquely determine $\mathscr{H}$, Σ, $\mathscr{R}$ and hence $\mathscr{G}$ and hence H, F, L, Γ, $\Gamma_u(0)$, where $\Gamma_u(0)$ is the covariance matrix of the $u(t)$. On the other hand, F, Γ determine $\mathscr{G}$ and hence F, Γ, $\Gamma_u(0)$ uniquely determine the spectrum of $u(t)$ and hence k, Σ and hence K, Σ and thus H, F, L, Γ, $\Gamma_u(0)$ uniquely determine H, F, L, K, Σ. [The determination of the spectrum of $u(t)$ from F, Γ, $\Gamma_u(0)$ does not involve spectral factorization.] We state this result as a theorem (see Solo, 1984, 1986).

Theorem 6.2.1. *Let $(k, l) \in U_\alpha$. Then an alternative set of coordinates to the freely varying elements in F, H, K, L, Σ in the canonical state-space representation are the freely varying elements in F, H, L, Γ, $\Gamma_u(0)$, where $\Gamma_u(j) = E\{u(t)u(t+j)'\}$ and Γ contains the same rows of the column of blocks $[\Gamma_u(1)', \Gamma_u(2)', \ldots]'$ as K does those of $[K(1)', K(2)', \ldots]'$.*

The Kalman filter, applied to our state-space representation of the ARMAX system, is of the following form (see Chapter 3), where the θ subscripts are omitted, for simplicity.

$$\hat{x}(t+1|t) = F\hat{x}(t|t-1) + Lz(t) + K(t)e(t), \quad \hat{x}(1|0) = 0;$$

$$\text{(6.2.7)} \quad \begin{aligned} y(t) &= H\hat{x}(t|t-1) + e(t), \\ K(t) &= \{FP(t)H' + K\Sigma\}\Sigma(t)^{-1}, \quad \Sigma(t) = HP(t)H' + \Sigma, \end{aligned}$$

$$P(t+1) = FP(t)F' + K\Sigma K' - K(t)\Sigma(t)K(t)', \quad P(1) = FP(1)F' + K\Sigma K'.$$

However, Son and Anderson have shown (see Anderson and Moore, 1979) that the last two lines may be replaced by

$$\text{(6.2.8)} \quad \begin{aligned} K(t) &= \{\Gamma - FQ(t)H'\}\Sigma^{-1}(t), \qquad \Sigma(t) = \{\Gamma_u(0) - HQ(t)H'\} \\ Q(t+1) &= FQ(t)F' + K(t)\Sigma(t)K(t)', \qquad Q(1) = 0, \end{aligned}$$

so that $e(t)$, $\Sigma(t)$ may be fully calculated from the first two lines of (6.2.7) together with (6.2.8), in terms of the new parameterization. [$Q(t)$ has no relation to the Q below (3.2.1) and, in particular, is not a time-dependent form of this. We realize that $Q(t)$ was used in a different sense in Section 3.2.] To see that (6.2.8) is valid, recall that if $x_u(t|t-1)$, $\hat{x}_u(t|t-1)$, correspond to $u(t)$

as $x(t|t-1)$, $\hat{x}(t|t-1)$ correspond to $y(t)$, then $\hat{x}_u(t+1|t) = F\hat{x}_u(t|t-1) + K(t)e(t)$. Moreover, $K(t)e(t)$ is the projection of $x_u(t+1|t)$ on $e(t)$ and $K(t)$ may be calculated as

$$E[\{Fx_u(t|t-1) + K\varepsilon(t)\}\{u(t) - H\hat{x}_u(t|t-1)\}']\Sigma(t)^{-1}$$

and the first factor is

$$-E\{F\hat{x}_u(t|t-1)\hat{x}_u(t|t-1)'H'\} + E\{x_u(t+1|t)u(t)'\}$$

because $F\{x_u(t|t-1) - \hat{x}_u(t|t-1)\} + K\varepsilon(t)$ is orthogonal to $e(t)$. This last expression is

$$-FQ(t)H' + \Gamma; \qquad Q(t) = E\{\hat{x}_u(t|t-1)\hat{x}_u(t|t-1)'\}$$

because $x_u(t+1|t)$ is composed of the rows of $(u(t+1|t)', u(t+2|t)', \ldots)'$ that occur in K, Γ and $E\{u_j(t+a|t)u_k(t)\} = E\{u_j(t+a)u_k(t)\}$ since $u_j(t+a) - u_j(t+a|t)$ is a linear combination of the elements of $\varepsilon(t+1), \varepsilon(t+2), \ldots, \varepsilon(t+a)$ and thus orthogonal to $u_k(t)$. Moreover,

$$\begin{aligned}\Sigma(t) &= E[\{u(t) - H\hat{x}_u(t|t-1)\}\{u(t) - H\hat{x}_u(t|t-1)\}'] \\ &= \Gamma_u(0) + HQ(t)H' - E\{u(t)\hat{x}_u(t|t-1)'H'\} - E\{H\hat{x}_u(t|t-1)u(t)'\} \\ &= \Gamma_u(0) - HQ(t)H'\end{aligned}$$

since $u(t) = H\hat{x}_u(t|t-1) + H\{x_u(t|t-1) - \hat{x}_u(t|t-1)\} + \varepsilon(t)$ and the last two terms are orthogonal to $\hat{x}_u(t|t-1)$. Finally,

$$Q(t+1) = E[\{F\hat{x}_u(t|t-1) + K(t)e(t)\}\{F\hat{x}_u(t|t-1) + K(t)e(t)\}']$$

and since $e(t)$ is orthogonal to $\hat{x}_u(t|t-1)$ the last line of (6.2.8) follows. We state all of this as a theorem.

Theorem 6.2.2. *If $(k, l) \in U_\alpha$ and the Kalman filter is constructed corresponding to the canonical prediction error state-space form, then the last two lines of (6.2.7) may be replaced by (6.2.8), where Γ, $\Gamma_u(0)$ have the same meaning as in Theorem 6.2.1.*

Thus the new parameterization, combined with (6.2.8), avoids spectral factorization in the calculation of the likelihood. Of course, if the $K(j)$, $L(j)$ are needed, finally, to construct a predictor, then spectral factorization will be needed at the end of the calculation. Since this is a complex task in general, some other method that avoids direct spectral factorization (see Section 6.4) may be preferable.

We now consider the important problem of finding initial estimates. In a sense this is the main question, whether the parameterization be by H, F, K, L, Σ or H, F, L, Γ, $\Gamma_u(0)$ or by other means (e.g., the ARMAX parameterization of the echelon forms). To illustrate how an initial estimate that even is consistent of optimal order, $T^{-1/2}$ (i.e., that differs in probability from the true values by quantities of order $T^{-1/2}$) may be inadequates consider the simplest nontrivial case [i.e., that where $s = n = 1$, $m = 0$ and $u(t) = \varepsilon(t) + \beta\varepsilon(t-1)$]. Then $F = 0$, $K = \beta$, $L = 0$, $H = 1$, $\Gamma = \beta$, $\Gamma(0) = 1 + \beta^2$, choosing $\sigma^2 = 1$ for convenience. Since, in general (see Chapter 3),

$$\hat{x}(t+1|t) = \{F - K(t)H\}\hat{x}(t|t-1) + u(t),$$

then in the present case the critical quantity determining the stability of the filtering operation is $|K(t)|$. Let $\hat{\gamma}(1)$, $\hat{\gamma}(0)$, be initial estimates of γ, $\Gamma_u(0)$. Then

$$K(t) = c_0\{1 - Q(t)/\hat{\gamma}(0)\}^{-1}, \qquad c_0 = \hat{\gamma}(1)/\hat{\gamma}(0)$$

and putting $a(t) = Q(t)/\hat{\gamma}(0)$,

$$a(t+1) = c_0^2\{1 - a(t)\}^{-1}, \qquad a(0) = 0.$$

This continued fraction may be evaluated by putting $a(t) = p(t)/q(t)$, whence

$$\begin{pmatrix} p(t) \\ q(t) \end{pmatrix} = \begin{bmatrix} 0 & c_0^2 \\ -1 & 1 \end{bmatrix} \begin{pmatrix} p(t-1) \\ q(t-1) \end{pmatrix}; \qquad p(0) = 0, \quad q(0) = 1.$$

It is evident that values $c_0^2 > \frac{1}{4}$ may cause trouble, for then the estimate of the spectrum of $u(t)$ based on $\hat{\gamma}(0)$, $\hat{\gamma}(1)$ will not factor since this is $\{\hat{\gamma}(0) + 2\hat{\gamma}(1)\cos\omega\}/(2\pi)$ and for $|\hat{\gamma}(1)/\hat{\gamma}(0)| > \frac{1}{2}$, this can be negative. From the above it is evident that $p(t)$, $q(t)$ satisfy the same difference equation $p(t) - p(t-1) + c_0^2 p(t-2) = 0$, with different initial conditions (because any matrix satisfies its own characteristic equation, which is $z^2 - z + c_0^2 = 0$ in the present case). Then if ω_0 is the argument of the complex zero of this characteristic equation (for $c_0^2 > \frac{1}{4}$), we find that

$$a(t) = c_0^2 \sin t\omega_0/\{\sin t\omega_0 - |c_0|\sin(t-1)\omega_0\}.$$

It is evident that this will oscillate wildly as the denominator is a sine function at the same frequency but out of phase with the numerator. Thus $K(t)$ will also oscillate wildly and the Kalman filter program will overflow. The natural estimators for Γ, $\Gamma_u(0)$ are $g(1)$, $g(0)$, but then $r(1) = g(1)/g(0)$ has, under (4.1.15), (4.1.16), the limiting distribution implied by treating $T^{1/2}\{r(1) - \rho(1)\}$ as normal with variance (see Theorem 6.2.6 below), $1 - 3\beta^2/(1 + \beta^2)^2 + 4\beta^4/(1 + \beta^2)^4$. At $\beta = 0.5$ this is 0.6224, and since then $\beta/(1 + \beta^2) = 0.4$, and then also, for $T = 100$, the standard deviation of $r(1)$ must be near $(0.62/100)^{1/2} = 0.08$ and values of $r(1)^2$ (i.e., c_0^2) greater than $\frac{1}{4}$ are quite likely. The maximum likelihood estimate of $\rho(1)$ has variance $0.1728T^{-1}$ from its limiting distribution (i.e., a standard deviation for $T = 100$ near 0.04) and thus is rather unlikely to give a value of c_0^2 greater than $\frac{1}{4}$ in these circumstances. Of course we find the ML estimates of $\rho(1)$ by finding the ML estimates of β.

Before going on to discuss the question of obtaining better initial estimates, let us first consider a general method for obtaining estimates of F, L (i.e., a, d). If these can be provided, the method based on (6.2.8) has a certain theoretical completeness. The ARMA case is discussed first. It will be convenient to associate an index α, such that $k \in U_\alpha$, with $f_u(\omega)$ also. Indeed, $f_u(\omega)$ determines k uniquely. Thus we shall say that $f_u(\omega)$ "has α associated with it." Let $\mathscr{G} = \mathscr{H}(I_\infty \otimes \Sigma)\mathscr{H}'$ as before. Write $\mathscr{G}_\alpha$ for the submatrix of $\mathscr{G}$ whose rows are determined by α, that is, having rows (u, j), $u = 1, \ldots, n_j$; $j = 1, \ldots, s$ if α indexes $(n_1, \ldots, n_s)$. $\mathscr{G}^\beta$ is similarly defined in terms of the columns of $\mathscr{G}$ and hence $\mathscr{G}_\alpha^\beta$ is defined.

Theorem 6.2.3. *Let $f_u(\omega)$ have α associated with it. Then $\mathscr{G}_\alpha^\beta$ is nonsingular if and only if $f_u(-\omega)$ has β associated with it.*

This and the next theorem are from An, Chen, and Hannan (1983). The proof here is as follows. $\mathscr{G}$ has rank $n = |\alpha|$, as it is the same as the rank of $\mathscr{H}$. $\mathscr{G}'$ has the same rank as $\mathscr{G}$ and $\mathscr{G}'$ corresponds to $f_u(-\omega)$, as $\mathscr{G}$ does to $f_u(\omega)$. Also, $2\pi f_u(-\omega) = \lambda\Omega\lambda^*$, let us say, where $\lambda(z)$ is the canonical factor and $\Omega > 0$. If $\lambda \in U_\beta$, where β indexes $(m_1, \ldots, m_s)$, then $\mathscr{G}_\alpha^\beta$ is of full rank since otherwise $\mathscr{G}_\alpha$ could not have been of full rank since the remaining columns of $\mathscr{G}_\alpha$ (other than those in $\mathscr{G}_\alpha^\beta$) are linear combinations of those in $\mathscr{G}_\alpha^\beta$. Conversely, if $\mathscr{G}_\alpha^\beta$ is not of full rank, then $\mathscr{G}^\beta$ is not of full rank, for the same reason, and $\lambda \notin U_\beta$.

Of course, the condition in the theorem holds trivially for $s = 1$, for then $k = \lambda$. Since each U_α is dense and open in $M(n)$, the condition of the theorem is, in any case, one that might be assumed to hold for $\beta = \alpha$.

Now let $k \in V_\alpha \subset U_\alpha$ and $k(z^{-1}) = \tilde{a}(z)^{-1}\tilde{b}(z)$, as in Section 5.5. The theorem is applicable since f_u has α associated with it. Now from Theorems 2.5.1 and 2.5.2,

$$\sum_{j=1}^{s}\sum_{v=0}^{n_i} \tilde{\alpha}_{ij}(v)u_j(t - n_i + v) = \sum_{j=1}^{s}\sum_{v=0}^{n_i} \tilde{\beta}_{ij}(n_i - v)\varepsilon_j(t - v), \qquad i = 1, \ldots, s \tag{6.2.9}$$

and multiplying by $u_l(t-w)$ and taking expectations, we obtain

$$\sum_{j=1}^{s}\sum_{v=0}^{n_i} \tilde{\alpha}_{ij}(v)\gamma_{lj}(w+v-n_i)=0, \qquad w>n_i; \qquad i,l=1,\ldots,s.$$

Putting $x=w-n_i-1$, and remembering that $\tilde{\alpha}_{ii}(n_i)=1$, this may be rewritten as

$$-\gamma_{lj}(x+n_i+1)=\sum_{j=1}^{s}\sum_{v=0}^{n_i-1}\tilde{\alpha}_{ij}(v)\gamma_{lj}(x+v+1), \qquad x\geq 0; \qquad i,l=1,\ldots,s. \tag{6.2.10}$$

Let γ_α^β be constituted from the (n_i, i), $i=1,\ldots,s$, rows of $\mathscr{G}^\beta$ (i.e., the rows numbered i in the n_ith blocks). Then γ_α^β is $s\times n$. Let A have ith row having s blocks, of which the jth has $\tilde{\alpha}_{ij}(v-1)$, $v=1,\ldots,n_j$ in the jth place. Some of these entries will be zero, unless α indexes $U(n)$. Then (6.2.10) may be rewritten as follows, taking $x=0,1,\ldots,m_l-1$ in (6.2.10):

$$A\mathscr{G}_\alpha^\beta=-\gamma_\alpha^\beta. \tag{6.2.11}$$

Now it follows immediately that if consistent estimators of $\mathscr{G}_\alpha^\beta, \gamma_\alpha^\beta$ are available, the solution, $\hat{A}$, of (6.2.11) corresponding to these will also be consistent. The obvious consistent estimators are the $g_{ij}(v)$. We state this as a theorem.

Theorem 6.2.4. *Let $u(t)$ be stationary and ergodic with zero mean and finite mean square and be generated by a process with rational spectrum and let $k(z)\in V_\alpha$. Let $\hat{\mathscr{G}}_\alpha^\beta, \hat{\gamma}_\alpha^\beta$ be obtained by replacing $\gamma_{ij}(v)$ by $g_{ij}(v)$ and let*

$$\hat{A}\hat{\mathscr{G}}_\alpha^\beta=-\hat{\gamma}_\alpha^\beta. \tag{6.2.12}$$

Then $\hat{A}\to A$, a.s. if $f_u(-\omega)$ has β associated with it.

After solving the equations the elements of A that are prescribed to zero should be set equal to zero. Of course, the conclusion of the theorem will still, then, remain true.

We briefly consider the case of $M(p,q)$ (see Chapter 2), that is, all ARMA systems having a representation

$$\sum_{0}^{p} A(j)u(t-j)=\sum_{0}^{q} B(j)\varepsilon(t-j), \qquad A(0)=B(0)=I_s \tag{6.2.13}$$

with $a(z)$, $b(z)$ left prime and having determinants with no zeros in $|z|\leq 1$ and with $[A(p), B(q)]$ of rank s. The analogue of (6.2.10) is now

$$\sum_{0}^{p} A(j)\Gamma(j-k) = 0, \qquad k \geq q+1, \tag{6.2.14}$$

and the matrix analogue of $\mathscr{G}_\alpha$ is

$$\Gamma_{p,q} = [\Gamma(j+k-q)]_{j=1,\ldots,p;k=1,2,\ldots}.$$

Since $A(0) = I_s$, the matrix of the system of equations (6.2.13) is that composed of the first p blocks of columns in $\Gamma_{p,q}$, which we call Γ_{pq}^{p}. Again we say that $f_u(\omega)$ is "associated with $M(p,q)$" if $k(z)$ in the unique spectral factorisation lies in $M(p,q)$. The following theorem is established in An, Chen, and Hannan (1983).

Theorem 6.2.5. *If $f_u(\omega)$ is associated with $M(p,q)$, then $\Gamma_{p,q}^{p}$ is of full rank if and only if $f_u(-\omega)$ is associated with $M(p,q)$. This condition always holds for $s = 1$ or $q = 0$.*

We do not give the rather lengthy proof from the reference cited because the theorem can, in fact, be proved in the same way as for Theorem 6.2.3. For $M(p,p)$ this is obvious since $M(p,p) = U(sp)$. However, all $M(p,q)$ can be obtained as "generic neighborhoods" in relation to an appropriately chosen $\mathscr{H}$. For example, if $\mathscr{H}$ has $K(i+j-2)$ as the (i,j)th block, the first n rows define a set of systems, having these rows as basic to $\mathscr{H}$ and this set is just $M(p+1,q)$ if $n = (p+1)s$, as is easily checked. Similarly, for any case $p \geq q$ we take the Hankel matrix with $K(i+j-p+q-1)$ as the (i,j)th block, i, $j = 1, 2, \ldots$, with $K(j) = 0$, $j < 0$, of course. If $p < q$, then the Hankel matrix must be similarly constructed from $k(z)^{-1}$. The fact that the condition holds for $s = 1$ or $q = 0$ is obvious, for then $f_u(-\omega) = f_u(\omega)$ (when $s = 1$) and when $q = 0$, $f_u(-\omega)^{-1}$ is again a matrix of trigonometric polynomials of degree p.

As a consequence, when the condition of theorem 6.2.5 holds, then $G(j-k)$ may be inserted in place of $\Gamma(j-k)$ in (6.2.14) and taking $k = q+1, \ldots, q+p$ the resulting equations solved to obtain consistent estimators of the $A(j)$, just as in Theorem 6.2.4. Theorems 6.2.3, 6.2.4, and 6.2.5 relate to the ARMA case. In the ARMAX case we can begin from

$$\begin{aligned}\sum_{j=1}^{s}\sum_{v=0}^{n_i} \tilde{\alpha}_{ij}(v) y_j(t-n_i+v) &= \sum_{j=1}^{m}\sum_{v=1}^{n_i} \tilde{\delta}_{ij}(n_i-v) z_j(t-v) \\ &\quad + \sum_{j=1}^{s}\sum_{v=0}^{n_i} \tilde{\beta}_{ij}(n_i-v)\varepsilon_j(t-v).\end{aligned} \tag{6.2.15}$$

Now choose a suitable set of instrumental variables [i.e., variables uncorrelated with the $\varepsilon_j(t-v)$ occurring in (6.2.15)] multiply by these and take expectations. Next replace the covariances, of the $y(t)$, $z(t)$, by estimates and solve the equations for estimates of the $\tilde{\alpha}_{ij}(v)$, $\tilde{\delta}_{ij}(v)$. A natural set of instrumental variables to choose is

$$y_j(t-w), \qquad w = n_i + x + 1, \qquad x = 0, 1, \ldots, n_j - 1;$$
$$z_j(t-v), \qquad v = 1, 2, \ldots, n_j.$$

Given that the estimates of the covariances are consistent, it is necessary to know that the matrix of the system of equations is nonsingular. It is not easy to give neat conditions ensuring that this will be so.

Even in the ARMA case there is a defect, for as we have seen in the example after Theorem 6.2.1, the estimates of the covariances in Theorem 6.2.4 may be too inefficient to be useful. This question has been considered in the scalar ARMA case by Walker (1962) (see also Godolphin, 1976; Anderson, 1971, pp. 226–268, 237–241). Walker explicitly deals with the Gaussian case, but the results are more general and follow from the following theorem.

Theorem 6.2.6. *Let $s = 1$ and*

$$u(t) = \sum_0^\infty \kappa(j)\varepsilon(t-j), \qquad \sum_1^\infty j^{1/2}\kappa(j)^2 < \infty$$

and let (4.1.15), (4.1.16) hold. Then $T^{1/2}\{r(j) - \rho(j)\}$, $j = 1, 2, \ldots, m$ are jointly asymptotically normal, for any m, with zero mean and covariance matrix with (i,j)th element.

(6.2.16)

$$\sum_{t=1}^\infty \{\rho(t+i) + \rho(t-i) - 2\rho(t)\rho(i)\}\{\rho(t+j) + \rho(t-j) - 2\rho(t)\rho(j)\}.$$

The proof is given in Hannan and Heyde (1972) and is reproduced in Hall and Heyde (1980, pp. 188–194). As can be seen from Theorem 6.2.1, translating that into the scalar ARMA context, the parameters may be chosen to be $\alpha(1), \ldots, \alpha(p)$, $\gamma(1), \ldots, \gamma(q)$, $\gamma(0)$, taking the basic model to be in $M(p,q)$ for $s = 1$. Indeed, now H is constant, F depends only on the $\alpha(j)$, L is null, Γ depends only on the $\gamma(j)$, $j = 1, 2, \ldots, \max(p,q)$, and $\Gamma_u(0) = \gamma(0)$. If $p > q$, so that certain constraints are imposed [as they also are for $p < q$ and are taken account of by putting $\alpha(j) = 0, j > p$], then

$$\sum_0^p \alpha(j)\gamma(j-k) = 0, \qquad k > q,$$

so that $\gamma(q+1), \gamma(q+2), \ldots, \gamma(p)$ are functions of the other parameters. Equivalently, we may use $\alpha(1), \ldots, \alpha(p)$, $\rho(1), \ldots, \rho(q)$, $\gamma(0)$. For that matter an equivalent set is

$$\rho(j), \quad 1 \le j \le q; \quad \sum_0^p \alpha(j)\rho(j-k), \quad k = q+1, \ldots, q+p; \quad \gamma(0).$$

Indeed, the mapping from $\alpha(1), \ldots, \alpha(p)$, $\rho(1), \ldots, \rho(q)$, $\gamma(0)$ to this set is one-to-one, since the matrix with entries $\rho(j-k)$, $j = 1, \ldots, p; k = q+1, \ldots, q+p$ is just the matrix that is shown to be nonsingular in Theorem 6.2.5 (for $s = 1$). Now Walker's clever idea is to consider the joint distribution of

$$T^{1/2}\{r(j) - \rho(j)\}, \quad 1 \le j \le q; \quad T^{1/2}\sum_0^p \alpha(j)r(j-k),$$
$$k = q+1, \ldots, q+p,$$

together with m additional $T^{1/2}\sum \alpha(j)r(j-k)$ $j = q+p+1, \ldots, q+p+m$ and to act as if that is the normal distribution derived from Theorem 6.2.6. Thus the quantities in (6.2.16) are treated as known, the likelihood is written down, and the "maximum likelihood" estimators are derived for $\alpha(1), \ldots, \alpha(p)$, $\rho(1), \ldots, \rho(q)$, which are the parameters. Of course, these estimators depend on the true $\rho(j)$ through (6.2.16). Walker truncates these sums appropriately and uses the $r(j)$ to estimate the $\rho(j)$ in them. As Godolphin points out, this procedure should be iterated with the estimates used to obtain better estimates of the quantities in (6.2.16) for the next iteration. Elements of subjective judgment are involved in determining a truncation point for the sums in (6.2.16) and in choosing m. Walker shows that if first $T \to \infty$ and then m is increased, the estimates by his method become asymptotically efficient in the sense that they have the same limiting distribution as those obtained from the Gaussian likelihood. We refer to Walker (1962) and Godolphin (1976) for details. The method differs from others in that the critical element is the location of the poles of $k(z)$, for this determines the location of the truncation point in (6.2.16) and also how large m needs to be. Of course, when spectral factorization is needed to obtain estimates of the $\beta(j)$, then the location of zeros will also be important. Walker's method has problems, nevertheless. It contains subjective elements. It is so far restricted to the scalar ARMA case. It is not easy to see how to reduce the calculations when p and q have also to be determined from the data. For these reasons we go on to study other methods.

3. AKAIKE'S METHOD

Again only the ARMA case is here considered, for the same reason as was referred to in Chapter 3, namely that

$$y(t+v|t) = \sum_{v}^{\infty} L(j)z(t+v-j) + \sum_{v}^{\infty} K(j)\varepsilon(t+v-j), \qquad v \geq 1,$$

is not, in general, the best linear predictor of $y(t+v)$ from observations to time t since the autocorrelation structure of $z(t)$ is neglected. The structure theory based on the Hankel matrix therefore has an artificial quality about it, and this makes it difficult to generalize Akaike's procedure. As a consequence, we shall call $u(t)$ the vector of observables under consideration throughout this section.

Let us construct the (infinite-dimensional) vectors $u^{(t)} = (u(t)', u(t+1)', \ldots)'$, $u_t = (u(t)', u(t-1)', \ldots)'$ with corresponding definitions for $\varepsilon^{(t)}$, ε_t. From

$$u(t) = \sum_{0}^{\infty} K(j)\varepsilon(t-j) \tag{6.3.1}$$

we obtain

$$u^{(t+1)} = \mathscr{H}\varepsilon_t + \mathscr{K}\varepsilon^{(t+1)}, \tag{6.3.2}$$

where $\mathscr{K}$ has $K(i-j)$ as the (i,j)th block, $i, j = 1, 2, \ldots$. This equation exhibits the relation between the "future" $u^{(t+1)}$ and the present and past ε_t and again shows the central part played by $\mathscr{H}$. The idea in Akaike (1976) is to find the structure of the ARMA system via a canonical correlation analysis procedure. Canonical correlation analysis was discussed in Section 2.8 in a theoretical way and will further be discussed in Chapter 7. We seek to find random variables ξ_j, η_j so that ξ_j is a linear function of the present and past and η_j of the future, so that

$$E(\xi_j\xi_k) = E(\eta_j\eta_k) = \delta_{jk}, \qquad E(\xi_j\eta_j) = \rho_j$$

with $\rho_1 \geq \rho_2 \geq \cdots \geq \rho_n > 0$, n being the McMillan degree. If $u^{(t+1)}$, ε_t were observable, we should find canonical correlations by finding the singular-value decomposition of

(6.3.3)

$$[E\{u^{(t+1)}u^{(t+1)'}\}]^{-1/2}E\{u^{(t+1)}\varepsilon_t'\}[E\{\varepsilon_t\varepsilon_t'\}]^{-1/2} = [\Gamma_u(k-j)]^{-1/2}\mathscr{H}(I_\infty \otimes \Sigma^{1/2}),$$

where the block in row j, column k is shown in the first matrix. There would be n of these singular values that correspond to nonnull canonical correlations. Of course, (6.3.3) is not available and approximations to $u^{(t+1)}$, ε_t have to be constructed. We discuss that below. In place of ε_t, we may as well use u_t since $u_t = \mathscr{S}\varepsilon_t$, where $\mathscr{S}$ has $K(j-i)$ in row i, column j of blocks and each ξ_j can equally be expressed as a linear function of u_t.

Akaike (1976) sets out to find the structure of the system by finding null functions $y_j = \tilde{a}_j' u^{(t+1)}$ such that $E\{y_j u_t\} = 0$. Assume that $u(t)$ is generated by an ARMA process with $k(z) \in V_\alpha$, where α indexes the Kronecker indices $(n_1, n_2, \ldots, n_s)$. Let a_j' be the jth row of the matrix A in (6.2.11). Then put $\tilde{a}_j' = (a_j', o_\infty')$, where o_∞' is an infinite row of zeros. Then $\tilde{a}_j'\mathscr{H} = 0$ and $y_j = \tilde{a}_j' u^{(t+1)}$ is evidently a null function, from (6.3.2). Having this in mind, we proceed as follows. Replace u_t by

$$u_{(h)}(t) = [u(t)', u(t-1)', \ldots, u(t-h+1)']',$$

where h is determined by fitting an autoregression to $u(t)$ and determining h by AIC (see Section 5.6 and also the next section for these procedures). Replace $u^{(t+1)}$ by

(6.3.4)

$$u_{p,q}(t) = [u(t+1)', u(t+2)', \ldots, u(t+p)', u_1(t+p+1), \ldots, u_q(t+p+1)]'$$
$$p = 1, 2, 3, \ldots; \qquad q = 0, 1, 2, \ldots, s-1.$$

If $n_1, n_2, \ldots, n_q \leq p+1$; $n_{q+1}, \ldots, n_s \leq p$, then the first p rows of blocks plus the first q elements of the $(p+1)$ row of blocks certainly span the rows of $\mathscr{H}$, which may as well be replaced by these rows. Call the resulting matrix $\mathscr{H}_{p,q}$. Thus we are led to consider, in place of (6.3.3),

$$X_T = \frac{1}{T}\sum_h^{T-p-1} u_{(h)}(t)u_{(h)}(t)', \qquad Y_T = \frac{1}{T}\sum_h^{T-p-1} u_{p,q}(t)u_{p,q}(t)',$$

so that the canonical correlations and corresponding "discriminant functions" are found by solving

$$\hat{a}_j'[\hat{\rho}_j Y_T - B_{p,q,T} X_T B_{p,q,T}'] = 0, \qquad \hat{\rho}_1 \geq \cdots \geq \hat{\rho}_{ps+q} \geq 0 \tag{6.3.5}$$
$$B_{p,q,T} = \{T^{-1} \textstyle\sum u_{pq}(t)u_{(h)}(t)'\} X_T^{-1}.$$

We refer the reader to Anderson (1958) for details. The $\hat{\rho}_j$ are the sample canonical correlations and $\hat{\eta}_j = \hat{a}_j' u_{pq}(t)$ are the corresponding functions of the future having correlation $\hat{\rho}_j$ with a corresponding function of the present and past, that being $\hat{a}_j' B_{p,q,T} u_{(h)}(t)$.

Let us say that $n_q = p, n_1, n_2, \ldots, n_{q-1} \geq p + 1; n_{q+1}, n_{q+2}, \ldots, n_s \geq p$. Then the first "true" null function we shall find by the series of canonical correlation analyses obtained by varying p, q in (6.3.4) is [see (6.2.9)]

$$\text{(6.3.6)} \quad \sum_{j=1}^{p} \sum_{v=0}^{p} \tilde{\alpha}_{q,j}(v) u_j(t + 1 + v) = \sum_{j=1}^{p} \sum_{v=0}^{p} \tilde{\beta}_{q,j}(p - v) \varepsilon_j(t + p + 1 - v),$$

which is certainly orthogonal to the past. This null function is unique up to a scalar since $n \geq (q - 1)(p + 1) + (s - q + 1)p = sp + q - 1$, so that there can be at most one null function. By the same argument there can be no null function for smaller p, or for the same p and smaller q, in (6.3.4). We need a criterion to decide if a null function has been found, and this clearly should be based on $\hat{\rho}_{ps+q}$, which is the smallest sample canonical correlation. Akaike (1976) uses the criterion

$$\text{(6.3.7)} \quad (T - v_{p,q}) \log(1 - \hat{\rho}_{ps+q}^2) + v_{p,q}, \qquad v_{p,q} = s(h - p) - q + 1.$$

When, *for the first time* [as we proceed through (p, q) values in dictionary order], (6.3.7) is positive, a null function is declared. Say that at $q = j$, $p = \hat{n}_j$ a null function is found; then we put $n_j = \hat{n}_j$ and take [see (6.3.5)] $\hat{a}_u$, $u = \hat{n}_j s + j$ as the estimate of the vector of values $\tilde{\alpha}_{j,i}(v)$, $i = 1, \ldots, s$, $v = 0, 1, \ldots, \hat{n}_j$ in (6.3.6), subject to the normalization $\tilde{\alpha}_j(\hat{n}_j) = 1$. *Now all $u_j(t + k)$, $k \geq \hat{n}_j$, are eliminated from later $u_{p,q}(t)$* (i.e., later in the dictionary ordering). *Strictly speaking, a new notation should then be introduced, but to avoid complexity we refrain and ask the reader to understand that this will have been done.* Now the procedure is continued until all $\hat{n}_k$ have been determined, and the estimates of the corresponding $\tilde{\alpha}_{k,i}(v)$, $i = 1, \ldots, s$; $v = 0, 1, \ldots, \hat{n}_k$, have been determined. In general (i.e., after the first time), $v_{p,q}$ in (6.3.7) will be $sh + 1$ less the dimension of the vector $u_{p,q}(t)$ [which dimension will change due to the deletion of $u_j(t + k)$, $k \geq \hat{n}_j$ for j such that $\hat{n}_j$ has previously been determined]. This agrees with (6.3.7) since there $v_{p,q}$ is $sh + 1 - (sp + q)$.

At the completion of the procedure $\hat{n}_1, \ldots, \hat{n}_s$ and $\hat{a}(z)$ will have been determined, the latter being the autoregressive part of the reversed echelon form for these Kronecker indices. Thus $\hat{F}$, $\hat{H}$ in the canonical state-space representation are also determined. Since we have ready made estimates of Γ, $\Gamma(0)$ given the $\hat{n}_j$ (see Theorem 6.2.1), from the $G(j)$, we have initial values for the procedure based on Theorems 6.2.1 and 6.2.2. (Now L does not occur because we consider only the ARMA case.) Thus the procedure is technically complete.

As the procedure actually is used by Akaike (1976), in the examples in that paper, the $u_j(t + k)$, $k \geq \hat{n}_j$, are *not* removed from later $u_{p,q}(t)$. In that case there should be an *additional* null function occurring at $(p, q) = (\hat{n}_j + v, j)$ for every $v > 0$ (e.g., if $v = 2$ and all other n_k are greater than $\hat{n}_j + v$, there should be three null functions). Then the criterion is also adjusted so that the first term involves the sum of the logarithms of the $1 - \hat{\rho}_u^2$, for the largest u values. This procedure can lead to contradictory results if the criterion does not indicate, for example, an additional null function at $(\hat{n}_j + v, j)$. Of course, such a contradiction has a virtue in suggesting some faults in the model.

No theory in relation to Akaike's method is available at the moment. There is no doubt that such a theory could be developed, along the lines of a later section, but it would be complicated. The matter is not pursued here for that reason and also for the following. (1) The method is restricted, so far, to the ARMA case. (2) The $\hat{n}_j$ determined by the canonical correlation procedure are inefficient since the calculation is based on sample autocovariances [see (6.3.5)] and it is known (see Section 6.2) that these are poor estimates. Of course, this will be true of all first estimates (e.g., those in Section 4), but here there is no simple procedure for improving those. Of course, one thing that might be done is to compute the maximum of the likelihood for a few sets of n_j, $j = 1, \ldots, s$ near $\hat{n}_j$, $j = 1, \ldots, s$, but this does not correspond to the program suggested in Section 6.1. (3) The procedure seems fairly costly, although no operation count has been made. However, finding the Kronecker indices is fairly certain to be costly in any case.

The procedure is an interesting one, proceeding from deep understanding. It could prove to be effective, but there is little evidence at the moment.

4. MODIFIED ALGORITHMS FOR AUTOREGRESSION

Throughout the next sections extensive use is made of the Levinson and Whittle algorithms (see Section 5.6). As explained earlier, these are Toeplitz calculations and involve the fiction that $y(t) = 0$ when either $t \leq 0$ or $t > T$. Consequently, biases are produced when T is not large (see Tjøstheim and Paulsen, 1983). There is an extensive discussion in the literature on systems and control, and also on signal processing, of modifications of these algorithms. This discussion uses a terminology unfamiliar to statisticians. Thus the partial correlations, $-\hat{\phi}_h(h)$ (see Section 5.6), are called reflection coefficients and the modified methods themselves are said to be lattice calculations or ladder forms, all of this terminology appearing to derive from the appearance of the flow diagrams describing the methods. We refer the reader to Friedlander (1982) and Makhoul (1981) for surveys of these methods and references. See also Section 7.5 below. Consider the recursion (5.6.13). We may use the $\hat{\Phi}_h(j)$, $\tilde{\Phi}_h(j)$ to compute forward and backward residuals.

$$\hat{\varepsilon}_h(t) = \sum_0^h \hat{\Phi}_h(j)y(t-j), \qquad \hat{r}_h(t) = \sum_0^h \tilde{\Phi}_h(j)y(t-h+j).$$

Then, from (5.6.13),

$$\begin{gathered}\hat{\varepsilon}_h(t) = \hat{\varepsilon}_{h-1}(t) + \hat{\Phi}_h(h)\hat{r}_{h-1}(t-1),\\ \hat{r}_h(t) = \hat{r}_{h-1}(t-1) + \tilde{\Phi}_h(h)\hat{\varepsilon}_{h-1}(t)\\ \hat{r}_h(0) = 0; \qquad \hat{\varepsilon}_0(t) = \hat{r}_0(t) = y(t), \qquad 1 \le t \le T;\\ y(t) = 0, \qquad t \le 0, \qquad t > T.\end{gathered} \tag{6.4.1}$$

Then (see Section 7.5 for details)

$$\Delta_h = \frac{1}{T}\sum_1^{T+h-1} \hat{\varepsilon}_{h-1}(t)\hat{r}_{h-1}(t-1)'. \tag{6.4.2}$$

Thus (5.6.13) could be effected by using (6.4.1) to calculate Δ_h and hence $\hat{\Phi}_h(h)$, $\tilde{\Phi}_h(h)$, via the last line of (5.6.13) and hence the $\hat{\Phi}_h(j)$, $\tilde{\Phi}_h(j)$ via the first line of (5.6.13). This would, for $s = 1$, involve about $3hT$ calculations, to order h, compared to about hT for (5.6.13) (taking $T \gg h$). The point of (6.4.1), however, is partly to allow for modifications of the procedure to mitigate the effects of the Toeplitz procedure. As we shall see in Chapter 7, and as can also be seen directly, the formulas (6.4.1) also lend themselves to real-time calculation. At the same time, it is desirable to maintain the truth of the conclusion of Theorem (5.6.2), that is, that

$$\det\left\{\sum_0^h \hat{\Phi}_h(j)z^j\right\}, \qquad \det\left\{\sum_0^h \tilde{\Phi}_h(j)z^j\right\} \neq 0, \qquad |z| \le 1. \tag{6.4.3}$$

(Actually only the former is proved in Theorem 5.6.2, but the latter follows in the same way.) The first person to recognize this seems to have been Burg (1975), who proposed, for $s = 1$, the following algorithm. Let us call $\bar{\phi}_h(j)$, $j = 0, 1, \ldots, h$, $\bar{\phi}_h(0) = 1$, the new coefficients to be calculated. Of course, for $s = 1$, $\tilde{\phi}_h(j) = \hat{\phi}_h(j)$ and no new backward coefficients need to be introduced. Having computed $\bar{\phi}_h(j)$, then (6.4.1) is used, with $\bar{\phi}_h(h)$ replacing $\hat{\phi}_h(h)$, $\tilde{\phi}_h(h)$ to compute (let us say) $\bar{\varepsilon}_h(t)$, $\bar{r}_h(t)$ and then

$$\bar{\phi}_{h+1}(h+1) = -\left\{\sum_{h+1}^T \bar{\varepsilon}_h(t)\bar{r}_h(t-1)\right\}\Bigg/\left\{\frac{1}{2}\sum_{h+1}^T \bar{\varepsilon}_h(t)^2 + \frac{1}{2}\sum_{h+1}^T \bar{r}_h(t-1)^2\right\}. \tag{6.4.4}$$

Then the first line of (5.6.6) is used to compute $\bar{\phi}_{h+1}(j)$, $j = 1, \ldots, h$. It is easily seen that $|\bar{\phi}_h(h)| < 1$. It is evident that the procedure is closely related to eliminating the first and last h rows of the array (5.6.3), thus mitigating the effects of the Toeplitz calculation. There is abundant evidence that it does that. It is almost obvious that the procedure is asymptotically equivalent to the Levinson procedure, since it is easily checked that

$$\sum_{1}^{T+h} \hat{\varepsilon}_h(t)^2 = \sum_{1}^{T+h} \hat{r}_h(t)^2,$$

so that if the sums in (6.4.4) were over $1 \leq t \leq T + h$, then the resulting procedure would just be the Levinson procedure. The effects of the elimination of $2h$ terms will, for h fixed, be asymptotically negligible.

In some ways a more natural definition of $\bar{\phi}_{h+1}(h+1)$ would be -1 by the coefficient of correlation between $\varepsilon_h(t), \bar{r}_h(t-1), h + 1 \leq t \leq T$. This, of course, would also ensure $|\bar{\phi}_h(h)| < 1$. This, in turn, ensures that (6.4.3) continues to hold, in this case $s = 1$. It is also this definition that generalizes, as we now show following Morf, Vieira, and Kailath (1978).

As has previously been pointed out (see immediately above Theorem 5.6.2), the Whittle recursion is completely determined by the $G(j)$ regarded as the Fourier coefficients of a spectral density $f(\omega)$, in this case $I(\omega) = w(\omega)w(\omega)^*$. Let us consider therefore any sequence $G(j)$ for which

$$G_h = [G(j-k)]_{j,k=1,\ldots,h} > 0, \qquad h = 1, 2, \ldots, H.$$

Then since, from (5.6.13),

$$\begin{bmatrix} I_s & \hat{\Phi}_h(1) & \cdots & \hat{\Phi}_h(h) \\ 0 & & I_{sh} & \end{bmatrix} G_{h+1} \begin{bmatrix} I_s & 0 \\ \hat{\Phi}_h(1)' & \\ \vdots & I_{sh} \\ \hat{\Phi}_h(h)' & \end{bmatrix} = \begin{bmatrix} \hat{\Sigma}_h & 0 \\ 0 & G_h \end{bmatrix},$$

it is evident that

$$\det G_{h+1} = \prod_{0}^{h} \det \hat{\Sigma}_j,$$

so that

$$(6.4.5) \qquad G_h > 0, \qquad h = 1, \ldots, H; \qquad \hat{\Sigma}_h > 0, \qquad h = 0, 1, \ldots, H - 1$$

are equivalent statements. (If $\det G_j > 0$, $j = 1, 2, \ldots, h$, then $G_h > 0$, as is easily seen.) By an obvious modification of these arguments, we find also that

(6.4.6) $G_h > 0, \quad h = 1, \ldots, H; \quad \tilde{\Sigma}_h > 0, \quad h = 0, 1, \ldots, H-1$

are equivalent and

$$\det G_{h+1} = \prod_0^h \det \tilde{\Sigma}_j, \qquad h = 0, 1, \ldots, H-1,$$

so that $\det \hat{\Sigma}_h = \det \tilde{\Sigma}_h$, $h = 0, 1, \ldots, H-1$. Since $G_h > 0$ implies [see (5.6.14) and following]

$$\det\left\{\sum_0^h \hat{\Phi}_h(j) z^j\right\} \neq 0, \qquad |z| \leq 1,$$

then so does $\hat{\Sigma}_j > 0$, $j = 0, 1, \ldots, h-1$ and similarly for $\tilde{\Sigma}_j > 0$, $j = 0, 1, \ldots, h-1$. Also, each ensures that

$$\det\left\{\sum_0^h \tilde{\Phi}_h(j) z^j\right\} \neq 0, \qquad |z| \leq 1.$$

Now, still following Morf, Vieira, and Kailath (1978), let us factor $\hat{\Sigma}_h$, $\tilde{\Sigma}_h$ as

$$\hat{\Sigma}_h = \hat{R}_h \hat{R}_h', \qquad \tilde{\Sigma}_h = \tilde{R}_h \tilde{R}_h',$$

where we may uniquely define $\hat{R}_h$, $\tilde{R}_h$ by requiring them to be lower triangular with positive diagonal elements. Thus we use the Choleski decomposition. Then put

$$P_{h+1}^{1/2} = \hat{R}_h^{-1} \hat{R}_{h+1}, \qquad Q_{h+1}^{1/2} = \tilde{R}_h^{-1} \tilde{R}_{h+1}, \qquad P_0^{1/2} = Q_0^{1/2} = \hat{R}_0$$

and

$$\Omega_h = \hat{R}_h^{-1} \Delta_h (\tilde{R}_h')^{-1},$$

so that, using the last line of (5.6.13),

$$\hat{R}_{h+1} \hat{R}_{h+1}' = \hat{R}_h \{I - \hat{R}_h^{-1} \Delta_h (\tilde{R}_h')^{-1} \tilde{R}_h^{-1} \Delta_h' (\hat{R}_h')^{-1}\} \hat{R}_h',$$

and thus

(6.4.7) $$P_{h+1} = \{I - \Omega_h \Omega_h'\}.$$

Note that, by this definition, $P_{h+1} = P_{h+1}^{1/2}(P_{h+1}^{1/2})'$, and this defines the unique Choleski decomposition of P_{h+1}. Similarly,

$$(6.4.8) \qquad Q_{h+1} = (I - \Omega_h'\Omega_h).$$

Now the first line of (5.6.13) can be rewritten as

$$(6.4.9) \quad \{\hat{R}_h^{-1}\hat{\Phi}_h(j)\} = P_h^{-1/2}\{\hat{R}_{h-1}^{-1}\hat{\Phi}_{h-1}(j) - \Omega_{h-1}\tilde{R}_{h-1}^{-1}\tilde{\Phi}_{h-1}(h-j)\},$$

$$(6.4.10) \quad \{\tilde{R}_h^{-1}\tilde{\Phi}_h(j)\} = Q_h^{-1/2}\{\tilde{R}_{h-1}^{-1}\tilde{\Phi}_{h-1}(j) - \Omega_{h-1}'\hat{R}_{h-1}^{-1}\hat{\Phi}_{h-1}(h-j)\}$$

$$j = 0, 1, \ldots, h; \qquad \hat{\Phi}_h(0) = \tilde{\Phi}_h(0) = I_s, \qquad \hat{\Phi}_h(j) = \tilde{\Phi}_h(j) = 0, \qquad j > h.$$

Moreover, Ω_h has all of its singular values less than unity in modulus, as is obvious from (6.4.7), since $P_{h+1} > 0$. Finally, since

$$P_0^{1/2}P_1^{1/2}\cdots P_h^{1/2} = \hat{R}_h, \qquad Q_0^{1/2}Q_1^{1/2}\cdots Q_h^{1/2} = \tilde{R}_h,$$

then

$$(6.4.11) \qquad \Omega_h = P_h^{-1/2}P_{h-1}^{-1/2}\cdots P_0^{-1/2}\Delta_h\tilde{R}_h^{-1}.$$

Thus equations (6.4.7) to (6.4.11) define a recursion that is alternative to (5.6.13). Moreover, say that a sequence, Ω_h, is given with each Ω_h, $h = 0, 1, \ldots, H-1$ having singular values less than unity, together with $G(0) > 0$. Then we may use these uniquely to define the quantities $\{\hat{R}_h^{-1}\hat{\Phi}_h(j)\}$, $\{\tilde{R}_h^{-1}\tilde{\Phi}_h(j)\}$, via (6.4.7) to (6.4.10). These are $\hat{R}_h^{-1}$, $\tilde{R}_h^{-1}$ for $j = 0$, and hence $\hat{\Phi}_h(j)$, $\tilde{\Phi}_h(j)$ and hence, also, the $G(j)$ are defined via

$$\Delta_h = \sum_0^h \hat{\Phi}_h(j)G(j-h-1), \qquad G(j) = G(-j)'.$$

Then the corresponding matrices G_h satisfy $G_h > 0$, $h = 1, \ldots, H$, because of the correspondences (6.4.5), (6.4.6) and the fact that $\hat{\Sigma}_h > 0$ is guaranteed by $\hat{R}_h$ being of the required form and this is guaranteed by $P_h > 0$, which follows from (6.4.7) and the condition on Ω_h. Thus we may use (6.4.7) to (6.4.10) to define the forward and backward autoregressions satisfying (6.4.3), once a sequence Ω_h is given having singular values less than unity. Now we may proceed as follows. We introduce, now, some confusion of notation to avoid a more complex notation by *using* $\hat{\Phi}_h(j)$, $\tilde{\Phi}_h(j)$, $\hat{\varepsilon}_h(t)$, $\hat{r}_h(t)$ *to stand for the quantities defined in terms of the* Ω_h *now to be introduced [see (6.4.12)], although they will be different from those defined via the* $G(j)$ *and (5.6.13), with the* $G(j)$ *as in Chapter 4. Given* $\hat{\Phi}_h(h)$, $\tilde{\Phi}_h(h)$ we may form the $\hat{\varepsilon}_h(t)$, $\hat{r}_h(t)$ recursively as in

(6.4.1) and then Choleski factorize

$$\sum_h^T \hat{\varepsilon}_h(t)\hat{\varepsilon}_h(t)', \qquad \sum_h^T \hat{r}_h(t-1)\hat{r}_h(t-1)',$$

as $R_h^\varepsilon(R_h^\varepsilon)'$ and $(R_h^r)(R_h^r)'$ followed by the formation of

$$\Omega_h = (R_h^\varepsilon)^{-1} \sum_h^T \hat{\varepsilon}_h(t)\hat{r}_h(t-1)'[(R_h^r)']^{-1}. \tag{6.4.12}$$

Then Ω_h has singular values as required since the solutions to

$$0 = \det\left[\lambda^2 \sum_h^T \hat{\varepsilon}_h(t)\hat{\varepsilon}_h(t)' - \sum_h^T \hat{\varepsilon}_h(t)\hat{r}_h(t-1)'\left(\sum_h^T \hat{r}_h(t-1)\hat{r}_h(t-1)'\right)^{-1}\left\{\sum_h^T \hat{\varepsilon}_h(t)\hat{r}_h(t-1)'\right\}'\right]$$

are the cosines of the angles between the planes in $(T-h+1)$-dimensional space spanned, respectively, by the rows of

$$[\hat{\varepsilon}_h(h), \ldots, \hat{\varepsilon}_h(T)], \qquad [\hat{r}_h(h-1), \ldots, \hat{r}_h(T-1)].$$

See Anderson (1958, p. 301) and Section 6.3.

This procedure would be computationally complex compared to the Levinson recursion and, for $T \gg H$, would involve more calculations than a direct solution of the equations of least squares, for each h, using rows $h+1$ to T of the array (5.6.3). This last procuedure is sometimes called the "covariance" calculation. The Toeplitz calculation is sometimes said to be "pre- and postwindowed" because of a hypothetical window that allows us to see only $y(1), \ldots, y(T)$ and makes us see the remainder of the series as null. Friedlander (1982) mainly deals with the prewindowed form and provides recursions without the postwindowed requirement. These calculations are really aimed at the real time, adaptive, situation and thus a "forgetting factor" may be introduced that weights down early $y(t)$ in a geometric fashion, depending on T, so that prewindowing is not important. Of course the elimination of prewindowing causes more trouble as increasing h, then, alters the previous columns in the array [see (5.6.3)]. In any case the calculations involved in the above are not very great by modern standards and might form a convenient way of calculating the $\hat{\Phi}_h(j)$. We discuss them further in Chapter 7 (i.e., Section 7.5).

An alternative to (6.4.7)–(6.4.10) together with (6.4.12) that is designed to reduce the effects of the pre- and postwindowing is implicit in Dahlhaus (1983), where $s = 1$ is considered in connection with the use of autoregression procedures in spectral estimation. In connection with the direct use of the

$$w(\omega) = T^{-1/2} \sum_{1}^{T} y(t)e^{it\omega}$$

for spectral estimation, a taper is often used (see Hannan, 1970, pp. 265–267), whereby $y(t)$ is replaced by $a(t/T)y(t)$, where $a(x)$ is a function that is zero outside $[0, 1]$ and at 0, 1 has high-order contact to the horizontal axis and is appropriately smooth. Dahlhaus considers the commonly used form

$$a(x) = \begin{cases} 1/2\{1 - \cos(2\pi x/\rho)\}, & 0 \leq x \leq \rho/2 \\ 1, & \rho/2 \leq x \leq 1 - \rho/2 \\ 1/2\{1 - \cos[2\pi(1 - x)/\rho]\}, & 1 - \rho/2 \leq x \leq 1. \end{cases}$$

It is clear that this will have effects similar to the elimination of pre- and postwindowing, and the simulations and theory in Dahlhaus (1983) support this. Once ρ is prescribed, this procedure would be easy to effect since the Levinson-Whittle procedure would be applied to the new vector series $a(t/T)y(t)$. Dahlhaus (1983) uses $\rho = 0.6$, for example. A rough guide based on (5.6.3) would seem to suggest $\rho = 4H/T$ (which would have been about 0.2 in the simulated example in Dahlhaus, 1983). However, ρ should clearly depend on h. The subjective element in the choice of ρ is a disadvantage.

As stated earlier, there is an extensive literature concerning algorithms such as (6.4.7)–(6.4.11), and we refer the reader to Friedlander (1982) for a survey and additional references. Further discussion appears in Section 7.5.

5. RECURSIVE REGRESSION-AUTOREGRESSION PROCEDURE

The procedure to be discussed here has sometimes been called the Hannan-Rissanen procedure, because it was presented in a paper by those authors in *Biometrika* (1982). However, the method, for the scalar ARMA case, with the orders assumed known and with a subjective choice of the initial autoregressive order, was suggested earlier by Durbin (1960), so that terminology will not be used. The extension to the vector case was presented in Hannan and Kavalieris (1984).

The idea is a simple one and it will be motivated first by considering the scalar ARMA case. In accordance with the developments in Sections 6.2 and 5.3, the criterion to be minimized can be taken to be

$$\text{(6.5.1)}\qquad \log\left\{\frac{1}{2\pi}\int |w(\omega)|^2|k_\tau|^{-2}\,d\omega\right\} + (p+q)C_T/T, \qquad p, q \le H_T,$$

where C_T and H_T are prescribed sequences. Assume that $\hat{p}$, $\hat{q}$, $\hat{k}$ have been calculated at a previous stage. Linearize k_τ^{-1} about $\hat{k}$. Then the expression whose logarithm is being taken in (6.5.1) is

$$\text{(6.5.2)}\qquad \frac{1}{2\pi}\int |w(\omega)|^2|\hat{k}^{-1} + (\partial_\tau \hat{k}^{-1})(\tau - \hat{\tau})|^2\,d\omega.$$

Here $\partial_\tau \hat{k}^{-1}$ is $\partial_\tau k_\tau^{-1}$ evaluated at $\hat{k}$. Now (6.5.2) is linear in τ and minimizing (6.5.1), using (6.5.2), we obtain

$$\text{(6.5.3)}\qquad \begin{aligned} \hat{\tau}^{(1)} &= -\left[\frac{1}{2\pi}\int |w(\omega)|^2(\partial_\tau \hat{k}^{-1})^*(\partial_\tau \hat{k}^{-1})\,d\omega\right]^{-1} \\ &\quad \times \frac{1}{2\pi}\int |w(\omega)|^2\{\hat{k}^{-1} - \partial_\tau \hat{k}^{-1}\hat{\tau}\}(\partial_\tau \hat{k}^{-1})^*\,d\omega. \end{aligned}$$

Thus a Gauss-Newton procedure is being used.

This expression (6.5.3) can be rewritten as a Toeplitz regression (see Section 5.6) of $(\hat{k}^{-1} - \partial_\tau \hat{k}^{-1}\hat{\tau})u(t)$ on $(\partial_\tau \hat{k}^{-1})'u(t)$. Here, of course, $(\hat{k}^{-1} - \partial_\tau \hat{k}^{-1}\hat{\tau})$ is interpreted as a lag operator operating on $u(t)$. Recall moreover that in so evaluating (6.5.3), $u(t)$ is to be treated as null outside $1 \le t \le T$. If the same convention is applied to all quantities, (6.5.3) is easily interpreted. Thus $(\hat{k}^{-1} - \partial_\tau \hat{k}^{-1}\hat{\tau})u(t)$ will also be null for $t \le 0$, evidently. Of course, putting $u(t) = 0$ for $t \le 0$ or $t > T$ is a fiction with bad effects, even though those effects are asymptotically negligible. That will be discussed below. Now

$$\partial_\tau \hat{k}^{-1}\hat{\tau} = (\hat{a} - 1)\hat{b}^{-1} - (\hat{a}/\hat{b}^2)(\hat{b} - 1) = -\hat{b}^{-1} + \hat{a}\hat{b}^{-2}.$$

Put [see also Section 5.6(ii)]

$$\hat{\varepsilon}(t) = \hat{k}^{-1}u(t), \qquad \hat{\xi}(t) = \hat{a}\hat{b}^{-2}u(t), \qquad \hat{\eta}(t) = \hat{b}^{-1}u(t).$$

In forming $(\partial_\tau \hat{k}^{-1})'u(t)$ put the derivatives with respect to the coefficients of $a(z)$ first down the vector and those with respect to the coefficients of b second. Then

$$\text{(6.5.4)}\qquad \begin{aligned} -\partial_\tau \hat{k}^{-1}u(t) &= (-\hat{\eta}(t-1), \ldots, -\hat{\eta}(t-p), \hat{\xi}(t-1), \ldots, \hat{\xi}(t-q)) \\ \{\hat{k}^{-1} - (\partial_\tau \hat{k}^{-1})\hat{\tau}\}u(t) &= \hat{\eta}(t) - \hat{\xi}(t) + \hat{\varepsilon}(t). \end{aligned}$$

Thus (6.5.3) is the vector of, Toeplitz, regression coefficients of $\hat{\eta}(t) - \hat{\xi}(t) + \hat{\varepsilon}(t)$ on $-(\partial_\tau \hat{k}^{-1})' u(t)$. In evaluating $\hat{\varepsilon}(t)$, $\hat{\xi}(t)$, $\hat{\eta}(t)$ recall the convention about values before $t = 1$. Then

$$\hat{b}\hat{\varepsilon}(t) = \hat{a}u(t),\ \hat{b}\hat{\xi}(t) = \hat{\varepsilon}(t),\ \hat{b}\hat{\eta}(t) = u(t), \qquad t = 1, \ldots, T;$$
$$u(t) = \hat{\varepsilon}(t) = \hat{\xi}(t) = \hat{\eta}(t) = 0, \qquad t \leq 0. \tag{6.5.5}$$

Thus the required quantities are easily formed recursively.

What is next required is an initial value. The key idea is to begin with

$$\hat{k}^{-1} = \hat{\phi}_{\hat{h}}(z) = \sum_0^{\hat{h}} \hat{\phi}_{\hat{h}}(j) z^j, \qquad \hat{\phi}_{\hat{h}}(0) = 1,$$

where the $\hat{\phi}_{\hat{h}}(j)$ are autoregression coefficients and $\hat{h}$ is itself determined by a criterion, (6.5.1), for $p = h, q = 0$. There are notational difficulties here, because $\hat{\phi}_{\hat{h}}(z)$ and $\hat{\phi}_{\hat{h}}(j)$ are so easily confused, but it is hoped that the context and the difference in the argument variable (z or j) will avoid confusion. *A convention will be adopted that will often be used below.* Although we write $\hat{\phi}_h(z)$, $\hat{\phi}_h(j)$, in general, when $h = \hat{h}$ we shall, in future write $\hat{\phi}_T(z)$, $\hat{\phi}_T(j)$, for simplicity. Of course, at the first iteration [i.e., the seond stage (the first stage being the autoregression)] $\hat{\eta}(t) = u(t)$, $\hat{\xi}(t) = \hat{\varepsilon}(t)$, so that the regression is of $u(t)$ on $-y(t-j), j = 1, \ldots, p; \hat{\varepsilon}(t-j), j = 1, \ldots, q$. However, at subsequent iterations the full detail of (6.5.4), (6.5.5) is needed. *At each iteration $\hat{p}$, $\hat{q}$ are determined anew, using (6.5.1).* (This was not emphasized sufficiently in Hannan and Rissanen, 1982.) Of course, (6.5.1) is, at the jth iteration,

$$\log\{\hat{\sigma}_{p,q}^{(j)}\}^2 + (p + q)C_T/T, \qquad p, q \leq H_T.$$

What is now necessary is to describe the recursive calculations at each iteration. *Here some terminology is first emphasized that is designed to help understanding.* We have spoken above of iterative "stages." Iteration is over j in $\hat{\eta}^{(j)}(t)$, and so on. Recursion is over p, q and is done for each j. We speak of recursive "steps." Thus $\hat{\alpha}_p^{(j)}(v)$, $v = 1, \ldots, p$, is produced at each j, recursively on p. Capital Roman numerals will be used to indicate iterative stages.

However, before going on to describe the recursions, let us generalize to the ARMAX case. Proceeding from the analogue of (6.4.1), for the scalar ARMAX case, the quantities needed are

$$\hat{b}\hat{\varepsilon}(t) = \hat{a}y(t) - \hat{d}z(t), \qquad \hat{b}\hat{\xi}(t) = \hat{\varepsilon}(t), \qquad \hat{b}\hat{\eta}(t) = y(t),$$
$$\hat{b}\hat{\zeta}(t) = z(t), \qquad t = 1, \ldots, T; \tag{6.5.6}$$
$$\hat{\varepsilon}(t) = \hat{\xi}(t) = \hat{\eta}(t) = y(t) = z(t) = 0, \qquad t \leq 0,$$

and the procedure is to regress $\hat{\eta}(t) - \hat{\xi}(t) + \hat{\varepsilon}(t)$ on $-\hat{\eta}(t-j)$, $j = 1, \ldots, p$; $\hat{\xi}(t-j)$, $j = 1, \ldots, q$; $\hat{\zeta}(t-j)$, $j = 1, \ldots, r$. Of course, at the first iteration (the second stage) once more we regress $y(t)$ on $-y(t-j)$, $\hat{\varepsilon}(t-j)$, $z(t-j)$. The initial values of $\hat{\varepsilon}(t)$ are provided again by a regression, autoregression procedure. Thus we calculate

$$\sum_{0}^{h} \hat{\phi}_h(j) y(t-j) = \sum_{1}^{h} \hat{\psi}_h(j) z(t-j) + \hat{\varepsilon}_h(t) \tag{6.5.7}$$

by a (block) Toeplitz regression. At this first stage the order of the regression, autoregression is called h and the same order is used for each part. At this first stage k, l are being estimated, putting $b = 1$, as the generating functions of, respectively, the autoregressive and regression parts of (6.5.7), so that p or r would be inappropriate notations. For the purposes of calculating this block Toeplitz regression, the algorithm (5.6.12), the Whittle recursion, is used, but it is necessary to present it in a new notation, for it will be used repeatedly with different inputs, replacing $y(t)$, as the input in (5.6.12), and the coefficient matrices will have elements with different interpretations, including the coefficients in (6.5.7). Thus in general we call the input $w(t)$ [at the risk of some confusion with $w(\omega)$, but hoping the context will avoid that confusion]. From these inputs covariance matrices

$$C(t) = \frac{1}{T} \sum_{s=1}^{T-t} w(s) w(s+t)' = C(-t)', \qquad t \geq 0$$

are computed and then the recursive algorithm is, from (5.6.12), in a new notation, letting the dimension of $w(t)$ be v.

$$\begin{aligned}
&F_h(j) = F_{h-1}(j) + F_h(h) \tilde{F}_{h-1}(h-j),\\
&\tilde{F}_h(j) = \tilde{F}_{h-1}(j) + \tilde{F}_h(h) F_{h-1}(h-j),\\
&F_h(0) = \tilde{F}_h(0) = I_v, \qquad j = 1, \ldots, h-1\\
&F_h(h) = -\Delta_{h-1} \tilde{S}_{h-1}^{-1}, \qquad \tilde{F}_h(h) = -\Delta_{h-1}' S_{h-1}^{-1},\\
&\qquad\qquad \Delta_h = \sum_{0}^{h} F_h(j) C(j-h-1),\\
&S_h = \{I_v - F_h(h) \tilde{F}_h(h)\} S_{h-1}, \qquad \tilde{S}_h = \{I_v - \tilde{F}_h(h) F_h(h)\} \tilde{S}_{h-1},\\
&S_0 = \tilde{S}_0 = C(0).
\end{aligned} \tag{6.5.8}$$

I. Now (6.5.7) is calculated, recursively on h, by identifying $w(t)$ as

$$w(t) = \begin{pmatrix} y(t) \\ -z(t) \end{pmatrix}, \qquad v = m + 1$$

and calling the first row of $F_h(j)$, $(\hat{\phi}_h(j), \hat{\psi}_h(j))$. Since the first row of $F_h(j)$ is the coefficient vector in the Toeplitz regression of $y(t)$ on $-y(t-j)$, $z(t-j)$, $j = 1, \ldots, h$, this is clearly correct. Also,

$$\hat{\sigma}_h^2 = \frac{1}{T} \sum \hat{\varepsilon}_h(t)^2$$

is the top left-hand corner element in S_h. At $h = \hat{h}$, determined from a criterion (6.5.1), we write $\hat{\sigma}_T^2$, $\hat{\phi}_T(j)$, $\hat{\psi}_T(j)$. Of course, $\hat{\varepsilon}_T(t)$ is obtained from (6.5.7) at $h = \hat{h}$.

II. At the first iteration, stage 2, we regress $y(t)$ on $-y(t-j)$, $\hat{\varepsilon}_T(t-j)$, $z(t-j)$. Consider first the case $p = q = r$. Then this regression may be calculated recursively by taking

$$w(t) = \begin{pmatrix} y(t) \\ -\hat{\varepsilon}_T(t) \\ -z(t) \end{pmatrix}, \qquad v = m + 2$$

as the input and taking the first row of F as $(\hat{\alpha}_p^{(1)}(j), \hat{\beta}_p^{(1)}(j), \hat{\delta}_p^{(1)}(j))$ while $\{\hat{\sigma}_p^{(1)}\}^2$ is the top left-hand corner element in S_p. *Observe that in calculating $C(t)$ for $p \leq h$ the covariances between $\hat{\varepsilon}_T(t)$ and $y(t-j)$, $z(t-j)$, $j = 1, \ldots, h$ need not be computed, but can be set equal to zero, since $\hat{\varepsilon}_T(t)$ is Toeplitz orthogonal to these. It is not quite true that the $\hat{\varepsilon}_T(t)$ are Toeplitz serially uncorrelated. However, if, in using (6.5.8), the autocovariances of the $\hat{\varepsilon}_T(t)$ are treated as $\delta_{0j}\hat{\sigma}_T^2$, then, from Theorem 6.6.1 below, it follows that the actual autocovariance will differ from this quantity by $O(Q_T)$, so that all theorems will continue to hold. With the autocovariances so defined, Franke (1985) gives a modified algorithm that requires fewer calculations than (6.5.8). All variances of the $\hat{\varepsilon}_T(t-j)$ are equal, of course, as they are for $z(t-j)$, $y(t-j)$.* The $\hat{\varepsilon}_h(t)$ need not themselves be computed, only their covariances with $\hat{\varepsilon}_h(t-j)$, $y(t-j)$, $z(t-j)$. Details concerning this, in the more complicated vector case, are given in Section 6.7. Again we use (6.5.1), that is,

$$\log\{\hat{\sigma}_p^{(1)}\}^2 + p(m+2)C_T/T, \qquad p \leq H_T,$$

to choose $\hat{p}$ and then write $\hat{\alpha}_T^{(1)}(j)$, $\hat{\beta}_T^{(1)}(j)$, $\hat{\delta}_T^{(1)}(j)$, $\{\hat{\sigma}_T^{(1)}\}^2$. The value of $\hat{p}$ will, under appropriate conditions to be delineated below, converge to $\max(p_0, q_0, r_0)$. It might be argued that probably p, q, r are not equal, so that parameters can be economized by allowing for this possibility. It is not difficult to see that this can be done by an extension of the procedure used many times before to include (or exclude) a variable from the Toeplitz regression by making it orthogonal to those already in the regression (or to be left in it). Indeed, replacing the lag covariances, for lags greater than zero, of the $\hat{\varepsilon}(t)$, by zero (see the italicized comment above), Franke (1985) produces an algorithm for the ARMA case for all p, q. *However, it seems most economical to effect the search to find a more economical parameterisation only after a final iteration, treating* $p = q = r$ *to that point.* Again at $p = \hat{p}^{(1)}$ we write $\hat{a}_T^{(1)}$, $\hat{b}_T^{(1)}$, $\hat{d}_T^{(1)}$ for the estimated transfer functions.

III. Now $\hat{\eta}^{(1)}(t)$, $\hat{\varepsilon}^{(1)}(t)$, $\hat{\xi}^{(1)}(t)$, $\hat{\zeta}^{(1)}(t)$, are formed via (6.5.6), using $\hat{a}_T^{(1)}(z)$, $\hat{b}_T^{(1)}(z)$, $\hat{d}_T^{(1)}(z)$ from stage II (i.e., the first iterate, hence the superscipt). Now put

$$w(t) = \begin{pmatrix} \hat{\eta}^{(1)}(t) \\ -\hat{\xi}^{(1)}(t) \\ -\hat{\zeta}^{(1)}(t) \end{pmatrix}$$

and use this to calculate (6.5.8). Then, as is easily seen,

(6.5.9)

$$\begin{pmatrix} \hat{\alpha}_p^{(2)}(j) \\ \hat{\beta}_p^{(2)}(j) \\ \hat{\delta}_p^{(2)}(j) \end{pmatrix} = \begin{pmatrix} \hat{\alpha}_{p-1}^{(2)}(j) \\ \hat{\beta}_{p-1}^{(2)}(j) \\ \hat{\delta}_{p-1}^{(2)}(j) \end{pmatrix} + \tilde{F}_{p-1}(p-j)' \begin{pmatrix} \hat{\alpha}_p^{(2)}(p) \\ \hat{\beta}_p^{(2)}(p) \\ \hat{\delta}_p^{(2)}(p) \end{pmatrix}, \qquad j = 1, 2, \ldots, p-1,$$

$$\begin{pmatrix} \hat{\alpha}_p^{(2)}(p) \\ \hat{\beta}_p^{(2)}(p) \\ \hat{\delta}_p^{(2)}(p) \end{pmatrix} = \tilde{S}_{p-1}^{-1}\left[\sum_0^{p-1} \tilde{F}_{p-1}(j)\frac{1}{T}\sum_t w(t-p+j)\{\hat{\eta}^{(1)}(t) + \hat{\varepsilon}^{(1)}(t) - \hat{\xi}^{(1)}(t)\}\right],$$

$$\hat{\alpha}_0(0) = \hat{\beta}_0(0) = 1, \qquad \hat{\delta}_0(0) = 0,$$

$$\{\hat{\sigma}_p^{(2)}\}^2 = \{\hat{\sigma}_{p-1}^{(2)}\}^2 - (\hat{\alpha}_p^{(2)}(p), \hat{\beta}_p^{(2)}(p), \hat{\delta}_p^{(2)}(p)')\tilde{S}_{p-1}(\hat{\alpha}_p^{(2)}(p), \hat{\beta}_p^{(2)}(p), \hat{\delta}_p^{(2)}(p)')',$$

$$\{\hat{\sigma}_0^{(2)}\}^2 = \frac{1}{T}\sum_t \{\hat{\eta}^{(1)}(t) - \hat{\xi}^{(1)}(t) + \hat{\varepsilon}^{(1)}(t)\}^2.$$

Now $\hat{p}^{(2)}$ is chosen by minimizing (6.5.1) again at $p = q = r$, that is,

(6.5.10) $$\log\{\hat{\sigma}_p^{(2)}\}^2 + 3pC_T/T, \qquad p \le H_T.$$

The definition of $\{\hat{\sigma}_p^{(2)}\}$ may seem strange, but recall (6.5.1), (6.5.2). It will be shown that for suitable C_T, H_T, stage II gives strongly consistent estimates of p_0, a_0, b_0, d_0. Thus examining (6.5.2) it is evident that for $p \ll \hat{p}^{(1)}$, then $\{\hat{\sigma}_p^{(2)}\}^2$ will be relatively large. At $p = \hat{p}^{(1)}$, $\{\hat{\sigma}_p^{(2)}\}^2 \le \{\hat{\sigma}_p^{(1)}\}^2$. [Indeed, at $p \ge \hat{p}^{(1)}$, and only then, $\{\hat{\sigma}_p^{(2)}\}^2$ is the residual mean square for the regression of $\hat{\varepsilon}^{(1)}(t)$ on $-\hat{\eta}^{(1)}(t-j)$, $\hat{\xi}^{(1)}(t-j)$, $\hat{\zeta}^{(1)}(t-j)$, $j = 1, \ldots, \hat{p}^{(1)}$ because $\hat{\eta}^{(1)}(t) - \hat{\xi}^{(1)}(t)$ is exactly linearly expressible in terms of $\hat{\eta}(t-j)$, $\hat{\xi}(t-j)$, $j = 1, 2, \ldots, \hat{p}^{(1)}$.] Thus really (6.5.10) need only be examined for values near $\hat{p}^{(1)}$ and certainly will not be minimized at very much smaller values. *On this account, if T, p_0, q_0, r_0 are large, it might be computationally cheaper to calculate the regression directly at some initial p value, not much smaller than $\hat{p}^{(1)}$, and then use (6.5.9).*

IV. At the third (and later) iterates, the procedure is the same as at stage III, beginning at $\hat{a}^{(2)}$, $\hat{b}^{(2)}$, $\hat{d}^{(2)}$, and no further description is needed.

The essential quantity that governs the cessation of iteration is (6.5.10) (i.e., its minimum value at any iteration). When this ceases to change appreciably, iteration should cease. It seems that more than three iterations will be rare.

A number of other remarks need to be made.

(i) Once iteration has ceased, the full likelihood could be obtained at the chosen p, q, r of the last iteration, initiating from the a, b, d found at these p, q, r at the last iteration, using one of the procedures described in Section 6.2.

(ii) There is one problem with the procedure above, which occurs for the first time at stage III. *It is not necessary for $\hat{b}^{(1)}(z)$ to have all its zeros outside* $|z| \le 1$, and if this does occur, then, in general, stage III will not run since the recursion forming $\hat{\eta}^{(1)}(t)$, for example, will explode. This problem troubles almost all procedures. [An exception is the procedure of Walker (1962), based on Theorem 6.2.6. However, this technique is for fixed p, q and is not easy to extend beyond the scalar ARMA case.] One way to proceed to solve this problem is to put

$$|\hat{b}^{(1)}(e^{i\omega})|^2 = \sum \hat{\gamma}(j) e^{ij\omega}, \tag{6.5.11}$$

the sum on the right being from $-\hat{p}^{(1)}$ to $\hat{p}^{(1)}$, and then to factorize this as

$$|\hat{b}^{(1)}(e^{i\omega})|^2 = |\tilde{b}^{(1)}(e^{i\omega})|^2, \qquad \tilde{b}(z) \ne 0, \qquad |z| \le 1.$$

[We can discount the possibility that $\hat{b}^{(1)}(z)$ has a zero precisely on $|z| = 1$.] An algorithm for this purpose is due to Wilson (1969). Put $\hat{\rho}(j) = \hat{\gamma}(j)/\hat{\gamma}(0)$. The algorithm factors $\sum \hat{\rho}(j) e^{ij\omega}$ and the coefficients, $\delta(j)$, finally obtained must be divided by $\delta(0)$ to obtain $\tilde{b}^{(1)}(j)$. Given any polynomial

$$\sum_{-p}^{p} \hat{\rho}(j)e^{ij\omega} > 0, \qquad \hat{\rho}(0) = 1,$$

the algorithm at the kth step finds

$$\delta_k = (\delta_k(0), \delta_k(1), \dots, \delta_k(p))'$$

$$\tfrac{1}{2}\delta_k + d_k = \delta_{k+1}, \qquad \delta_0 = (1, 0, \dots, 0)', \qquad d_k = A(\delta_k)^{-1}\hat{\rho}$$

$$\hat{\rho} = (1, \hat{\rho}(1), \dots, \hat{\rho}(p))',$$

$$A(\delta) = \begin{bmatrix} \delta(0) & \delta(1) & \cdots & \delta(p) \\ \delta(1) & \delta(2) & \cdots & 0 \\ \vdots & \vdots & \vdots & \vdots \\ \delta(p) & 0 & \cdots & 0 \end{bmatrix} + \begin{bmatrix} \delta(0) & \delta(1) & \cdots & \delta(p) \\ 0 & \delta(0) & \cdots & \delta(p-1) \\ \vdots & \vdots & \vdots & \vdots \\ 0 & 0 & \cdots & \delta(0) \end{bmatrix}.$$

At each step the polynomial $\sum \delta_k(j)z^j \neq 0$, $|z| \leq 1$. The iteration is continued to convergence.

It may be that it will be computationally cheaper to check the stability of $\hat{b}^{(1)}(z)$ (i.e., the location of the zeros), by a Schur-Cohn (Routh-Hurwitz) criterion and only to use Wilson's algorithm when there are zeros inside $|z| = 1$. This procedure of Wilson's effectively reflects zeros that are inside $|z| = 1$ in the unit circle. Of course, asymptotically, as we shall see below, $\hat{b}^{(1)}(z)$ will be stable and therefore it is reasonable to expect that at worst the zeros, if not stable, are at least near true values that are near $|z| = 1$, so that the reflection procedure will produce, again, zeros near the true values. The same problem with stability occurs at all stages after stage I and can be dealt with in the same way.

(iii) After the last iteration it may be thought necessary to see if parameters can be reduced in number by looking for p, q, r that are not equal. In principle it should only be necessary to try to reduce the common value, $\hat{p}^{(j)}$, since $\hat{p}^{(j)}$ will be shown, under appropriate conditions, to converge to $\max(p_0, q_0, r_0)$. If only a few possibilities, for (p, q, r) near $(\hat{p}^{(j)}, \hat{p}^{(j)}, \hat{p}^{(j)})$, are to be examined it will be easiest to do the regression individually, and for this reason no further detail will be given here. A recursion can be established, but it is more complicated. (See Monden, Yamada, and Arimoto, 1982, for a related case.)

(iv) It will be seen in the next section that there is a bias at state II, if $C_T = \log T$ at stages I and II, that leads to some tendency to overestimate p_0. This bias is due to the fact that, as will be seen, $\hat{h}$ at stage I increases as $\log T$. The following procedure corrects that bias. *Whether it is worthwhile doing this is not fully apparent since the bias disappears at stage III in any case.* The procedure is as follows. Form

$$\hat{\rho}_{p,q,r}(k) = \sum_{0}^{q-k} \hat{\beta}_q^{(1)}(j)\hat{\beta}_q^{(1)}(j+k), \qquad k = 0, 1, \ldots, q.$$

Then use the Levinson recursion (5.6.6) with these $\hat{\rho}_{p,q,r}(k)$ in place of $g(k)$ to generate $\hat{\omega}^2_{p,q,r}(k)$, $k = 0, 1, \ldots$, where these are interpreted as $\hat{\sigma}_k^2$ in the recursion. Then put

$$\begin{aligned} BIC_{p,q,r}(i,j,k) &= BIC(i,j,k) = \log(\hat{\sigma}^{(1)}_{i,j,k})^2 + (i + j + mk)\log T/T, \\ &\qquad\qquad i \le p \text{ or } j \le q \text{ or } k \le r \\ BIC_{p,q,r}(i,j,k) &= BIC(i,j,k) + \hat{h}\{\hat{\omega}^2_{p,q,r}(0) - \hat{\omega}^2_{p,q,r}(v)\}, \\ v &= \min(i-p, j-q, k-r), \qquad i > p, j > q, k > r. \end{aligned} \tag{6.5.12}$$

Here, of course, the constant $\hat{h}$ is the order of autoregression chosen at stage I. We now consider, for each p, q, r, the values of $BIC_{p,q,r}(i,j,k)$ minimizing this, always taking p, q, r, i, j, k less than $H_T = O\{(\log T)^a\}$, $a < \infty$. It may be that $BIC_{p,q,r}(i,j,k)$ is minimized, over i, j, k, at $(i,j,k) = (p,q,r)$. If so, let us say that (p,q,r) is a "marked" point. We now show that $\hat{p}^{(1)}, \hat{q}^{(1)}, \hat{r}^{(1)}$ is always a marked point and there is at most one other with $p \le \hat{p}^{(1)}$, $q \le \hat{q}^{(1)}$ and $r \le \hat{r}^{(1)}$. We put $\hat{p}_c^{(1)}, \hat{q}_c^{(1)}, \hat{r}_c^{(1)}$ as this, then, or if there is no other, we put it at $\hat{p}^{(1)}, \hat{q}^{(1)}, \hat{r}^{(1)}$.

For brevity call $BIC_{p,q,r}(i,j,k)$ at $(p,q,r) = (\hat{p}^{(1)}, \hat{q}^{(1)}, \hat{r}^{(1)})$, just $\hat{B}(i,j,k)$. Then $\hat{B}(i,j,k) \le BIC(i,j,k)$ and $\hat{B}(\hat{p}^{(1)}, \hat{q}^{(1)}, \hat{r}^{(1)}) = BIC(\hat{p}^{(1)}, \hat{q}^{(1)}, \hat{r}^{(1)})$, so that $\hat{p}^{(1)}, \hat{q}^{(1)}, \hat{r}^{(1)}$ is a marked point. If $p \ge \hat{p}^{(1)}$, $q \ge \hat{q}^{(1)}$, $r \ge \hat{r}^{(1)}$, then

$$BIC_{p,q,r}(i,j,k) \ge BIC(\hat{p}^{(1)}, \hat{q}^{(1)}, \hat{r}^{(1)})$$

for the same reason, so that any other marked point must be in the region $p < \hat{p}^{(1)}$, $q < \hat{q}^{(1)}$, $r < \hat{r}^{(1)}$. We now consider only such points. For them

$$BIC_{p,q,r}(i,j,k) \ge BIC_{p,q,r}(\hat{p}^{(1)}, \hat{q}^{(1)}, \hat{r}^{(1)}), \qquad i \ge \hat{p}^{(1)}, j \ge \hat{q}^{(1)}, k \ge \hat{r}^{(1)},$$

so that we need examine only $i < \hat{p}^{(1)}$, $j < \hat{p}^{(1)}$, $k < \hat{r}^{(1)}$. Let p, q, r be a marked point and assume that $p_1 < p$ or $q_1 < q$ or $r_1 < r$. Then p_1, q_1, r_1 is not a marked point because

$$\begin{aligned} BIC_{p_1,q_1,r_1}(p_1,q_1,r_1) &= BIC(p_1,q_1,r_1) = BIC_{p,q,r}(p_1,q_1,r_1) > BIC_{p,q,r}(p,q,r) \\ &= BIC(p,q,r) = BIC_{p_1,q_1,r_1}(p,q,r). \end{aligned}$$

This shows that there is at most one marked point other than $\hat{p}^{(1)}$, $\hat{q}^{(1)}$, $\hat{r}^{(1)}$, as required. *This argument also shows that the calculation to discover* $\hat{p}_c^{(1)}$, $\hat{q}_c^{(1)}$,

$r_c^{(1)}$ *is not large*. Indeed, we need example only $p \le \hat{p}^{(1)}, q \le \hat{q}^{(1)}, r \le \hat{r}^{(1)}$. Once another marked point is found, we stop. Moreover, for each p, q, r, we need consider only $i \le \hat{p}^{(1)}, j \le \hat{q}^{(1)}, k \le \hat{r}^{(1)}$. This procedure is due to Hannan and Kavalieris (1984).

(V) Although all procedures have been expressed in terms of the Levinson-Whittle algorithm, needless to say any of the modified procedures introduced in Section 4 could be used.

(VI) There is a very large literature developing around methods of the kind discussed in this section. For example, Spliid (1983) suggests, for fixed p, q, r, iterating stage II, using at state I an autoregression of fixed order, say $h = (p + q)$. However, Hannan and McDougall (1988) show that this procedure may not converge. [For $r = 0$, $p = q = 1$, the critical condition is $\alpha_0(1)\beta_0(1) \le \frac{1}{2}$.] If it does converge, the limiting distribution, for $r = 0$, is the same as that for stage II using an autoregression at stage I as there described. Saikonnen (1987) suggests two methods for $r = 0$. They could be generalized to $r > 0$. For example, Saikonnen uses the $\hat{\phi}_T(j)$ and the quantities

$$\tilde{g}(i) = \sum_0^{\hat{h}-i} \hat{\phi}_T(j)\hat{\phi}_T(j+i).$$

The latter are, except for a scale factor independent of i, the estimates of the "inverse autocorrelations" (Bhansali, 1980, 1983), that is, the coefficients in the Laurent expansion of $f(\omega)^{-1}$. The Whittle recursion may be used with $w(t)' = (\hat{\phi}_T(t-1), -\delta_{0,t-1})$, $t = 1, 2, \ldots, \hat{h} + 1$ to obtain a sequence of estimators $\check{\tau}'_p = (\check{\beta}_p(1), \ldots, \check{\beta}_p(p), \check{\alpha}_p(1), \ldots, \check{\alpha}_p(p))$, $p = 0, 1, 2, \ldots$. A criterion (6.5.1) could be used to choose $\check{p}$. It may be observed that now $\sum \check{\beta}_p(j)z^j \ne 0$, $|z| \le 1$. Indeed, the form of $w(t)$ ensures that, recalling that $F_h(j)$ are the quantities in (6.5.8) for these $w(t)$,

$$\sum_0^p F_p(j)z^j = \begin{bmatrix} \sum \check{\beta}_p(j)z^j & \sum \check{\alpha}(j)z^j \\ 0 & 1 \end{bmatrix}.$$

Since the sequence $\delta_{0,t-1}$ is Toeplitz orthogonal to $\hat{\phi}_T(t-k-1)$, $k > 0$, and to $\delta_{0,t-k-1}$, the required result follows from Theorem 5.6.2. It may be pointed out that the same will be true of $\hat{a}(z) = \sum \hat{\alpha}_p(j)z^j$ if the conventions adopted by Franke (1985) are used [i.e., that $\hat{\varepsilon}_T(t)$ is treated as Toeplitz orthogonal to $y(t-j)$, $z(t-j)$, $\hat{\varepsilon}_T(t-j)$, $j > 0$]. Thus $\hat{a}(z)$ is very likely to be stable even without this convention since $\hat{\varepsilon}_T(t)$ is in any case Toeplitz orthogonal to $y(t-j)$, $z(t-j)$ and will be nearly so, orthogonal to $\hat{\varepsilon}_T(t-j)$. However, we need to filter by a filter whose z-transform is an estimate of $b(z)^{-1}$, so that it is the result for Saikonnen's method that is interesting.

The rationale behind Saikonnen's procedure is that, putting $\phi(z) = a(z)/b(z)$, then

$$\sum_0^q \beta(j)\phi(k-j) = \sum_1^p \alpha(j)\delta_{j,k};$$

$$\sum_0^q \beta(j)\left\{\sum_{u=0}^{\infty} \phi(u)\phi(u+k-j)\right\} = \sum_1^p \alpha(j)\phi(j-k); \qquad k = 1, 2, \ldots .$$

It is these equations that are being solved, putting $\phi(k) = \hat{\phi}_T(k)$, $k = 0, 1, \ldots, \hat{h}$, $\phi(k) = 0$, $k > \hat{h}$.

A stage III procedure is suggested by Saikonnen (1987) and the method, when p_0, q_0 are known, is shown to give estimates having the same asymptotic distribution as for the procedure at stage III above, used, again, for p_0, q_0 known.

6. SOME ASYMPTOTIC THEORY

Throughout this section the zero subscript on Σ, k, l, and so on, indicating a true value, will be omitted, but it will be retained for order parameters, p_0, q_0, r_0. Much of the theory in this section will be for the vector case but some, relating to stages II, III of Section 5, will be for the scalar case. The work leans heavily on Hannan and Kavalieris (1984) and, especially, on the thesis of Kavalieris (1984).

The section is divided into three parts. Initially, we state some conditions and then some theorems relating to Section 5. The remaining two subsections give proofs of these. *The second subsection deals with some results of a general kind, of interest beyond this section.* The last proves the theorems stated in the first subsection.

(i) Conditions and Theorems

It has been found necessary to add some conditions on $z(t)$. That will be required to be generated by a stationary process with zero mean and absolutely continuous spectrum with density, $f_z(\omega)$, satisfying

$$\underline{c}I_m \leq f_z(\omega) \leq \overline{c}I_m, \qquad 0 < \underline{c} \leq \overline{c} < \infty. \tag{6.6.1}$$

Here and below, as usual, c and also $\underline{c}$, $\overline{c}$, will stand for arbitrary finite positive constants, not always the same ones. The condition (6.6.1) implies that

$$z(t) = \sum_0^{\infty} K_z(j)\varepsilon_z(t-j), \quad E\{\varepsilon_z(s)\varepsilon_z(t)'\} = \delta_{st}\Sigma_z, \quad \Sigma_z > 0, \quad K_z(0) = I_m.$$

However, it is additionally required that

$$\sum_{1}^{\infty} j^{1/2}\|K_z(j)\| < \infty. \tag{6.6.2}$$

It is a consequence of (6.6.1), (6.6.2) that, putting

$$k_z(e^{i\omega})^{-1} = \sum_{0}^{\infty} \Phi_z(j)e^{ij\omega}, \qquad \Phi_z(0) = I_s, \qquad k_z(e^{i\omega}) = \sum_{0}^{\infty} K_z(j)e^{ij\omega},$$

then

$$\sum_{1}^{\infty} j^{1/2}\|\Phi_z(j)\| < \infty. \tag{6.6.3}$$

The converse also holds, so that (6.6.3) might as well be imposed as a condition. That (6.6.1), (6.6.2) imply (6.6.3) will be established in Chapter 7.

Finally we require that, for some sequence H_T, later to be specified, $H_T \uparrow \infty$,

$$\sup_{0 \le t \le H_T} \|G_z(t) - \Gamma_z(t)\| = e_T H_T^{-1}, \qquad e_T \to 0. \tag{6.6.4}$$

Here $\Gamma_z(t) = E\{z(s)z(s+t)'\}$, needless to say. This condition is not onerous, as can be seen from Chapter 5, which for $H_T \le (\log T)^a$ would suggest that we may take $e_T \le cT^{-1/2}H_T(\log\log T)^{1/2}$. However, much weaker conditions than are used in Section 5 would, as will be seen in Chapter 7, ensure that $e_T \le cT^{-1/2}H_T \log T$. Indeed, the conditions of Chapter 7 are overly strong for (6.6.4). For example, if $\mathscr{F}_{z,-\infty}$ is a trivial σ-algebra [i.e., consists only of the empty set and the full space over which $z(t)$ is defined], then $z(t)$ is "regular" and (Ibragimov and Linnik, 1971) $z_u(s)z_v(s+t) - \gamma_{z,uv}(t)$ has an absolutely continuous spectrum for all u, v, t. If those spectra are uniformly bounded in u, v, t, we may take $e_T = cH_T T^{-b}$ for all $b < \frac{1}{2}$ (see Hannan, 1978, for example). We omit further discussion here but repeat that (6.6.4) does not seem unduly restrictive.

These conditions are gathered together.

Conditions A: The conditions of Theorem 5.3.1 together with (6.6.1), (6.6.3), (6.6.4) as well as $\det k(z) \neq 0$, $|z| \le 1$.

For some purposes a strengthening is needed.

$$E\{\varepsilon(s)\varepsilon(t)'|\mathscr{F}_{t-1}\} = \delta_{st}\Sigma, \quad E\{\varepsilon_j(t)^4 \log_+|\varepsilon_j(t)|\} < \infty, \quad j = 1,\ldots,s. \tag{6.6.5}$$

Of course, the first part of this is just (4.1.16). The second part is minor. This condition is needed for certain "in probability" results where variances have to be evaluated. *If (6.6.5) is added to Conditions A, we speak of Conditions B.*

Some further preliminaries will be needed in relation to stage I. This is a mixed regresssion-autoregression, or ARX, procedure, and here it will be convenient to deal with the vector case. The defining equations are

$$\sum_0^h \hat{\Phi}_h(j)G_y(j-k) - \sum_1^h \hat{\Psi}_h(j)G_{zy}(j-k) = \delta_{0k}\hat{\Sigma}_h, \qquad k = 0, 1, \ldots, h,$$

$$\sum_0^h \hat{\Phi}_h(j)G_{yz}(j-k) - \sum_1^h \hat{\Psi}_h(j)G_z(j-k) = 0, \qquad k = 1, \ldots, h.$$

Analogous "true" quantities, $\Phi_h(j)$, $\Psi_h(j)$, Σ_h are defined by replacing $G_y(j)$, and so on, by $\Gamma_y(j)$ and so on. Put, in extension of the notation used in Section 4,

$$G_h = \begin{bmatrix} G_{y,h} & G_{yz,h} \\ G_{zy,h} & G_{z,h} \end{bmatrix},$$

with $G_{y,h}$ a block Toeplitz matrix having $G_y(j-k)$ as the block in the (j,k)th place, j, $k = 1,\ldots,h$. Similarly, $G_{yz,h}$ has $G_{yz}(j-k)$ in that place, and so on. Define Γ_h similarly in terms of $\Gamma_{y,h}$, $\Gamma_{yz,h}$, and so on. Put

$$g_h = [G_y(-1), \ldots, G_y(-h), G_{yz}(-1), \ldots, G_{yz}(-h)]$$

with γ_h defined analogously in terms of $\Gamma_y(-1), \ldots, \Gamma_{yz}(-1), \ldots$. Put

$$[-\hat{\Phi}_h, \hat{\Psi}_h] = [-\hat{\Phi}_h(1), \ldots, -\hat{\Phi}_h(h), \hat{\Psi}_h(1), \ldots, \hat{\Psi}_h(h)]$$

with $[-\Phi_h, \Psi_h]$ defined analogously. Then

$$(6.6.6) \qquad [-\hat{\Phi}_h, \hat{\Psi}_h]G_h = g_h, \qquad [-\Phi_h, \Psi_h]\Gamma_h = \gamma_h,$$

and Σ_h is defined as for $\hat{\Sigma}_h$.

Because it is desirable to avoid (4.1.16) as far as possible, we introduce $\dot{\Gamma}_h$, $[\dot{\Phi}_h, \dot{\Psi}_h]$, $\dot{\gamma}_h$ defined exactly as for the second part of (6.6.6) but with $\Gamma_y(t) = \Gamma_u(t) + \Gamma_v(t)$ replaced by $\dot{\Gamma}_y(t) = \dot{\Gamma}_u(t) + \Gamma_v(t)$:

$$\dot{\Gamma}_u(t) = \sum_{j=0}^{\infty} K(j)\dot{\Sigma}K(j+t)', \qquad \dot{\Sigma} = \frac{1}{T}\sum_{s=1}^{T} \varepsilon(s)\varepsilon(s)'.$$

Finally, put

$$\phi(z) = k(z)^{-1} = \sum_0^\infty \Phi(j)z^j, \qquad \psi(z) = k(z)^{-1}l(z) = \sum_1^\infty \Psi(j)z^j,$$

so that

$$\begin{aligned}\sum_0^\infty \Phi(j)\Gamma_y(j-k) - \sum_1^\infty \Psi(j)\Gamma_{zy}(j-k) &= \delta_{0k}\Sigma, \qquad k = 0, 1, \ldots \\ \sum_0^\infty \Psi(j)\Gamma_{yz}(j-k) - \sum_1^\infty \Psi(j)\Gamma_z(j-k) &= 0, \qquad k = 1, 2, \ldots .\end{aligned} \tag{6.6.7}$$

Observe also that (6.6.7) holds with $\Gamma_y(j-k)$ replaced by $\dot{\Gamma}_y(j-k)$, and Σ by $\dot{\Sigma}$.

Theorem 6.6.1. *Let $y(t)$ be generated by an ARMAX process safisfying conditions A for $H_T \le (\log T)^a$, $a < \infty$. Then*

$$\begin{aligned}&\sup_{h \le H_T} \sup_{j \le h} \|\hat{\Phi}_h(j) - \dot{\Phi}_h(j), \hat{\Psi}_h(j) - \dot{\Psi}_h(j)\| \\ &\quad = O(Q_T) + o\left\{H_T^{-1} \sum_{h+1}^\infty \|[\Phi(j), \Psi(j)]\|\right\}.\end{aligned} \tag{6.6.8}$$

In the ARMA case the last term on the right side of (6.6.8) may be omitted. For $s = 1$, or for $s > 1$ when (4.1.16) holds, $\dot{\Phi}_h(j)$, $\dot{\Psi}_h(j)$ may be replaced by $\Phi_h(j)$, $\Psi_h(j)$. This may be done in any case if the factor H_T^{-1} is omitted from the last term.

We now discuss this theorem. In the first place let ρ_0^{-1}, $0 \le \rho_0 \le 1$, be the modulus of a zero of $\det b(z)$ [i.e., $\det k(z)$] nearest $|z| = 1$. Then

$$\begin{aligned}&\limsup_{j\to\infty} \rho_0^{-j}\|[\Phi(j), \Psi(j)]\| = c_1, \qquad 0 < c_1 < \infty \\ &\exists N < \infty, \qquad n_j \to \infty, \qquad n_{j+1} - n_j \le N,\end{aligned} \tag{6.6.9}$$

such that

$$\lim_{j\to\infty} \rho_0^{-n_j}\|[\Phi(j), \Psi(j)]\| = c_1.$$

The proof follows from a partial fraction decomposition of $\{\det b(z)\}^{-1}$, in a left prime m.f.d for $k(z)$. Thus if there are two zeros of modulus ρ_0^{-1} that are complex conjugates and are of the form $\rho_0^{-1}\exp(-i\theta)$, $\rho_0^{-1}\exp(i\theta)$, the dominant term in the partial fraction decomposition will be

$$c\{1 - z\rho_0 e^{i\theta}\}^{-1} + \bar{c}\{1 - z\rho_0 e^{-i\theta}\}^{-1}$$

and the coefficient of z^k in the expansion of this will be

$$|c|\rho_0^k \cos(k\theta + \phi), \qquad c = |c|e^{i\phi}.$$

For example,

$$\rho_0^{-j}\Phi(j) = \sum_0^j |c|\rho_0^{-k}\cos\{(j-k)\theta + \phi\}D(k) + \rho_0^{-j}X(j). \tag{6.6.10}$$

Here $D(k)$ is the typical coefficient matrix in $\det b(z)\cdot b(z)^{-1}a(z)$, which is a matrix of polynomials, while $X(j)$ comes from other terms in the partial fraction decomposition so that $\rho_0^{-j}X(j) \to 0$. Thus (6.6.9) follows from (6.6.10), in this case. We leave the reader to consider the case of multiple zeros, for example.

Put

$$l(T) = \log T/(-2\log\rho_0). \tag{6.6.11}$$

Then from (6.6.9) we obtain

$$\sum_{h+1}^{\infty} \|[\Phi(j), \Psi(j)]\| \le c\exp(-h\log\rho_0) \le cT^{-1/2}, \; c < \infty, \; h \ge l(T). \tag{6.6.12}$$

Below we shall also generalize a result due to Baxter (1962) which shows, in particular, that, for some c, $H < \infty$,

$$\sum_1^h \|[\Phi_h(j) - \Phi(j), \Psi_h(j) - \Psi(j)]\| \le c\sum_{h+1}^{\infty} \|[\Phi(j), \Psi(j)]\|, \quad h \ge H \tag{6.6.13}$$

with the same result holding when $\Phi_h(j)$, $\Psi_h(j)$ *are replaced by* $\dot{\Phi}_h(j)$, $\dot{\Psi}_h(j)$. Thus the following corollary to Theorem 6.6.1 holds.

Corollary 6.6.2. *If* $y(t)$ *is as in Theorem 6.6.1, then*

$$\sup_{l(T)\le h\le H_T} \sup_{j\le h} \|[\hat{\Phi}_h(j) - \Phi(j), \hat{\Psi}_h(j) - \Psi(j)]\| = O(Q_T).$$

Indeed, the replacement of $\dot{\Phi}_h(j)$, $\dot{\Psi}_h(j)$ by $\Phi(j)$, $\Psi(j)$ can, because of (6.6.13), be made at the expense of eliminating the factor H_T^{-1} in the last term in (6.6.8), which is $o(T^{-1/2})$ by (6.6.12). This is, of course, the best result possible, except for evaluation of a best constant in the $O(Q_T)$ term. The result is not empty, as the next theorem shows.

Theorem 6.6.3. *Let $y(t)$ be as in Theorem 6.6.1 for $q_0 > 0$. Let $\hat{h}$ minimize*

$$BIC(h) = \log\det\Sigma_h + h(s^2 + sm)\log T/T, \qquad h \le H_T.$$

Then

$$\lim_{T\to\infty} \hat{h}/l(T) = 1, \qquad a.s.$$

Actually, the result holds with [see (5.2.1)] $C_T = c_0 \log\log T$, for c_0 sufficiently large, but c_0 is difficult to prescribe and $\log\log T$ is not a very relevant number unless T is very large, so we do not emphasize that. Of course, this theorem says that $\hat{h} = l(T)\{1 + o(1)\}$ and since the $o(1)$ term could be negative, it seems that if Corollary 6.6.2 is to be available, then h would have to be taken as $c\hat{h}$, $c > 1$. Of course, even with $\hat{h}$ [see (6.6.17)] we obtain, in place of $O(Q_T)$ in Corollary 6.6.2, $O(T^{-b})$ for all $b < \frac{1}{2}$.

In practice it is likely that $AIC(h)$ will often be used, as some experience suggests that $BIC(h)$ tends to give somewhat too low a value for some purposes. That is covered by the next result.

Theorem 6.6.4. *Let $y(t)$ be ARMAX for $q_0 > 0$ and satisfy conditions B for $H_T = (\log T)^a$, $a < \infty$, and let $\hat{h}$ minimize*

$$AIC(h) = \log\det\Sigma_h + 2h(s^2 + sm)/T, \qquad h \le H_T.$$

Then

$$\operatorname{plim}_{T\to\infty} \hat{h}/l(T) = 1.$$

This theorem is essentially due to Shibata (1980), in the Gaussian, ARMA case for $s = 1$. Using $h = c\hat{h}$, $c > 1$, $\hat{h}$ determined from $AIC(h)$, then Corollary 6.6.2 would hold with $O(Q_T) = O_P(Q_T)$. One feels that here Q_T should be replaceable by something smaller (e.g., $T^{-1/2}$), but it is not easy to do that.

Simulations suggest that $l(T)$ overestimates $\hat{h}$ unless T is very large, especially when ρ_0 is near unity. Obvious candidates for better approximations are

$$l_A(T) = \arg\min\{\log\det\Sigma_h + 2h(s^2 + sm)/T\},$$

$$l_B(T) = \arg\min\{\log\det\Sigma_h + h(s^2 + sm)\log T/T\}.$$

Of course, $l_A(T)/l(T) \to 1$, $l_B(T)/l(T) \to 1$.

If $\hat{h}$ minimizes $\log\det\hat{\Sigma}_h + hC_T/T$ and $\hat{h}'$ minimizes $\log\det\hat{\Sigma}_h + hC'_T/T$, then for $C'_T \ge C_T$, $\hat{h} \ge \hat{h}'$. Indeed,

$$\log\det\hat{\Sigma}_h - \log\det\hat{\Sigma}_{\hat{h}} + (h-\hat{h})(s^2+sm)C_T/T \geq 0, \qquad h \geq \hat{h},$$

and thus the same holds with C'_T replacing C_T so that $\hat{h}' \leq \hat{h}$. *Thus $\hat{h}$ is a nonincreasing function of C_T.*

So far, in relation to stage I we have established theorems about the accuracy of the $\hat{\Phi}_h(j)$, $\hat{\Psi}_h(j)$ and about the behavior of $\hat{h}$ from AIC, BIC. Finally, a result is stated concerning the accuracy of the $\hat{\varepsilon}_h(t)$,

$$\hat{\varepsilon}_h(t) = \sum_0^h \hat{\Phi}_h(j)y(t-j) - \sum_1^h \hat{\Psi}_h(j)z(t-j).$$

Theorem 6.6.5. *Let $y(t)$ be as in Theorem 6.6.4. Then uniformly in h, $l(T) \leq h \leq H_T$, for $j \geq 0$,*

$$\frac{1}{T}\sum_1^T \{\hat{\varepsilon}_h(t) - \varepsilon(t)\}\{\hat{\varepsilon}_h(t-j) - \varepsilon(t-j)\}' = \delta_{0j}\frac{h}{T}(s+m)\Sigma + o_P(h/T).$$

$$\frac{1}{T}\sum_1^T \varepsilon(t)\{\hat{\varepsilon}_h(t-j) - \varepsilon(t-j)\}' = -\delta_{0j}\frac{h}{T}(s+m)\Sigma + o_P(h/T).$$

Again Theorems 6.6.3 and 6.6.4 show the relevance of the condition on h. The calculation shows, taking $j = 0$, how accurate the $\hat{\varepsilon}_h(t)$ are for the mean-square error when AIC or BIC is used is essentially of order $\log T/T$.

Now we consider stage II, but for $s = 1$. In the recommended procedure we put $p = q = r$, but that is for computational reasons, and theorems are stated as if p, q, r are individually determined. If only $p = q = r$ is to be considered, the theorems can be interpreted by taking $p_0 = q_0 = r_0 = \max(p_0, q_0, r_0)$.

There is another technical point that comes from the discussion below Theorem 6.6.3, where it is pointed out that the theorem merely ensures that $\hat{h} = l(T)\{1 + o(1)\}$. As a consequence, for example, (6.6.13) would be $O(T^{-1/2})$ if $h = c_1\hat{h}$, $c_1 > 1$, but might not be of that order for $h = \hat{h}$. *For this reason in all theorems below where AIC or BIC are used at stage I, we interpret $\hat{h}$ as $c_1\hat{h}_1$, $c_1 > 1$, where $\hat{h}_1$ is the value determined by AIC or BIC at stage I.* This saves repeated statement of this requirement in the theorems.

Theorem 6.6.6. *Let $y(t)$ be as in Theorem 6.6.1 for $s = 1$. Let $C_T = \log T$ at stage I and $0 < c_0 \leq C_T/\{\log T \log\log T\} \leq (\log T)^b$, $b < \infty$, at stage II, for c_0 sufficiently large and let H_T at stage I satisfy $l(T) \leq H_T \leq (\log T)^a$, $a < \infty$, and, at stage II, $H_T \uparrow \infty$, $H_T \leq (\log T)^a$, $a < \infty$. Then $\hat{p}^{(1)}, \hat{q}^{(1)}, \hat{r}^{(1)} \to p_0, q_0, r_0$, a.s. and $\hat{a}_T^{(1)}(z), \hat{b}_T^{(1)}(z), \hat{d}_T^{(1)}(z) \to a(z), b(z), d(z)$, a.s.*

In interpreting this result, recall that $\hat{a}_T^{(1)}(z)$, and so on, are the estimates at $\hat{p}^{(1)}, \hat{q}^{(1)}, \hat{r}^{(1)}$. The value of c_0 depends on unknown quantities, and this aspect of the theorem is therefore not useful in practice. The theorem is stated thus so as to show how large C_T must be. In fact, it is desirable to keep C_T as small as possible to reduce underestimation of order. For this reason we go on to consider $C_T = \log T$ at stage II. Let $x(t) = b\varepsilon(t)\sigma^{-1} = \sigma^{-1}\sum \beta(j)\varepsilon(t-j)$ and let ω_v^2 be the variance of the error from the best linear predictor of $x(t)$ from $x(t-1), x(t-2), \ldots, x(t-v)$. Let $v_0(z) - 1$ be the z-transform of that best predictor [i.e., $(v_0(z) - 1)x(t)$ is the best predictor].

Theorem 6.6.7. *Let $y(t)$ be as in Theorem 6.6.4 for $s = 1$ and AIC or BIC be used at stage I for $l(T) \le H_T \le (\log T)^a$, $a < \infty$. If $C_T = \log T$ at stage II and $H_T \le (\log T)^a$, $a < \infty$, $H_T \uparrow \infty$, then*

$$\operatorname{plim}_{T\to\infty}(\hat{p}^{(1)}, \hat{q}^{(1)}, \hat{r}^{(1)}) = (p_0 + v_0, q_0 + v_0, r_0 + v_0),$$

where v_0 minimises

$$(m+1)c_1\omega_v^2 + v(m+2)(-2\log\rho_0)$$

where c_1 is defined above Theorem 6.6.6. Also, for $p \ge p_0, q \ge q_0, r \ge r_0, q_0 > 0$,

$$(\hat{a}_p^{(1)}(z), \hat{b}_q^{(1)}(z), \hat{d}_r^{(1)}(z)) = v_0(z)\{1 + o_P(1)\}(a(z), b(z), d(z)) + O(Q_T), \qquad a.s.,$$

for $v = \min(p - p_0, q - q_0, r - r_0)$, in the definition of $v_0(z)$.

Note that $v_0(z)$ has zeros outside $|z| = 1$, from Theorem 5.6.2. Theorem 6.6.7 shows that for $C_T = \log T$ at stage II, the method may overestimate the order at stage II. As will be seen, this is of no great consequence since this overestimation will be corrected at stage III. Of course, v_0 may be zero and indeed for $m = 0$, $b(z) = 1 + \beta z$, $|\beta| < 0.89$, when c_1 is very near 1, some detailed calculations show that $v_0 = 0$. Also, the overestimation tends to be less than the theorem indicates, for the following reason. It will be shown in the proof that

$$(\hat{\sigma}_{p,q,r}^{(1)})^2 = \sigma^2 + \{\sigma^2(m+1)\hat{h}/T\}(\omega_v^2 - 1) + o_P(\log T/T), \qquad p \ge p_0, \quad q \ge q_0, \quad r \ge r_0, \quad q_0 > 0. \tag{6.6.14}$$

The need for $q_0 > 0$ in Theorem 6.6.7 and (6.6.14) arises from the fact that the theorem uses the increase of $\hat{h}$ at the rate $\log T$, from Theorems 6.6.3 and 6.6.4. The result in the theorem is obtained by replacing $\hat{h}$ by $c_1 l(T)$. However, for T not very large and ρ_0 near unity, then $\hat{h}$, for $c_1 = 1$, is usually much smaller than $l(T)$, so that (6.6.14) overestimates the bias in the right side. Some further details and simulations are given in Hannan and Kavalieris (1984). The procedure in remark (iv) at the end of Section 6.5 corrects for the over-estimation shown in (6.6.14).

Theorem 6.6.8. *Under the same conditions as for Theorem 6.6.7, for $q_0 > 0$, the expression (6.6.14) is valid, and using the procedure in (6.5.12) and below,*

$$\operatorname{plim}_{T\to\infty}(\hat{p}_c^{(1)}, \hat{q}_c^{(1)}, \hat{r}_c^{(1)}) = (p_0, q_0, r_0).$$

Finally, consider stage III, again for $s = 1$.

Theorem 6.6.9. *(a) If the conditions of Theorem 6.6.6 hold at stages I and II and $C_T = \log T$ at stage III, $H_T \leq (\log T)^a$, $a < \infty$, then $(\hat{p}^{(2)}, \hat{q}^{(2)}, \hat{r}^{(2)}) \to (p_0, q_0, r_0)$, a.s. and $(\hat{a}^{(2)}, \hat{b}^{(2)}, \hat{d}^{(2)}) \to (a, b, d)$, a.s. and $T^{1/2}(\hat{\tau}^{(2)} - \tau)$ obeys the central limit theorem with the covariance matrix given by Theorem 4.3.1.*

(b) If the conditions of Theorem 6.6.7 hold at stages I and II and at later stages $C_T = \log T$ and $H_T \leq (\log T)^a$, $a < \infty$, then $(\hat{p}^{(2)}, \hat{q}^{(2)}, \hat{r}^{(2)}) \to (p_0, q_0, r_0)$, $(\hat{a}^{(2)}, \hat{b}^{(2)}, \hat{d}^{(2)}) \to (a, b, d)$, in probability. Also, $(\hat{p}^{(3)}, \hat{q}^{(3)}, \hat{r}^{(3)}) \to (p_0, q_0, r_0)$ in probability and $T^{1/2}(\hat{\tau}^{(3)} - \tau)$ obeys the central limit theorem with the covariance matrix as in Theorem 4.3.1, for the case where (4.1.16) holds.

REMARK. $T^{1/2}(\hat{\tau}^{(2)} - \tau)$ will also obey the central limit theorem but will not be asymptotically efficient unless $v_0 = 0$. If $\hat{p}_c^{(1)}$, $\hat{q}_c^{(1)}$, $\hat{r}_c^{(1)}$ are used in place of $\hat{p}^{(1)}$, $\hat{q}^{(1)}$, $\hat{r}^{(1)}$ and the associated estimates of a, b, d at these orders, then again part (b) of the theorem holds but with $\hat{\tau}^{(2)}$ replacing $\hat{\tau}^{(3)}$.

(ii) Preliminary Results

First some results are given concerning the norms of matrices. In accordance with earlier notation, $\Gamma_{u,h}$ is the matrix of h^2 blocks, the typical one being $\Gamma_u(j-k)$. Recall that $\lambda_m(A)$, $\lambda_M(A)$ are the smallest and largest singular values of A. We also need the evaluations, for $H_T = (\log T)^a$, $a < \infty$,

$$\max_{0 \leq |t| \leq H_T} \|G_v(t) - \Gamma_v(t)\| = o(H_T^{-1}),$$
(6.6.15)
$$\max_{0 \leq |t| \leq H_T} \|G_{vz}(t) - \Gamma_{vz}(t)\| = o(H_T^{-1}).$$

Take the second expression. Then

$$\begin{aligned} G_{vz}(t) - \Gamma_{vz}(t) &= \sum_0^\infty L(k)\left\{\frac{1}{T}\sum_{s=1}^{T-t} z(s-k)z(s+t)' - \Gamma_z(t+k)\right\} \\ &= \sum_0^{c \log T} L(k)\left\{\frac{1}{T}\sum_1^{T-t} z(s-k)z(s+t)' - \Gamma_z(t+k)\right\} + o(H_T^{-1}), \end{aligned}$$

from which the result follows immediately from (6.6.4) and $\|z(t)\| = o(t^{1/4})$. The first part follows similarly.

Theorem 6.6.10. *Under the same conditions as for Theorem 6.6.1,*

$$0 < \underline{c} \le \lambda_m(G_h) \le \lambda_M(G_h) \le \bar{c} < \infty, \qquad h \le H_T.$$

Proof:

$$G_h = \begin{bmatrix} G_{u,h} & 0 \\ 0 & 0 \end{bmatrix} + \begin{bmatrix} G_{v,h} & G_{vz,h} \\ G_{zv,h} & G_{z,h} \end{bmatrix} + \begin{bmatrix} G_{uv,h} + G_{vu,h} & G_{uz,h} \\ G_{zu,h} & 0 \end{bmatrix}. \tag{6.6.16}$$

The last matrix has elements that are $O(Q_T)$ by Theorem 5.3.4. Also using (6.6.3), (6.6.4), (6.6.15), the second matrix on the right differs by quantities that are $o(H_T^{-1})$ from its limit

$$\begin{bmatrix} \Gamma_{v,h} & \Gamma_{vz,h} \\ \Gamma_{zv,h} & \Gamma_{z,h} \end{bmatrix}.$$

Thus it is sufficient to consider

$$\begin{bmatrix} G_{u,h} & 0 \\ 0 & 0 \end{bmatrix} + \begin{bmatrix} \Gamma_{v,h} & \Gamma_{vz,h} \\ \Gamma_{zv,h} & \Gamma_{z,h} \end{bmatrix}. \tag{6.6.17}$$

By (5.3.7) the elements of $G_{u,h} - \Gamma_{uh}$ are $O(Q_T)$. However,

$$0 < \underline{c} \le \lambda_m(\Gamma_{u,h}) \le \lambda_M(\Gamma_{u,h}) \le \bar{c} < \infty, \quad \text{a.s.},$$

since $f_u(\omega) = (1/2\pi)/k(e^{i\omega})\Sigma k(e^{i\omega})^*$ satisfies

$$\underline{c}I_s \le f_u(\omega) \le \bar{c}I_s, \quad \text{a.s.}, \quad 0 < \underline{c} \le \bar{c} < \infty,$$

so that

$$x^*\Gamma_{uh}x = \int \left(\sum x_j e^{ij\omega}\right)^* f_u(\omega) \left(\sum x_j e^{ij\omega}\right) d\omega,$$

x_j being the jth block of elements in x,

$$\le \bar{c}\frac{1}{2\pi} \int \left\|\sum x_j e^{ij\omega}\right\|^2 d\omega = \bar{c} \sum \|x_j\|^2.$$

A similar argument establishes the lower bound. However, the Hermitian form from the second term in (6.6.17) is

$$\int \left\{ \sum_1^h x_j^* e^{ij\omega} l^*(e^{i\omega}) + \sum_1^h y_j^* e^{ij\omega} \right\} f_z(\omega) \left\{ l(e^{i\omega}) \sum_1^h x_j e^{ij\omega} + \sum_1^h y_j e^{ij\omega} \right\} d\omega,$$

where x_j and y_j are the typical component blocks in v and z parts of the vector. Because of (6.6.1) this is bounded and bounded away from zero for $\sum(\|x_j\|^2 + \|y_j\|^2) = 1$. This proves the theorem.

However, this result is not sufficient for all purposes, since theorems below relate to results uniform in the lag, as in Theorem 5.3.1. Thus the vector space norm is then the uniform norm

$$|x|_\infty = \max_j |x_j|, \qquad x' = (x_1, x_2, \ldots, x_n).$$

(Now the x_j are scalars.) The corresponding matrix norm in this Banach space of column vectors is the "row sum norm"

$$|A|_\infty = \max_i \sum_j |a_{ij}|, \qquad |Ax|_\infty \le |A|_\infty |x|_\infty. \tag{6.6.18}$$

This satisfies $|AB|_\infty \le |A|_\infty |B|_\infty$ (i.e., is a Banach norm). Also, for a symmetric nonnegative matrix, A,

$$\lambda_M(A) \ge x'Ay/(x'xy'y)^{1/2},$$

and if i_0 is the row maximizing the row sum in (6.6.18), we may choose x_i to be null except at $i = i_0$, where $x_{i_0} = 1$ and $y_j = \operatorname{sign} a_{ij}$, $i = i_0$. Then

$$\lambda_M(A) \ge \max_i \sum_j |a_{ij}|/n^{1/2},$$

so that

$$|A|_\infty \le (\dim A)^{1/2} \lambda_M(A) = (\dim A)^{1/2} \|A\|. \tag{6.6.19}$$

Now $|\Gamma_h|_\infty \le c < \infty$, uniformly in h, because

$$\sum_0^\infty \|\Gamma_z(k)\| < \infty, \qquad \sum_0^\infty \|\Gamma_y(k)\| < \infty$$

so that the row sum norms are uniformly bounded. [Here and below the equivalence of norms in a finite-dimensional space is used. This cannot be used for $|\Gamma_h|_\infty$, but only for $\|\Gamma_z(k)\|$, $\|\Gamma_y(k)\|$, since h is not fixed.] Hence

$$|G_h|_\infty \le c < \infty, \qquad h \le H_T = (\log T)^a, \qquad a < \infty,$$

by an argument of the same nature as was used to prove Theorem 6.6.10. However, a uniform norm bound for G_h^{-1} is also needed.

Theorem 6.6.11. *Under Conditions A*

$$|\Gamma_h^{-1}|_\infty \le c < \infty, \qquad |G_h^{-1}|_\infty \le c < \infty, \qquad h \le H_T = (\log T)^a, \qquad a < \infty.$$

First it will be shown that $|\Gamma_{z,h}^{-1}|_\infty \le c < \infty$, $h = 1, 2, \ldots$. Let A_h be the matrix with $\Phi_z(i-j)$ as the block in row i, column j, $i, j = 1, \ldots, h$. Put $z_{t,h}' = (z(t+1)', z(t+2)', \ldots, z(t+h)')$, $A_h z_{t,h} = \varepsilon_{z,t,h} + \mathcal{N}_h z_t$, where $\varepsilon_{z,t,h}$ is as for $z_{t,h}$ but with $\varepsilon_z(t+j)$ replacing $z(t+j)$, $\mathcal{N}_h$ is the part of the Hankel matrix composed from the $\Phi_z(j)$ consisting of the first h rows of blocks and $z_t' = (z(t+1)', z(t+2)', \ldots)$. Then

$$A_h \Gamma_{z,h} A_h' = I_h \otimes \Sigma + \mathcal{N}_h \Gamma_{z,\infty} \mathcal{N}_h'.$$

Let M_{ij} be the typical block in the second matrix on the right.

$$n^{1/2} \sum_{i=n}^{\infty} |M_{ij}|_\infty \le \sum_{u=0}^{\infty} \left[\left\{ n^{1/2} \sum_{i=n}^{\infty} |\Phi_z(i+u)|_\infty \sum_{v=0}^{\infty} |\Gamma_z(u-v)\Phi_z(j+v)'|_\infty \right\} \right]$$

because

$$M_{ij} = \sum_{u,v=0}^{\infty}\sum \Phi_z(i+u)\Gamma_z(u-v)\Phi_z(j+v)'.$$

Thus

$$n^{1/2} \max_j \sum_{i=n}^{\infty} |M_{ij}|_\infty \to 0, \qquad n^{1/2} \max_i \sum_{j=n}^{\infty} |M_{ij}|_\infty \to 0 \tag{6.6.20}$$

because of (6.6.3) (and the symmetry of $\mathcal{N}_h \Gamma_\infty \mathcal{N}_h'$) and

$$\sum_{u,v=0}^{\infty}\sum |\Gamma_z(u-v)\Phi_z(j+v)'|_\infty \le c < \infty,$$

which follows from (6.6.3) and $|\Gamma_{z,\infty}|_\infty < \infty$ since $f_z(\omega)$ has an absolutely convergent Fourier series. Now partition $A_h\Gamma_{z,h}A_h'$ after the nth block of rows and columns as

$$\begin{bmatrix} A & B \\ B' & C \end{bmatrix}.$$

Then

$$\begin{bmatrix} A & B \\ B' & C \end{bmatrix}^{-1} = \begin{bmatrix} A^{-1} + A^{-1}B(C - B'A^{-1}B)B'A^{-1} & -A^{-1}B(C - B'A^{-1}B)^{-1} \\ -(C - B'A^{-1}B)^{-1}B'A^{-1} & (C - B'A^{-1}B)^{-1} \end{bmatrix},$$

where

$$C = (I_{h-n} \otimes \Sigma_z) + M_n, \qquad M_n = [M_{ij}]_{ij=n+1,\ldots,h}.$$

Taking n large but fixed $|M_n|_\infty$ may be made arbitrarily small by (6.6.20) and hence $|(I_{h-n} \otimes \Sigma_z^{-1})M_n|_\infty$ may be made arbitrarily small. Thus expanding $\{I_{(h-n)m} + (I_{h-n} \otimes \Sigma_z^{-1})M_n\}^{-1}$ in a Neumann series, we see that this matrix is $I_{(h-n)m} + C_1$ where $|C_1|_\infty$ is bounded uniformly in h and hence so is $(I_{h-n} \otimes \Sigma_z)(I_{(h-n)m} + C_1)$, so that $|C^{-1}|_\infty \le c \le \infty$. Also,

$$|B'A^{-1}B|_\infty \le |B'|_\infty |A^{-1}|_\infty |B_\infty| \le n^{1/2}\lambda_M(A)|B_\infty|\,|B'|_\infty$$

and since $|B_\infty|$, $|B'|_\infty$ can be made arbitrarily small by taking n large, from (6.6.20) again, then $|B'A^{-1}B|_\infty$ can be made arbitrarily small. Thus so can $|C^{-1}B'A^{-1}B|_\infty$ and again using a Neumann expansion of $(I - C^{-1}B'A^{-1}B)$, we obtain $|I - C^{-1}B'A^{-1}B|_\infty \le c < \infty$. Hold n fixed but large. Then $|A|_\infty \le c < \infty$ and hence all elements in $(A_h\Gamma_{z,h}A_h')^{-1}$ (i.e., in its partition) are row sum bounded. As $|A_h|_\infty \le c < \infty$, it follows that $|\Gamma_{z,h}^{-1}|_\infty \le c < \infty$. Now Γ_h^{-1} corresponds to

$$\begin{bmatrix} k(z) & l(z)k_z(z) \\ 0 & k_z(z) \end{bmatrix}^{-1} = \begin{bmatrix} k(z)^{-1} & -k(z)^{-1}l(z) \\ 0 & k_z(z)^{-1} \end{bmatrix} \tag{6.6.21}$$

as $\Gamma_{z,h}$ corresponds to $k_z(z)^{-1}$ and $k(z)^{-1}$, $k(z)^{-1}l(z)$ have coefficients decreasing geometrically to zero while (6.6.2) holds for $k_z(z)$. Thus the same proof applies to $|\Gamma_h^{-1}|_\infty$ as to $|\Gamma_{z,h}^{-1}|_\infty$, and thus $|\Gamma_h^{-1}|_\infty \le c < \infty$, $h = 1, 2, \ldots$. The same is evidently true of $|\hat{\Gamma}_h^{-1}|_\infty$, a.s., for T large enough. Consider

$$\begin{aligned} |G_h^{-1}|_\infty &\le |\Gamma_h^{-1}|_\infty + |G_h^{-1} - \Gamma_h^{-1}|_\infty \\ &\le |\Gamma_h^{-1}|_\infty + |G_h^{-1}|_\infty |G_h - \Gamma_h|_\infty |\Gamma_h^{-1}|_\infty. \end{aligned}$$

Then

$$|G_h^{-1}|_\infty\{1 - |G_h - \Gamma_h|_\infty |\Gamma_h^{-1}|_\infty\} \le |\Gamma_h^{-1}|_\infty.$$

From (5.3.7), $|G_h - \Gamma_h|_\infty = O(hQ_T) \to 0$, so that $|G_h^{-1}|_\infty \le c < \infty$, which establishes the theorem.

Finally in this subsection we generalize the theorem of Baxter (1963).

Theorem 6.6.12. *Let* $2\pi f(\omega) = k\Sigma k^*$, $k(z) = \sum K(j)z^j$ *being analytic for* $|z| \le 1$ *with* $\det(k) \ne 0$, $|z| \le 1$. *Put* $\Phi(z) = k(z)^{-1} = \sum \Phi(j)z^j$. *Let* $\Phi_h(j)$, Σ_h *be as in (6.6.6) and below. If* $\sum |j|^\lambda \|K(j)\| < \infty$, $\lambda > 0$, *there are* $H < \infty$, $c < \infty$, *depending only on* $f(\omega)$, *so that*

$$\sum_1^h (2^\lambda + j^\lambda)\|\Sigma_h^{-1}\Phi_h(j) - \Sigma^{-1}\Phi(j)\| \le c \sum_{h+1}^\infty (2^\lambda + j^\lambda)\|\Phi(j)\|, \qquad h \ge H.$$

The proof is given in the outline necessary to extend the proof in Baxter (1962). We shall call the result in this theorem "Baxter's inequality." As will be shown in Chapter 7 (see Theorem 7.4.2), $\sum j^\lambda \|K(j)\| < \infty$ implies $\sum j^\lambda \|\Phi(j)\| < \infty$ because $\det\{k(z)\} \ne 0$, $|z| \le 1$. Since

$$\int \sum_0^h \Sigma_h^{-1}\Phi_h(j)e^{ij\omega}f(\omega)e^{-ik\omega}\,d\omega = \delta_{0k}I,$$

$$k = 0, 1, \ldots, h$$

$$\int \sum_0^\infty \Sigma^{-1}\Phi(j)e^{ij\omega}f(\omega)e^{-ik\omega}\,d\omega = \delta_{0k}I,$$

$$k = 0, 1, \ldots$$

then

$$\int \sum_0^h \{\Sigma_h^{-1}\Phi_h(j) - \Sigma^{-1}\Phi(j)\}e^{ij\omega}f(\omega)e^{-ik\omega}\,d\omega$$

$$= \int \sum_{h+1}^\infty \Sigma^{-1}\Phi(j)e^{ij\omega}f(\omega)e^{-ik\omega}\,d\omega, \qquad k = 0, 1, \ldots, h. \tag{6.6.22}$$

Call $F_h(k)$ the right side. It is shown in Baxter (1963) that if for $f(\omega)$ as in the theorem and $s = 1$,

$$\int H_h(e^{i\omega})f(\omega)e^{-ik\omega}\,d\omega = F_h(k), \qquad k = 0, 1, \ldots, h,$$

then

$$\text{(6.6.23)}\qquad \sum_0^h (2^\lambda + j^\lambda)\|H_h(j)\| \le c \sum_0^h (2^\lambda + j^\lambda)\|F_h(j)\|, \qquad h \ge H,$$

where the $H_h(j)$ are the Fourier coefficients of $H_h(z)$. Baxter's proof depends on factorizing $f(\omega)$ as $A(e^{i\omega})B(e^{i\omega})$, where $A(z)$ is analytic for $|z| \le 1$ and $B(z)$ for $|z| \ge 1$. If, in the vector context, $A(z)$ is taken as $k(z)\Sigma^{1/2}$, $B(z)$ as $\Sigma^{1/2}k(z^{-1})'$, then the proof in Baxter (1963) is hardly altered so that (6.6.23) again holds. Put $v(j) = 2^\lambda + |j|^\lambda$. Then $v(j) \le v(k)v(j-k)$, for all j, k. [This is why the $v(j)$ are so defined.] Thus applying (6.6.23) to (6.6.22), we obtain

$$\sum_{j=1}^h (2^\lambda + j^\lambda)\|\Sigma_h^{-1}\Phi_h(j) - \Sigma^{-1}\Phi(j)\| \le c \sum_{k=0}^h (2^\lambda + k^\lambda)\| \sum_{h+1}^\infty \Sigma^{-1}\Phi(j)\Gamma(k-j)\|$$

$$\le c \sum_{h+1}^\infty v(j)\|\Phi(j)\| \sum_{k=0}^\infty v(k-j)\|\Gamma(k-j)\| \le c \sum_{h+1}^\infty (2^\lambda + j^\lambda)\|\Phi(j)\|$$

since it is easily seen [again using $v(j) \le v(k)v(j-k)$] that $\Sigma_0^\infty v(j)\|\Gamma(j)\| < \infty$. This establishes the theorem.

REMARK 1. The case of general λ has been dealt with because of applications in Chapter 7.

REMARK 2. If

$$f(\omega) = \begin{bmatrix} f_y(\omega) & f_{yz}(\omega) \\ f_{zy}(\omega) & f_z(\omega) \end{bmatrix}$$

and we call $[\Phi_h(j), \Psi_h(j)]$ the first s rows of what was *previously* $\Phi_h(j)$, then

$$\sum_1^h (2^\lambda + j^\lambda)\|\Sigma_h^{-1}[\Phi_h(j), \Psi_h(j)] - \Sigma^{-1}[\Phi(j), \Psi(j)]\|$$

$$< c \sum_{h+1}^\infty (2^\lambda + j^\lambda)\|[\Phi(j), \Psi(j)]\|, \qquad h \ge H.$$

REMARK 3. The factors Σ_h^{-1}, Σ^{-1} can be eliminated as follows. It is evident that

$$\sum_1^h (2^\lambda + j^\lambda)\|\Phi_h(j) - \Sigma_h\Sigma^{-1}\Phi(j)\| \le C \sum_{h+1}^\infty (2^\lambda + j^\lambda)\|\Phi(j)\|, \qquad h \ge H,$$

since $\Sigma_h \ge \Sigma > 0$, $\Sigma_h < \infty$. Thus it is sufficient to show that

$$\sum_{1}^{h} (2^{\lambda} + j^{\lambda}) \|(\Sigma_h - \Sigma)\Sigma^{-1}\Phi(j)\| \leq c \sum_{h+1}^{\infty} (2^{\lambda} + j^{\lambda}) \|\Phi(j)\|, \qquad h \geq H.$$

However, much more is true because

$$\begin{aligned} \Sigma_h &= \int \left\{ \sum_{0}^{h} \Phi_h(j) e^{ij\omega} \right\} f(\omega) \left\{ \sum_{0}^{h} \Phi_h(j) e^{ij\omega} \right\}^* d\omega \\ &\leq \int \left\{ \sum_{0}^{h} \Phi(j) e^{ij\omega} \right\} f(\omega) \left\{ \sum_{0}^{h} \Phi(j) e^{ij\omega} \right\}^* d\omega \end{aligned}$$

since the $\Phi_h(j)$ are minimizing. But

$$\begin{aligned} &\int \left\{ \sum_{0}^{h} \Phi(j) e^{ij\omega} \right\} f(\omega) \left\{ \sum_{0}^{h} \Phi(j) e^{ij\omega} \right\}^* d\omega \\ &\qquad = \Sigma - \int \left\{ \sum_{h+1}^{\infty} \Phi(j) e^{ij\omega} \right\} f(\omega) \left\{ \sum_{h+1}^{\infty} \Phi(j) e^{ij\omega} \right\}^* d\omega, \end{aligned}$$

so that

$$\|\Sigma_h - \Sigma\| \leq C \left\{ \sum_{h+1}^{\infty} \|\Phi(j)\| \right\}^2.$$

Thus for suitable c, H depending only on $f(\omega)$,

$$\sum_{1}^{h} (2^{\lambda} + j^{\lambda}) \|\Phi_h(j) - \Phi(j)\| \leq c \sum_{h+1}^{\infty} (2^{\lambda} + j^{\lambda}) \|\Phi(j)\|, \qquad h \geq H,$$

and in particular, (6.6.13) follows, and similarly for $\dot{\Phi}_h(j)$, $\dot{\Psi}_h(j)$.

(iii) Proofs of Theorems

Theorem 6.6.1. Put, for example,

$$[-\Phi, \Psi]_h = [-\Phi(1), \ldots, -\Phi(h), \Psi(1), \ldots, \Psi(h)].$$

Then [see above (6.6.6) for the notation] from (6.6.6),

$$\begin{aligned} \{[-\hat{\Phi}_h, \hat{\Psi}_h] - [-\dot{\Phi}_h, \dot{\Psi}_h]\} G_h = & -[-\Phi, \Psi]_h G_h \\ & + (g_h + \{[-\Phi, \Psi]_h - [-\dot{\Phi}_h, \dot{\Psi}_h]\} \dot{\Gamma}_h) \\ & + \{[-\Phi, \Psi]_h - [\dot{\Phi}_h, \dot{\Psi}_h]\} (G_h - \dot{\Gamma}_h). \end{aligned} \tag{6.6.24}$$

The last term is, by (5.3.7), (6.6.4), and (6.6.15), which ensure that $G_h - \Gamma_h = o(H_T^{-1})$,

$$o\left\{H_T^{-1} \sum_1^h \|[\Phi(j), \Psi(j)] - [\dot{\Phi}_h(j), \dot{\Psi}_h(j)]\|\right\} = o\left\{H_T^{-1} \sum_{h+1}^{\infty} \|[\Phi(j), \Psi(j)]\|\right\}.$$

The right side follows from Theorem 6.6.12. The second term in (6.6.24) has as ith, $(i + h)$th blocks [see (6.6.7), which also holds with $\dot{\Gamma}_y$ replacing Γ_y]

$$(6.6.25)\qquad \begin{aligned} &\sum_{h+1}^{\infty} \Phi(j)\dot{\Gamma}_y(j-i) - \sum_{h+1}^{\infty} \Psi(j)\Gamma_{zy}(j-i) + G_y(-i) \\ &\sum_{h+1}^{\infty} \Phi(j)\Gamma_{yz}(j-i) - \sum \Psi(j)\Gamma_z(j-i) + G_{yz}(-i). \end{aligned}$$

Now replace $\dot{\Gamma}_y$, Γ_{zy}, Γ_z by G_y, G_{zy}, G_z in (6.6.25) and then add these blocks to the corresponding blocks in the first term in (6.6.24) to obtain for these blocks in (6.6.24), using (5.3.7), (6.6.4), (6.6.15) again,

$$(6.6.26)\qquad \begin{aligned} &\sum_0^{\infty} \Phi(j)G_y(j-i) - \sum_1^{\infty} \Psi(j)G_{zy}(j-i) + o\left\{H_T^{-1} \sum_{h+1}^{\infty} \|[\Phi(j), \Psi(j)]\|\right\} \\ &\sum_0^{\infty} \Phi(j)G_{yz}(j-i) - \sum_1^{\infty} \Psi(j)G_z(j-i) + o\left\{H_T^{-1} \sum_{h+1}^{\infty} \|[\Phi(j), \Psi(j)]\|\right\}. \end{aligned}$$

Now, $G_y(j - i)$ may, for example, be replaced by

$$\tilde{G}_y(j-i) = \frac{1}{T}\sum_1^T y(t-j)y(t-i)',$$

introducing an error that is uniformly in i, j, that is, $o(T^{-1/2})$. Indeed, for $j \le i$, for example,

$$G_y(j-i) - \tilde{G}_y(j-i) = \frac{1}{T}\sum_{t=T+1}^{T+j} y(t-j)y(t-i)' - \frac{1}{T}\sum_{t=1}^{i} y(t-j)y(t-i)',$$

which is $= o\{jT^{-1/2} + i^{3/2}T^{-1}\}$ by the finiteness of the fourth moment. For $j > i$ we obtain a bound that is $o\{iT^{-1/2} + j^{3/2}T^{-1}\}$, needless to say. Since the $\Phi(j)$ converge geometrically, the error in (6.6.26) from this replacement is $o(T^{-1/2})$. Similarly, $G_{zy}(j - i)$ may be replaced by

$$\tilde{G}_{zy}(j-i) = \frac{1}{T}\sum_1^T z(t-j)y(t-i)'$$

and we obtain for (6.6.26) the expression

$$\frac{1}{T}\sum_{1}^{T}\varepsilon(t)y(t-i)' + o\left\{H_T^{-1}\sum_{h+1}^{\infty}\|[\Phi(j),\Psi(j)]\|\right\} + o(T^{-1/2})$$

$$\frac{1}{T}\sum_{1}^{T}\varepsilon(t)z(t-i)' + o\left\{H_T^{-1}\sum_{h+1}^{\infty}\|[\Phi(j),\Psi(j)]\|\right\} + o(T^{-1/2}).$$

Now the first part of the theorem is completed using Theorems 6.6.11 and 5.3.1. In the ARMA case $G_h - \Gamma_h = O(Q_T)$, so H_T^{-1} may be replaced by Q_T, giving the desired result. For $s = 1$ autocorrelations may replace autocovariances and, from Corollary 5.3.2, the use of the $\Phi(j)$, $\Psi(j)$ is avoided and the same is true when (4.1.16) holds for $s > 1$. The last statement follows from Baxter's inequality.

Theorems 6.6.3 and 6.6.4. We follow Kavalieris (1984). Put O_∞ for the matrix of infinitely many blocks of zeros. Then, as will be shown,

$$\hat{\Sigma}_h = \dot{\Sigma} + ([(\dot{\Phi}_h \vdots O_\infty),(\dot{\Psi}_h \vdots O_\infty)] - [\Phi,\Psi]_\infty)\dot{\Gamma}_\infty([(\dot{\Phi}_h \vdots O_\infty),(\dot{\Psi}_h \vdots O_\infty)] - [\Phi,\Psi]_\infty)'\{1+o(1)\} + O(hQ_T^2). \tag{6.6.27}$$

Call $\dot{\Omega}$ the second term on the right in (6.6.27) ignoring the factor $\{1 + o(1)\}$. Put $S_1, S_2, \ldots, S_5$ for the expressions

$$\frac{1}{T}\sum_{1}^{T}\{\dot{\varepsilon}_h(t)-\varepsilon(t)\}\{\dot{\varepsilon}_h(t)-\varepsilon(t)\}', \quad \frac{1}{T}\sum_{1}^{T}\varepsilon(t)\{\dot{\varepsilon}_h(t)-\varepsilon(t)\}',$$

$$\frac{1}{T}\sum\{\hat{\varepsilon}_h(t)-\dot{\varepsilon}_h(t)\}\{\dot{\varepsilon}_h(t)-\varepsilon(t)\}',$$

$$\frac{1}{T}\sum_{1}^{T}\varepsilon(t)\{\hat{\varepsilon}_h(t)-\dot{\varepsilon}_h(t)\}', \quad \frac{1}{T}\sum_{1}^{T}\{\hat{\varepsilon}_h(t)-\dot{\varepsilon}_h(t)\}\{\hat{\varepsilon}_h(t)-\dot{\varepsilon}_h(t)\}',$$

where

$$\dot{\varepsilon}_h(t) = \sum_{0}^{h}\dot{\Phi}_h(j)y(t-j) - \sum_{1}^{h}\dot{\Psi}_h(j)z(t-j).$$

Then $\hat{\Sigma}_h = \dot{\Sigma} + S_1 + (S_2 + S_2') + (S_3 + S_3') + (S_4 + S_4') + S_5$. Let $\tilde{G}_h$ have entries $\tilde{G}_y(j)$, $\tilde{G}_{yz}(j)$, and so on. Then

$$S_1 = ([(\dot{\Phi}_h \vdots O_\infty),(\dot{\Psi}_h \vdots O_\infty)] - [\Phi,\Psi]_\infty)\tilde{G}_\infty([(\dot{\Phi}_h \vdots O_\infty),(\dot{\Psi}_h \vdots O_\infty)] - (\Phi,\Psi)]_\infty)'.$$

In this expression $\tilde{G}_\infty$ may be replaced by $\dot{\Gamma}_\infty$ up to a factor $\{1 + o(1)\}$ because the infinite series may be truncated at $c \log T = K$ (say) for c large, and $\tilde{G}_K - G_K = o(H_T T^{1/2})$, $G_K - \Gamma_K = o(H_T^{-1})$, so that $\tilde{G}_K - \Gamma_K$ has spectral norm, that is, $o(1)$. Then $\dot{\Gamma}_K$ may be replaced by $\dot{\Gamma}_\infty$. With this replacement we obtain $\dot{\Omega}$. Thus it remains to show that the remaining S_j, $j = 2, \ldots, 5$, give a contribution that is $O(hQ_T^2)$. The term S_2' may be similarly manipulated to give

$$\sum_1^T [\{\dot{\Phi}_h(j) - \Phi(j)\} y(t-j) + \{\dot{\Psi}_h(j) - \Psi(j) z(t-j)] \varepsilon(t)' + o(T^{-1}),$$

and using Theorem (5.3.1),

$$S_2 = O\left\{Q_T \sum_{h+1}^{\infty} \|[\Phi(j), \Psi(j)]\|\right\} + o(T^{-1}).$$

In much the same way,

$$S_3 = O\left\{Q_T \sum_{h+1}^{\infty} \|[\Phi(j), \Psi(j)\|\right\} + o\left[h^{-1}\left\{\sum_{h+1}^{\infty} \|[\Phi(j), \Psi(j)]\|\right\}^2\right] + o(T^{-1}).$$

Put

$$F = [\hat{\Phi}_h, \hat{\Psi}_h] - [\dot{\Phi}_h, \dot{\Psi}_h] + \frac{1}{T}\sum_1^T \varepsilon(t)[y(t,h)', z(t,h)']\hat{\Gamma}_h^{-1}, \tag{6.6.28}$$
$$y(t,h)' = (y(t-1)', \ldots, y(t-h)'), \qquad z(t,h)' = (z(t-1)', \ldots, z(t-h)').$$

Then

$$\begin{aligned}\hat{\varepsilon}_h(t) - \dot{\varepsilon}(t) &= \sum_1^h \{[\hat{\Phi}_h(j) - \dot{\Psi}_h(j)] y(t-j) - [\hat{\Phi}_h(j) - \dot{\Psi}_h(j)] z(t-j)\} \\ &= -\frac{1}{T}\sum_1^T \varepsilon(s)[y(s,h)', z(s,h)'] G_h^{-1} \begin{bmatrix} y(t,h) \\ z(t,h) \end{bmatrix} + F \begin{bmatrix} y(t,h) \\ z(t,h) \end{bmatrix},\end{aligned}$$

so that

$$\begin{aligned} S_5 = &-\frac{1}{T}\sum_1^T \varepsilon(t)[y(t,h)', z(t,h)'] G_h^{-1} \frac{1}{T} \begin{bmatrix} y(t,h) \\ z(t,h) \end{bmatrix} \varepsilon(t)' \\ &- F \frac{1}{T}\sum_1^T \begin{bmatrix} y(t,h) \\ z(t,h) \end{bmatrix} \varepsilon(t)' - \frac{1}{T}\sum_1^T \varepsilon(t)[y(t,h)', z(t,h)'] F' \\ &+ F G_h F' + o(T^{-1}).\end{aligned}$$

The $o(T^{-1})$ term comes from replacing $\tilde{G}_h$ by G_h. In the same way,

$$S_4 = -\frac{1}{T}\sum_1^T \varepsilon(t)[y(t,h)', z(t,h)']G_h^{-1}\frac{1}{T}\sum_1^T \begin{bmatrix} y(t,h) \\ z(t,h) \end{bmatrix}\varepsilon(t)'$$
$$+\frac{1}{T}\sum_1^T \varepsilon(t)[y(t,h)', z(t,h)']F' + o(T^{-1}).$$

Now the elements of F are

$$o\left\{h^{-1}\sum_{h+1}^{\infty} \|[\Phi(j), \Psi(j)]\|\right\} + o(T^{-1/2})$$

from the proof of Theorem 6.6.1 [see below (6.6.24)], so that

$$FG_hF' = o(h/T) + o\left[h^{-1}\left\{\sum_{h+1}^{\infty} \|[\Phi(j), \Psi(j)]\|\right\}^2\right],$$

from which it follows that $FG_hF' = o(h/T) + o(\dot{\Omega})$, because

$$\underline{c}\sum_{h+1}^{\infty} \|[\Phi(j), \Psi(j)]\|^2 \leq \|\dot{\Omega}\| \leq \bar{c}\sum_{h+1}^{\infty} \|[\Phi(j), \Psi(j)]\|^2, \tag{6.6.29}$$
$$\left\{\sum_{h+1}^{\infty} \|[\Phi(j), \Psi(j)]\|\right\}^2 = O\left\{\sum_{h+1}^{\infty} \|[\Phi(j), \Psi(j)]\|^2\right\}.$$

The latter follows from (6.6.9) and the former from $o < \lambda_m(\Gamma_\infty) < \infty$ together with Baxter's inequality and the evident fact that

$$\|[\dot{\Phi}_h \vdots O_\infty), (\dot{\Psi}_h \vdots O_\infty)] - [\Phi, \Psi]_\infty\|^2 \geq \sum_{h+1}^{\infty} \|[\Phi(j), \Psi(j)\|^2.$$

Also using these methods for S_3, we obtain, noticing the cancellations in $S_4 + S_4' + S_5$,

$$\hat{\Sigma}_h = \dot{\Sigma} + \dot{\Omega}\{1 + o(1)\} + O\left\{Q_T\sum_{h+1}^{\infty} \|[\Phi(j), \Psi(j)]\|\right\} \tag{6.6.30}$$
$$+\frac{1}{T}\sum_1^T \varepsilon(t)[y(t,h)', z(t,h)']G_h^{-1}\frac{1}{T}\sum_1^T \begin{bmatrix} y(t,h) \\ z(t,h) \end{bmatrix}\varepsilon(t)' + o\left(\frac{h}{T}\right).$$

To complete the proof it is necessary to show that

$$P(h, T) = Q_T \left\{ \sum_{h+1}^{\infty} \|[\Phi(j), \Psi(j)]\| \right\} \Big/ \max \left\{ \frac{h}{T}, \sum_{h+1}^{\infty} \|[\Phi(j), \Psi(j)]\|^2 \right\} \to 0.$$

Let $h(T)$ be any sequence satisfying $h(T) \le H_T$ and put $P(T) = P(h(T), T)$. If $\limsup P(T) = c > 0$, then, along some subsequence, T_i, $P(T_i) \to c$. If $\limsup h(T_i)/l(T_i) \ge 1$, there is a sub-subsequence, $\tilde{T}_i$, because of (6.6.9), so that

$$\tilde{T}_i^{1/2} \sum_{h(\tilde{T}_i)+1}^{\infty} \|[\Phi(j), \Psi(j)]\| \to c_1 < \infty,$$

so that we obtain the contradiction

$$\begin{aligned} P(\tilde{T}_i) &\le Q_{\tilde{T}_i} \left\{ \sum_{h(\tilde{T}_i)+1}^{\infty} \|[\Phi(j), \Psi(j)]\| / \{h(\tilde{T}_i) \tilde{T}_i^{-1}\} \right\} \\ &= O\{Q_{\tilde{T}_i} \tilde{T}_i^{1/2} / h(\tilde{T}_i)\} \to 0. \end{aligned}$$

Thus $\limsup h(T_i)/l(T_i) < 1$ and then along some sub-subsequence, $\tilde{T}_i$, $h(\tilde{T}_i)/l(\tilde{T}_i) \to c_1$ and, again using (6.6.9),

$$\tilde{T}_i^b \left\{ \sum_{h(\tilde{T}_i)+1}^{\infty} \|[\Phi(j), \Psi(j)]\| \right\} \to c > 0, \qquad b < \tfrac{1}{2}.$$

Thus

$$\begin{aligned} P(\tilde{T}_i) &\le Q_{\tilde{T}_i} \sum_{h(\tilde{T}_i)+1}^{\infty} \|[\Phi(j), \Psi(j)]\| \Big/ \left\{ \sum_{h(\tilde{T}_i)+1}^{\infty} \|[\Phi(j), \Psi(j)]\|^2 \right\} \\ &= O(Q_{\tilde{T}_i} \tilde{T}_i^b) \to 0 \end{aligned}$$

and we have a contradiction to $\limsup P(T) = c > 0$. This establishes (6.6.27) since the fourth term on the right in (6.6.30) is $O(hQ_T^2)$, from Theorems 5.3.1 and 6.6.11. Now from this

$$BIC(h) = \log\det \dot{\Sigma} + \operatorname{tr}\{\Sigma^{-1}\dot{\Omega}\}\{1 + o(1)\} + \{hs(s+m)\log T/T\}\{1 + o(1)\}$$

since $\dot{\Sigma} \to \Sigma$. Also,

$$\text{(6.6.31)} \quad \operatorname{tr}\{\Sigma^{-1}\dot{\Omega}\} + hs(s+m)\log T/T = c_h \rho_0^{2h} + hs(s+m)\log T/T,$$

where the c_h sequence is bounded and bounded away from zero. This follows from (6.6.29) and (6.6.9). Thus if $\hat{h}$ minimizes $BIC(h)$, then $\hat{h}/l(T) \to 1$. Indeed, the value of h, say $l(T)$, minimizing (6.6.31) is evidently $l(T)\{1 + o(1)\}$ and Theorem 6.6.3 follows easily.

To prove Theorem 6.6.4 we first observe that because $\hat{h}$ is a nonincreasing function of C_T, $\hat{h}$ for AIC, $\hat{h}_A$ let us say, satisfies $\liminf \hat{h}_A/l(T) \geq 1$, a.s. Now we shall show, in the proof of Theorem 6.6.5, that the fourth term on the right in (6.6.30) is $(h/T)\Sigma(s + m) + o_P(h/T)$, for $h \geq l(T)$. Thus *for such* h,

$$\begin{aligned} AIC(h) &= \log\det\dot{\Sigma} + \operatorname{tr}\{\Sigma^{-1}\dot{\Omega}\}\{1 + o(1)\} + \{hs(s + m)/T\}\{1 + o_P(1)\} \\ &= \log\det\Sigma + \{c_h\rho_0^{2h} + hs(s + m)/T\}\{1 + o_P(1)\}, \end{aligned}$$

which implies, since the h minimizing this is $l(T)\{1 + o_P(1)\}$,

$$\lim_{T\to\infty} P\{\hat{h}/l(T) > 1)\} = 0.$$

This proves Theorem 6.6.4.

Theorem 6.6.5. Now (4.1.16) holds, as it also does for Theorem 6.6.4. Put, for $a \geq 0$ and each i, j,

$$\tilde{G}_y(i - j - a) = \frac{1}{T}\sum_1^T y(t - i)y(t - j - a)', \qquad i, j = 0, 1, \ldots$$

and correspondingly $\tilde{G}_{y,h}(a)$ for the matrix with $\tilde{G}_y(i - j - a)$ as the (i, j)th block. Defining $\tilde{G}_{yz,h}(a)$, $\tilde{G}_{z,h}(a)$ analogously, we then put

$$\tilde{G}_h(a) = \begin{bmatrix} \tilde{G}_{y,h}(a) & \tilde{G}_{yz,h}(a) \\ \tilde{G}_{zy,h}(a) & \tilde{G}_{z,h}(a) \end{bmatrix}.$$

We need these expressions only for $a = 0$ for Theorem 6.6.5 but for $a > 0$ later. For $a = 0$, then $\tilde{G}_h(a) = \tilde{G}_h$. Consider

$$\text{(6.6.32)} \quad \frac{1}{hT}\sum\sum_{s,t=1}^T \varepsilon(s)[y(s, h)', z(s, h)']\tilde{G}_h^{-1}\tilde{G}_h(a)\tilde{G}_h^{-1}\begin{bmatrix} y(t, h) \\ z(t, h) \end{bmatrix}\varepsilon(t)'.$$

We shall show that this is, *for* $h \geq c\log T$, $c > 0$,

$$\text{(6.6.33)} \quad \begin{aligned} &\frac{1}{hT}\sum_{j=1}^{h-a}\left\{\sum_{s=1}^T \varepsilon(s)\varepsilon(s - j)'\right\}\Sigma^{-1}\left\{\sum_{t=1}^T \varepsilon(t - j - a)\varepsilon(t)'\right\} \\ &+ \frac{1}{hT}\sum\sum_{s,t=1}^T \varepsilon(s)z(s, h)'\tilde{G}_{z,h}^{-1}\tilde{G}_{z,h}(a)\tilde{G}_{z,h}^{-1}z(t, h)\varepsilon(t)' + O\left(\frac{\log\log T}{\log T}\right). \end{aligned}$$

Put

$$B_h = \begin{bmatrix} I_s & \Phi(1) & \cdots & \Phi(h-1) & 0 & \Psi(1) & \cdots & \Psi(h-1) \\ 0 & I_s & \cdots & \Phi(h-2) & 0 & 0 & \cdots & \Psi(h-2) \\ \vdots & \vdots & \vdots & \vdots & \vdots & \vdots & \vdots & \vdots \\ 0 & 0 & \cdots & I_s & 0 & 0 & \cdots & 0 \end{bmatrix}.$$

Put

$$D_h = \begin{bmatrix} \Phi(h+1) & \Psi(h+1) & \Phi(h+2) & \Psi(h+2) & \cdots \\ \Phi(h) & \Psi(h) & \Phi(h+1) & \Psi(h+1) & \cdots \\ \vdots & \vdots & \vdots & \vdots & \vdots \\ \Phi(1) & \Psi(1) & \Phi(2) & \Psi(2)\ldots & \end{bmatrix}.$$

Then

$$B_h \begin{bmatrix} y(t,h) \\ z(t,h) \end{bmatrix} = \varepsilon(t,h) + D_h[y(t-h-1)', z(t-h-1)', y(t-h-2)', \ldots]'.$$

Call $w_h(t)'$ the vector on which D_h operates. [This definition of D_h, $w_h(t)$ has been adopted so as more conveniently to define these infinite matrices and vectors.] Now define $\hat{B}_h$ as for B_h but with $\hat{\Phi}_{h-u}(j)$, $\hat{\Psi}_{h-u}(j)$ replacing $\Phi(j)$, $\Psi(j)$ in the uth row, $u = 1, \ldots, h$. Then from the defining relations for $\hat{\Phi}_h(j)$, $\hat{\Psi}_h(j)$, we have

$$\hat{B}_h G_h \hat{B}_h' = S_h, \qquad \hat{B}_h'[G_{zy,h}, G_{z,h}] = 0,$$

where S_h is block diagonal with Σ_{h-u} as the uth diagonal block. Thus

$$G_h^{-1} = \begin{bmatrix} & \vdots & 0 \\ \hat{B}_h' & & \\ & \vdots & I_{hm} \end{bmatrix} \begin{bmatrix} S_h^{-1} & 0 \\ 0 & G_{z,h}^{-1} \end{bmatrix} \begin{bmatrix} \hat{B}_h & \\ \cdots & \cdots \\ 0 & I_{hm} \end{bmatrix}.$$

Now the sum of the moduli of all the elements of $\hat{B}_h - B_h$ is $o(1)$ and the same is true of $S_h - (I_h \otimes \Sigma)$. These follow from Theorems 6.6.1 and 6.6.11 and (6.6.27). Thus

$$G_h^{-1} = \begin{bmatrix} & \vdots & 0 \\ B_h' & & \\ & \vdots & I_{hm} \end{bmatrix} \begin{bmatrix} I_h \otimes \Sigma^{-1} & 0 \\ 0 & G_{z,h}^{-1} \end{bmatrix} \begin{bmatrix} B_h & \\ \cdots & \cdots \\ 0 & I_{hm} \end{bmatrix} + E_h,$$

where E_h has the sum of the moduli of its elements bounded uniformly. The same is true for $\tilde{G}_h$ using Theorem 6.6.11, since $\tilde{G}_h - G_h$ has elements $o(hT^{-1/2})$. Also,

$$\begin{bmatrix} B_h & \vdots & 0 \\ 0 & \vdots & I_{hm} \end{bmatrix} \tilde{G}_h(a) \begin{bmatrix} & \vdots & 0 \\ B_h' & & \\ & \vdots & I_{hm} \end{bmatrix} = \begin{bmatrix} I_h(a) \otimes \Sigma & 0 \\ 0 & \tilde{G}_{z,h}^{(a)} \end{bmatrix} + F_h,$$

with F_h having the same properties as E_h and $I_h(a)$ has units in the first diagonal below the main diagonal and zeros elsewhere. *We shall continue to call E_h the matrix of errors in the expression for $\tilde{G}_h^{-1}$.* Now

$$\begin{aligned}
&\sum_t \varepsilon(t)[y(t,h)', z(t,h)']\tilde{G}_h^{-1} \\
&\quad = \sum \varepsilon(t)[y(t,h)', z(t,h)'] \\
&\quad\quad \times \left\{ \begin{bmatrix} & \vdots & 0 \\ B_h' & & \\ & \vdots & I_{hm} \end{bmatrix} \begin{bmatrix} I_h \otimes \Sigma^{-1} & 0 \\ 0 & \tilde{G}_{z,h}^{-1} \end{bmatrix} \begin{bmatrix} & B_h & \\ \cdots & & \cdots \\ 0 & & I_{hm} \end{bmatrix} + E_h \right\} \\
&\quad = \sum \varepsilon(t)[\varepsilon(t,h)' + w_h(t)D_h', z(t,h)'] \\
&\quad\quad \left\{ \begin{bmatrix} I_h \otimes \Sigma^{-1} & 0 \\ 0 & \tilde{G}_{z,h}^{-1} \end{bmatrix} \begin{bmatrix} & B_h & \\ \cdots & & \cdots \\ 0 & & I_{hm} \end{bmatrix} + E_{1h} \right\},
\end{aligned}$$

where E_{1h} again has the sum of the moduli of its elements uniformly bounded. Then (6.6.33) is

$$\begin{aligned}
&\sum_t \sum_u \varepsilon(t)[\varepsilon(t,h)', z(t,h)'] \begin{bmatrix} I_h \otimes \Sigma^{-1} & 0 \\ 0 & \tilde{G}_{z,h}^{-1} \end{bmatrix} \begin{bmatrix} I_h(a) \otimes \Sigma & 0 \\ 0 & \tilde{G}_{z,h}(a) \end{bmatrix} \\
&\quad \times \begin{bmatrix} I_h \otimes \Sigma^{-1} & 0 \\ 0 & \tilde{G}_{z,h}^{-1} \end{bmatrix} \begin{bmatrix} \varepsilon(u,h) \\ z(u,h) \end{bmatrix} + F_h.
\end{aligned}$$

This is (6.3.33) except for the remainder term, F_h. The latter is made up of expressions of the form $x_1 X x_2$, where x_i' is $\sum \varepsilon(t)[y(t,h)', z(t,h)']$ or $\sum \varepsilon(t) w_h(t) D_h'$ and thus are matrices whose entries are $O(TQ_T)$. The matrix X is a product of three factors chosen from

$$\begin{bmatrix} I_h \otimes \Sigma^{-1} & 0 \\ 0 & \tilde{G}_{z,h}^{-1} \end{bmatrix}, \quad \begin{bmatrix} I_h(a) \otimes \Sigma & 0 \\ 0 & \tilde{G}_{z,h}(a) \end{bmatrix}, \quad E_{1h}, \quad F_h.$$

If either E_{1h} or F_h occurs, then X has the sum of the moduli of its elements bounded since if $|A|_\infty \le c$ and E has this property, then AE has the property. Thus if E_{1h} or F_h occurs, then $x_1' X x_2 = O(T^2 Q_T^2) = O(T \log\log T)$. Otherwise, we obtain $O(hT^2 Q_T^2) = O(hT \log\log T)$. Since $(hT)^{-1} O(hT \log\log T) = O(\log\log T/\log T)$, because $h \ge c \log T$, then (6.6.33) is established. Next we establish

$$\frac{1}{hT}\sum_{j=1}^{h-a}\left\{\sum_{t=1}^{T}\varepsilon(t)\varepsilon(t-j)'\right\}\Sigma^{-1}\left\{\sum_{t=1}^{T}\varepsilon(t-j-a)\varepsilon(t)'\right\} = s\delta_{0a}\Sigma + o_P(1)$$

(6.6.34)

$$\frac{1}{hT}\sum_{t,u=1}^{T}\sum \varepsilon(t)z(t,h)'\tilde{G}_{z,h}^{-1}\tilde{G}_{z,h}(a)\tilde{G}_{z,h}^{-1}z(u,h)\varepsilon(u) = m\delta_{0a}\Sigma + o_P(1).$$

We give the proof of the first of these formulas for $a = 0$ and $s = 1$, for brevity. The general proof is almost the same but a little more complicated. Take $\sigma^2 = 1$, again for notational simplicity. Then

$$(hT)^{-1}\sum_{j=1}^{h}\left\{\sum_{t=1}^{T}\varepsilon(t)\varepsilon(t-j)\right\}^2 - 1 \tag{6.6.35}$$

is to be shown to converge in probability to zero. The squared terms in this are

$$h^{-1}\sum_j\left[\frac{1}{T}\sum_1^T\{\varepsilon(t)^2-1\}\varepsilon(t-j)^2 + \frac{1}{T}\sum_1^T\{\varepsilon(t-j)^2-1\}\right].$$

The second term in brackets converges to zero uniformly in j by ergodicity. Consider therefore

$$\sum_1^T\{\varepsilon(t)^2-1\}\varepsilon(t-j)^2,$$

which is martingale. Of course, this is $o(T)$ for each fixed j, but it is necessary to show this uniformly in j. We use a proof that will be used again in other contexts. First replace $\varepsilon(t)^2$ by

$$g(t) = \varepsilon(t)^2 I\{\varepsilon(t)^2 \le a(t)\} + a(t)I\{\varepsilon(t)^2 > a(t)\}$$

where $a(t) = t^a$, $a < \frac{1}{2}$. Put

$$Y_j(t) = [g(t) - E\{g(t)|\mathscr{F}_{t-1}\}]g(t-j).$$

To show that these truncation effects are negligible it is sufficient to show, in the first place, that

$$\frac{1}{T}\sum_1^T \{g(t) - \varepsilon(t)^2\}^2 = \frac{1}{T}\sum_1^T \{\varepsilon(t)^2 - a(t)\}^2 I\{\varepsilon(t)^2 > a(t)\}$$

converges to zero, which will follow, via Kronecker's lemma, from

$$\sum_1^\infty \frac{1}{t}\int_{a(t)}^\infty x^2\,dF(x) < \infty.$$

The left side is not greater than

$$c\sum_1^\infty \log t \int_{a(t)}^{a(t+1)} x^2\,dF(x) \le c\int_0^\infty x^2 \log_+ x\,dF(x) < \infty,$$

by condition (6.6.5). Also,

$$\frac{1}{T}\sum_1^T E|E\{g(t) - 1|\mathscr{F}_{t-1}\}g(t-j)| = \frac{1}{T}\sum_1^T E|E\{\varepsilon(t)^2 - g(t)|\mathscr{F}_{t-1}\}g(t-j)|$$

and this converges to zero uniformly in j by the same argument. Thus truncation effects may be ignored.

Put $\lambda(T) = (\log T)^d/T$, $d > 1$, so that $\lambda(T)a(T)^2 \to 0$, and consider

$$\tau_j(T) = \lambda(T)\sum_1^T Y_j(t) - \tfrac{1}{2}\lambda(T)^2\{1 + a(T)\lambda(T)/2\}\sum_1^T E\{Y_j(t)^2|\mathscr{F}_{t-1}\}.$$

Then by Stout (1974, p. 299), for arbitrary $b > 0$,

$$P\{\tau_j(T) > b\log T\} \le T^{-b},$$

so that

$$\sum_{T=1}^\infty P\left\{\max_j \tau_j(T) > b\log T\right\} \le \sum_{T=1}^\infty hT^{-b} < \infty, \qquad b > 1,$$

so that, considering also $-Y_j(T)$, for which the same result holds, then

$$\max_{1\le j\le h} \frac{(\log T)^{d-1}}{T}\left|\sum_1^T Y_j(t)\right| \to 0, \quad \text{a.s.},$$

and we shall have established what is required, provided that

$$\max_j \frac{\lambda(T)^2}{\log T} \sum_1^T E\{Y_j(t)^2 | \mathscr{F}_{t-1}\} \to 0, \quad \text{a.s.}$$

The left side is bounded above by

$$\frac{(\log T)^{2d-1}}{T^2} T^{2a} \sum_1^T E[\{\varepsilon(t)^2 - 1 | \mathscr{F}_{t-1}] \tag{6.6.36}$$

since $E\{Y_j(t)^2 | \mathscr{F}_{t-1}\} = g(t-j)^2 E([g(t) - E\{g(t)|\mathscr{F}_{t-1}\}]^2) \leq a(T)^2 E([g(t) - E\{g(t)|\mathscr{F}_{t-1}]^2) \leq a(T)^2 E[\{\varepsilon(t)^2 - 1\}^2]$, because of Stout (1970), where it is shown that the expectation is an increasing function of the truncation point. Then (6.6.36) converges to zero for some $d > 1$ since $2a < 1$. Thus (6.6.35) converges to zero if this is true of

$$(hT)^{-1} \sum_1^h \sum_{t \neq u} \sum \varepsilon(t)\varepsilon(t-j)\varepsilon(u)\varepsilon(u-j). \tag{6.6.37}$$

The terms there for $t = u - j$ or $u = t - j$ are of the form

$$\frac{1}{h} \sum_1^h \frac{1}{T} \sum_1^T \varepsilon(t)\varepsilon(t-j)^2 \varepsilon(t-2j),$$

and this may be shown to converge to zero by the same technique as was used for the squared terms in (6.6.35), since the summands are again martingale differences. Thus we must consider

$$(hT)^{-1} \sum_1^h \sum \sum \varepsilon(t)\varepsilon(t-j)\varepsilon(u)\varepsilon(u-j), \tag{6.6.38}$$

where now the inner double sum is over $1 \leq t, u \leq T, t \neq u, t \neq u - j, u \neq t - j$. We now show that the variance of the double sum is $O(T^2)$ uniformly in j while the co-covariance between two of these double sums, for j, k, is $O(T)$, uniformly in (j, k). Then the variance of (6.6.38) is $(hT)^{-2}\{hO(T^2) + h^2 O(T)\} = O(h^{-1}) + O(T^{-1})$ and the first part of (6.6.34) is established for $s = 1$, $a = 0$. As has been said, the general case is hardly more difficult. To prove the assertions about (6.6.38), we repeatedly use (4.1.16). Consider $E\{\varepsilon(t)\varepsilon(t-j)\varepsilon(u)\varepsilon(u-j)\varepsilon(v)\varepsilon(v-j)\varepsilon(w)\varepsilon(w-j)\}$. We may take $t > u$, $v > w$. Then the expression vanishes unless $t = v$ and then is $E\{\varepsilon(u)\varepsilon(u-j)\varepsilon(w)\varepsilon(w-j)\}$. This in turn will vanish unless $u = w$, so that we see that the variance of the double sum is $O(T^2)$ uniformly in j, as stated. Next take $E\{\varepsilon(t)\varepsilon(t-j)\varepsilon(u)\varepsilon(u-j)\varepsilon(v)\varepsilon(v-k)\varepsilon(w)\varepsilon(w-k)\}$, again considering $t > u$, $v > w$, $j < k$. Again this vanishes unless $t = v$, so that we consider $E\{\varepsilon(t-j)\varepsilon(t-k)\varepsilon(u)\varepsilon(u-j)\varepsilon(w)\varepsilon(w-k)\}$ for $t-j \neq u, t \neq u-j, t-k \neq w, w-k \neq t, t > u, t > w, j < k$. Then the largest argument is $t-j$ or u or w and since $t-j \neq u$, then we may consider only terms for which $u = w$ is the largest index or $t-j = w$ is the largest index. In the latter case we obtain $E\{\varepsilon(t-k)\varepsilon(u)\varepsilon(u-j)\varepsilon(t-j-k)\}$. Then we must have $t-k = u$ and thus have terms for $t = v$, $t-j = w$, $t-k = u$, of which there are only $O(T)$ terms in the variance, for each $j < k$. Similarly, for $u = w > t-j$ we obtain $E\{\varepsilon(t-j)\varepsilon(t-k)\varepsilon(u-j)\varepsilon(u-k)\}$, which vanishes unless $t = u$, so that again there are only $O(T)$ terms. This finishes the proof of the first part of (6.6.34).

The argument in the case of the second part is simpler. It will be sufficient to show that the expectation converges to the right side of (6.6.34) and the variance converges to zero, treating $z(t)$ as fixed and a.s. in these $z(t)$ values. We take $s = 1$ and put $z(t,h)'\tilde{G}_{z,h}^{-1}\tilde{G}_{z,h}(a)\tilde{G}_{z,h}^{-1}z(u,h) = P(t,u)$. Then $E\{\varepsilon(t)P(t,u)\varepsilon(u)\varepsilon(v)P(v,w)\varepsilon(w)|z(t), t = 1,\ldots,T\}$ is null unless $t = v$, $u = w$, when it is $P(t,u)^2$. However,

$$\frac{1}{(hT)^2}\sum_{t,u=1}^{T}\sum P(t,u)^2 = \frac{1}{h^2}\operatorname{tr}\{\tilde{G}_{z,h}^{-1}\tilde{G}_{z,h}(a)\tilde{G}_{z,h}^{-1}\tilde{G}'_{z,h}(a)\} = O(h^{-1}).$$

Thus we need to consider only

$$\frac{1}{hT}\sum_{1}^{T}\varepsilon(t)^2 P(t,t), \tag{6.6.39}$$

whose expectation is $h^{-1}\operatorname{tr}\{\tilde{G}_{z,h}^{-1}\tilde{G}_{z,h}(a)\}$, which converges to m for $a = 0$ and to zero otherwise, since $\tilde{G}_{z,h}^{-1}\tilde{G}_{z,h}(a)$ has only a nonnull elements in this main diagonal for $a \neq 0$ since all except the a last columns of $\tilde{G}_{z,h}(a)$ are columns of $\tilde{G}_{z,h}$. The variance of (6.6.39), conditional on $z(t)$ fixed, is fairly easily calculated as

$$\begin{aligned} E\{\varepsilon(t)^4 - 1\}(hT)^{-2}\sum_{1}^{T}P(t,t)^2 &= h^{-2}E\{\varepsilon(t)^4 - 1\}\operatorname{tr}\{\tilde{G}_{z,h}^{-1}\tilde{G}_{z,h}(a)\tilde{G}_{z,h}^{-1}\tilde{G}'_{z,h}(a)\} \\ &= O(h^{-1}), \end{aligned}$$

which completes the proof for $s = 1$. The case $s > 1$ is only marginally more complex.

Now the proof of Theorem 6.6.5 may be completed. Again we consider the case $s = 1$, $j = 0$ and take the first of the two formulas. Now

$$\hat{\varepsilon}_h(t) - \varepsilon(t) = \sum_1^\infty [\{\hat{\phi}_h(j) - \phi(j)\}y(t-j) + \{\hat{\psi}_h(j) - \psi(j)\}z(t-j)], \tag{6.6.40}$$

using lowercase letters, as usual, in this scalar case. Of course, $\hat{\phi}_h(j), \hat{\psi}_h(j) = 0$, $j > h$. Now

$$\begin{aligned} \frac{1}{T}\sum_1^T \{\hat{\varepsilon}_h(t) - \varepsilon(t)\}^2 &= \{[\hat{\phi}_h, \hat{\psi}_h] - [\phi, \psi]_h\}\tilde{G}_h\{[\hat{\phi}_h, \hat{\psi}_h] - [\phi,\psi]_h\}' \\ &\quad + O(T^{-1/2}Q_T). \end{aligned} \tag{6.6.41}$$

This follows by truncating (6.6.40) at $j = h$, observing that, for $j > h$, $[\phi(j), \psi(j)] = O(T^{-1/2})\rho_0^{j-h}$, $0 < \rho_0 < 1$ and that $\hat{\phi}_h(j) - \phi(j)$, $\hat{\psi}_h(j) - \psi(j)$ are $O(Q_T)$, $j \le h$. Now from (6.6.24) and the formulas below that [see following (6.6.27)], since now (4.1.16) holds,

$$[\hat{\phi}_h, \hat{\psi}_h] - [\phi, \psi]_h = -\frac{1}{T}\sum_1^T \varepsilon(t)[y(t,h)', z(t,h)']G_h^{-1} + o(T^{-1/2}),$$

so that (6.6.41) becomes

$$\frac{1}{T^2}\sum_1^T \varepsilon(t)[y(t,h)', z(t,h)']G_h^{-1}\tilde{G}_h G_h^{-1}\sum_1^T \begin{bmatrix} y(t,h) \\ z(t,h) \end{bmatrix}\varepsilon(t) + o_P\left(\frac{h}{T}\right).$$

Now we may replace G_h in this expression by $\tilde{G}_h$ since $G_h - \tilde{G}_h$ has elements that are $o(hT^{-1/2})$, so that the overall error is $o(hT^{-1/2}h^2Q_T^2) = o(h/T)$. Now we may use (6.6.33) and (6.6.34) to complete the proof of Theorem 6.6.5.

The last four theorems of this section are explicitly for the case $s = 1$, *all previous theorems having been for the general case.* Consider in relation to Theorem 6.6.6 the expression

$$\frac{1}{T}\sum \{ay(t) - dz(t) - (b-1)\hat{\varepsilon}_T(t)\}^2, \tag{6.6.42}$$

where $\hat{\varepsilon}_T(t)$ is obtained from stage I, the formula is Toeplitz, and the expression is to be minimized with respect to the vector, τ, of coefficients $\alpha(j)$, $j = 1, \ldots, p$; $\beta(j)$, $j = 1, \ldots, q$ and $\delta(j)$, $j = 1, 2, \ldots, r$. The minimized value is $\{\hat{\sigma}_{p,q,r}^{(1)}\}^2$. We now point out that if $p < p_0$ or $q < q_0$ or $r < r_0$, then for large T $\{\hat{\sigma}_{p,q,r}^{(1)}\}^2 \ge \sigma_0^2 + c$, $c > 0$. To see this, observe that (6.6.42) converges to

$$\int_{-\pi}^{\pi} f_u(\omega)|a - (b-1)k_0^{-1}|\,d\omega + \int_{-\pi}^{\pi} (al_0 - d)'\,dF_z(\omega)(al_0 - d). \tag{6.6.43}$$

(Here, and in using σ_0^2 for the true variance, we have reverted to using a zero subscript since now a, b, d are trial values and need to be distinguished from true values.) For example,

$$\begin{aligned}\frac{1}{T}\sum y(t-j)\hat{\varepsilon}_T(t-k) &= \frac{1}{T}\sum y(t-j)\varepsilon(t-k) \\ &\quad + \frac{1}{T}\sum y(t-j)\{\hat{\varepsilon}_T(t-k)-\varepsilon(t-k)\},\end{aligned} \tag{6.6.44}$$

and this converges to $E\{y(t-j)\varepsilon(t-k)\}$ using Theorem 6.6.5. Moreover, the covariance matrix of the quantities $y(t-j)$, $\hat{\varepsilon}_T(t-j)$, $z(t-j)$, $j > 0$, consisting of quantities such as (6.6.44), is nonsingular and converges to a nonsingular limit, since otherwise there would be some relation $(a-1)y(t) - dz(t) - (b-1)\varepsilon(t) = 0$, which is impossible for $p \le p_0$, $q \le q_0$, $r \le r_0$. Thus the minimum of (6.6.42) converges to that of (6.6.43), which is strictly greater than σ_0^2 if $p < p_0$ or $q < q_0$ or $r < r_0$. Thus the criterion cannot, for sufficiently large T, be minimized at such values of p, q, r since, as we shall see,

$$\log\{\hat{\sigma}^{(1)}_{p_0,q_0,r_0}\}^2 \le \frac{1}{T}\sum \varepsilon(t)^2 + O(Q_T^2 \log T). \tag{6.6.45}$$

Indeed, an upper bound to (6.6.42) is then obtained at a_0, b_0, d_0, at which values (6.6.42) becomes

$$\frac{1}{T}\sum [\varepsilon(t) - (b_0 - 1)\{\hat{\varepsilon}_T(t) - \varepsilon(t)\}]^2$$

and this is equal to the right side of (6.6.45), using Theorem 6.6.5. However, that theorem uses (4.1.16), which is not required in the theorem we are proving. However, we can use the evaluations of S_1 to S_5 obtained below (6.6.27) to establish that $T^{-1}\sum \varepsilon(t)\{\hat{\varepsilon}_T(t) - \varepsilon(t)\}$, $T^{-1}\sum\{\hat{\varepsilon}_T(t)-\varepsilon(t)\}^2$ are $O(\hat{h}Q_T^2)$, taking account of the fact that $\hat{h} \ge l(T)$ (see the remarks at the beginning of stage II, just above the statement of Theorem 6.6.6) to show that S_1 is of the required order.

Next it will be shown that, for $p \ge p_0$, $q \ge q_0$, $r \ge r_0$,

$$\{\hat{\sigma}^{(1)}_{p,q,r}\}^2 \ge \frac{1}{T}\sum \varepsilon(t)^2 + O(Q_T^2 \log T). \tag{6.6.46}$$

The expression $(a-1)y(t) - dz(t) - (b-1)\hat{\varepsilon}_T(t)$ is a linear function of $y(t-1)$,

$\ldots, y(t-\hat{h}-q), z(t-1), \ldots, z(t-r)$ and thus a lower bound will be obtained by a direct regression of $y(t)$ on those quantities. Now (6.6.30) may be used, to bound (6.6.46) below by $T^{-1}\sum \varepsilon(t)^2 + O(\hat{h}Q_T^2)$, which establishes what is needed. Thus from (6.6.45), (6.6.46), for $p \geq p_0$, $q \geq q_0$, $r \geq r_0$,

$$\log\{\hat{\sigma}^{(1)}_{p,q,r}\}^2 + (p+q+rm)C_T/T - \log\{\hat{\sigma}^{(1)}_{p_0,q_0,r_0}\}^2 - (p_0+q_0+r_0m)C_T/T$$
$$\geq O(Q_T^2 \log T) + \{p-p_0+q-q_0+(r-r_0)m\}C_T/T$$

and since $C_T \geq c_0 \log T(\log\log T)$, then for c_0 sufficiently large the result follows that $(\tilde{p}^{(1)}, \tilde{q}^{(1)}, \tilde{r}^{(1)}) \to (p_0, q_0, r_0)$. The last statement of the theorem follows from the fact, previously asserted, that at p_0, q_0, r_0, the minimizing values of (6.6.42) converge to those of (6.6.43).

We turn now to Theorem 6.6.7. Introduce the new parameterization of Section 5.4. Then $\tau' = (\chi', \psi')X_0$, where X_0 depends only on a_0, b_0, d_0 and hence

$$(a, b, d) = v(z)(a_0, b_0, d_0) + \sum_1^N \chi_j \xi_j(z). \tag{6.6.47}$$

[Here $v(z)$ is the same as what is called $h(z)$ in Section 5.4, but that notation is now avoided since h is used with a different meaning at stage I.] Now use (6.6.47) in (6.6.42) to obtain, after some rearrangements, and using $a_0 y(t) - d_0 z(t) - b_0 \varepsilon(t) = 0$,

$$\begin{aligned} ay(t) &- dz(t) - (b-1)\hat{\varepsilon}_T(t) \\ &= (v-1)a_0 y(t) - (v-1)b_0\hat{\varepsilon}_T(t) - (v-1)d_0 z(t) \\ &\quad + a_0 y(t) - d_0 z(t) - (b_0 - 1)\hat{\varepsilon}_T(t) \\ &\quad + \sum_1^N \chi_j \xi_j(y(t), -\hat{\varepsilon}_T(t), -z(t)')' \\ &= -(v-1)b_0\{\hat{\varepsilon}_T(t) - \varepsilon(t)\} + \sum_1^N \chi_j \xi_j(y(t), -\hat{\varepsilon}_T(t), -z(t)')' \\ &\quad + \varepsilon(t) - (b_0 - 1)\{\hat{\varepsilon}_T(t) - \varepsilon(t)\}. \end{aligned}$$

Here, of course, ξ_j is a polynomial lag operator (a row vector of such) operating on $(y(t), -\hat{\varepsilon}_T(t), -z(t)')'$. Thus the minimization of (6.6.42) consists of the Toeplitz regression of $\varepsilon(t) - (b_0 - 1)\{\hat{\varepsilon}_T(t) - \varepsilon(t)\}$ on $\hat{\varepsilon}_T(t-j) - \varepsilon(t-j)$, $j = 1, \ldots, v = \min(p-p_0, q-q_0, r-r_0)$ and on $w_j(t) \overset{\mathrm{d}}{=} \xi_j(y(t), -\hat{\varepsilon}_T(t), z(t)')'$, $j = 1, \ldots, N = p+q+rm-v$. Call $w(t)$ the vector composed of the $w_j(t)$. Then $T^{-1}\sum w(t)w(t)'$ converges to a nonsingular limit, for otherwise there would be

$\tilde{\chi}_j, \sum \tilde{\chi}_j^2 = 1$, so that $\sum \tilde{\chi}_j w_j(t)$ would converge to zero and since, from Theorems 6.6.5 and 5.3.1,

$$\frac{1}{T}\sum \{\hat{\varepsilon}_T(t) - \varepsilon(t)\}^2 \to 0, \quad \frac{1}{T}\sum \varepsilon(t)\{\hat{\varepsilon}_T(t) - \varepsilon(t)\} \to 0, \quad \frac{1}{T}\sum \varepsilon(t)w(t) \to 0$$

[the latter because $\xi_j(0) = 0, j = 1, \ldots, N$], then (6.6.42) would converge to σ_0^2, which is impossible since (6.6.43), which is the limit of (6.6.42), would then be σ_0^2 at values of a, b, d that cannot be va_0, vb_0, vd_0 since $\tilde{\chi} \neq 0$. Now carry out the regression of the expression $\zeta(t) = \varepsilon(t) - (b_0 - 1)\{\hat{\varepsilon}_T(t) - \varepsilon(t)\}$ on the $b_0\{\hat{\varepsilon}_T(t-j) - \varepsilon(t-j)\}$, $j = 1, \ldots, v$ after they have been made orthogonal to the $w_j(t)$ and on the $w_j(t)$, $j = 1, \ldots, N$. The mean square for regression on the latter is $O(Q_T^2)$, as we now show. To show this we have only to show that $T^{-1}\sum w_j(t)\zeta(t) = O(Q_T)$. It is evident, from Theorems 5.3.1 and 6.6.5, that $T^{-1}\sum w_j(t)\varepsilon(t) = O(Q_T)$. Consider, for example,

$$\frac{1}{T}\sum y(t-j)\{\hat{\varepsilon}_T(t-k) - \varepsilon(t-k)\}$$

$$= \sum_1^{\hat{h}} \{\hat{\phi}_{\hat{h}}(u) - \phi(u)\}\frac{1}{T}\sum y(t-j)y(t-u-k) + O(T^{-1/2}),$$

the last coming from the geometric rate of decrease of the $\phi(j)$. The first term is

$$\sum_1^{\hat{h}} \{\hat{\phi}_{\hat{h}}(u) - \phi(u)\}[E\{y(t-j)y(t-u-k)\} + o(\hat{h}^{-1})] = O(Q_T)$$

from (6.6.4) and Theorem 6.6.1 since

$$\sum_1^{\hat{h}} |E\{y(t-j)y(t-u-k)\}| \leq c < \infty.$$

The same applies if $y(t-j)$ is replaced by $z(t-j)$, and finally,

$$\frac{1}{T}\sum \hat{\varepsilon}_T(t-j)\{\hat{\varepsilon}_T(t-k) - \varepsilon(t-k)\}$$

$$= \frac{1}{T}\sum \varepsilon(t-j)\{\hat{\varepsilon}_T(t-k) - \varepsilon(t-k)\} + O(\log T/T),$$

from Theorem 6.6.5. The first term is $O(Q_T)$, by the same proof as for $y(t-j)$.

Also, by Theorem 6.6.5,

$$
\begin{aligned}
T^{-1}\sum \zeta(t)^2 &= T^{-1}\sum \varepsilon(t)^2 + (m+1)\frac{\hat{h}}{T}\sigma_0^2 \sum_1^{q_0} \beta_0(j)^2 + o_P(\log T/T) \\
&= T^{-1}\sum \varepsilon(t)^2 + \sigma_0^2(m+1)\frac{\hat{h}}{T}(\omega_0^2 - 1) + o_P(\log T/T)
\end{aligned}
\tag{6.6.48}
$$

in the notation of the theorem.

Finally, we calculate the regression of $\zeta(t)$ on $b_0\{\hat{\varepsilon}_T(t-j) - \varepsilon(t-j)\}$, $j = 1,\ldots,v$, after these have been made orthogonal to the $w_j(t)$. The coefficients in regressions of the $b_0\{\hat{\varepsilon}_T(t-j) - \varepsilon(t-j)\}$ on the $w_j(t)$ are $O(Q_T)$, by the same argument as was just used, and it is easily seen that the regression mean square in the regression of $\zeta(t)$ on the $b_0\{\hat{\varepsilon}_T(t-j) - \varepsilon(t-j)\}$ after correction for the $w_j(t)$ is $O(Q_T^2)$ plus that for the regression of $\zeta(t)$ on the $b_0\{\hat{\varepsilon}_T(t-j) - \varepsilon(t-j)\}$. Now we use Theorem 6.6.5 once again, from which it follows that

$$
\begin{aligned}
&\frac{1}{T}\sum b_0\{\hat{\varepsilon}_T(t-j) - \varepsilon(t-j)\} b_0\{\hat{\varepsilon}_T(t-k) - \varepsilon(t-k)\} \\
&\qquad = (m+1)\frac{\hat{h}}{T}\sigma_0^2 \sum \beta_0(u)\beta_0(u+j-k) + O_P(\hat{h}/T)
\end{aligned}
\tag{6.6.49}
$$

using the convention that $\beta_0(u) = 0$ unless $0 \le u \le q_0$. In the same way,

$$
\begin{aligned}
&\frac{1}{T}\sum \zeta(t) b_0\{\hat{\varepsilon}_T(t-k) - \varepsilon(t-k)\} \\
&\qquad = -(m+1)\frac{\hat{h}}{T}\sum_u \beta_0(u)\beta_0(u-k) + o_p(\hat{h}/T).
\end{aligned}
\tag{6.6.50}
$$

Thus neglecting the terms that are $O_P(\hat{h}/T)$, it is readily seen that the mean square for regression is $(m+1)(\hat{h}/T)\sigma_0^2$, by the mean square for regression of $x(t)$ on $x(t-1),\ldots,x(t-v)$. [See above the statement of the theorem for the definition of $x(t)$.] Indeed, to see this one has only to check that the variances and covariances involved are as in (6.6.49) and (6.6.50), except for the minus sign in (6.6.50), which does not affect the mean square. Thus taking account of (6.6.48), we obtain, for $p \ge p_0$, $q \ge q_0$, $r \ge r_0$,

$$
\{\hat{\sigma}_{p,q,r}^{(1)}\}^2 = \frac{1}{T}\sum \varepsilon(t)^2 + \sigma_0^2(m+1)\frac{\hat{h}}{T}(\omega_v^2 - 1) + o_p(\log T/T)
$$

since $\omega_v^2 = \omega_0^2 -$(mean square for regression). Now recalling that $T^{-1}\sum\varepsilon(t)^2 = \sigma_0^2 + O(Q_T)$ and so that

$$\log\{\hat{\sigma}_{p,q,r}^{(1)}\}^2 = \log\{T^{-1}\sum\varepsilon(t)^2\} + (m+1)\frac{\hat{h}}{T}(\omega_v^2 - 1) + o_p(\log T/T),$$

we obtain, recalling that $\hat{h}/\{c_1 \log T/(-2\log\rho_0)\} \to 1$,

$$\begin{aligned}\log\{\hat{\sigma}_{p,q,r}^{(1)}\}^2 &+ (p+q+rm)\log T/T - \log\{\hat{\sigma}_{p_0,q_0,r_0}^{(1)}\}^2 - (p_0+q_0+r_0 m)\log T/T \\ &= (m+1)c_1\log T(\omega_v^2 - \omega_0^2)/\{T(-2\log\rho_0)\} + \{(p-p_0) \\ &\quad + (q-q_0) + (r-r_0)m\}\log T/T + o_p(\log T/T).\end{aligned}$$

The first part of the theorem now follows. The second part follows from (6.6.47) and the evaluation of $\hat{\chi}_j$ as $O(Q_T)$ accomplished above, together with the proof that has been given that the coefficients in the estimate of $v_0(z)$ will converge to those of $v_0(z)$.

Theorem 6.6.8. If $p < p_0$ or $q < q_0$ or $r < r_0$, then $BIC_{p,q,r}(i,j,k)$ cannot be minimized, eventually, at $i = p$, $j = q$, $k = r$, because then $BIC_{p,q,r}(p,q,r) = BIC(p,q,r) > BIC_{p,q,r}(p_0,q_0,r_0) = BIC(p_0,q_0,r_0)$. For $p = p_0$, $q = q_0$, $r = r_0$, then $\tilde{\omega}_v^2(p_0,q_0,r_0)$ converges to ω_v^2 by Theorem 6.6.7, so that, for $i \geq p_0$, $j \geq q_0$, $k \geq r_0$,

$$\begin{aligned}BIC_{p_0,q_0,r_0}(i,j,k) &= \log\left\{\frac{1}{T}\sum\varepsilon(t)^2\right\} + (m+1)\frac{\hat{h}}{T}(\omega_0^2 - 1) \\ &\quad + (i+j+km)\log T/T + o_p(\log T/T),\end{aligned}$$

which will eventually be minimized at $i = p_0$, $j = q_0$, $k = r_0$. This p_0, q_0, r_0 will eventually be a "marked" point [see below (6.5.12)], and since there is at most one such for $p \leq \hat{p}^{(1)}$, $q \leq \hat{q}^{(1)}$, $r \leq \hat{r}^{(1)}$, other than $\hat{p}^{(1)}$, $\hat{q}^{(1)}$, $\hat{r}^{(1)}$, the theorem is established.

Theorem 6.6.9. We give the proof of part (b) since the proof of part (a) is more direct. We know that

$$(\hat{a}_T^{(1)}, \hat{b}_T^{(1)}, \hat{d}_T^{(1)}) \to v_0(z)\{1 + o_p(1)\}(a_0, b_0, d_0) + O(Q_T),$$

where we now use v_0 for v at $\upsilon = \upsilon_0$. The function $\hat{v}$ [see (6.6.47)] has, for sufficiently large T, its zeros bounded away from $|z| = 1$ with arbitrarily high probability since $\hat{v}$ converges in probability to v_0, which has this property. Thus the coefficients of z^i in

$$\{\hat{b}_T^{(1)}\}^{-1} - (\hat{v}b_0)^{-1} = \{\hat{v}b_0 - \hat{b}_T^{(1)}\}\{\hat{b}_T^{(1)}\}^{-1} b_0^{-1} \hat{v}^{-1}$$

are $O(Q_T \rho^i)$, where $0 < \rho < 1$, with arbitrarily high probability for T sufficiently large. Thus $\hat{\eta}^{(1)}(t) - \hat{v}^{-1} b_0^{-1} y(t)$ is a linear combination of $y(t-j)$, $j > 0$, with coefficients decreasing to zero at a geometric rate that are also $O(Q_T)$, with arbitrarily high probability for T large. The same is true of $\hat{\xi}^{(1)}(t)$, $\hat{\varepsilon}^{(1)}(t)$, $\hat{\zeta}^{(1)}(t)$ in relation to $\hat{v}^{-1} b_0^{-1} \varepsilon(t)$, $\varepsilon(t)$, $\hat{v}^{-1} b_0^{-1} z(t)$. Thus, for example, with arbitrarily high probability for T large,

$$\frac{1}{T} \sum \varepsilon(t) \hat{\xi}^{(1)}(t) = \frac{1}{T} \sum \varepsilon(t) \hat{v}^{-1} b_0^{-1} \varepsilon(t) + O(Q_T^2).$$

Moreover, for $p < p_0$ or $q < q_0$ or $r < r_0$ or for $p = p_0$, $q = q_0$, $r = r_0$, then $\hat{\eta}^{(1)}(t-j)$, $j = 1, \ldots, p$, $\hat{\xi}^{(1)}(t-j)$, $j = 1, \ldots, q$, $\hat{\zeta}^{(1)}(t-j)$, $j = 1, \ldots, r$ have a covariance matrix that will be, for T sufficiently large, nonsingular with arbitrarily high probability. Now the fact that $\hat{p}^{(2)} \to p_0$, $\hat{q}^{(2)} \to q_0$, $\hat{r}^{(2)} \to r_0$, in probability is almost an in the proof of Theorem 6.6.6. Thus for $p = p_0$, $q = q_0$, $r = r_0$ we take $a = a_0$, $b = b_0$, $d = d_0$ and evaluate

$$\frac{1}{T} \sum \{a\hat{\eta}^{(1)}(t) - b\hat{\xi}^{(1)}(t) - d\hat{\zeta}^{(1)}(t) + \hat{\varepsilon}^{(1)}(t)\}^2$$

$$\leq \frac{1}{T} \sum \{a_0 \hat{\eta}^{(1)}(t) - b_0 \hat{\xi}^{(1)}(t) - d_0 \hat{\zeta}^{(1)}(t) + \hat{\varepsilon}^{(1)}(t)\}^2 \tag{6.6.51}$$

$$= \frac{1}{T} \sum \{\varepsilon(t) + R(t)\}^2,$$

where $R(t)$ is $a_0\{\hat{\eta}^{(1)}(t) - \hat{v}^{-1} b_0^{-1} y(t)\} - b_0\{\hat{\xi}^{(1)}(t) - b_0 \hat{v}^{-1} b_0^{-1} \varepsilon(t)\} - d_0\{\hat{\zeta}^{(1)}(t) - \hat{v}^{-1} b_0^{-1} z(t)\} + \hat{\varepsilon}^{(1)}(t) - \varepsilon(t)$. This is, with arbitrarily high probability for T large,

$$\frac{1}{T} \sum \varepsilon(t)^2 + O(Q_T^2).$$

Similarly, for $p \geq p_0$, $q \geq q_0$, $r \geq r_0$ we obtain a lower bound by considering the regression on $\hat{b}^{-2} y(t-j)$, $j = 1, \ldots, \max(\hat{p}^{(1)} + q, \hat{q}^{(1)} + p)$, $\hat{b}^{-2} z(t-j)$, $j = 1, \ldots, r + \hat{q}$, since $\hat{\eta}^{(1)}(t-j)$, $\hat{\xi}^{(1)}(t-j)$, $\hat{\zeta}^{(1)}(t-j)$ are linear combinations of these. To $O(Q_T^2)$, as before, we may replace these regressors by $\hat{v}^{-2} b_0^{-1} y(t-j)$, $\hat{v}^{-2} b_0^{-2} z(t-j)$ and also $\hat{\eta}^{(1)}(t) - \hat{\xi}^{(1)}(t) + \hat{\varepsilon}^{(1)}(t)$ by $\hat{v}^{-1} b_0^{-1} y(t) - \hat{v}^{-1} b_0^{-1} \varepsilon(t) + \varepsilon(t)$. The residual mean square is of the form

$$\frac{1}{T}\sum\left\{\varepsilon(t)-\sum e_j\hat{v}^{-2}b_0^{-2}y(t-j)-\sum f_j\hat{v}^{-2}b_0^{-2}z(t-j)+R_1(t)\right\}^2$$

with $R_1(t)$ of the same nature as $R(t)$, since $\hat{v}^{-1}b_0^{-1}y(t)-\hat{v}^{-1}b_0^{-1}\varepsilon(t)=(b_0-a_0)\hat{v}(\hat{v}^{-2}b_0^{-2}y(t)\}$ and thus is a linear combination of the regressors. Now, as for (6.6.51), we obtain the lower bound, with arbitrarily high probability for T large,

$$\frac{1}{T}\sum\varepsilon(t)^2+O(Q_T^2).$$

Again, as before, it is established that we cannot have $p<p_0$ or $q<q_0$ or $r<r_0$, and as before we see that $(\hat{p}^{(2)},\hat{q}^{(2)},\hat{r}^{(2)})-(p_0,q_0,r_0)$ converges in probability to zero. As above, $a_0\hat{\eta}^{(1)}(t)-b_0\hat{\xi}^{(1)}(t)-d_0\hat{\zeta}^{(1)}(t)=O(Q_T)$, with arbitrarily high probability for T large. Since $(\hat{\eta}^{(1)}(t),\hat{\varepsilon}^{(1)}(t),\hat{\xi}^{(1)}(t),\hat{\zeta}^{(1)}(t))=(\hat{v}^{-1}b_0^{-1}y(t),\varepsilon(t),\hat{v}^{-1}b_0^{-1}\varepsilon(t),\hat{v}^{-1}b_0^{-1}z(t)+O(Q_T)$, in the same sense, then the assertion that $(\hat{a}^{(2)},\hat{b}^{(2)},\hat{d}^{(2)})\to(a_0,b_0,d_0)$ in probability is equivalent to the assertion that, for $p=p_0$, $q=q_0$, $r=r_0$ the vector of regression coefficients of $\varepsilon(t)$ on $(\hat{v}^{-1}b_0^{-1}y(t),\hat{v}^{-1}b_0^{-1}\varepsilon(t),\hat{v}^{-1}b_0^{-1}z(t))$ converge to zero, which is clearly true. In the same way it may be shown that $T^{1/2}(\hat{\tau}^{(2)}-\tau_0)$ converges in distribution to a normally distributed random vector with zero mean. Indeed, the distribution is that of the vector of regression coefficients of $\varepsilon(t)$ on $(v_0^{-1}b_0^{-1}y(t),v_0^{-1}b_0^{-1}\varepsilon(t),v_0^{-1}b_0^{-1}z(t))$.

The last assertion in part (b) of the theorem has a direct but fairly tedious proof. The proof that $(\hat{p}^{(3)},\hat{q}^{(3)},\hat{r}^{(3)})\to(p_0,q_0,r_0)$, in probability is, of course, almost the same as for $\hat{p}^{(2)}$, $\hat{q}^{(2)}$, $\hat{r}^{(2)}$ save that $\hat{v}\equiv 1$. The central limit result remains. This depends in the first place on showing that, for $p=p_0$, $q=q_0$, $r=r_0$, quantities such as

$$\text{(6.6.52)}\qquad \begin{aligned}&T^{-1}\sum\hat{\eta}^{(2)}(t-j)\hat{\xi}^{(2)}(t-k),\\&T^{-1/2}\sum\hat{\eta}^{(2)}(t-j)\{\hat{\varepsilon}(t)+a_0\hat{\eta}^{(2)}(t)-b_0\hat{\xi}^{(2)}(t)-d_0\hat{\zeta}^{(2)}(t)\}\end{aligned}$$

converge in probability to, respectively,

$$E\{\eta_0(t-j)\xi_0(t-k)\},\qquad T^{-1/2}\sum\eta_0(t-j)\varepsilon(t),$$

where $\eta_0(t)=b_0^{-1}y(t)$, $\xi_0(t)=b_0^{-1}\varepsilon(t)$. If that is true, then $T^{1/2}(\hat{\tau}^{(3)}-\tau_0)$ is distributed asymptotically as $T^{1/2}$ by the vector of regression coefficients of $\varepsilon(t)$ on $\eta_0(t-j)$, $j=1,\ldots,p_0$, $\xi_0(t-j)$, $j=1,\ldots,q_0$, $\zeta_0(t-j)$, $j=1,\ldots,r_0$. It remains only to check that the covariance matrix of this is, in its limiting distribution, as stated in the theorem. Consider the second term in (6.6.52), for example. This is of the form

$$T^{-1/2}\sum[\{b_0^{-1}-(\hat{b}^{(2)}b_0)^{-1}(\hat{b}^{(2)}-b_0)\}y(t-j)]$$
$$[\varepsilon(t)+\hat{b}^{-2}b_0^{-1}(\hat{b}^{(2)}-b_0)\{(\hat{a}^{(2)}b_0-\hat{b}^{(2)}a_0)y(t)+(\hat{d}^{(2)}b_0-\hat{b}^{(2)}d_0)z(t)\}].$$

Since $\hat{b}^{(2)}-b_0$, $\hat{a}^{(2)}b_0-\hat{b}^{(2)}a_0$, $\hat{d}^{(2)}b_0-\hat{b}^{(2)}d_0$ are $O(T^{-1/2})$ in probability, since the vector $T^{1/2}(\hat{\tau}^{(2)}-\tau_0)$ obeys the central limit theorem, while $(\hat{b}^{(2)}b_0)^{-1}$ has coefficients decreasing geometrically in its expansion, with arbitrarily high probability for T large and since $T^{-1/2}\sum\varepsilon(t)y(t-u)$ also obeys the central limit theorem [or for that matter is $O\{(\log\log T)^{1/2}\}$], we see that the assertion concerning the second component of (6.6.52) holds. That for the first component is more easily established. Thus the theorem follows.

7. RECURSIVE ALGORITHMS FOR THE VECTOR CASE

(i) The Use of U_α

One approach to the construction of an algorithm for the general, vector, case is that of van Overbeek and Ljung (1982) and van Overbeek (1982). They consider the optimization of a criterion over $M(n)$ such as the likelihood function or some approximation to it (see Section 6.2). The numerical problem will be easier if the Hessian of the expression being optimized, in the coordinate system being used, is well conditioned near the true optimum. In the case of the likelihood function this condition number is closely related to the degree to which the rows of the Hankel matrix defining the coordinate system being used are "very linearly independent" (e.g., nearly orthogonal). This can be judged from H_0H_0', where H_0 contains the rows, spanning the rows of $\mathscr{H}$, that define the coordinate system being used. Of course, H_0 is a submatrix of that constituted by the first ns rows of $\mathscr{H}$ and if the first s rows of $\mathscr{H}$ are always in H_0, as is only reasonable, then H_0 is a submatrix of the matrix constituted by the first $(n-s+1)s$ rows of $\mathscr{H}$. It is argued by van Overbeek and Ljung that it is therefore best to work in terms of the U_α rather than the V_α since one can move from U_α to U_β, with U_α, U_β adjacent in some sense, if H_0H_0' for U_α becomes ill-conditioned. ("Adjacent" appears to mean with most rows in the basis for $\mathscr{H}$ held in common.) Corresponding to H_0 will be the state-space model

$$x(t+1|t)=Fx(t|t-1)+K\varepsilon(t),\qquad y(t)=Hx(t|t-1)+\varepsilon(t). \tag{6.7.1}$$

As has been seen in Chapter 2, the vector $x(t+1|t)$ is made up of components of the $y(t+j|t)$ corresponding to the rows in H_0, $y_i(t+j|t)$ corresponding to row $h(j,i)$, this being the ith row in the jth block. Put $\mathcal{O}' = [H', F'H', \ldots, (F')^{n-1}H']$. Then $x(t|t-1)$ corresponding to any coordinate neighborhood leads to a state vector chosen from the elements of $\mathcal{O}x(t|t-1)$, since $HF^k x(t|t-1) = y(t+k|t-1)$. The rows of the Hankel matrix $\mathscr{H}$ providing the basis corresponding to the representation (6.7.1) are given by the matrix $[K, FK, F^2K \ldots] = \mathscr{C}_\infty$. Since $E\{x(t|t-1)x(t|t-1)'\} = \mathscr{C}_\infty \mathscr{C}_\infty'$ and any state vector consists of elements of $\mathcal{O}x(t|t-1)$, then $\mathcal{O}\mathscr{C}_\infty \mathscr{C}_\infty' \mathcal{O}'$ should be examined. If LDL' is a Choleski decomposition of $\mathscr{C}_\infty \mathscr{C}_\infty'$, L being lower triangular with units in the main diagonal and with D diagonal, $W = \mathcal{O}L$ is to be examined and these rows chosen that give a new $x(t|t-1)$, say $\tilde{x}(t|t-1)$, with $E\{\tilde{x}(t|t-1)\tilde{x}(t|t-1)'\} = \tilde{L}\tilde{D}\tilde{L}'$ having $\tilde{D}$ with diagonal elements as nearly as possible the same. Of course, the rows have to be chosen to give a "nice" selection from $\mathscr{H}$, that is, so that they are of the form $h(u,j)$, $u = 1, \ldots, n_j$; $j = 1, \ldots, s$, $\sum n_j = n$. Thus having estimated the structure at some iterative stage and found an F, H, K, Σ, and having determined that the condition number is not good, a procedure of the following kind is recommended. Rows of $W = \mathcal{O}L$ are chosen so that when orthogonalized with respect to D as weight function, they will have the greatest length possible, the squared length being $w_i' D w_i$, w_i' being a typical row of W. The first s rows are always chosen. Then a next row is chosen from rows numbered $s+1, s+2, \ldots, 2s$, after these are made orthogonal to the first s rows, which maximizes the length. (At least one of these rows must be in the basis.) This is adjoined to the original set of s. Then the procedure continues, the next set of rows being examined being determined by the set so far chosen. Thus if the $(s+1)$st row chosen was the jth in the second block, we next examine rows numbered $s+1$, $s+2, \ldots$, $s+j-1$, $2s+j$, $s+j+1, \ldots, 2s$, one of which must be in the basis. The procedure continues until n rows are found and thus a new coordinate basis is selected. As indicated above, this somewhat arduous, procedure would be used only when the existing basis became ill-conditioned and this could be judged from the information matrix.

The notion that one should choose n and then τ, the vector of system parameters given n, is appealing. We have heard arguments against the use of the U_α, and in favor of the examination of the V_α, that proceed as follows. It is pointed out that the dimension of the state vector in (6.7.1) is $n(2s+m)$, so that as n changes, the number of system parameters changes by $2s+m$. However, if all V_α are considered, their dimensions (see Theorem 2.4.5) are much closer together, so that a more parsimonious parameterization may be achieved. The choice between the various V_α would be based on a criterion of the form of (5.2.1). Arguments against this point of view might be the following. Why are the $M(n)$ being subdivided in this arbitrary way? How, indeed, does one know which ordering of the output vector to use? (Of course, one could attempt to discover the n_{0j} as an ordered sequence.) Does this subdivision into

V_α have any physical meaning? The actual maximum of the likelihood over $M(n)$ will surely be found in the intersection of the U_α, which is open and dense in $M(n)$. Is it not, then, true that any saving in parameters must be illusory? The question is not an easy one to resolve. Nevertheless, the approach of van Overbeek and Ljung has, as already stated, something to commend it. However, it will clearly be a computationally costly business, particularly as $\hat{n}$ has to be found. An alternative is to try to locate the maximum fairly accurately via a first calculation based on a regression-autoregression procedure and to use only one coordinate system. The numerical problem due to the use of a bad coordinate system may then be reduced, since one will, it is hoped, begin to search close to the true maximum.

It is, of course, possible to determine the Kronecker indices, and this will be dealt with in the theorems below. However, for brevity, no great detail concerning algorithms for finding these will be given.

(ii) A Gauss-Newton Procedure

Now we described the vector analogue of the method of Section 6.5. The method extends to the determination of the Kronecker indices, but as has just been said, we give few details there. Instead, we shall seek first only to determine n and the best point in $U(n)$. We therefore consider the ARMAX representation in the reversed echelon form for $U(n)$, where $A(j)$, $B(j)$, $D(j)$, the coefficient matrices in $a(z)$, $b(z)$, $d(z)$, have the form given almost immediately. For $U(n)$, see Remarks below Theorem 2.6.5. We exhibit these matrices with all partitions after the row q and column q, where $n = ps + q$, $0 \leq q < s$. An asterisk indicates a submatrix of freely varying elements.

$$A(0) = B(0) = \begin{bmatrix} I_q & 0 \\ * & I_s - q \end{bmatrix}, \qquad A(1) = \begin{bmatrix} * & 0 \\ * & * \end{bmatrix}$$

$$(6.7.2) \qquad A(p+1), \qquad B(p+1), \qquad D(p+1) = \begin{bmatrix} * & * \\ 0 & 0 \end{bmatrix}.$$

All other $A(j)$, $B(j)$, $D(j)$, $j = 0, 1, \ldots, p+1$, are unconstrained. It is *not* implied that $A(p+1) = B(p+1) = D(q+1)$. The technique used below is a Gauss-Newton procedure for optimizing the likelihood and is the natural generalization of that presented in Section 6.5. However, at each iterative stage the recursive calculation is in two parts. The first of these seeks to determine p_0 in $n_0 = p_0 s + q_0$, $0 \leq q_0 < s$. The second determines q_0. The reason for this procedure is that the Levinson-Whittle recursion can be used to calculate the estimate $\hat{p}^{(1)}$ of p_0, and it will be shown, under appropriate conditions, that $\hat{p}^{(1)} \to p_0 + 1$ if $q_0 > 0$ and to p_0 otherwise. (We state this loosely here, but precise results will be given in theorems below.) Then a further step determines $\hat{p}$, $\hat{q}$. This second stage involves $(s-1)$ regressions. As will be seen, the overall procedure will be very arduous if s is large and is likely to be used only for $s \leq 5$. These additional regressions do not therefore constitute a large task. There are three stages, of which only the third is iterated.

I. This is effectively the same as I of Section 6.5. Thus (6.5.8) is used to calculate, recursively on h, the regression-autoregression

$$\sum_0^h \hat{\Phi}_h(j)y(t-j) = \sum_1^h \hat{\Psi}_h(j)z(t-j) + \hat{\varepsilon}_h(t),$$

taking $w(t)' = (y(t)', z(t)')$. Thus the first s rows only of $F_h(j)$ are directly used and constitute $[\hat{\Phi}_h(j), \hat{\Psi}(j)]$ while $\hat{\Sigma}_h$ is the top left-hand $s \times s$ submatrix of S_h. Now $\hat{h}$ is chosen to minimize

$$\log\det\hat{\Sigma}_h + h(s^2 + sm)C_T/T, \qquad h \leq H_T. \tag{6.7.3}$$

Of course, C_T, H_T, and especially the former, must be specified. Write $\hat{\Phi}(j)$, $\hat{\Psi}(j)$, $\hat{\varepsilon}(t)$, $\hat{\Sigma}$ for $\hat{\Phi}_h(j)$, and so on, at $h = \hat{h}$. The $\hat{\varepsilon}(t)$ are not directly used at the later stages but only the autocovariance matrices, namely $G_y(t)$, $G_z(t)$, $G_{yz}(t)$ and

$$\begin{aligned}
\hat{G}_{\varepsilon y}(t) &= \sum_0^{\hat{h}} \hat{\Phi}(j)G_y(j+t) + \sum_1^{\hat{h}} \hat{\Psi}(j)G_{zy}(j+t), \qquad t>0;\\
&= \hat{\Sigma}, \qquad t=0;\\
&= 0, \qquad -\hat{h} \leq t < 0.\\
\hat{G}_{\varepsilon z}(t) &= \sum_0^{\hat{h}} \hat{\Phi}(j)G_{yz}(j+t) + \sum_1^{\hat{h}} \hat{\Psi}(j)G_z(j+t), \qquad t \geq 0;\\
&= 0, \qquad -\hat{h} \leq t < 0.\\
\hat{G}_\varepsilon(t) &= \sum\sum_{u,v=0}^{\hat{h}} [\hat{\Phi}(u), \hat{\Psi}(u)]G(u+t-v)\begin{bmatrix}\hat{\Phi}(v)'\\ \hat{\Psi}(v)'\end{bmatrix}, \qquad t>0\\
&= \hat{\Sigma}, \qquad t=0,\\
&= \hat{G}_\varepsilon(-t)', \qquad t<0.
\end{aligned} \tag{6.7.4}$$

Here, as before,

$$G(u) = \begin{bmatrix} G_y(u) & G_{yz}(u)\\ G_{zy}(u) & G_z(u)\end{bmatrix}.$$

Since $\hat{\varepsilon}(t)$ is a good approximation to $\varepsilon(t)$, it would be reasonable to put $\hat{G}_\varepsilon(t) = \delta_{0,t}\hat{\Sigma}$. Theorem 6.6.5 shows that $\hat{G}_\varepsilon(t)$, as defined above, is $\delta_{0,t}\hat{\Sigma} + O(Q_T)$ and, indeed, that it is $T^{-1}\sum \varepsilon(s)\varepsilon(s+t)' + O_P(\log T/T)$. When $\hat{G}_\varepsilon(t) = \delta_{0,t}\hat{\Sigma}$ is used, it will be possible, following Franke (1985), to use a more economical procedure than that given below at stage II. We shall not give the details here. *All theroems below will hold if $\hat{G}_\varepsilon(t) = \delta_{0,t}\hat{\Sigma}$ is used.*

IIa. Now (6.5.8) is again used, with $w(t)' = (y(t)', \hat{\varepsilon}(t)', z(t)')$, that is, with $C(t)$ obtained from (6.7.4), together with $G_y(t)$, $G_z(t)$, $G_{yz}(t)$. Then the first s rows of $F_p(j)$ constitute $[\hat{A}_p^{(1)}(j), -\hat{B}_p^{(1)}(j), -\hat{D}_p^{(1)}(j)]$, $j = 1, \ldots, p$ and $\hat{\Sigma}_p^{(1)}$ is the top left-hand $s \times s$ submatrix of S_p. Then $\hat{p}^{(1)}$ is chosen to minimize a criterion

$$\log\det \hat{\Sigma}_p^{(1)} + p(2s^2 + sm)C_T/T, \qquad p \leq P_T. \tag{6.7.5}$$

At $p = \hat{p}^{(1)}$ we write $\hat{A}^{(1)}(j)$, $\hat{B}^{(1)}(j)$, $\hat{D}^{(1)}(j)$, $\hat{\Sigma}^{(1)}$.

IIb. Now $\hat{p}$, $\hat{q}$ are determined. Here $\hat{p} = \hat{p}^{(1)} - 1$, $0 < \hat{q} < s$ or $\hat{p} = \hat{p}^{(1)}$, $\hat{q} = 0$. The choice is based on the quantity

$$\log\det \hat{\Sigma}_{p,q}^{(1)} + (ps + q)(2s + m)C_T/T, \qquad p = \hat{p}^{(1)} - 1, \qquad 0 < q \leq s, \tag{6.7.6}$$

where at $q = s$ we set $\hat{\Sigma}_{p,q}^{(1)}$ at $\hat{\Sigma}^{(1)}$. Thus we need to describe the calculation of $\hat{\Sigma}_{p,q}^{(1)}$ for $0 < q < s$, $p = \hat{p}^{(1)} - 1$. This is to be the covariance matrix of the residuals in the Toeplitz regression as follows. [See (6.7.2) to understand the regression.]

(i) For $i = 1, \ldots, q$ regress $y_i(t)$ on $-y_j(t-1)$, $j = 1, \ldots, q$ and on all components of $z(t-1)$, $\hat{\varepsilon}(t-1)$, $-y(t-k)$, $\hat{\varepsilon}(t-k)$, $z(t-k)$, $k = 2, \ldots, \hat{p}^{(1)}$. These regressions yield the first q rows of (6.7.2) and the first q components in the residual vector.

(ii) For $i = q+1, \ldots, s$, regress $y_i(t)$ on $-y_j(t) + \hat{\varepsilon}_j(t)$, $j = 1, \ldots, q$ and on all components of $-y(t-k)$, $\hat{\varepsilon}(t-k)$, $z(t-k)$, $k = 1, \ldots, \hat{p}^{(1)} - 1$. These regressions yield the last $s - q$ rows of (6.7.2) and the last $s - q$ components in the residual vector.

Of course, the $\hat{\varepsilon}(t)$ are not themselves needed for these regressions, nor the $y(t)$, $z(t)$ for that matter, but again, all that have to be used are the covariances computed via (6.7.4) together with $G_y(t)$, $G_z(t)$, $G_{yz}(t)$. The matrix $\hat{\Sigma}_{p,q}^{(1)}$, $p = \hat{p}^{(1)} - 1$, $0 < q \leq s$, is, of course, the matrix of variances and covariances of the residuals from these regesssions and again the computation of this does not require the formation of the residuals but only the use of (6.7.4), and so on, and the regression coefficient matrices computed. Indeed, if $\hat{A}_{p,q}^{(1)}(j)$, $\hat{B}_{p,q}^{(1)}(j)$, $\hat{D}_{p,q}^{(1)}(j)$ are the coefficient matrices estimating (6.7.2) for these p, q resulting from the regressions (i), (ii) above, then

$$\hat{\Sigma}_{p,q}^{(1)} = \sum \sum [\hat{A}_{p,q}^{(1)}(j), -\hat{B}_{p,q}^{(1)}(j) + \delta_{0,j}I_s, -\hat{D}_{p,q}^{(1)}(j)]\hat{G}(j-k)$$
$$\times [\hat{A}_{p,q}^{(1)}(k), -\hat{B}_{p,q}^{(1)}(k) + \delta_{0,k}I_s, -\hat{D}_{p,q}^{(1)}(k)]',$$

where

$$\hat{G}(t) = \begin{bmatrix} G_y(t) & \hat{G}_{y\varepsilon}(t) & G_{yz}(t) \\ \hat{G}_{\varepsilon y}(t) & \hat{G}_\varepsilon(t) & \hat{G}_{\varepsilon z}(t) \\ G_{zy}(t) & \hat{G}_{z\varepsilon}(t) & G_z(t) \end{bmatrix}.$$

It is possible to effect the regressions (i), (ii) more economically using the output from stage I since for each q the regression is obtained from the regression at stage IIa by excluding some of the regressors included in IIa and including some others. Moreover, it is only $\hat{\Sigma}_{p,q}^{(1)}$ that is required until $\hat{q}$ is determined and these regressions may be done in such a way as to compute the regression coefficient matrices only when $\hat{q}$ has been fixed. We shall not go into details of this algorithm here but refer the reader to Hannan and Kavalieris (1984). An alternative that may be better than the procedure prescribed there is to proceed by a recursion on q, beginning from $q = s - 1$ and proceeding to lower values. For example, at $q = s - 1$ the regressor $-y_s(t-1)$ must be removed from the regressions corresponding to the first $(s-1)$ rows that was computed at $p = \hat{p}^{(1)}$ in stage IIa and the regressors $-y_j(t) + \hat{\varepsilon}_j(t)$, $j = 1, \ldots, s-1$ have to be added into the last row of the regression computed at $p = \hat{p}^{(1)} - 1$ at stage IIa. Since the computations making $-y_j(t) + \hat{\varepsilon}_j(t)$ Toeplitz orthogonal to $-y(t-k)$, $z(t-k)$, $\hat{\varepsilon}(t-k)$, $k = 1, \ldots, \hat{p}^{(1)} - 1$ have already been computed at stage IIa, the latter is a relatively easy task. Again to remove $-y_s(t-1)$ we need to make this Toeplitz orthogonal to the regressions remaining in the regression, and some of the calculations necessary for this have already been done.

We indicate the final output from stage II by $\hat{A}(j)$, $\hat{B}(j)$, $\hat{D}(j)$ $\hat{\Sigma}_{\hat{n}}$ with $\hat{n} = \hat{p}s + \hat{q}$. Of course, the coefficient matrices are of the form (6.7.2).

III. It is only this stage that is iterated. At each iteration the input will be a set of coefficient matrices $\hat{A}(j)$, $\hat{B}(j)$, $\hat{D}(j)$, $\hat{\Sigma}_{\hat{n}}$, for some $\hat{n} = \hat{p}s + \hat{q}$ which will have been determined at the previous iteration. Of course, the coefficient matrices will be of the form (6.7.2). Needless to say, at the first use of stage III the input comes from stage II.

The calculations at state III are more costly than those for stage II or for stage III when $s = 1$. This is because all regressions now must be of dimension $s(2s + m)$, which even for $s = m = 5$ is 75. Although it is better to redetermine $\hat{n}$ at each iteration of stage III it would be computationally much cheapter, needless to say, to use the $\hat{n}$ from stage II and to use stage III only to determine asymptotically efficient estimates of the $A(j)$, $B(j)$, $D(j)$ matrices. As in Section

6.5 (see Theorems 6.6.6 and 6.6.7), if AIC or BIC are used at stage I and $C_T = \log T$ at stage II, there may be asymptotic overestimation of the order at stage II. This can be avoided by a repetition of stage II using the output at the first use of stage II to define the estimates of $\varepsilon(t)$. However, it would be preferable, aside from the computations involved, to redetermine the estimates, including $\hat{n}$, at stage III. Asymptotically, nothing will be gained from iterating stage III, but in practice some iteration may be worthwhile.

At the first step in stage III we form a matrix sequence $[\eta(t), \zeta(t), \xi(t)]$ of s rows and $s(2s+m)$ columns by solving

$$\sum_0^{\hat{p}} \hat{B}(j)[\eta(t-j), \zeta(t-j), \xi(t-j)] = (y(t)', z(t)', \hat{\varepsilon}(t)') \otimes I_s. \tag{6.7.7}$$

Here $\hat{\varepsilon}(t)$ is obtained from

$$\sum_0^{\hat{p}} \hat{B}(j)\hat{\varepsilon}(t-j) = -\sum_1^{\hat{p}} \hat{D}(j)z(t-j) + \sum_0^{\hat{p}} \hat{A}(j)y(t-j).$$

In (6.7.6), (6.7.7) we initiate by $y(t) = z(t) = \hat{\varepsilon}(t) = 0$, $t \le 0$ and set $y(t) = z(t) = 0$, $t > T$, for the Toeplitz calculation. Then form

$$\begin{aligned} \hat{C}(t) &= \frac{1}{T}\sum \begin{bmatrix} \eta(s)' \\ -\zeta(s)' \\ -\xi(s)' \end{bmatrix} \Sigma^{-1}[\eta(s+t), -\zeta(s+t), -\xi(s+t)] \\ &= \hat{C}(-t)', \qquad t \ge 0. \end{aligned} \tag{6.7.8}$$

It is $\hat{C}(t)$ that is now used in (6.5.8) [as the $C(t)$ there defined] to generate $F_p(j)$, and so on. However, a further step has to be performed, for what is needed is a different Toeplitz regression. Form $\hat{\eta}(t)$, $\hat{\zeta}(t)$, $\hat{\xi}(t)$ [that differ from $\eta(t)$, $\zeta(t)$, $\xi(t)$ and that correspond most closely to the correspondingly designated quantities in Section 6.5; see above (6.5.6)]. Thus

$$\sum_0^{\hat{p}} \hat{B}(j)[\hat{\eta}(t-j), \hat{\zeta}(t-j), \hat{\xi}(t-j)] = [y(t), z(t), \hat{\varepsilon}(t)],$$

with the usual initiation. Correspondingly, $\hat{\eta}(t)$ is obtained by adding columns numbered $i + s(i-1)$, $i = 1, \ldots, s$ in $\eta(t)$, and similarly for $\hat{\zeta}(t)$, $\hat{\xi}(t)$. Now using the output of the Levinson-Whittle recursion based on (6.7.8), form the $s(2s+m)$-dimensional vector

$$\hat{\tau}_{p,q} = \tilde{S}_{p-1}^{-1} \sum_{0}^{p-1} \tilde{F}_{p-1}(j) \frac{1}{T} \sum [\eta(t-p+j), -\zeta(t-p+j), -\xi(t-p+j)]'$$
$$\times (\hat{\eta}(t) - \hat{\xi}(t) + \hat{\varepsilon}(t)).$$

Finally, put

$$\hat{\tau}_{p,j} = \hat{\tau}_{p-1,j} + \tilde{F}_{p-1}(p-j)\hat{\tau}_{p,q}, \qquad j = 0, 1, \ldots, p-1.$$

To initiate take $\hat{\tau}_{0,0}$ to have zeros everywhere save in rows $i + s(i-1)$, $i = 1, \ldots, s$ and rows $i + s(s+m) + s(i-1)$, $i = 1, \ldots, s$, where it has units. The matrices $\hat{\tau}_{p,j}$ estimate the $A(j)$, $B(j)$, $C(j)$ in (6.7.2) for the case $n = ps$ (i.e., $q = 0$). Thus $\hat{\tau}_{p,j} = \text{vec}[\hat{A}_n(j), \hat{D}_n(j), \hat{B}_n(j)]$, where we remind the reader that by $\text{vec}\, X$ we mean the vector obtained from the matrix X by laying the columns one underneath the other, proceeding from left to right across the matrix. Next,

(6.7.9)
$$\hat{\Sigma}_n = \frac{1}{T} \sum \hat{\varepsilon}(t)\hat{\varepsilon}(t)', \qquad n = ps,$$
$$\sum_{0}^{p} \hat{B}_p^{(1)}(j)\hat{\varepsilon}(t-j) = \sum_{0}^{p} \hat{A}_n(j)y(t-j) - \sum_{1}^{p} \hat{D}_n(j)z(t-j).$$

Then $\hat{n} = \hat{p}^{(1)}s$ is chosen so that $\hat{p}^{(1)}$ minimizes

(6.7.10)
$$\log\det \hat{\Sigma}_{ps} + ps(2s+m)C_T/T, \qquad p \leq P.$$

Finally, we have to determine $\hat{p}$, $\hat{q}$ in $\hat{n} = \hat{p}s + \hat{q}$ by a set of $s-1$ regressions of the same nature as (i), (ii) in stage II. Consider $[-\eta(t-j), \zeta(t-j), \xi(t-j)]$, formed in (6.7.7). As in forming the $\hat{\tau}_{p,j}$, the columns of these are associated with various elements of the $A(j)$, $B(j)$, $D(j)$ for various p, q. Thus the column numbered $i + s(k-1)$, $i, k = 1, \ldots, s$, is associated with $a_{ik}(j)$, column numbered $s^2 + i + s(k-1)$, $i = 1, \ldots, s$, $k = 1, \ldots, m$ is associated with $d_{ik}(j)$ and column $s^2 + sm + i + s(k-1)$ with $b_{ik}(j)$, $i = 1, \ldots, s$, $k = 1, \ldots, s$. Now eliminate all columns for which the corresponding $a_{ik}(j)$ $b_{ik}(j)$ or $d_{ik}(j)$ in (6.7.2), for given p, q, is prescribed to be zero and, for $j \geq 1$, call the resulting matrix $X_j(t)$. For $X_0(t)$ we again eliminate all elements corresponding to prescribed zero elements in $A(0)$, $B(0)$, $D(0)$ but also further reduce the number of columns by adding column $s^2 + sm + i + s(k-1)$ to column $i + s(k-1)$, $i = q+1, \ldots, s$, $k = 1, \ldots, s$, eliminating the former columns. Thus $X_0(t)$ has $q(s-q)$ columns. Then put $X(t) = [X_0(t), X_1(t), \ldots, X_{p+1}(t)]$. Let $\hat{\tau}_n' = [\hat{\tau}_{n,0}', \hat{\tau}_{n,1}', \ldots,$

$\hat{\tau}'_{n,p+1}]$, where $\hat{\tau}_{n,j} = \text{vec}\,[\hat{A}_n(j), \hat{D}_n(j), \hat{B}_n(j)]$. Then, for $n = ps + q$, $p = \hat{p}^{(1)} - 1$, $q = 1, \ldots, s - 1$,

$$\hat{\tau}_n = \left\{\frac{1}{T}\sum X(t)'\hat{\Sigma}^{-1}X(t)\right\}^{-1}\left\{\frac{1}{T}\sum X(t)'\hat{\Sigma}^{-1}\{\hat{\eta}(t) + \hat{\varepsilon}(t) - \hat{\xi}(t)\}\right\}$$

$$\text{(6.7.11)}\quad \hat{\Sigma}_n = \frac{1}{T}\sum \hat{\varepsilon}_n(t)\hat{\varepsilon}_n(t)',$$

$$\sum_0^{p+1} \hat{B}_n(j)\hat{\varepsilon}_n(t-j) = \sum_0^{p+1} \hat{A}_n(j)y(t-j) - \sum_1^{p+1} \hat{D}_n(j)z(t-j)$$

with the usual initiations and Toeplitz assumptions. Of course, $\hat{\Sigma}_n$ for $n = \hat{p}^{(1)}s$ (i.e., $q = s$) has already been computed in (6.7.11). Finally, $\hat{q}^{(1)}$ is chosen to minimize

$$\text{(6.7.12)}\quad \log\det\hat{\Sigma}_n + n(2s+m)C_T/T, \qquad n = (\hat{p}^{(1)} - 1)s + q, \qquad q = 1, \ldots, s$$

(i.e., $\hat{p}$ is $\hat{p}^{(1)} - 1$ if $\hat{q}^{(1)} < s$ and $\hat{p}^{(1)}$ if $\hat{q}^{(1)} = s$ and $\hat{q} = \hat{q}^{(1)}$ if $\hat{q}^{(1)} < s$ and $\hat{q} = 0$ if $\hat{q}^{(1)} = s$).

This completes stage III. As stated previously, this stage may be iterated until (6.7.12), at its minimum, ceases to change appreciably. We now make some remarks concerning the procedure.

In the first place all the comments of Section 6.5 (see following stage IV) have obvious analogues here. Again the step (6.7.6) (see also the formation of the $\hat{\eta}$, $\hat{\zeta}$, $\hat{\xi}$) will fail if det $(\hat{b})$ has zeros for $|z| \leq 1$. Of course, this may be checked by a Schur-Cohn procedure and again, if that test fails, we may seek to factorize $\hat{b}(\exp(i\omega))\hat{\Sigma}\hat{b}(\exp(i\omega))^*$ as $\tilde{b}\tilde{\Sigma}\tilde{b}^*$, where $\tilde{b}$ satisfies the required condition. Such a procedure is described in Tunnicliffe Wilson (1972).

We emphasize again that all procedures involving the use of the Levinson-Whittle recursion could be replaced by procedures of the kind introduced in Section 4. Paulsen and Tjøstheim (1985) show that the techniques based on Morf, Vieira, and Kailath (1978) work better, especially when det (A) has zeros near $|z| = 1$, than does the Levinson-Whittle recursion and that the differences are quite large even for $T = 500$. *An alternative is to taper the data.* It remains to be seen how well that will work in the present context, but there is some evidence that it performs as well as do Burg-type algorithms (see Section 6.4).

The question of the examination of manifolds other than the $U(n)$ needs discussion. One procedure would be to examine the $M(p, q, r)$ by examining first $M(p, p, p) = U(ps)$ and then, having determined a best p value, to seek to economize parameters by choosing q, $r < p$. The modification of the procedures discussed above to effect these calculations is fairly obvious. If one seeks to so examine $M(p, q, r)$, one is, presumably, using prior information or, better, experience that suggests that an adequate treatment of the data will be obtained with, say, q quite small. Finally, there is the question of estimating the Kronecker indices, n_{0j}. As will be shown in theorems below $\hat{p}^{(1)}$ will, asymptotically, be equal to the greatest n_{0j}, under appropriate conditions, so that a procedure similar to that described above could be used to investigate all sets of Kronecker indices with $n_j < \hat{p}^{(1)}$. The procedure, *which we call stage (IIb)′ if it follows IIa and stage III′ if it replaces stage III*, could be as follows. First, all Kronecker indices $n_1, n_2, \ldots, n_{s-1}$ are set equal to $\hat{p}^{(1)}$ and $n_s = 1, 2, \ldots, \hat{p}^{(1)} - 1$ is examined. [Of course $n_s = \hat{p}^{(1)}$ leads to $U(\hat{p}^{(1)}s)$.] At stages II, III a regression will then be effected that is the full analogue of that described under (i), (ii) for stage IIb and in (6.5.11) for stage III. It can be determined from an ARMAX representation for the echelon form which will replace (6.7.2) and will derive from Theorems 2.5.1 and 2.5.2. It is, of course, much more complicated to write down. We give an example for $s = 3$, $m = 0$ and $n_{01} = 4$, $n_{02} = 2$, $n_{03} = 3$. Then Theorem 2.5.1 tells us that in the echelon form for $k_0(z) = \tilde{a}_0(z)^{-1}\tilde{b}_0(z)$ the degrees of the elements of $\tilde{a}_0(z)$, $\tilde{b}_0(z) - \tilde{a}_0(z)$ are

$$\begin{array}{cc} \tilde{a}_0(z) & \tilde{b}_0(z) - \tilde{a}_0(z) \\ \begin{bmatrix} 4 & 1 & 2 \\ 2 & 2 & 1 \\ 3 & 1 & 3 \end{bmatrix} & \begin{bmatrix} 3 & 3 & 3 \\ 1 & 1 & 1 \\ 2 & 2 & 2 \end{bmatrix} \end{array}$$

Thus $a_0(z) = Z\tilde{a}_0(z)$, with Z diagonal and having $z^{n_{0,i}}$ in the ith diagonal place has the form

$$A_0(0) = B(0) = \begin{bmatrix} 1 & 0 & 0 \\ * & 1 & 0 \\ * & 0 & 1 \end{bmatrix}, \qquad A_0(1) = \begin{bmatrix} * & 0 & 0 \\ * & * & * \\ * & 0 & * \end{bmatrix},$$

$$A_0(2) = \begin{bmatrix} * & 0 & * \\ * & * & * \\ * & * & * \end{bmatrix}, \qquad A_0(3) = \begin{bmatrix} * & * & * \\ 0 & 0 & 0 \\ * & * & * \end{bmatrix}, \qquad A_0(4) = \begin{bmatrix} * & * & * \\ 0 & 0 & 0 \\ 0 & 0 & 0 \end{bmatrix}.$$

Then also $B_0(1)$, $B_0(2)$ have all elements freely varying and $B_0(3)$ has elements freely varying except for a second row of zeros, while $B_0(4)$ has the first row freely varying and zeros elsewhere. For example, the regression at stage (IIb)′ for the second row will be of $y_2(t)$ on $-y_1(t) + \hat{\varepsilon}_1(t)$ and on $y_j(t-1)$, $y_j(t-2)$, $\hat{\varepsilon}_j(t-1)$, $\hat{\varepsilon}_j(t-2)$; $j = 1, 2, 3$. The value of $\hat{n}_s$ will be chosen from $1, 2, \ldots, \hat{p}^{(1)}$ by minimizing

$$\log\det\hat{\Sigma}_\alpha + d(\alpha)C_T/T, \qquad \alpha \leq A_T, \tag{6.7.13}$$

where $d(\alpha)$ is the dimension of the vector of system parameters for V_α from Theorem 2.4.5 and α corresponding to the Kronecker indices $n_j = \hat{p}^{(1)}$, $j = 1, \ldots, s-1$, $n_s \leq \hat{p}^{(1)}$. Of course, $\hat{\Sigma}_\alpha$ is the residual covariance matrix.

Having chosen $\hat{n}_s \leq \hat{p}^{(1)}$, we then fix $n_1 = n_2 = n_{s-2} = \hat{p}^{(1)}$, $n_s = \hat{n}_s$ and consider $n_{s-1} = 1, 2, \ldots, \hat{p}^{(1)}$ by doing $\hat{p}^{(1)} - 1$ further regressions, and so on, until all $\hat{n}_j$ are determined. The total number of regressions is $(\hat{p}^{(1)} - 1)s$. In principle this procedure should be used at stage II and every iteration of stage III, but in practice, once the $\hat{n}_j$ have been calculated at stage II, some judgment might be used as to which values to examine at stage III since the theorems immediately below show that these initial estimates already have good properties.

Although the procedure here described has good asymptotic properties, as we shall show below, in practice, needless to say $\hat{p}^{(1)}$ might be less than the greatest n_{0j} and it might pay to examine sets of Kronecker indices near those found by the method above, but with $\max n_j > \hat{p}^{(1)}$, using (6.7.13) to make the final choice.

(iii) Theorems

We shall now state and prove two theorems. We continue to adopt the convention stated in italics immediately before Theorem 6.6.6, namely that $\hat{h}$ found at stage I will be replaced by $h = c_1\hat{h}$, $c_1 > 1$, at stage II.

Theorem 6.7.1. *Let Conditions A hold, C_T in (6.7.3) be $\log T$ and $H_T \leq (\log T)^a$, $a < \infty$, and let $C_T = c_0 \log T \log\log T$ in (6.7.5), (6.7.6), (6.7.13) for $c_0 < \infty$ but sufficiently large. Let P_T, A_T be sufficiently large. If $[k_0, l_0] \in U(n_0)$, $n_0 = p_0 s + q_0$, $0 \leq q_0 < s$, then at stage IIa*

$$\lim_{T\to\infty} \hat{p}^{(1)} = p_0, \qquad q_0 = 0; \tag{6.7.14}$$

$$= p_0 + 1, \qquad q_0 > 0, \qquad \textit{a.s.,}$$

and at stage IIb $(\hat{p}, \hat{q}) \to (p_0, q_0)$, a.s. If $[k_0, l_0] \in V_{\alpha_0}$, α_0 indexing $(n_{0j}, j = 1, \ldots, s)$ and $\|\tau_0\| \leq c_2 < \infty$, where c_2 is known, then if in stage IIa minimization is over $\|\tau\| \leq c_2$,

$$\lim_{T\to\infty} \hat{p}^{(1)} = \max_j n_{0j}, \qquad \textit{a.s.} \tag{6.7.15}$$

and at stage (IIb)′ $\hat{n}_j \to n_{0j}$, a.s. In either case the limiting distribution of $T^{1/2}(\hat{\tau} - \tau_0)$ at each iteration of stage III, or stage III′, using the $\hat{n}_j$ from stage II or (IIb)′, is that of the first part of Theorem 4.3.1, assuming that (4.1.16) also holds. If Conditions B replace Conditions A, this limiting distribution is Gaussian with zero mean vector and covariance matrix given by (4.3.9).

This theorem is unsatisfactory for many reasons. First c_0 is not prescribed. That could be done, but the value would depend on unknowns. The occurrence of $\log \log T$ in C_T is also unsatisfactory. Third, the restriction to minimization over $\|\tau\| \le c_2$ for some arbitrarily large c_2 is unsatisfactory. We give the theorem because it is fairly simple to prove and because (6.7.15) enables a restricted evaluation of the n_{0j} to be considered which reduces the labor involved in finding the Kronecker indices. In fact, there may be no true Kronecker indices, so that theorems of this kind have, in any case, value only in suggesting that the associated methods have some meaning. *The theorem also assumes that the n_{0j} are not reestimated at stage III but* the values from stage II (or II″) are used. Since these are strongly consistent, under the conditions of the theorem, this is reasonable. However, it is certain that the estimates of the n_{0j} at stage III (or III′) will be superior. We discuss this further below.

Because of the unsatisfactory nature of Theorem 6.7.1, we give a second theorem.

Theorem 6.7.2. *Assume that $[k_0, l_0] \in V_{\alpha_0}$, α_0 indexing $\{n_{0j}, j = 1, \ldots, s\}$, $|\alpha_0| \le P < \infty$ and let Conditions B hold. Let AIC or BIC be used at stage I with $H_T \le (\log T)^a$, $a < \infty$. At stage II let α be chosen to minimize*

$$\log \det \hat{\Sigma}_\alpha + d(\alpha) c_0 \log T/T, \qquad |\alpha| \le P. \tag{6.7.16}$$

Then for c_0 sufficiently large at stage IIa, $\operatorname{plim}_{T\to\infty} \hat{p}^{(1)} = \max_j n_{0j}$ and

$$\operatorname{plim}_{T\to\infty} \hat{n}_j = n_{0j}, \qquad j = 1, \ldots, s.$$

Also, $T^{1/2}(\hat{\tau} - \tau_0)$, at stage III, using these $\hat{n}_j$ from stage II, has the same limiting distribution as in the last statement in Theorem 6.7.1.

This theorem is somewhat more satisfactory, but c_0 remains. *It should be possible to prove a full analogue of Theorem 6.6.9, that is, one for which $C_T = \log T$ at stage II and stage III [i.e., in (6.7.5), (6.7.6), (6.7.10), (6.7.12), (6.7.13) and convergence in probability of the $\hat{n}_j$ occurs at stage III].* The proof would be long and involved and cannot be given here. There is also a problem in proving the analogue of the last statement in Theorem 6.6.7.

(iv) Proofs of Theorems

We may make use of Theorems 6.6.1 and 6.6.5 since these were proved for the vector case. We first establish (6.7.14), (6.7.15). At stage IIa we are minimizing the trace of

$$\frac{1}{T} \sum \{ay(t) - dz(t) - (b - I_z)\hat{\varepsilon}(t)\}\{ay(t) - dz(t) - (b - I_s)\hat{\varepsilon}(t)\}'. \tag{6.7.17}$$

This is minimized by carrying out a set of regressions, one for each "row" [i.e., element of $y(t)$]. The regressors are components of $-y(t-j)$, $z(t-j)$, $\hat{\varepsilon}(t-j)$, and some components of $-y(t)+\hat{\varepsilon}(t)$. However, $\hat{\varepsilon}(t)$ is a linear combination of $y(t-j)$, $z(t-j)$, $j=1,\dots,h=c_1\hat{h}$ (see above Theorem 6.6.6 for c_1). Thus a lower bound to (6.7.17) will be found by replacing the $\hat{\varepsilon}(t-j)$ by $-y(t-j), z(t-j), j=1,2,\dots,n_j+h$, for the jth row. [Note that $y(t)-\hat{\varepsilon}(t)$ depends only on $y(t-j)$, $j>0$.] Now from (6.6.27) it follows that (6.7.17) is bounded below by

$$\frac{1}{T}\sum \varepsilon(t)\varepsilon(t)' + O(Q_T^2 \log T), \tag{6.7.18}$$

because $\hat{\Omega}$ in that formula (6.6.27) is $O(T^{-1})$ because

$$\sum_{h+1}^{\infty} \|[\Phi(j), \Psi(j)]\|^2 = O(T^{-1}).$$

(It is for this reason that $h=c_1\hat{h}$, $c_1>1$.) Of course, (6.6.27) refers to a regression where the regressors are the same in each row but we could make that so since we are seeking a lower bound, and in any case, the result would be the same when the regressor set is row dependent. In proving (6.7.14), (6.7.15), we take α corresponding to $n_j \equiv p$, but we have given the proof more generally for later use.

At the true Kronecker indices, n_{0j}, an upper bound is obtained by putting $(a,b,d)=(a_0,b_0,d_0)$ in (6.7.17). Then that expression is

$$\begin{aligned}&\frac{1}{T}\sum [\varepsilon(t)-(b_0-I_s)\{\hat{\varepsilon}(t)-\varepsilon(t)\}][\varepsilon(t)-(b_0-I_s)\{\hat{\varepsilon}(t)-\varepsilon(t)\}]'\\ &\quad=\frac{1}{T}\sum \varepsilon(t)\varepsilon(t)' + O(Q_T^2\log T).\end{aligned} \tag{6.7.19}$$

This again follows from the proof of Theorems 6.6.3, 6.6.4 (see the evaluation of S_1 to S_5). Now if it can be shown that for $p<\max n_{0j}$, (6.7.17) has a lower bound strictly greater than Σ_0, the upper bound (6.7.19) shows that eventually $p \ge \max n_{0j}$, while the lower bound (6.7.18) used for $p>\max n_{0j}$ then establishes (6.7.14), (6.7.15). For example, we cannot have $p>\max n_{0j}$ because for $n_j \equiv p$, writing $\alpha(ps)$ for the corresponding α,

$$\begin{aligned}&\log\det\hat{\Sigma}_{\alpha(ps)} + ps(2s+m)c_0\log T\cdot\log\log T/T - \log\det\hat{\Sigma}_{\alpha 0}\\ &\quad - d(\alpha_0)c_0\log T\cdot\log\log T/T\\ &\quad \ge c_0\{ps(2s+m)-d(\alpha_0)\}\log T\cdot\log\log T/T + O(Q_T^2\log T)\end{aligned}$$

and since $ps(2s+m) > d(\alpha_0) = \sum n_{0j}$, then for sufficiently large c_0 this expression is positive. [Of course, (6.7.14) is a special case of (6.7.15). It is singled out because it can be proved under more general conditions.] Now for (6.7.17) at $n_j \equiv p$ we are doing the (Toeplitz) regression of $y(t)$ on $-y(t-j)$, $z(t-j)$, $\hat{\varepsilon}(t-j)$, $j = 1,\ldots,p$. First consider $(k_0, l_0) \in U(n_0)$, $n_0 = p_0 s + q_0$. Then the vector $\hat{w}(t)$, $\hat{w}(t)' = (-y(t-1)',\ldots,-y(t-p)', z(t-1)',\ldots,\hat{\varepsilon}(t-p)'$, satisfies

$$\liminf_{T\to\infty} \frac{1}{T} \sum \hat{w}(t)\hat{w}(t)' \geq cI, \qquad c > 0.$$

Indeed, putting $w(t)$ for $\hat{w}(t)$ with $\hat{\varepsilon}(t)$ replaced by $\varepsilon(t)$, then $T^{-1}\sum w(t)w(t)'$ converges to $E\{w(t)w(t)'\}$. Putting $y_p^{(t+1)} = (y(t+1)', y(t+2)',\ldots,y(t+p)')$, and similarly $z_p^{(t+1)}$, $\varepsilon_p^{(t+1)}$, then if $E\{w(t)w(t)'\}$ were singular, we should have $e'y_p^{(t+1)} = f'z_p^{(t+1)} + g'\varepsilon_p^{(t+1)}$ and, from the ARMAX analogue of (6.3.3) it follows that $e'_\infty \mathscr{H} = 0, f = 0$, where $e_\infty = (e', o'_\infty)'$. Indeed, $y^{(t+1)} = \mathscr{H}_1 \varepsilon_t + \mathscr{H}_2 z_t + \mathscr{K}\varepsilon^{(t+1)}$, where $y^{(t+1)}$, and so on, have the meanings given above (6.3.3) and $\mathscr{H}_1$ is the Hankel matrix composed from the $K_0(j)$ while $\mathscr{H}_2$ is composed of the $L_0(j)$. But then $e'_\infty y^{(t+1)} = e'_\infty \mathscr{H}_1 \varepsilon_t + e'_\infty \mathscr{H}_2 z_t + e'_\infty \mathscr{K}\varepsilon^{(t+1)} = f'z_p^{(t+1)} + g'\varepsilon_p^{(t+1)}$, which implies that $e'_\infty \mathscr{H} = f = 0$ [using (6.6.1)]. However, for $p < p_0$, $q_0 = 0$, or $p < p_0 + 1$, $q_0 > 0$, this is impossible. Thus $E\{w(t)w(t)'\}$ is nonsingular and hence $T^{-1}\sum \hat{w}(t)\hat{w}(t)'$ eventually has its smallest eigenvalue bounded away from zero. [It is possible to give a simpler proof than the above. We have used this one because of its relation to the proof of (6.7.15).] Thus the solutions to the minimization of (6.7.17) are uniformly bounded and hence, using the ergodic properties of the autocovariances involved, including their uniform convergence properties established in Chapter 5, then (6.7.17) evaluated at $\hat{a}$, $\hat{b}$, $\hat{d}$, the optimizing values, converges to

$$\begin{aligned}
&\frac{1}{2\pi}\int (\hat{a}k_0 - \hat{b} + I_s)\Sigma_0(\hat{a}k_0 - b_0 + I_s)^* \, d\omega \\
&\qquad + \int (\hat{a}l_0 - \hat{d}) f_z(\omega)(\hat{a}l_0 - \hat{d})^* \, d\omega \\
&= \Sigma_0 + \frac{1}{2\pi}\int (\hat{a}k_0 - \hat{b})\Sigma_0(\hat{a}k_0 - b_0)^* \, d\omega \\
&\qquad + \int (\hat{a}l_0 - \hat{d}) f_z(\omega)(\hat{a}l_0 - \hat{d})^* \, d\omega.
\end{aligned} \tag{6.7.20}$$

However, if $p < \max n_{0j}$, then the sum of the second and third terms on the right of (6.7.20) cannot be zero and indeed the lower bound must be a nonzero matrix (see Section 2.5). This completes the proof of (6.7.14). Now (6.7.15) also follows since (6.7.20), *under the boundedness condition* $\|\tau\| \leq c_2$, again follows. Unfortunately, the argument involving the nonsingularity of $T^{-1}\sum \hat{w}(t)\hat{w}(t)'$ now does not follow since we could have $e_\infty \mathscr{H} = 0$ for $p < \max n_{0j}$, which is why the boundedness condition is needed. Now it remains to prove that when $(k, l) \in U(n)$, $n < n_0$, (6.7.17) has a lower bound greater than Σ_0 and that for $\|\tau\| < c_2 < \infty$ $(k, l) \in V_\alpha$ the same is true if any $n_j < n_{0j}$. The proof of these propositions is the same, essentially, as that just given and is omitted. The last part of Theorem 6.7.1 is almost the same as the proof of that part of Theorem 6.6.9.

We turn to Theorem 6.7.2. To prove this we establish the following results. We first establish

(a) $\alpha \not\geq \alpha_0 \Rightarrow \hat{\Sigma}_\alpha \geq \Sigma_\alpha + o_p(1), \qquad \Sigma_\alpha > \Sigma_{\alpha_0}$.
(b) $n_j \equiv p,\ p_j \geq n_{0j},\ p > n_{0j}$ for some $j \Rightarrow \hat{\Sigma}_\alpha > \frac{1}{T}\sum \varepsilon(t)\varepsilon(t)' + o_p(\log T/T)$.
(c) $\alpha > \alpha_0 \Rightarrow \hat{\Sigma}_\alpha \geq \frac{1}{T}\sum \varepsilon(t)\varepsilon(t)' + K_1(\log T)/T + o_p(\log T/T)$.
(d) $\alpha = \alpha_0 \Rightarrow \hat{\Sigma}_\alpha = \frac{1}{T}\sum \varepsilon(t)\varepsilon(t)' + K_0(\log T)/T + o_p(\log T/T)$.

Here K_0, K_1 depend only on τ_0. Σ_α is the minimum of (6.7.20) when $\hat{a}, \hat{b}, \hat{d}$ vary over all of V_α. If we do not take c_0 sufficiently large then we may overestimate at stage II because, writing $BIC(\alpha)$ for BIC obtained for Kronecker indices indexed by α,

$$\text{(6.7.21)} \quad BIC(\alpha_0) - BIC(\alpha) = [\operatorname{tr}\{\Sigma_0^{-1}(K_0 - K_1) + d(\alpha_0) - d(\alpha)\}](\log T)/T + o_p(\log T/T),$$

which is positive when the factor multiplying $(\log T)/T$ on the right is positive, which can occur.

We can see that weak consistency of the $\hat{n}_j$ is obtained, once (a) to (d) are established, as follows. First that plim $\hat{p}^{(1)} = \max n_{0j}$ follows precisely as in the proof of Theorem 6.7.1 using (a) to (d). Since $O(Q_T^2 \log T)$ is now replaced by $K_j(\log T)/T + o_p(\log T/T)$, then $c_0 \log T$ may replace $c_0 \log T \log\log T$ to give the required convergence in probability. Now assume that the procedure (IIb)′ is used and we have shown plim $(\hat{n}_{q+1}, \ldots, \hat{n}_s) = (n_{0q+1}, \ldots, n_{0s})$ and at the next stage we put $n_j = \hat{n}_j$, $j = q + 1, \ldots, s$; $\hat{n}_1, \ldots, \hat{n}_{q-1} = \hat{p}^{(1)}$ and examine $n_q = 0, 1, 2, \ldots, \hat{p}^{(1)} - 1, \hat{p}^{(1)}$ using (6.7.16). Call α_p these Kronecker indices for $n_q = p$. If $p > n_{0q}$, then $\alpha_p > \alpha_0$ and (c) applies. This shows, by the same kind of comparison as is used in (6.7.12), that for c_0 sufficiently large we cannot, eventually, have $\hat{p} > n_{0q}$. If $p < n_{0q}$, then $\alpha_p \not\geq \alpha_0$ and (a) applies, so that again, eventually, such a value of p cannot be chosen. Thus $\hat{n}_q$ converges in probability to n_{0q} and the proof of the first part of the theorem follows by induction. The last part is then proved in much the same way as Theorem 6.6.9.

It remains to prove (a) to (d).

(a) Here $\hat{\Sigma}_\alpha$, $\alpha \not\geq \alpha_0$ is obtained by regressing $y_i(t)$, $i = 1, \ldots, s$, on certain components of $y(t) - \hat{\varepsilon}(t)$, $y(t-j)$, $z(t-j)$, $\hat{\varepsilon}(t-j)$. Now where $\hat{\varepsilon}_j(t-k)$ occurs in this set, replace it by $\varepsilon_j(t-k)$ to obtain a vector $R_i(t)$. Let $S(t)$ be the vector composed of the components of $\hat{\varepsilon}(t-k) - \varepsilon(t-k)$, $k = 0, 1, \ldots, \max n_j$. Then regressing on the vector composed of $R_i(t)$ and $S(t)$ can only reduce the residual matrix of variances and covariances below $\hat{\Sigma}_\alpha$. The vector $R_i(t)$ is generated by a stationary process, but its covariance matrix could be singular. Moving to a new basis, we remove all null components so that the new vector, which we again call $R_i(t)$ for notational simplicity, has a nonsingular covariance matrix. Then the residual covariance matrix for the set of s residuals from the regressions of the $y_i(t)$ on the $R_i(t)$ will converge to a matrix $\Sigma_\alpha > \Sigma_{\alpha_0}$, by the same argument as was used in proving Theorem 6.7.1. We have yet to regress on $S(t)$ after that has been made orthogonal to $R_i(t)$. By the argument used in proving Theorems 6.6.3 and 6.6.4 [see the proofs for S_1 to S_5 below (6.6.27)] we see that, calling $S_i(t)$ the residual from the regression of $S(t)$ on $R_i(t)$.

$$S_i(t) = S(t) + AR_i(t),$$

where A has elements that are $O(Q_T)$. By the same argument,

$$\frac{1}{T} \sum y_i(t) S_i(t) = O(Q_T)$$

while

$$\frac{1}{T} \sum S_i(t) S_i(t)' = \frac{1}{T} \sum S(t) S(t)' + O(Q_T^2),$$

and from Theorem 6.6.5 this is

$$(s + m)(I_p \otimes \Sigma_0) h/T + o_p(h/T), \qquad p = \max n_j.$$

Thus the reduction in the residual matrix of variances and covariances is $O_p(\log \log T/\log T)$ and (a) holds.

(b), (c), (d) We use the parameterization of Section 5.5. We now are considering V_α, $\alpha \geq \alpha_0$. However, the technique is the same, the only thing changing being N, which is evaluated in Theorem 2.5.3(v), where the necessary facts to institute the parameterization are also established. Then we have

$$(a, b, d) = v(z)(a_0, b_0, d_0) + \sum_1^N \chi_j \xi_j(z).$$

Here v and ξ_j are matrices. (We now use v in place of h, as in Section 5.5, to avoid confusion with h, the order determined at stage I.) Put $w_j(t) = \xi_j(t)(y(t)', -\hat{\varepsilon}(t)', z(t)')'$, using z as a backward shift, as usual. Then, referring to (6.7.17),

$$ay(t) - (b - I_s)\hat{\varepsilon}(t) - dz(t)$$
$$= -(v - I_s)b_0\{\hat{\varepsilon}(t) - \varepsilon(t)\} + \sum_1^N \chi_j w_j(t) + \varepsilon(t) - (b_0 - I_s)\{\hat{\varepsilon}(t) - \varepsilon(t)\}.$$

The proof from this point on is not essentially different from that of Theorem 6.6.7 in Section 6.6. [see below (6.6.47)]. The constant K_1 in (c) corresponds, of course, to the second term on the right in (6.6.14). The only essential difference between the two treatments is due to the fact that $v(z) - I_s$ may be linearly constrained. [Of course, $v(0) - I_s$ is lower triangular with zeros on the main diagonal, since both $b(0)$ and $v(0)b(0)$ are lower triangular with unit on the main diagonal.] Thus evaluating, using Theorem 6.6.5, the contribution to the minimum value of (6.7.17) from the regression of $\varepsilon(t) - (b_0 - I_s)\{\hat{\varepsilon}(t) - \varepsilon(t)\}$, on $b_0\{\hat{\varepsilon}(t) - \varepsilon(t)\}$, one has to take account of these constraints, which are complicated, except in special cases. Thus K_1 has not been evaluated precisely. K_0 and K_1 are of course nonnegative matrices and $K_0 \geq K_1$. The proof for (d), is, again, essentially as in the proof of Theorem 6.6.7. Now K_0 is easily evaluated by evaluating

$$\frac{1}{T}\sum [\varepsilon(t) - (b_0 - I_s)\{\hat{\varepsilon}(t) - \varepsilon(t)\}][\varepsilon(t) - (b_0 - I_s)\{\hat{\varepsilon}(t) - \varepsilon(t)\}]'.$$

Using Theorem 6.6.5, we obtain

$$K_0 = (s + m)\left[\sum_l \{B_0(l) - \delta_{0l}I_s\}\Sigma_0\{B_0(l) - \delta_{0l}I_s\}' - \{B_0(0) - I_s\}\Sigma_0 - \Sigma_0\{B_0(0) - I_s\}'\right].$$

CHAPTER 7

Approximation by Rational Transfer Functions

1. Introduction

Now the problem to be considered is that of approximating some true structure by one with a rational transfer function. The traditional development of statistics has been on the basis of the "Fisherian paradigm," namely, that the data are generated by a stochastic process, specified except for a finite number of parameters. This paradigm has been used, of course, recognizing its falsity, in a wide range of circumstances, but recognizing also that it leads to useful analyses, adroitly used (see Brillinger, 1984, pp. 1145–1147, for a discussion). In Chapters 5 and 6 the severity of the paradigm has been moderated by allowing the number of parameters to be unknown and hence estimated. However, in this chapter we abandon the paradigm altogether and explicitly admit that the data are not generated by any model in the model set.

In the next section a deterministic situation is considered where $k(z)$ is known (albeit not necessarily rational) and a rational approximation to that is to be constructed. Such a situation could conceivably arise in an experimental situation. Again $k(z)$ might be rational but of very high McMillan degree (as with a large number of rational transfer function systems in series) and a lower-degree approximation might be required.

The approximation theory delt with in Section 2 provides insight into the situation, which is why it is presented, but in fact $k(z)$ will not be known. However, Theorems 5.4.1, 5.4.2, 5.5.1, and 5.5.2 have no meaning when there is no true structure in the model set and a new justification has to be provided for AIC, BIC, and so on. This is taken up in the third section.

In Section 4 some theory of the type given in Chapters 5 and 6 is presented for the situation of this chapter. This will include the a.s. behavior of $\hat{k}$, $\hat{l}$, and some theory about autoregressive approximation for n determined from the data. For this theory some theorems, analogous to those in Section 5.3, about autocovariances are needed.

Finally, we deal briefly with on-line calculation, the situation where the estimation procedure proceeds on an adaptive basis as each time point is passed. Such calculations could be in "real time" (e.g., to be completed in a fraction of a second). This field has been expertly surveyed recently by Ljung and Söderström (1983) and Young (1984) and we shall deal only with procedures related to those of Section 6.5. The discussion will allow us to introduce certain algorithms to the reader, especially the so-called lattice algorithms, which are of great importance.

In much of this chapter the cases considered will be those where there are no inputs. This is partly because, once a rational transfer function context is abandoned, we may as well regard inputs and outputs as generated by the same system. [Of course, there would be some means and periodic effects (e.g., seasonal effects) to be accounted for, but these can be removed by regression before the analysis of the system is begun.] We shall, however, then, call the output $y(t)$ rather than $u(t)$ as we have done previously, when it was necessary to stress the fact that the ARMA case was being considered.

2. Hankel Norm Approximation*

Here only the case is discussed where the input is unobserved white noise and constitutes the innovation sequence, for the same reason as was given in Section 6.3. Let

$$(7.2.1) \qquad y(t) = \sum_0^\infty K(j)\varepsilon(t-j), \qquad \sum_1^\infty j\|K(j)\|^2 < \infty, K(0) = I$$

The second part of (7.2.1) ensures that $\operatorname{tr}(\mathscr{H}\mathscr{H}') < \infty$, so that the sum of squares of the singular values of $\mathscr{H}$ is finite. Here $\mathscr{H}$ is being considered as an operator in l_2 (i.e., the space of all square summable sequences). Put

$$H_1 = [K(1), K(2), K(3), \ldots]$$

and let H_0 consist of rows of $\mathscr{H}$ spanning the space spanned by all of the rows of $\mathscr{H}$. (All rows of $\mathscr{H}$ lie in l_2.) Put

$$\varepsilon_t = (\varepsilon(t)', \varepsilon(t-1)', \ldots,)', \qquad \varepsilon^{(t+1)} = (\varepsilon(t+1)', \varepsilon(t+2)', \ldots,)'.$$

Then (see Sections 2.5 and 2.8) ε_t represents the present and past (for brevity we again call this "the past") and $\varepsilon^{(t+1)}$ the future. Put $H_0 = [K, H_2]$, where K is the first column of blocks in H_0. Then H_1 and H_2 are composed of rows of $\mathscr{H}$ (see Theorem 2.4.1) so that there are matrices F, H so that

*This section may be omitted. Later sections do not use it.

$H_1 = HH_0$, $H_2 = FH_0$. Put $x(t|t-1) = H_0\varepsilon_{t-1}$. Since $y(t) = H_1\varepsilon_{t-1} + \varepsilon(t)$, $x(t+1|t) = H_2\varepsilon_{t-1} + K\varepsilon(t)$, we have

$$(7.2.2)\quad y(t) = Hx(t|t-1) + \varepsilon(t), \qquad x(t+1|t) = Fx(t|t-1) + K\varepsilon(t).$$

This is the general form of the prediction error state-space representation. Alternatively, we could have begun from $\tilde{\mathscr{H}}$, having $K(i+j-2)$ as the (i,j)th block. Defining $\tilde{H}_0$, $\tilde{H}_1$, $\tilde{K}$, $\tilde{F}$, $\tilde{H}$ accordingly, we obtain

$$(7.2.3)\qquad y(t) = \tilde{H}\tilde{x}(t+1|t), \qquad \tilde{x}(t+1|t) = \tilde{F}\tilde{x}(t|t-1) + \tilde{K}\varepsilon(t).$$

We mainly refer to (7.2.2), although (7.2.3) is also relevant. Defining $y^{(t)}$, y_t as for $\varepsilon^{(t)}$, ε_t, we have [see (6.3.2)]

$$(7.2.4)\qquad y^{(t+1)} = \mathscr{H}\varepsilon_t + \mathscr{K}\varepsilon^{(t+1)},$$

where $\mathscr{K}$ has $K(i-j)$ as the (i,j)th block, $K(j) = 0$, $j < 0$. Of course, $E\{\varepsilon^{(t+1)}\varepsilon_t'\} = 0$. As in Section 6.3, the equation (7.2.4) describes the relation between future and past.

The problem is to approximate to $k(z)$, corresponding to (7.2.1), by a rational matrix function $\hat{k}(z)$ when $k(z)$ is known. A natural measure of the approximation of k is

$$(7.2.5)\qquad \sup_{\omega} \|k(e^{i\omega}) - \hat{k}(e^{i\omega})\| = \|\hat{k} - k\|_{\infty}.$$

This L_∞ norm is connected with the Hankel norm, which is the largest singular value of $\mathscr{H}$ (i.e., the square root of the largest eigenvalue of $\mathscr{H}'\mathscr{H}$). Let us write $\mathscr{H}_\phi$ for the Hankel matrix corresponding to the function

$$\phi(z) = \sum_{-\infty}^{\infty} \Phi(j)z^j.$$

Only $\Phi(j)$ for $j > 0$ enter into $\mathscr{H}_\phi$, but $\phi(z)$ need not be analytic at $z = 0$. The function ϕ is said to be the symbol of $\mathscr{H}_\phi$. Write $\|\mathscr{H}_\phi\|$ for the largest singular value of $\mathscr{H}_\phi$. Let ϕ be essentially bounded on $|z| = 1$ (i.e., $\phi \in L_\infty$) and write $\|\phi\|_\infty$ for the essential uniform bound to $|\phi(e^{i\omega})|$. By H_∞ we mean the functions analytic for $|z| > 1$ that are essentially bounded on $|z| = 1$. (Usually, H_∞ would be said to be the functions analytic for $|z| < 1$ that are almost uniformly bounded on $|z| = 1$, but our conventions earlier in this book force us to our present definition.) Then $\mathscr{H}_\phi$ may be associated with the coset $\phi + H_\infty$ (i.e., the set of all functions $\phi + h$, $h \in H_\infty$). This coset is a point in L_∞/H_∞ whose norm is

$$\|\phi + H_\infty\|_{L_\infty/H_\infty} = \inf\{\|\phi - h\|_\infty, h \in H_\infty\}, \tag{7.2.6}$$

and which makes L_∞/H_∞ into a Banach space. A theorem due to Nehari (1957) asserts that the correspondence $\mathscr{H}_\phi \leftrightarrow \phi + H_\infty$ is a linear isometry, that is, that

$$\|\mathscr{H}_\phi\| = \|\phi + H_\infty\|_{L_\infty/H_\infty} \tag{7.2.7}$$

and that there is an h for which the equality in (7.2.6), (7.2.7) is attained. Adamyan, Arov, and Krein (1971) extend this result. Let $H_\infty^{(\kappa)}$, $\kappa > 0$, κ an integer, be the functions essentially bounded on $|z| = 1$ which have exactly κ poles, counting multiplicity (i.e., order), in $|z| > 1$. Let ρ_k be the kth singular value of $\mathscr{H}_\phi$, $\rho_1 \geq \rho_2 \geq \cdots$, and κ be such that $\rho_{\kappa-1} > \rho_\kappa = \rho_{\kappa+1} = \cdots = \rho_{\kappa+r}$. Then there is a Hankel matrix $\mathscr{H}_{(\kappa)}$, *of rank* κ, and only one of rank not exceeding $\kappa + r$, so that $\|\mathscr{H}_\phi - \mathscr{H}_{(\kappa)}\| = \rho_\kappa$. Moreover, if $h^{(\kappa)}$ is the symbol of $\mathscr{H}_{(\kappa)}$, that function uniquely *attains* the minimum

$$\inf\{\|\phi - h\|_\infty, h \in H_\infty^{(\kappa)}\} = \rho_\kappa. \tag{7.2.8}$$

Adamyan, Arov, and Krein (1971) also show how to construct $h^{(\kappa)}$. Let $\xi^{(\kappa)}$, $\eta^{(\kappa)}$ be a "Schmidt pair" for $\mathscr{H}_\phi$ associated with ρ_κ, that is,

$$\mathscr{H}_\phi \xi^{(\kappa)} = \rho_\kappa \eta^{(\kappa)}.$$

There may be more than one such pair, but the result is independent of the choice. Let $\xi_j^{(\kappa)}$ and $\eta_j^{(\kappa)}$ be the jth components of the infinite vectors. Then

$$\begin{aligned} h^{(\kappa)}(z) &= \phi(z) - \rho_\kappa \eta_-^{(\kappa)}(z)/\xi_+^{(\kappa)}(z), \\ \xi_+^{(\kappa)}(z) &= \sum_1^\infty \xi_j^{(\kappa)} z^{-j+1}, \qquad \eta_-^{(\kappa)}(z) = \sum_1^\infty \eta_j^{(\kappa)} z^j. \end{aligned} \tag{7.2.9}$$

Since for $s = 1$, $\mathscr{H}_\phi$ will be real (and symmetric) in circumstances relevant to us, then $\xi^{(\kappa)} = \pm\eta^{(\kappa)}$.

This remarkable result seems, however, of limited use to us, for although $\mathscr{H}_{(\kappa)}$ is of finite rank, $h^{(\kappa)}(z)$ is not analytic for $|z| \leq 1$, nor will it be rational. (Remember that only coefficients of positive powers of z occur in $\mathscr{H}_{(\kappa)}$.) Thus if $\phi(z)$ is taken as $k(z)$, then certainly we have approximated satisfactorily to $\mathscr{H}_k$ by $\mathscr{H}_{(\kappa)}$, but (7.2.8) merely says that $h^{(\kappa)}$ approximates, to a known accuracy, to $k(z)$. However, what would be required is an approximation to $k(z)$ that is rational and analytic for $|z| \leq 1$.

It is not apparent that $\|\mathscr{H}\|$ is a relevant criterion. Return to (7.2.4) and recall the considerations of Sections 2.8 and 6.3. It seems more relevant to replace $\mathscr{H}$ by the matrix obtained when $y^{(t+1)}$, ε_t are linearly transformed so that the new vectors are composed of orthonormal random variables. One such transformation may be achieved as follows. For ε_t this means that we change to $(I_\infty \otimes \Sigma^{-1/2})\varepsilon_t$, where the matrix is block diagonal with all diagonal blocks $\Sigma^{1/2}$. Thus $\mathscr{H}$ is replaced by $\mathscr{H}(I_\infty \otimes \Sigma^{1/2})$. Also,

$$E\{y^{(t+1)}y^{(t+1)\prime}\} = [\Gamma(k-j)]_{j,k=1,2,\ldots},$$

where the (j,k)th block is shown. This is the block covariance matrix corresponding to the spectrum $f(-\omega) = f(\omega)'$, the change to $-\omega$ being due to the time reversal inherent in the fact that the time argument increases as you go down the column $y^{(t+1)}$. Now [see (1.3.6) and below (2.8.4)] we have

$$2\pi f(-\omega) = \lambda(e^{i\omega})\Omega\lambda(e^{i\omega})^*, \qquad \lambda(z) = \sum_0^\infty \Lambda(j)z^j, \qquad \Lambda(0) = I_s.$$

Let $\mathscr{L}$ have $\Lambda(k-j)$ as the (j,k)th block, $\Lambda(j) = 0, j < 0$. Then

$$E\{y^{(t+1)}y^{(t+1)\prime}\} = \mathscr{L}(I_\infty \otimes \Omega)\mathscr{L}',$$

so that $(I_\infty \otimes \Omega^{-1/2})\mathscr{L}^{-1}y^{(t+1)}$ is of the required form and $\mathscr{H}$ finally may be replaced by

$$\mathscr{E} = (I_\infty \otimes \Omega^{-1/2})\mathscr{L}^{-1}\mathscr{H}(I_\infty \otimes \Sigma^{1/2}). \tag{7.2.10}$$

The matrix $\mathscr{E}$ is again a Hankel matrix. Indeed, any upper triangular block Toeplitz matrix has an inverse of the same form and when such a matrix multiplies a Hankel matrix on the left the result is again a Hankel matrix. Moreover, the symbols obey the same algebra, so that $\mathscr{E}$ has the symbol

$$e(z) = \Omega^{-1/2}\lambda(z^{-1})^{-1}k(z)\Sigma^{1/2}. \tag{7.2.11}$$

Indeed, since

$$\frac{1}{2\pi}\int \lambda(e^{i\omega})e^{-i(j-k)\omega}\,d\omega = \Lambda(j-k),$$

then $\lambda(z^{-1})$ is the symbol of the Toeplitz matrix $\mathscr{L}$ and $\lambda(z^{-1})^{-1}$ of $\mathscr{L}^{-1}$. The matrix $e(z)$ is not analytic for $z < 1$ but it is unitary on $|z| = 1$, since, for $z = \exp(i\omega)$,

$$e(z)e(z)^* = 2\pi\Omega^{-1/2}\lambda(e^{-i\omega})^{-1}f(\omega)\lambda(e^{-i\omega})^{*^{-1}}\Omega^{-1/2} = I_s.$$

The singular values of $\mathscr{E}$ are the canonical correlations between future and past discussed in Section 2.8 in the rational case. Indeed,

$$\mathscr{E} = \sum_1^\infty \rho_j \xi^{(j)}\eta^{(j)^*}, \qquad \xi^{(j)}\xi^{(k)^*} = \eta^{(j)}\eta^{(k)^*} = \delta_{jk}, \quad \rho_1 \geq \rho_2 \geq \cdots \geq 0.$$

The discriminant functions are

(7.2.12) $y_j = \eta^{(j)^*}(I_\infty \otimes \Omega^{-1/2})\mathscr{L}^{-1}y^{(t+1)}, \qquad x_j = \xi^{(j)^*}(I_\infty \otimes \Sigma^{-1/2})\varepsilon_t.$

We bring this together as a theorem.

Theorem 7.2.1. *Let $y(t)$ be generated by (7.2.1) with $|\det k| \geq c > 0$, $|z| = 1$. Then (7.2.10) is well defined and has symbol (7.2.11). The canonical correlations between future and past are the singular values of $\mathscr{E}$ and the corresponding discriminant functions are given by (7.2.12).*

REMARKS

1. All of the conditions are by no means necessary.
2. Green (1987) has derived the theorem of Section 2.8 directly from $e(z)$.
3. When $s = 1$, then $f(\omega) = f(-\omega)$, so that $\lambda(z) = k(z)$, $\Omega = \Sigma$ and $e(z) = k(z)/k(z^{-1})$, which is of modulus unity on $|z| = 1$.

For $s = 1$, Jonckheere and Helton (1985) produce a rational approximation to $k(z)$ using this theory. Beginning from (7.2.3) and $\mathscr{H}$ we attain $\tilde{\mathscr{E}} = (I_\infty \otimes \Omega^{-1/2})\mathscr{L}^{-1}\mathscr{H}(I_\infty \otimes \Sigma^{1/2})$. Then $\tilde{e}(z) = zk(z)/k(z^{-1})$, since $zk(z)$ is the symbol of $\mathscr{H}$. The introduction of z is connected with the fact that $\xi_+^{(\kappa)}(z) = \pm z\eta_-^{(\kappa)}(z)$. Let $\tilde{h}^{(k)}(z)$ solve (7.2.8) for $\phi = \tilde{e}(z)$. Next let $\tilde{h}_0^{(\kappa)}(z)$ solve (7.2.6) for $\phi = \tilde{h}^{(\kappa)}(z^{-1})$ with $\tilde{\rho}_0$ the minimized value (i.e., the norm of the Hankel matrix $\mathscr{H}^{(\kappa)}$). Now letting ρ_κ be the singular values of $\mathscr{H}$, since $\tilde{h}_0^{(\kappa)} = \tilde{h}^{(\kappa)} - \tilde{\rho}_0\tilde{\eta}_-/\tilde{\xi}_+$, where $\tilde{\eta}_-$ and $\tilde{\xi}_+$ are constructed from the Schmidt pair for $\mathscr{H}^{(\kappa)}$ corresponding to $\tilde{\rho}_0$, then

$$\rho_\kappa = \sup_\omega |e^{i\omega}\{k(e^{i\omega})/k(e^{-i\omega})\} - \tilde{h}^{(\kappa)}(e^{i\omega})|$$

$$= \sup_\omega |e^{i\omega}\{k(e^{i\omega})/k(e^{-i\omega})\} - \tilde{h}_0^{(\kappa)}(e^{i\omega}) - \tilde{\rho}_0\tilde{\eta}_-(e^{i\omega})/\tilde{\xi}_+(e^{i\omega})|.$$

Since $\tilde{h}_0^{(\kappa)} \in H_\infty$, its coefficients do not enter into the Hankel matrix corresponding to $\tilde{h}_0^{(\kappa)} + \tilde{\rho}_0\tilde{\eta}_-/\tilde{\xi}_= \tilde{h}^{(k)}$ [see (7.2.9)], which determines ρ_κ and finally,

$$\rho_\kappa = \sup_\omega |e^{i\omega}\{k(e^{i\omega})/k(e^{-i\omega})\} - \tilde{\rho}_0\tilde{\eta}_-(e^{i\omega})/\tilde{\xi}_+(e^{i\omega})|.$$

Put

$$\hat{k}(z) = cz^{-1}\tilde{\eta}_-(z) = c\sum_1^\infty \tilde{\eta}_j z^{j-1}. \tag{7.2.13}$$

Then $\tilde{\rho}_0\tilde{\eta}_-/\tilde{\xi}_+ = \tilde{\rho}_0 z\hat{k}(z)/\hat{k}(z^{-1})$. If $\tilde{\rho}_0 = 1$, it is natural to take $\hat{k}(z)$, which is rational, as the approximation to $k(z)$. [The fact that $\hat{k}(z)$ is rational depends on further theory for the Schmidt pairs that we have not presented here. It also depends on assuming that the singular values of $\tilde{\mathscr{H}}$ are simple.] It does not seem to be established that $\tilde{\rho}_0 = 1$, nor is a choice of c in (7.2.13) well motivated. It could be chosen to make the spectrum corresponding to $\hat{k}$ equal to $f(\omega)$ at one point. Certainly, we are not achieving a minimum to (7.2.5). Finally, the case $s = 1$, only, is being treated.

The vector case is treated in Adamyan, Arov, and Krein (1978). However, here we discuss only Glover (1984). Glover (1984) begins from a $k(z)$ of McMillan degree n, which may be high. Let a minimal state-space representation of the system have system matrices F, H, K and state vector $x(t|t-1)$ [see (7.2.2)]. Put

$$\mathscr{O}'_\infty = (H', F'H', (F')^2H', \ldots), \qquad \mathscr{C}_\infty = (K, FK, F^2K, \ldots),$$

where now there are infinitely many blocks. Then $\mathscr{H} = \mathscr{O}_\infty\mathscr{C}_\infty$. Put $P = \mathscr{C}_\infty\mathscr{C}'_\infty$, $Q = \mathscr{O}'_\infty\mathscr{O}_\infty$. These are both $n \times n$. Factor Q as $R'R$, for example, choosing R to be a Choleski factor. Put $RPR' = VS^2V'$, $V'V = I_n$ where S is diagonal with positive diagonal elements. Since $RPR' > 0$ (see Theorem 2.3.3) this is possible. Now in (7.2.2) make the transformation (2.3.9) with $T = S^{-1/2}V'R$. Then $P \to TPT' = S^{-1/2}V'RPR'VS^{-1/2} = S^{-1/2}V'VS^2V'VS^{-1/2} = S$. Similarly, $Q \to (T')^{-1}QT^{-1} = S$. Thus we have now changed to a new state vector, $Tx(t|t-1)$, so that in the new coordinates F, H, K are such that $P = Q = S$ and S is diagonal and positive definite. This transformation is due to Moore (1981) and is said to constitute a "balanced realization" of the system. *To avoid a more complex notation, let us assume that the initial minimal realization is balanced* so that F, H, K, $\mathscr{O}_\infty$, $\mathscr{C}_\infty$ are such that $\mathscr{O}'_\infty\mathscr{O}_\infty = \mathscr{C}_\infty\mathscr{C}'_\infty = S$. We may assume that the diagonal elements of S, namely σ_i, are such that $\sigma_1 \geq \sigma_2 \geq \cdots \geq \sigma_n > 0$. Now since $\mathscr{H} = \mathscr{O}_\infty\mathscr{C}_\infty$, then

$$\mathscr{H} = \sum_1^n \sigma_i \eta^{(i)}\xi^{(i)'}, \qquad \eta^{(i)'}\eta^{(j)} = \xi^{(i)'}\xi^{(j)} = \delta_{ij}, \tag{7.2.14}$$

where

$$\eta^{(i)} = \sigma_i^{-1/2} u_i, \qquad \xi^{(i)} = \sigma_i^{-1/2} v_i$$

and u_i is the ith column of $\mathcal{O}_\infty$ and v_i' is the ith row of $\mathscr{C}_\infty$. Since $\mathcal{O}_\infty' \mathcal{O}_\infty = S$, the $\eta^{(i)}$ are orthonormal, as are the $\xi^{(i)}$ because $\mathscr{C}_\infty \mathscr{C}_\infty' = S$ and (7.2.14) follows. Thus (7.2.14) constitutes the singular-value decomposition of $\mathscr{H}$.

We might just as well have begun from $\mathscr{E}$ in place of $\mathscr{H}$. Of course, $\mathscr{E}$ has the same rank (i.e., McMillan degree) as $\mathscr{H}$. Again we can find a balanced realization and a singular-value decomposition (7.2.14). Glover (1984) then shows how to find an optimal Hankel norm approximation to $\mathscr{H}$ or $\mathscr{E}$ of prescribed McMillan degree [i.e., how to solve (7.2.8)]. We shall not give the details here. Of course, (7.2.14) cannot merely be truncated to retain only the first κ terms, as the result may not be a Hankel matrix. However, a simple thing to do is to omit all states of $x(t|t-1)$ except the first κ (i.e., to omit all except the first κ rows of F, K and all save the first κ columns of F, H). Then we retain only the first κ rows of $\mathscr{C}_\infty$, $\mathcal{O}_\infty'$ and achieve a system of McMillan degree κ. Let $\mathscr{H}_{(\kappa)}$ be the corresponding Hankel matrix. Then Glover (1984) shows that $\|k - k_k\|_\infty \leq 2(\sigma_{k+1} + \cdots + \sigma_n)$, where $k_k(z)$ is the transfer function corresponding to the truncated system. If this technique is to be used, one needs to know $k(z)$. This will not be so in practice but one procedure would be to commence from, say, an autoregressive estimate of the transfer function and then proceed to reduce the system order by this balanced truncation method. We go on to discuss autoregressive approximation in Section 7.4, after first discussing approximation criteria.

3. APPROXIMATION CRITERIA

The classical procedure for choosing a model is the method of maximum likelihood, ML. When the dimension of the parameter vector varies, this is in general inadequate, as can be seen from Chapter 5, since ML will usually uncritically choose the largest dimension allowed. Classically, this problem is handled via significance tests, but they have a vague meaning when the truth belongs to no model set considered.

The first person to reconsider this problem recently appears to be Akaike (1969). Shimizu (1978), Ogata (1980), Hosoya (1984), and Bhansali (1986) have discussed Akaike's ideas, and Shibata (1980, 1981) has established properties relating to his methods. Here we follow Findley (1985). Put $L_T(\theta)$ for the log likelihood, where θ indexes the model. Put $E_T(\theta) = E\{L_T(\theta)\}$, the expectation being with respect to the true probability structure, θ_0, assuming such to exist. Then a theorem, essentially due to Shannon, asserts that $E_T(\theta)$ is maximized at θ_0. If $\hat{\theta}_1$, $\hat{\theta}_2$ are two ML estimates for two model sets, possibly of different dimensions, it is suggested that $\hat{\theta}_1$ be preferred if $E_T(\hat{\theta}_1) - E_T(\hat{\theta}_2) > 0$.

However, the left side is unknown. What Ogata (1980) and Findley (1985) set out to establish is that

$$L_T(\hat{\theta}_1) - L_T(\hat{\theta}_2) - \dim \hat{\theta}_1 + \dim \hat{\theta}_2$$

is asymptotically unbiased for $E_T(\hat{\theta}_1) - E_T(\hat{\theta}_2)$. The basic theorem in Findley (1985) assumes that $\hat{\theta}_j$, $j = 1, 2$, converges as $T \to \infty$ to a point $\theta_{j,\infty}$ that lies in the interior of the set over which optimization was effected to get $\hat{\theta}_j$. The point $\theta_{j,\infty}$ need not be the "true" point. If the model sets are Gaussian ARMA models, of different McMillan degrees, n_1, n_2, then from Section 5.4, taking the scalar case, for example,

$$-\frac{2}{T}\{L_T(\hat{\theta}_1) - L_T(\hat{\theta}_2) - n_1 + n_2\} = \log \hat{\sigma}_{n_1}^2 - \log \hat{\sigma}_{n_2}^2 - \frac{2n_1}{T} + \frac{2n_2}{T}$$
$$- T \log \det P_T(\hat{\theta}_1) + T^{-1} \log \det P_T(\hat{\theta}_2).$$

In a wide range of cases the last two terms provide a contribution that is $o(T^{-1})$, so that these results lead to the use of AIC.

The argument is not fully convincing, but neither can be any argument, perhaps, relating to any criterion. There is the jump from $E_T(\theta_0) > E_T(\theta)$ to $\hat{\theta}_1$ being preferred if $E_T(\hat{\theta}_1) - E_T(\hat{\theta}_2) > 0$, and then the use of an asymptotically unbiased estimator of the left side. Of course, the same objection might be raised to ML. For ML there is a wealth of theory showing the method to have desirable properties in certain circumstances. Shibata (1980, 1981) has provided some support of the same kind for *AIC*. He introduces, for an autoregressive model set being fitted to data generated by a stationary process with innovation variance σ^2, the criterion

$$M_T(h) = h\sigma^2/T + \frac{\sigma_h^2 - \sigma^2}{\sigma^2}. \qquad (7.3.1)$$

Here σ_h^2 is the minimal prediction error from h previous values. This criterion is arrived at in a slightly suspect way. An autoregression of order h is first fitted to the data. Then that is used to predict, one step forward, *an independent realization* of the same process. The expectation of the squared prediction error is then evaluated along a sequence of orders $h_T \to \infty$, assuming that

$$\sum_0^\infty \phi(j)y(t-j) = \varepsilon(t), \qquad \sum |\phi_j| < \infty, \qquad \phi(0) = 1$$

$$\phi(j) \neq 0 \quad \text{for infinitely many } j$$

$$\phi(z) = \sum \phi(j)z^j \neq 0, \qquad |z| \leq 1, \qquad h \leq H_T = o(T^{1/2}).$$

This evaluation leads to (7.3.1). Such questions will be considered further in Section 4. If h_T^* minimizes (7.3.1) for each T, then Shibata (1980) shows, *essentially*, that $\hat{h}_T/h_T^* \to 1$, in probability, where $\hat{h}_T$ minimizes $AIC(h)$. Again see Section 4.

The criterion (7.3.1) is, in this context, related to the way in which AIC was originally introduced in Akaike (1969). Nevertheless, the result indicates a virtue in AIC. This is reinforced by a result in Shibata (1981). The fitted autoregression

$$\sum_0^h \hat{\phi}_h(j)y(t-j) = \hat{\varepsilon}_h(t), \qquad \hat{\phi}_h(0) = 1$$

might be used to estimate the true spectrum $f(\omega)$ by

$$\hat{f}_h(\omega) = (\hat{\sigma}_h^2/2\pi)\left|\sum \hat{\phi}_h(j)e^{ij\omega}\right|^{-2}.$$

Then a measure of the virtue of this estimate might be

$$\hat{E}(h) = \frac{1}{2\pi}\int_{-\pi}^{\pi}\left|\frac{\hat{f}_h(\omega) - f(\omega)}{f(\omega)}\right|^2 d\omega. \tag{7.3.2}$$

Shibata shows that, under the same conditions, for any $\varepsilon > 0$,

$$\lim_{T\to\infty} P\left\{\hat{E}(h_T)/M_T(h_T^*) \geq \frac{2}{\sigma^2} - \varepsilon\right\} = 1$$

and that the lower bound $2/\sigma^2$ is attained by $\hat{h}_T$.

This result is impressive since (7.3.2) is an appealing figure of merit and autoregressive spectral estimation is important. Consider, however, the situation studied in Franke, Gasser, and Steinberg (1985). They used autoregressive estimates of spectra to classify EEG records choosing the orders by AIC and compared this with an expert's classification. It was found that BIC worked better. The reason can be seen from (7.3.2), for that uses *relative* squared error. The EEG spectra are usefully classified on the basis of a narrow range of frequencies, others appearing almost irrelevant. AIC is evidently trying to obtain a good relative fit even where $f(\omega)$ is very small and thus is taking $\hat{h}_T$ too high for the purpose. This is introducing variation between individuals that is making classification difficult.

Consider now the important ideas of Rissanen (1978, 1983, 1986). Put

$$y^t = (y(1)', y(2)', \ldots, y(t)')'$$

for the data to time t. (There might also be inputs observed, but that does not affect the discussion below, so we keep to the simpler situation notationally.) It is assumed that $y(t)$ is recorded to a certain fixed accuracy (e.g., four decimal digits). Let $P(y^t|\theta)$ be a model for the (discrete) probability density for y^t. When this is the true density, $-E\{\log_2 P(y^t|\theta)\}$ is an almost achievable lower bound to the average code length for y^t, in bits. This is essentially the same theorem as that which says that $E_T(\theta) \leq E_T(\theta_0)$. Thus, given the model θ, we might reasonably encode y^t so that $-\log_2 P(y^t|\theta)$ bits (or the smallest integer greater than that) are used. [This argument is again of the same nature as that which says "Prefer $\hat{\theta}_1$ to $\hat{\theta}_2$ when $E_T(\hat{\theta}_1) > E_T(\hat{\theta}_2)$."] However, to decode we must also transmit θ to the decoder. There is little point in recording θ more accurately than it can be known from the data. If $cT^{-1/2}$ is the accuracy for θ_j, then $\log_2 c + \frac{1}{2}\log_2 T$ bits, approximately, are needed, so $\frac{1}{2}d\log_2 T + O(1)$ bits for θ, where d is the dimension of θ. Thus

$$-\log_2 P(y^t|\theta) + \tfrac{1}{2}d\log_2 T + O(1) \tag{7.3.3}$$

bits are needed in all. We refer the reader to Rissanen (1983) for a more detailed discussion. Rissanen (1986) shows essentially that for any reasonable code length $l(y^t)$, θ being the true model, for every $\varepsilon > 0$,

$$E\{l(y^t)\} \geq -\log_2 P(y^t|\theta) + d(\tfrac{1}{2} - \varepsilon)\log_2 T$$

for all but finitely many T. [This is an extension of Shannon's theorem that $E_T(\theta)$ is maximized at θ_0.] Thus (7.3.3) does lead to a minimal encoding.

Proceeding in much the same way as we did when we showed that for the Gaussian case $E_T(\theta)$ leads to AIC, we now see that the use of the first two terms in (7.3.3) leads to BIC. Indeed, multiplying by $\log_e 2$ so as to revert to natural logarithms and taking the Gaussian case then, after multiplying the first two terms in (7.3.3) by $2/T$ and using the results in Chapter 4, we arrive at

$$\log\det\Sigma_\theta + \{d(\theta)\log T\}/T,$$

after neglecting lower-order terms. [We have also interpreted $P(y^t|\theta)$ as the probability density of a continuously varying random variable.] Minimizing with respect to θ for fixed $d(\theta)$ we arrive at BIC, that is,

$$\log\det\hat{\Sigma}_d + d\log T/T.$$

Rissanen speaks of this as the minimum description length (MDL) principle. It again has been justified by the theorems of Chapter 5, which show that BIC

has virtues, although these relate to the case where there is a true order (as with ML). Rissanen's MDL principle is an appealing one. It rests on somewhat the same foundations as does the idea underlying *AIC*.

Rissanen (1986) introduces a predictive minimum description length (PDML) principle that also seems important. We may write

(7.3.4)

$$-\log P(y^t|\theta) = -\sum_{s=0}^{t-1} \log f_{\theta,d}(y(s+1)|y^s), \quad y^s = (y(1)', y(2)', \ldots, y(s)')',$$

where $f_{\theta,d}(y(s+1)|y^s)$ is the conditional probability density for $y(s+1)$ given y^s. Now d is made explicit, although it is also inherent in θ. For fixed d put $\hat{\theta}(t)$ for the value of θ minimizing (7.3.4) (i.e., for the ML estimator). Thus Rissanen (1986) suggests that $\hat{d}(t)$ be chosen to minimize

$$-\sum_{0}^{t-1} \log f_{\hat{\theta}(s),d}(y(s+1)|y^s). \tag{7.3.5}$$

This is not the same as choosing θ, d simultaneously to minimize (7.3.4) because (7.3.5) depends on all $\hat{\theta}(s)$, not only $\hat{\theta}(t)$. Now no transmission of parameter values is needed since at time $t+1$ the decoder will have decoded $y(s)$, $s \leq t$ and hence, using the same algorithm as the encoder, have determined $\hat{d}(t)$, $\hat{\theta}(t)$ and hence be able to decode $y(t+1)$.

To illustrate, consider a scalar autoregressive model set

$$\sum_{0}^{h} \phi_h(j)y(t-j) = \varepsilon_h(t), \qquad \phi_y(0) = 1, \qquad E\{\varepsilon_h(s)\varepsilon_h(t)\} = \delta_{st}\sigma_h^2. \tag{7.3.6}$$

We now use h in place of d to agree with earlier usage. Of course, θ also includes σ^2, so that $d = h + 1$. Then, ignoring a constant,

$$-\log f_{\theta,d}(y(s+1)|y^s) = \tfrac{1}{2}\log\sigma_h^2 + \tfrac{1}{2}\varepsilon_h(s+1)^2/\sigma_h^2.$$

Of course $\varepsilon_h(s+1)$ depends on θ through its definition, (7.3.6). Call $e_h(s+1)$ the value of $\varepsilon_h(s+1)$ when the $\phi_h(j)$ (and σ^2) are estimated by some least-squares procedure, such as the Toeplitz procedure of Section 5.6. The value of $e_h(s+1)$ will depend on the actual procedure, but the notation will not be encumbered with that detail. Then, ignoring the factor $\frac{1}{2}$, (7.3.5) becomes, treating $y(t)$ as zero before $t = 0$,

$$\sum_{0}^{t-1} \left\{\log\hat{\sigma}_h^2(s) + \frac{e_h(s+1)^2}{\hat{\sigma}_h^2(s)}\right\}, \tag{7.3.7}$$

where $\hat{\sigma}_h^2(s)$ is the least-squares estimator of σ^2. *Of course, (7.3.7) is not restricted to the AR case but holds for any ARMAX system with appropriate initial conditions.* It is illuminating to consider the case where $y(t)$ is generated by an autoregression of order h_0. If $h \geq h_0$, then, from (6.6.33), we have

$$\hat{\sigma}_h^2(t) = \dot{\sigma}^2(t) + \frac{h}{t}c_t, \qquad \dot{\sigma}^2(t) = \frac{1}{t}\sum_1^t \varepsilon(s)^2, \qquad |c_t| = O(Q_t^2).$$

Then (7.3.7) is, dividing by t,

$$\frac{1}{t}\sum_0^{t-1}\left[\log\dot{\sigma}^2(s) + \frac{hc_s}{\sigma^2 s}\{1+o(1)\} + \frac{1}{\sigma^2}e_h(s+1)^2\left\{1 + \frac{hc_s}{s\sigma^2}(1+o(1))\right\}\right]$$

$$= \frac{1}{t}\sum_0^{t-1}\dot{\sigma}^2(s) - \frac{1}{\sigma^2 t}\sum_1^t\{e_h(s)^2 - \sigma^2\}\frac{c_s}{s}\{1+o(1)\} + \frac{1}{\sigma^2 t}\sum_1^t e_h(s)^2 + o(t^{-1}).$$

The first term is independent of h. The second can be shown to be negligible compared to the third. [This is fairly evident, both because of the subtraction of σ^2 and because of the factor c_s/s, which is $O(\log\log s/s^2)$.] Thus it is the second last term that is of consequence. It can be shown that, for $h \geq h_0$,

$$\text{(7.3.8)} \qquad \frac{1}{t}\sum_1^t e_h(s)^2 = \dot{\sigma}_{(t)}^2 + \frac{\sigma^2 h\log t}{t}\{1+o(1)\} + o(t^{-1}), \text{ a.s.}$$

under (4.1.15), (4.1.16), (4.1.20). This expression involves the recursive residuals [i.e., $e_h(s)^2$ involves only the estimate of $\phi_h(j)$ using data to time $s-1$] as compared to $\hat{\sigma}_h^2(t)$, which is the mean square of the off-line residuals, each computed using estimates of $\phi_h(j)$ from data to time t. Of course, $\hat{\sigma}_h^2(t)$ decreases monotonically with h for t fixed but, somewhat remarkably, (7.3.8), to order $\log t/t$, increases with h, for $h \geq h_0$. It is also somewhat remarkable that again $\log t/t$ emerges, as in MDL. The derivation of (7.3.8) is far from trivial but will not be given here. It will appear in forthcoming work by Hannan, McDougall, and Poskitt.

It is possible to compute (7.3.7) in real time (and also the MDL criterion, for that matter). However, (7.3.8) can be computed with fewer calculations at each time point, and since it is the essential component in (7.3.7), might itself be used to choose h at each time point. In practice the past will have to be forgotten or weighted down. This introduces essential changes. Such forgetting will be discussed in Section 7.5.

It seems impossible to assert that any particular principle is supreme. Both the principle based on $E_T(\theta)$ and the various minimum description length principles have appeal. The notion that statistics is concerned with the "reduction of data" is an old one (Fisher, 1944, p. 1) and has great force. What is needed is to investigate the properties of such procedures, particularly when the model set does not contain the truth. This we go on to discuss in the next section.

4. STATISTICAL PROPERTIES OF APPROXIMATION METHODS

In this section first some essential properties of autocovariances are discussed under general circumstances. In the fourth subsection the a.s. behavior of $\hat{k}$, $\hat{l}$, for given McMillan degree, is investigated when the data are generated by a more general process. A detailed investigation of autoregressive approximation is presented in the second and third subsections.

(i) Autocovariances

The theorem referred to in (5.3.2) is first established. Recall the condition (1.4.5) defining a purely nondeterministic process. Such a process is evidently also linearly purely nondeterminstic (linearly regular) and thus has an absolutely continuous spectrum (see Section 1.3). Thus $\Gamma(t)$ converges to zero as $t \to \infty$ for purely nondeterministic $y(t)$. For $t \geq T$ we put $G(t) = 0$. [See (4.1.7) for the definition of $G(t)$. We write $\Gamma(t)$, $G(t)$ in place of $\Gamma_y(t)$, $G_y(t)$, for simplicity.]

Theorem 7.4.1. *Let $y(t)$ be generated by a vector stationary process with finite variance and be ergodic and purely nondeterministic. Then*

$$\lim_{T \to \infty} \sup_{0 \leq t < \infty} \|G(t) - \Gamma(t)\| = 0, \quad a.s.$$

As in Section 1.4, we have, with $\mathscr{F}_t$ the σ-algebra of events to time t,

$$y(t) = \sum_0^\infty u(t, t-j), \qquad u(t, t-j) = E\{y(t)|\mathscr{F}_{t-j}\} - E\{y(t)|\mathscr{F}_{t-j-1}\}.$$

Since the proof is only notationally more complex for the vector case, we take $s = 1$. Then we write $g(t)$, $\gamma(t)$ for $G(t)$, $\Gamma(t)$. Put

$$w(t) = \sum_{j=0}^{r} u(t, t-j), \qquad e(t) = y(t) - w(t).$$

Let $\tilde{g}(t)$, $\tilde{\gamma}(t)$ correspond to $w(t)$. Then

$$|\tilde{g}(t) - g(t)| \leq 2\left[\left\{T^{-1} \sum_1^T w(s)^2\right\}\left\{T^{-1} \sum_1^T e(s)^2\right\}\right]^{1/2} + T^{-1} \sum_1^T e(s)^2$$

and uniformly in t the right side converges a.s. to a quantity that may be made arbitrarily small by taking r large, by ergodicity, since $e(t)$ has a variance that may be made arbitrarily small by increasing r. The same is true of $|\tilde{\gamma}(t) - \gamma(t)|$. Thus it is sufficient to prove the theorem for $\tilde{g}(t)$, $\tilde{\gamma}(t)$ for fixed r.

$$\tilde{g}(t) - \tilde{\gamma}(t) = \sum_{j,k=0}^{r}\sum \left\{T^{-1} \sum_{s=1}^{T-t} u(s, s-j)u(s+t, s+t-k)\right\} - \tilde{\gamma}(t), \qquad 0 \leq t < T. \tag{7.4.1}$$

Evidently only these values of t need to be considered since $\tilde{\gamma}(t) \to 0$, $\tilde{g}(t) = 0$, $t \geq T$. There are $(r + 1)$ values of t, only, for which $t - k$ is not positive, $k = 0, 1, \ldots, r$. Omit these since certainly $\tilde{g}(t) - \tilde{\gamma}(t)$ converges, a.s., to zero for each fixed t. Equally, a single pair of j, k values, only, need be considered, for $t > k$. Thus consider

$$T^{-1} \sum_{s=1}^{T-t} u(s, s-j)u(s+t, s+t-k), \qquad t = k+1, \ldots, T-1. \tag{7.4.2}$$

These have zero expectation since $E\{u(s, s-j)u(s+t, s+t-k)|\mathscr{F}_{s+t-k-1}\} = u(s, s-j)\, E\{u(s+t, s+t-k)|\mathscr{F}_{s+t-k-1}\} = 0$. Next put

$$u'(t, t-j) = \begin{cases} u(t, t-j), & |u(t, t-j)| \leq A \\ 0, & |u(t, t-j)| > A, \end{cases}$$

and $u''(t, t-j) = u(t, t-j) - u'(t, t-j)$. Then (7.4.2) computed for $u(t)$ and $u'(t)$ differs by a quantity bounded in modulus by

$$\left[\left\{T^{-1} \sum_{1}^{T} u'(s, s-j)^2\right\}\left\{T^{-1} \sum_{1}^{T} u''(s, s-k)^2\right\}\right]^{1/2}$$
$$+ \left[\left\{T^{-1} \sum_{1}^{T} u''(s, s-j)^2\right\}\left\{T^{-1} \sum_{1}^{T} u'(s, s-k)^2\right\}\right]^{1/2}$$
$$+ [\{T^{-1} \sum_{1}^{T} u''(s, s-j)^2\}\{T^{-1} \sum_{1}^{T} u''(s, s-k)^2\}]^{1/2},$$

which converges a.s. to a quantity that may be made arbitrarily small by taking A large. Thus $u(t, t-j)$ may be replaced by $u'(t, t-j)$, and similarly for $u(s+t, s+t-k)$. Put

$$\bar{u}(t, t-j) = u'(t, t-j) - E\{u'(t, t-j)|\mathscr{F}_{t-j-1}\}.$$

Now

$$\left|\frac{1}{T}\sum_{1}^{T-t}[E\{u'(s,s-j)|\mathscr{F}_{s-j-1}\}E\{u'(s+t,s+t-k)|\mathscr{F}_{s+t-k-1}\}]\right|$$
$$=\left|\frac{1}{T}\sum_{1}^{T-t}[E\{u''(s,s-j)|\mathscr{F}_{s-j-1}\}E\{u''(s+t,s+t-k)|\mathscr{F}_{s+t-k-1}\}]\right|$$

since $E\{u(s,s-j)|\mathscr{F}_{s-j-1}\}=0$. This is bounded e.g. for $j=k$ by

$$\frac{1}{T}\sum_{1}^{T}E\{u''(s,s-j)|\mathscr{F}_{s-j-1}\}^2\leq\frac{1}{T}\sum_{1}^{T}E\{u''(s,s-j)^2|\mathscr{F}_{s-j-1}\},$$

which converges a.s. to $E\{u''(0-j)^2\}$, which may be made arbitrarily small by taking A large. In the same way,

$$\frac{1}{T}\sum_{1}^{T-t}u'(s,s-j)\mathrm{E}\{u'(s+t,s+t-k)|\mathscr{F}_{s+t-k-1}\},$$
$$\frac{1}{T}\sum_{1}^{T-t}E\{u'(s,s-j)|\mathscr{F}_{s-j-1}\}u'(s+t,s+t-k)$$

may be shown to be negligible. Thus $u'(t,t-j)$ may be replaced by $\bar{u}(t,t-j)$. The quantities $\bar{u}(s,s-j)\bar{u}(s+t,s+t-k)$, $t=k+1,\ldots,T-1$ are for each t martingale differences with respect to $\mathscr{F}_{s+t-k}$. Now Burkholder's inequality (Hall and Heyde, 1980, p. 23) shows that [recalling that $\bar{u}(s,s-j)$ is uniformly bounded]

$$E\left[\left\{\sum_{1}^{T-t}\bar{u}(s,s-j)\bar{u}(s+t,s+t-k)\right\}^6\right]$$
$$\leq cE\left[\left\{\sum_{1}^{T-t}\bar{u}(s,s-j)^2\bar{u}(s+t,s+t-k)^2\right\}^3\right]$$
$$\leq cT^3,$$

so that

$$P\left\{\sup_{r+1\leq t<T}\left|\frac{1}{T}\sum_{1}^{T-t}\bar{u}(s,s-j)\bar{u}(s+t,s+t-k)\right|>\varepsilon\right\}$$
$$\leq\sum_{t=r+1}^{T-t}P\left\{\left|\frac{1}{T}\sum_{1}^{T-t}\bar{u}(s,s-j)\bar{u}(s+t,s+t-k)\right|>\varepsilon\right\}$$
$$\leq\sum_{t}c\varepsilon^{-6}T^{-6}T^3\leq cT^{-2},$$

for each $\varepsilon > 0$. Thus by the Borel-Cantelli lemma,

$$\lim_{T\to\infty} \sup_{r<t<T} \left| \frac{1}{T} \sum_{1}^{T-t} \bar{u}(s, s-j)\bar{u}(s+t, s+t-k) \right| = 0, \quad \text{a.s.}$$

and Theorem 7.4.1 is proved.

We now wish to prove a stronger result, under stricter conditions, analogous to the result of Section 5.3, but now not for ARMA processes. These stronger results are needed in connection with the theorems concerning autoregressive approximation in the third part of this section. We restrict attention to the case where are no inputs; that is, we consider

$$(7.4.3) \quad y(t) = \sum_{0}^{\infty} K(j)\varepsilon(t-j), \qquad E\{\varepsilon(s)\varepsilon(t)'\} = \delta_{st}\Sigma, \qquad K(0) = I_s.$$

Stronger conditions will be imposed on the $\varepsilon(t)$, including at least (4.1.15), but some conditions are needed also on the $K(j)$, namely,

$$(7.4.4) \qquad \sum_{1}^{\infty} j^{1/2} \|K(j)\| < \infty.$$

In fact, more relevant than (7.4.3) is the autoregressive representation

$$(7.4.5) \qquad \sum_{0}^{\infty} \Phi(j)u(t-j) = \varepsilon(t), \qquad \Phi(0) = I_s.$$

Put

$$\Phi(z) = \sum_{0}^{\infty} \Phi(j)z^j = k(z)^{-1}$$

and require that

$$(7.4.6) \qquad \det \Phi(z) \neq 0, \qquad |z| \leq 1,$$

$$(7.4.7) \qquad \sum_{1}^{\infty} j^{1/2} \|\Phi(j)\| < \infty.$$

Of course, (7.4.6) is necessary in order that $\varepsilon(t)$ be the innovation sequence for $y(t)$. The condition (7.4.4), combined with $\det k(z) \neq 0$, $|z| \leq 1$, implies (7.4.6), (7.4.7), and conversely. We state this well-known result as a theorem and give a brief proof, generalizing the result to cover other cases.

Theorem 7.4.2. *If* $\det k(z) \neq 0, |z| \leq 1$ *and*

$$\sum_1^\infty j^\lambda \|K(j)\| < \infty, \qquad \lambda > 0,$$

then $\Phi(z) = k(z)^{-1} = \sum_0^\infty \Phi(j) z^j$ *also satisfies*

$$\sum_1^\infty j^\lambda \|\Phi(j)\| < \infty.$$

Proof. If $\sum j^\lambda |a_j|, \sum j^\lambda |b_j|$ converge, $j = 1, 2, \ldots$, then $\sum j^\lambda |(a * b)_j|$ also converges, where $(a * b)_j = \sum_{k=0}^\infty a_k b_{j-k}$. Thus putting $k(z)^{-1} = \{\det k(z)\}^{-1} \operatorname{adj} k(z)$, we see that it is sufficient to prove the result for $\det k$ (i.e., for the scalar case). Consider the sequences $a = \{a_j, j = 0, 1, 2, \ldots\}$ with convolution as multiplication. They form an algebra over the field of complex numbers, and if we put $\|a\| = \sum (2^\lambda + j^\lambda)|a_j|$, this is a commutative Banach algebra (see Rudin, 1973) with this norm. The essential thing to check is that $\|a * b\| \leq \|a\| \, \|b\|$ and this follows, putting $v(j) = 2^\lambda + j^\lambda$, from $v(j + k) \leq v(j)v(k)$. Put $x = (0, 1, 0, 0, \ldots)$. Write x^2 for $x * x$, and so on, $x^0 = (1, 0, 0, \ldots)$. Then if $f(\cdot)$ is any homeomorphism of the algebra into the complex numbers, $|f(x^n)| = |f(x)|^n$ and $|f(x^n)| \leq \|x^n\| = (2^\lambda + n^\lambda)$, Rudin (1973, p. 231). Thus $|f(x)| \leq 1$. Now the theorem follows in the same way as for Wiener's lemma (Rudin, 1973, p. 266).

Next a stronger result than Theorem 7.4.1 will be proved under stronger conditions. Recall the definition of $\dot{\Gamma}(t)$ in Section 5.3, above (5.3.7).

Theorem 7.4.3. *Let* $\varepsilon(t)$ *satisfy (4.1.15), (4.1.17) and have finite fourth moments. Let* $y(t)$ *be generated by (7.4.3) with (7.4.4) holding and*

$$\sup_j j \|K(j)\| < \infty. \tag{7.4.8}$$

Then, putting $G(t) = 0, |t| \geq T$,

$$\sup_{0 \leq t < \infty} \|G(t) - \dot{\Gamma}(t)\| = 0\{(\log T/T)^{1/2}\}.$$

If (4.1.16) holds, then $\dot{\Gamma}(t)$ *may be replaced by* $\Gamma(t)$ *in this statement.*

REMARK. Since

$$\dot{\Gamma}(t) - \Gamma(t) = \sum_{j=0}^\infty K(j)\{\dot{\Sigma}_T - \Sigma\}K(j + t)', \qquad \dot{\Sigma}_T = \frac{1}{T}\sum_1^T \varepsilon(s)\varepsilon(s)',$$

then $G(t) - \Gamma(t)$ is, uniformly in t, $O\{(\log T/T)^{1/2}\} + O(\|\dot{\Sigma}_T - \Sigma\|)$. If the vector stationary process composed of the $\varepsilon_j(t)\varepsilon_k(t), j, k > 1, \ldots, s$ has an absolutely continuous spectrum with uniformly bounded spectral density, then $\|\dot{\Sigma}_T - \Sigma\| = o\{(\log T)^a/T^{1/2}\}$, for all $a > \frac{3}{2}$ (see Hannan, 1978, Theorem 1). Under slightly stronger, but quite realistic conditions, but without (4.1.16), $\|\dot{\Sigma}_T - \Sigma\| = o\{(\log T)^\delta/T^{1/2}\}$, $\delta > \frac{1}{2}$. In particular, if $\mathscr{F}_{-\infty}$ is trivial, that is, contains only sets of measure 0 or 1 [in which case $y(t)$ is said to be regular, Ibragimov and Linnik, 1971, p. 301, and Section 1.4], the following condition is sufficient for $\|\dot{\Sigma}_T - \Sigma\| = o\{(\log T)^\delta/T^{1/2}\}$, $\delta > \frac{1}{2}$.

(7.4.9)

$$\sum_{j=1}^{\infty} E\{|E\{\varepsilon_a(t)\varepsilon_b(t)|\mathscr{F}_{t-j}\} - E\{\varepsilon_a(t)\varepsilon_b(t)|\mathscr{F}_{t-j-1}\}|^2\}^{1/2} < \infty, \qquad a, b = 1, \ldots, s$$

(see Hannan, 1978, Theorem 2). This seems a relatively mild condition.

Proof. Again the proof is essentially the same for the vector and the scalar case, so we take $s = 1$. Then, using lowercase letters for this scalar case, as usual,

$$\begin{aligned} g(t) - \gamma(t) &= \sum\sum \kappa(j)\kappa(k)\left\{\frac{1}{T}\sum_1^{T-t} \varepsilon(s-j)\varepsilon(s+t-k) - \sigma^2\delta_{j,k-t}\right\} \\ &= \sum_{k-j\neq t}\sum \kappa(j)\kappa(k)\left\{\frac{1}{T}\sum_1^{T-t} \varepsilon(s-j)\varepsilon(s+t-k) - \sigma^2\delta_{j,k-t}\right\} \\ &\quad + \sum_j \kappa(j)\kappa(j+t)\left\{\frac{1}{T}\sum_1^{T-t} \varepsilon(s-j)^2 - \sigma^2\right\} \qquad (7.4.10) \\ &= \sum_{k-j\neq t}\sum \kappa(j)\kappa(k)\left\{\frac{1}{T}\sum_1^{T-t} \varepsilon(s-j)\varepsilon(s+t-k)\right\} + \dot{\gamma}(t) - \gamma(t) \\ &\quad + o(T^{-1/2}), \end{aligned}$$

uniformly in t, exactly as in the proof of Theorem 5.3.2. [There only $\sum j\kappa(j)^2 < \infty$ was required, but $\sum j^{1/2}|\kappa(j)| < \infty$ is needed here, for later purposes.] Thus it is necessary to consider only the first term in (7.4.10). Put $T^{-1}S_{j,k,t}(T) = T^{-1}\sum_1^{T-t}\{\varepsilon(s-j)\varepsilon(s+t-k)\}$. Exactly as in Section 5.3, since $t\gamma(t)/T \to 0$, this may replace the bracketed expression in the first term in (7.4.10). Now

$$x_t(T) = T^{-1}\sum_0^{\infty}\sum \kappa(j)\kappa(k)S_{j,k,t}(T) - T^{-1}\sum_0^{T^a}\sum \kappa(j)\kappa(k)S_{j,k,t}(T)$$

is made up of three terms, of which one is

$$T^{-1} \sum_{T^a+1}^{\infty} \sum_{0}^{\infty} \kappa(j)\kappa(k)S_{j,k,t}(T),$$

whose mean square is bounded by

$$cT^{-1}\left\{ \sum_{T^a+1}^{\infty} |\kappa(j)| \sum_{0}^{\infty} |\kappa(j)| \right\}^2$$

since $E\{S_{j,k,t}^2\} \leq cT$. This last expression is $O(T^{-(1+a)})$ because of (7.4.4), uniformly in j, k, t. The same is true of the other two terms. Since, because of (7.4.4), $\gamma(t) = o(T^{-1/2})$, $t \geq T$ and $g(t) = 0$, $t \geq T$, we need consider only $t \leq T$ and thus

$$\sum_T P\left\{ \max_{0 \leq t \leq T} |x_t(T)| \geq cT^{-1/2} \right\} \leq c \sum_T T^2 T^{-(1+a)}.$$

Here T^2 comes from the fact that the maximum is taken over T values of t combined with the bound $T^{-1/2}$, which gives a factor T via Chebyshev's inequality. This series converges for $a > 2$, so that, uniformly in t, we need consider only

$$T^{-1} \sum\sum \kappa(j)\kappa(k)S_{j,k,t}(T), \qquad 0 \leq t < T,$$

where the double sum is over $0 \leq j, k \leq T^a$, $j + t \neq k$. Because of (7.4.4) we may consider

$$f_{j,k}^{-1} T^{-1} S_{j,k,t}(T), \qquad f_{j,k} = \{(1+j)(1+k)\}^{1/2}$$

for j, k, t running over the same values. If we show that uniformly in j, k, t this is $O\{(\log T/T)^{1/2}\}$, we shall have shown what is required.

The remainder of the proof is much as that for Theorem 5.3.1, but is simpler since no subsequences [i.e., the T_k below (5.3.8)] need to be considered. First the $\varepsilon(t)$ are truncated at $\lambda(t)^{1/2}$, $\lambda(t) = (t/\log t)^{1/2}$. Thus $d(t) = \varepsilon(t)I\{|\varepsilon(t)| \leq \lambda(|t|)^{1/2}\}$, $e(t) = \varepsilon(t) - d(t)$. Then exactly as in the proof just above the definition of $s_t^2(T)$ [see (5.3.12)] in Section 5.3, we see that

$$\sum_{1}^{\infty} E\{e(t)^2\}/\phi(t) < \infty, \qquad \phi(t) = (t \log t)^{1/2}. \tag{7.4.11}$$

Then we consider the martingales

$$\sum_{j,k,t} [d(s)d(s - |j - k + t|) - E\{d(s)d(s - |j - k + t|)|\mathscr{F}_{s-1}\}], \tag{7.4.12}$$

where the sum is over $t - k + 1 \leq s \leq T - k$ for $t + j - k > 0$ and over $j + 1 \leq s \leq T + j - t$, $t + j - k < 0$. As in Section 5.3, using (7.4.11) it is shown that the truncation and the subtraction of the second term in the summand in (7.4.12) produces a negligible effect, uniformly in j, k, t. Put

$$s_{j,k,t}^2(T) = \sum_{j,k,t} (E\{d^2(s)d^2(s - |j - k + t|)|\mathscr{F}_{s-1}\} - [E\{d(s)d(s - |j - k + t|)|\mathscr{F}_{s-1}\}]^2). \tag{7.4.13}$$

Consider [see (7.4.10)]

$$P\left\{\max_{j,k,t} S_{j,k,t}(T) > b\phi(T)f_{jk}\right\} \leq \sum_{j,k,t} P\{S_{j,k,t}(T) > b\phi(T)f_{jk}\}.$$

If this series (in T) can be shown to converge for some $0 < b < \infty$, and the same with $S_{j,k,t}(T)$ replaced by $-S_{j,k,t}(T)$, then what is required will have been shown. The proof is essentially the same for both $S_{j,k,t}(T)$ and $-S_{j,k,t}(T)$, so we do not mention this again. Since there are at most T^{2a+1} values of j, k, t, then if the sequence

$$P\{S_{j,k,t}(T) > b\phi(T)f_{j,k}\} \tag{7.4.14}$$

can be shown to be bounded by T^{-f}, $f > 2a + 2$, uniformly in j, k, t, the proof will be completed since then $T^{2a+1}T^{-f}$ converges. Let $\lambda_{j,k,t}(T)$ be the largest value of $\lambda(s)$ occurring for any $\varepsilon(s)$ arising in $S_{j,k,t}(T)$. Then consider

$$\begin{aligned} P\{\lambda_{j,k,t}(T)^{-1}S_{j,k,t}(T) > b'\phi(T)f_{jk}\lambda_{j,k,t}(T)^{-1}\} \\ = P\{\lambda_{j,k,t}(T)^{-1}S_{j,k,t}(T) - (3/4)(\lambda_{j,k,t})^{-2}s_{j,k,t}^2 \\ > b'\phi(T)f_{jk}\lambda_{j,k,t}(T)^{-1} - (3/4)(\lambda_{j,k,t})^{-2}s_{j,k,t}^2\}. \end{aligned} \tag{7.4.15}$$

Now $b'\phi(T)f_{jk}\lambda_{j,k,t}(T)^{-1} \geq b\log T$, that is, $f_{jk}\lambda_{j,k,t}(T)^{-1} \geq (b/b')T^{-1/2} \times (\log T)^{1/2}$. For example, for $j = k = 0$, $f_{jk} = 1$, $\lambda_{j,k,t} \leq T^{-1/2}(\log T)^{1/2}$. For $j = T^a$, $k = 0$, $t = 0$, then $f_{jk} = (1 + T^a)^{1/2}$, $\lambda_{j,k,t}(T) \leq T^{a/2}/(a/2 \log T)$. In any case b may be chosen arbitrarily large by taking b' large. Moreover, as in Section 5.3, the second term on the right in the argument of $P(\cdot)$ in (7.4.15) may be shown to be negligible compared to the first, uniformly in j, k, t. Now the Corollary 5.4.1 in Stout (1970, p. 299) may be used to bound (7.4.15) by $\exp(-b\log T) = T^{-b}$, and thus the theorem is proved.

We finally have, exactly as in Corollary 5.3.3, the following result.

Theorem 7.4.4. *Let $y(t)$ be as in Theorem 7.4.3 for $s = 1$. Then*

$$\sup_{0 \le t \le \infty} |r(t) - \rho(t)| = O\{(\log T/T)^{1/2}\}.$$

The proof is precisely as for Corollary 5.3.3.

(ii) Autoregressive Approximation

We now consider autoregressive approximation under the more general conditions (7.2.1), (7.4.4). [Of course, (7.4.4) implies the second half of (7.2.1).] Again it is assumed that $\det(k) \neq 0, |z| \le 1$, so that (7.4.7) also holds. For some purposes (4.1.16) will be needed, but for others only (4.1.15). We shall need (4.1.20) and also sometimes

$$E\{\varepsilon_j(t)^4 \log_+ |\varepsilon_j(t)|\} < \infty, \qquad j = 1, \ldots, s. \tag{7.4.16}$$

The basic model, including inputs, could be

$$\sum_0^\infty \Phi(j)y(t-j) = \sum_1^\infty \Psi(j)z(t-j) + \varepsilon(t), \qquad \Phi(0) = I_s.$$

However, a model would be needed for $z(t)$, and this is, naturally,

$$\sum_0^\infty \Lambda(j)z(t-j) = \sum_1^\infty M(j)h(t-j) + \eta(t), \qquad \Lambda(0) = I_m$$

$$E\{\eta(s)\eta(t)'\} = \delta_{st}\Omega.$$

However, unless the conditions on the coefficients $\Lambda(j)$, $M(j)$ are essentially different from those on $\Phi(j)$, $\Psi(j)$ (as was the case in Chapters 5 and 6, when the latter decreased geometrically) there is little point in discriminating between $y(t)$ and $z(t)$ and they may as well as be aggregated into one vector, which we again call $y(t)$, and which is generated by (7.4.3).

Of course, as usual, mean corrections would customarily need to be made, and also, perhaps, corrections for periodic (i.e., seasonal) effects, but all results below would continue to hold when these are made. Thus we shall act as if such corrections can be ignored.

What are to be considered are the properties of the estimates of the system structure, (7.4.7), obtained by fitting autoregressions, using

$$\log\det \hat{\Sigma}_h + hs^2 C_T/T, \qquad h \le H_T, \qquad C_T/T \downarrow 0, \qquad C_T > 1, \tag{7.4.17}$$

to determine the order, h, of the fitted autoregression. The same notation will be used as in Chapters 5 and 6, namely $\hat{\Phi}_h(j)$, $\Phi_h(j)$, $j = 0, 1, \ldots, h$, will represent the solutions to the Yule-Walker equations using, respectively, the sample autocovariances, $G(t)$, or the true autocovariances, $\Gamma(t)$. Again, as in Chapter 6, the notation $\dot{\Gamma}(t)$ will be used when Σ in the definition of $\Gamma(t)$ is replaced by $\dot{\Sigma} = T^{-1}\sum \varepsilon(t)\varepsilon(t)'$.

Theorem 7.4.5. *Let $y(t)$ be generated by (7.4.3) with (7.4.7), (4.1.15), (4.1.17), (4.1.20) holding. Then, for $h \le H_T = o\{(T/\log T)^{1/2}\}$, uniformly in h,*

$$\max_{1\le j\le h} \|\hat{\Phi}_h(j) - \Phi_h(j)\| = O\{(\log T/T)^{1/2}\} + o(1)\sum_{h+1}^{\infty} \|\Phi(j)\|. \tag{7.4.18}$$

If (4.1.16) holds, the term that is $o(1)$ in (7.4.18) is also $O\{(\log T/T)^{1/2}\}$, and for $s = 1$ this is true without (4.1.16) if $j\|K(j)\| \le c < \infty$.

Exactly the same remarks may be made in relation to this theorem as were made below Theorem 7.4.3, namely that the right side of (7.4.18) is $O\{(\log T/T)^{1/2}\} + o\{(\log T)^{\delta}T^{-1/2}\}\sum_{h+1}^{\infty} \|\Phi(j)\|$, $\delta > \frac{1}{2}$, under very general conditions, not including (4.1.16). If $h = O\{(\log T)^a\}$, $a > \frac{1}{2}$, then (7.4.7) ensures that (7.4.18) is $O\{(\log T/T)^{1/2}\}$, for example.

The next theorem relates to the properties of

$$\hat{\varepsilon}_h(t) = \sum_0^h \hat{\Phi}_h(j)y(t-j)$$

as estimates of the $\varepsilon(t)$. *If h_T is a sequence of h values, we shall write $\hat{\varepsilon}_T(t)$ in place of $\hat{\varepsilon}_{h_T}(t)$, for simplicity.*

Theorem 7.4.6. *Let $y(t)$ be generated by (7.4.3) with (7.4.7), (4.1.15), (4.1.16), (4.1.20) holding and $H_T = o\{(T/\log T)^{1/2}\}$. Then the following hold.*

(a) Uniformly in $h \le H_T$

$$\frac{1}{T}\sum_1^T \{\hat{\varepsilon}_h(t) - \varepsilon(t)\}\{\hat{\varepsilon}_h(t) - \varepsilon(t)\}' = (\Sigma_h - \Sigma)\{1 + o(1)\} + O(h\log T/T),$$

$$\frac{1}{T}\sum_1^T \varepsilon(t)\{\hat{\varepsilon}_h(t) - \varepsilon(t)\}' = O(h\log T/T).$$

(b) Let $h = h_T$ satisfy

$$\liminf_{T\to\infty} h_T/(\log\log T)^2 = \infty; \tag{7.4.19}$$

then if (7.4.16) also holds,

$$\frac{1}{T}\sum_1^T \{\hat{\varepsilon}_T(t)-\varepsilon(t)\}\{\hat{\varepsilon}_T(t)-\varepsilon(t)\}' = \left(\Sigma_{h_T}-\Sigma+\frac{sh_T}{T}\Sigma\right)\{1+o(1)\} + o_P(h_T/T), \tag{7.4.20}$$

$$\frac{1}{T}\sum_1^T \varepsilon(t)\{\hat{\varepsilon}_T(t)-\varepsilon(t)\}' = -\frac{sh_T}{T}\Sigma\{1+o(1)\}+o_P(h_T/T).$$

If (7.4.19) is strengthened to

$$\liminf_{T\to\infty} h_T/(\log T)^2 > 0, \tag{7.4.21}$$

then the term that is $o_P(h_T/T)$ in (7.4.20) may be replaced by $o(h_T/T)$.

As is also known from the results in Chapter 5, even for an ARMA process the estimates $\hat{h}_T$ got by *BIC* increases as $\log T$. Since $\hat{h}_T$ increases as C_T decreases, then, for $C_T \le c\log T$, $\hat{h}_T$ can be expected to increase at least as $\log T$, so that the first part of (b) of the theorem relates to $h_T = \hat{h}_T$. This improves on part (a) by decreasing the error from $O(h_T \log T/T)$ to $O_P(h_T/T)$. In fact, as will be seen below, it can be expected that $\hat{h}_T$ will increase as some power of T under a fair range of circumstances, so that (7.4.21) will hold for $h_T = \hat{h}_T$ and the error is $O(h_T/T)$.

Theorem 7.4.7. *Let the conditions of Theorem 7.4.6 hold and $y(t)$ not be generated by a finite autoregression. Then the following hold.*

(a) If $C_T/\log T \to \infty$, uniformly in $h \le H_T$,

$$\log\det\hat{\Sigma}_h + hs^2C_T/T = \log\det\dot{\Sigma} + \frac{hs^2}{T}(C_T-1) + \mathrm{tr}\{\Sigma^{-1}(\Sigma_h-\Sigma)\}\{1+o(1)\}. \tag{7.4.22}$$

(b) If (7.4.16) also holds and $C_T \ge c > 1$, then (7.4.22) holds with $o(1)$ replaced by $o_P(1)$.

REMARK 1. The requirement that $y(t)$ be not generated by an autoregression is of no real importance since, from the results in Chapter 5, the autoregressive case is fairly fully understood.

REMARK 2. The second part of the theorem seems, at first, more relevant because of the weak condition on C_T and the fact that Rissanen's important theories require that $C_T = \log T$. It is likely, however, that part (a) holds if $\liminf C_T/\log T = C > 0$, although that has not been proved, in which case part (a) becomes the relevant part.

These results in Theorem 7.4.7 show that $\hat{h}_T$ is effectively determined by h_T^*, where this minimizes

$$L_T(h, C_T) = \frac{hs^2}{T}(C_T - 1) + \operatorname{tr}\{\Sigma^{-1}(\Sigma_h - \Sigma)\}. \tag{7.4.23}$$

Indeed, we know that if $\hat{h}_T$ minimizes the left side of (7.4.22), then $\hat{h}_T \to \infty$, a.s. Indeed, otherwise $\operatorname{tr}\{\Sigma^{-1}(\Sigma_h - \Sigma)\}\{1 + o(1)\}$ remain bounded away from zero for h values along the $\hat{h}_T$ sequence, and since $hs^2 C_T/T \to 0$, a minimum is not being achieved. For brevity, call the second term on the right in (7.4.23) just $\theta(h)$. Extend the definition of $\theta(h)$ to all $h \geq 0$, for example, as a step function changing only at integer values of h. Assume that there is a function $\tilde{\theta}(h)$ which is twice differentiable, with a second derivative $\tilde{\theta}''(h)$ and

$$\lim_{h\to\infty} \theta(h)/\tilde{\theta}(h) = 1.$$

Then put

$$\tilde{L}_T(h, C_T) = L_T(h, C_T) + \tilde{\theta}(h) - \theta(h) = L_T(h, C_T)\{1 + o_h(1)\},$$

where $o_h(1)$ decreases to zero as $h \to \infty$. Thus if $\hat{L}_T(h, C_T)$ is the left side of (7.4.22), less $\log\det\dot{\Sigma}$, then since $\hat{h}_T \to \infty$, if $o(1)$ holds in (7.4.22),

$$\hat{L}_T(\hat{h}_T, C_T) = \tilde{L}_T(\hat{h}_T, C_T)\{1 + o(1)\}.$$

Let h_T^* minimize $\tilde{L}_T(h, C_T)$. Then

$$\tilde{L}_T(\hat{h}_T, C_T) = \tilde{L}_T(h_T^*, C_T) + \tfrac{1}{2}\tilde{\theta}''(\bar{h}_T)(\hat{h}_T - h_T^*)^2,$$

where $\bar{h}_T$ lies between $\hat{h}_T$ and h_T^*. Thus

$$\left(\frac{\hat{h}_T}{h_T^*} - 1\right)^2 = 2\frac{\tilde{L}_T(\hat{h}_T, C_T) - \tilde{L}_T(h_T^*, C_T)}{(h_T^*)^2\tilde{\theta}''(\bar{h}_T)}.$$

But

$$\tilde{L}_T(h_T^*, C_T) \leq \tilde{L}_T(\hat{h}_T, C_T)$$

and also

$$\tilde{L}_T(\hat{h}_T, C_T) = \hat{L}_T(\hat{h}_T, C_T)\{1 + o(1)\}$$

and

$$\hat{L}_T(\hat{h}_T, C_T) \leq \hat{L}_T(h_T^*, C_T) = \tilde{L}_T(h_T^*, C_T)\{1 + o(1)\}.$$

Thus

$$\frac{\tilde{L}_T(\hat{h}_T, C_T)}{\tilde{L}_T(h_T^*, C_T)} \to 1$$

and $\hat{h}_T/h_T^* \to 1$, a.s., if $o(1)$ holds in (7.4.22) and

$$\liminf_{h\to\infty} \left|\frac{h^2\tilde{\theta}''(h)}{\tilde{\theta}(h)}\right| > 0. \tag{7.4.24}$$

If $o(1)$ must be replaced by $o_P(1)$ in (7.4.22) then, similarly, $\hat{h}_T/h_T^* \to 1$ in probability.

For example, if $y(t)$ is generated by an ARMA model, then $\tilde{\theta}(h)$ may be taken of the form $c_0\rho_0^{-2h}$, where ρ_0 is the modulus of a zero of $k_0(z)$ nearest $|z| = 1$. Then $\tilde{\theta}''(h) = (2\log\rho_0)^2\tilde{\theta}(h)$ and h_T^* solves

$$\frac{s^2}{T}(C_T - 1) = 2\log\rho_0(c_0\rho_0^{-2h}),$$

that is,

$$h_T^* = \frac{\log T}{2\log\rho_0}\{1 + o(1)\},$$

as we already know. Similarly, if we may take $\tilde{\theta}(h) = c_0 h^{-\beta}$, $\beta \geq 1$, then $\tilde{\theta}''(h) = C_0\beta(\beta + 1)h^{-(\beta+2)}$ and we find that

$$h_T^* = [C_0\beta T/\{s^2(C_T - 1)\}]^{1/(1+\beta)}.$$

However, the theorem requires that $h_T^* = o\{(T/\log T)^{1/2}\}$, so for part (b) of the theorem to be applicable, we need $\beta > 1$ unless $C_T/\log T$ increases to infinity. Theorems 7.4.5, 7.4.6, and 7.4.7 are due to Hannan and Kavalieris (1986). Earlier work, on Gaussian assumptions but requiring weaker conditions otherwise, that leads, in particular, to (7.4.22), is that of Shibata (1980).

Finally, a modification of a theorem due to Lewis and Reinsel (1985) will be presented. For this purpose put $\hat{\phi}_h = \operatorname{vec}\{\hat{\Phi}_h(1), \hat{\Phi}_h(2), \ldots, \hat{\Phi}_h(h)\}$, $\phi_h = \operatorname{vec}\{\Phi(1), \Phi(2), \ldots, \Phi(h)\}$. Note that ϕ_h is defined in terms of the $\Phi(j)$, not the $\Phi_h(j)$. We write $\hat{\phi}_T$, ϕ_T in place of $\hat{\phi}_h$, ϕ_h when $h = h_T$ is a function of T and also Γ_T for Γ_h at $h = h_T$.

Theorem 7.4.8. *Let $y(t)$ be generated by (7.4.3) with (7.4.4), (4.1.15), (4.1.16), (4.1.20) holding and assume that h_T satisfies*

$$h_T = o(T^{1/3}), \qquad T^{1/2} \sum_{h_T+1}^{\infty} \|\Phi(j)\| \to 0.$$

Then if α_T is a vector of $h_T s^2$ components satisfying

$$0 < c_1 \le \alpha_T' \alpha_T \le c_2 < \infty,$$

$$T^{1/2} \alpha_T'(\hat{\phi}_T - \phi_T)/\{\alpha_T'(\Gamma_T^{-1} \otimes \Sigma)\alpha_T\}^{1/2} \tag{7.4.25}$$

is asymptotically normal with zero mean and unit covariance matrix.

REMARK 1. It is possible to replace (4.1.16) with (4.1.17), but then a more complicated norming constant will be needed in (7.4.25).

REMARK 2. This result does *not* show that $\max_j \|\hat{\Phi}_h(j) - \Phi(j)\| = O_p(T^{-1/2})$ since α_h is to be chosen independently of the $y(t)$ sequence. Indeed, that result is not true, as h increases to infinity.

(iii) Proofs of Theorems

These proofs in some cases resemble closely the proofs in Chapter 6 and for this reason will be given in somewhat abbreviated form. The basic references are the works of Hannan and Kavalieris (1983, 1986).

We first point out that the analogue of Theorem 6.6.11 holds: namely,

Theorem 7.4.9. *Let $y(t)$ be generated by (7.4.3) and (7.4.4), (4.1.15), (4.1.17), (4.1.20) hold. Then $\sup_{0<h<\infty} |\Gamma_h^{-1}|_\infty \le c < \infty$; $|G_h^{-1}|_\infty \le c < \infty$, $h \le H_T = o\{(T/\log T)^{1/2}\}$.*

The first part of this theorem is proved in Theorem 6.6.11. The second part follows as below (6.6.21) since $|G_h - \Gamma_h|_\infty = O\{h(\log T/T)^{1/2}\} = o(1)$ for $h = o\{(T/\log T)^{1/2}\}$.

Proof of Theorem 7.4.5. We shall write $G(j-k)$ for $G_y(j-k)$, for simplicity, and similarly for $\Gamma(j-k)$.

$$\sum_1^h \{\hat{\Phi}_h(j) - \Phi_h(j)\} G(j-k) = -\sum_0^h \Phi(j)G(j-k)$$

$$(7.4.26) \qquad + \sum_1^h \{\Phi(j) - \Phi_h(j)\}\{G(j-k) - \Gamma(j-k)\}$$

$$+ \sum_0^h \Phi(j)\Gamma(j-k), \qquad k = 1, 2, \ldots, h,$$

because of the estimation equations. The second term is, in the uniform norm,

$$o(1) \sum_1^h \|\Phi(j) - \Phi_h(j)\| = o(1) \sum_{h+1}^{\infty} \|\Phi(j)\|$$

by Theorems 7.4.1 and 6.6.12. Using (7.4.7) for the last term we have

$$\sum_0^h \Phi(j)\Gamma(j-k) = -\sum_{h+1}^{\infty} \Phi(j)\Gamma(j-k) = -\sum_{h+1}^{T} \Phi(j)\Gamma(j-k) + o(T^{-1/2})$$

$$= -\sum_{h+1}^{\infty} \Phi(j)G(j-k) + o(T^{-1/2}) + o(1) \sum_{h+1}^{\infty} \|\Phi(j)\|,$$

where the error is again evaluated in the uniform norm and Theorem 7.4.1 is again used. Thus the right side of (7.4.26) is

$$(7.4.27) \qquad -\sum_0^T \Phi(j)G(j-k) + o(1) \sum_{h+1}^{\infty} \|\Phi(j)\| + o(T^{-1/2}).$$

If (4.1.16) holds, the $o(1)$ term is $O\{(\log T/T)^{1/2}\}$ by Theorem 7.4.3, as also is the first term, so that the second part of the theorem follows from Theorem 7.4.9. If $s = 1$, we may use Theorem 7.4.4 to establish the required result so that only the main part of the theorem remains. For this purpose if $G(j-k)$ is replaced by $\tilde{G}(j-k)$ [see below (6.6.26)], then (7.4.27) becomes, as below (6.6.26),

$$-\frac{1}{T}\sum_1^T \varepsilon(t)y(t-k)' + \frac{1}{T}\sum_{T+1}^{\infty} \Phi(j) \sum_1^T y(t-j)y(t-k)'$$

$$+ o(1) \sum_{h+1}^{\infty} \|\Phi(j)\| + o(T^{-1/2})$$

$$= -\frac{1}{T}\sum_1^T \varepsilon(t)y(t-k)' + o(1) \sum_{h+1}^{\infty} \|\Phi(j)\| + o(T^{-1/2})$$

using (7.4.7). Exactly as in the proof of Theorem 7.4.3, the first term is $O\{(\log T/T)^{1/2}\}$ uniformly in k, and using Theorem 7.4.9, the theorem then follows. It therefore remains to justify the replacement of $G(j-k)$ by $\tilde{G}(j-k)$. For this purpose it is only necessary [see below (6.6.26)] to show that

(7.4.28)

$$T^{-1}\sum_0^T \Phi(j)\sum_{jk}{}' y(t-j)y(t-k)' - T^{-1}\sum_0^T \Phi(j)\sum_{jk}{}'' y(t-j)y(t-k)'$$
$$= o(T^{-1/2}),$$

where $\sum_{jk}'$ is over $T < t \leq T + \min(j,k)$ and $\sum_{jk}''$ is over $1 \leq t < \max(j,k)$. The second term is bounded in norm (using the finiteness of the fourth moment) by

$$T^{-1}\sum_0^k \|\Phi(j)\| j^{3/2} + T^{-1}\sum_{k+1}^T \|\Phi(j)\| k^{3/4}j^{1/4} = o\left\{T^{-1}\left(\frac{T}{\log T}\right)^{1/2}\right\} = o(T^{-1/2})$$

since $j, k = o\{(T/\log T)^{1/2}\}$ and using (7.4.7). The first term remains. To show that this is negligible, we first show that all terms involving $\varepsilon(s)\varepsilon(s)'$ may be removed. These are

$$T^{-1}\sum_0^T \Phi(j)\sum_0^\infty \left\{K(u)\sum_{jk}{}' \varepsilon(t-j-u)\varepsilon(t-j-u)'K(u+j-k)'\right\}$$
$$= T^{-1}\sum_0^T \Phi(j)\sum_1^\infty \left\{K(u-j)\sum_{jk}{}' \varepsilon(t-u)\varepsilon(t-u)'K(u-k)'\right\} + o(T^{-1/2}),$$

where $K(u) = 0$, $u < 0$. This is, since $\sum_0^\infty \Phi(j)K(u-j) = 0$, $u \geq 1$,

$$T^{-1}\sum_0^k \Phi(j)\sum_1^\infty K(u-j)\sum_T^{T+j} \varepsilon(t-u)\varepsilon(t-u)'K(u-k)'$$
$$- T^{-1}\sum_0^k \Phi(j)\sum_1^\infty K(u-j)\sum_T^{T+k} \varepsilon(t-u)\varepsilon(t-u)'K(u-k)' + o(T^{-1/2}),$$

which is bounded in norm by

$$o(1)T^{-1/2}\sum_0^k \|\Phi(j)\|\sum_1^\infty \|K(u-j)\|\,\|K(u-k)\|(k-j) + o(T^{-1/2})$$
$$= o(1)T^{-1/2}\sum_0^k \|\Phi(j)\|\sum_k^\infty \|K(u)\|\,\|K(u+k-j)\|(k-j) + o(T^{-1/2})$$

$$\leq o(1)T^{-1/2}\left[\sum_{k}^{\infty}\|K(u)\|u^{1/2}\|K(u+k)\|k^{1/2}\right.$$

$$\left.+\sum_{1}^{\infty}\|\Phi(j)\|j^{1/2}\left\{\sum_{k}^{\infty}\|K(u)\|u^{1/2}\|K(u+k-j)\|(k-j)^{1/2}\right\}\right]$$

$$+\,o(T^{-1/2})=o(T^{-1/2}).$$

Thus we need to consider only

$$T^{-1}\sum_{0}^{T}\Phi(j)\sum_{v\neq u+j-k}\sum\left\{K(u)\sum_{jk}{}'\,\varepsilon(t-j-u)\varepsilon(t-v-k)K(v)'\right\},$$

so that finally it is sufficient to show that, uniformly in $a, b \leq 2T, j \leq H_T$,

$$(7.4.29)\qquad (T\log T)^{-1/2}\sum_{T+1}^{T+j}\varepsilon(t-a)\varepsilon(t-\mathrm{b})'\to 0,\quad \text{a.s.}\qquad a\neq b.$$

That this is so may be established as in the proof of Theorem (7.4.3). Thus consider

$$\sum_{T+1}^{T+j}\varepsilon_k(t-a)\varepsilon_l(t-b),\qquad a\neq b,\qquad a,b\leq H_T$$

and truncate the $\varepsilon_k(t)$, replacing these by

$$d_k(t)=\varepsilon_k(t)I\{|\varepsilon_k(t)|\leq a(t)\}+a(t)I\{\varepsilon_k(t)>a(t)\}-a(t)I\{\varepsilon_k(t)<-a(t)\}.$$

This mode of truncation has advantages (Chow and Studden, 1969) and might have been used previously since, putting $\mathscr{G}_t=\mathscr{F}_{t-a}, a<b$, $\mathscr{G}_t=\mathscr{F}_{t-b}, b<a$, we have, putting $Y_t=d_k(t-a)d_l(t-b)-E\{d_k(t-a)d_l(t-b)|\mathscr{G}_{t-1}\}$, that $E([d_k(t-a)-E\{d_k(t-a)/\mathscr{G}_{t-1}\}]^2|\mathscr{G}_{t-1})$ is an increasing function of the truncation point. Now for $a<b$, for example,

$$\begin{aligned}E\{Y_t^2|\mathscr{G}_{t-1}\}&=d(t-b)^2E([d_k(t-a)-E\{d_k(t-a)|\mathscr{G}_{t-1}\}]^2|\mathscr{G}_{t-1})\\&\leq\varepsilon_k(t-b)^2E\{\varepsilon_k(t-a)^2|\mathscr{G}_{t-1}\},\end{aligned}$$

so that

$$\frac{1}{T}\sum_{T+1}^{T+j}E\{Y_t^2|\mathscr{G}_{t-1}\}\leq\frac{1}{T}\sum_{T+1}^{T+H_T}E\{\varepsilon_k(t-a)^2\varepsilon_b(t-b)^2|\mathscr{G}_{t-1}\}$$

$$\le \left[\frac{1}{T}\sum_{T+1}^{T+H_T} E\{\varepsilon_k(t-a)^4|\mathscr{G}_{t-1}\}\frac{1}{T}\sum_{T+1}^{T+H_T} E\{\varepsilon_l(t-b)^4|\mathscr{G}_{t-1}\}\right]^{1/2}$$

$$\le c < \infty, \quad \text{a.s.}$$

and uniformly in a, b, since

$$\frac{1}{T}\sum_{T+1}^{T+H_T} E\{\varepsilon_k(t-a)^4|\mathscr{G}_{t-1}\} \le \frac{1}{T}\sum_{1}^{2T} E\{\varepsilon_k(t)^4|\mathscr{F}_{t-1}\} \le c < \infty.$$

The fact that truncation effects are negligible, if $a_t = (t/\log t)^{1/4}$, follows from

$$\frac{1}{(T\log T)^{1/2}}\sum_{1}^{T} \varepsilon_k(t)^2 \to 0, \qquad c_k(t) = \varepsilon_k(t) - d_k(t),$$

which is proved in almost the same way as in the proof of Theorem 5.3.1 [see (5.3.12)]. Thus put $\lambda_T = a_{T+H_T}^{-2}$ and consider

$$\mathscr{Z}_j = \lambda_T \sum_{T+1}^{T+j} Y_t - \tfrac{1}{2}\lambda_T^2(1 + \lambda_T a_{T+H_T}/2) \sum_{T+1}^{T+j} E\{Y_t^2|\mathscr{G}_{t-1}\}. \tag{7.4.30}$$

Here, of course, $\lambda_T a_{T+H_T}^2 = 1$. Then from Stout (1974, p. 299),

$$P\left\{\max_j \mathscr{Z}_j > b_T\right\} \le e^{-b_T}.$$

Choose $b_T = b\log T$ and observe that from what has just been shown, the second term on the right in (7.4.30) is bounded by $c\log T$ for some finite c. Choosing b sufficiently large, we see that

$$\sum_{T=1}^{\infty} P\left\{\max_{k,l,a,b,j} (T\log T)^{-1/2}\sum_{T+1}^{T+j} \varepsilon_k(t-a)\varepsilon_l(t-b) > \varepsilon\right\} \le c\sum_{1}^{\infty} T^{-b}T^{3/2} < \infty$$

since there are at most $T^{3/2}$ values of a, b, j to be considered. Replacing $\varepsilon_k(t-a)\varepsilon_l(t-b)$ by $-\varepsilon_k(t-a)\varepsilon_l(t-b)$, the result (7.4.29) is established and hence Theorem 7.4.5 is proved.

Proof of Theorem 7.4.6. Here we shall take the scalar case, for simplicity, and give the proof for $T^{-1}\sum\{\hat{\varepsilon}_h(t) - \varepsilon(t)\}^2$. We shall take $\sigma^2 = 1$. The proof for $T^{-1}\sum\varepsilon(t)\{\hat{\varepsilon}_h(t) - \varepsilon(t)\}$ is of the same form but a little simpler. We shall give the proof in outline, referring the reader to Hannan and Kavalieris (1986) for omitted details. Recalling the notation of Section 6.6, we put

$y(t,h)' = (y(t-1),\dots,y(t-h))$ and

$$\hat{u}_k = T^{-1}\sum_1^T \varepsilon(t)y(t-k), \qquad \hat{u}' = (\hat{u}_1,\dots,\hat{u}_n).$$

We observe that $\hat{u}_k = O\{(\log T/T)^{1/2}\}$, uniformly in k, by the same proof as for Theorem 7.4.3. Moreover, if $k < (\log T)^a$, $a < \infty$ and (4.1.16) holds, then $\hat{u}_k = O\{(\log\log T/T)^{1/2}\}$ uniformly in such k. The proof of this is the same as that for the second part of Theorem 5.3.1. Let $\tilde{G}$ be the infinite matrix with $\tilde{g}(j,k)$ in row j, column k, where $\tilde{g}(j,k)$ is defined in Section 6.6. Let $\tilde{G}_{11}$ have zero in place of $\tilde{g}(j,k)$ if $j > h$ or $k > h$. Similarly, $\tilde{G}_{22}$ has zero in place of $\tilde{g}(j,k)$ if $j \le h$ or $k \le h$. The definitions for $\tilde{G}_{12}$, and $\tilde{G}_{21}$ are now obvious. Γ, $\Gamma_{11}, \Gamma_{12}, \Gamma_{22}$ are defined similarly in terms of $\gamma(j-k)$. The first step is to prove that

$$\text{(7.4.31)} \quad \sum_1^h \{\hat{\phi}_h(j) - \phi(j)\}\tilde{g}(j,k) = -\hat{u}_k + \sum_{h+1}^\infty \phi(j)\tilde{g}(j,k) + o\{(\log T/T)^{1/2}\},$$

using Theorems 7.4.5 and 6.6.12 and (4.1.16).

Under the conditions of part (b) of the theorem being proved, the last term in (7.4.31) is $o(T^{-1/2})$ [i.e., when (7.4.16) holds]. The proof is essentially the same as that below (7.4.29) except that now, because of (7.4.16), we may put $a_t = \{t/(\log t)^2\}^{1/4}$, so that λ_T may be chosen as $c\{\log T/T^{1/2}\}$. Put

$$\phi = (\phi(1),\phi(2),\dots)', \qquad \phi_h^{(1)} = (\phi(1),\dots,\phi(h),0,0,\dots,)'$$

and $\phi_h^{(2)} = \phi - \phi_h^{(1)}$. Put $\hat{\phi}_h^{(1)}$, and so on, for the same quantities, with $\hat{\phi}(j)$ replacing $\phi(j)$. Thus

$$\hat{\phi}_h^{(1)} - \phi_h^{(1)} = -\tilde{G}_{11}^{-1}\hat{u} + \tilde{G}_{11}^{-1}\tilde{G}_{12}\phi_h^{(2)} + \tilde{G}_{11}^{-1}o\{(\log T/T)^{1/2}\},$$

where $\tilde{G}_{11}^{-1}$ is the obvious generalized inverse and where the last term is a vector that has elements uniformly of the indicated order and that order becomes $o(T^{-1/2})$ under (7.4.16). We shall write out results for this case only, for brevity.

$$\hat{\varepsilon}_h(t) - \varepsilon(t) = -y(t,\infty)'\{\tilde{G}_{11}^{-1}\hat{u} + \tilde{G}_{11}^{-1}\tilde{G}_{12}\phi_h^{(2)} + \tilde{G}_{11}^{-1}o(T^{-1/2}) - \phi_h^{(2)}\}.$$

Now $\hat{u}'\tilde{G}_{11}^{-1}\tilde{G}(I_\infty - \tilde{G}_{12})\phi_h^{(2)} = 0$, $o(T^{-1/2})'\tilde{G}_{11}o(T^{-1/2}) = o(h/T)$, so that

$$\frac{1}{T}\sum_1^T \{\hat{\varepsilon}_h(t) - \varepsilon(t)\}^2 = \hat{u}'\tilde{G}_{11}^{-1}\hat{u} + \phi_h^{(2)\prime}\{\tilde{G}_{22} - \tilde{G}_{21}\tilde{G}_{11}^{-1}\tilde{G}_{12}\}\phi_h^{(2)} + o\left(\frac{h}{T}\right).$$

Now $\phi_h^{(2)\prime}\{\Gamma_{22} - \Gamma_{21}\Gamma_{11}^{-1}\Gamma_{12}\}\phi_h^{(2)} = \sigma_h^2 - 1$ and replacing $\tilde{G}_{ij}$ by Γ_{ij} on the right, *now changing notation so that Γ_{11} is the nonnull part, only, of the original* Γ_{11}, arrive at

$$(7.4.32)\qquad \frac{1}{T}\sum_1^T \{\hat{\varepsilon}_h(t) - \varepsilon(t)\}^2 = \hat{u}'\Gamma_{11}^{-1}\hat{u} + (\sigma_h^2 - 1)\{1 + o(1)\} + o\left(\frac{h}{T}\right).$$

For example,

$$\phi_h^{(2)\prime}(\tilde{G}_{22} - \Gamma_{22})\phi_h^{(2)} = \frac{1}{T}\sum_1^T [\xi_h(t)^2 - E\{\xi_h(t)^2\}]$$

$$\xi_h(t) = \sum_{h+1}^{\infty} \lambda_h(j)\varepsilon(t-j), \qquad \lambda_h(j) = \sum_{h+1}^{\infty} \phi(u)K(j-u),$$

where the $[\xi_h(t)^2 - E\{\xi_h(t)^2\}]$ are martingale differences, because of (4.1.16), and $\xi_h(t)$ has variance $\le c(\sigma_h^2 - 1)$. This is proved as for Theorem 7.4.3. Now $\hat{u}'\Gamma_h^{-1}\hat{u} = O(h \log T/T)$ and using (7.4.32) with $o(h/T)$ replaced by $o(h \log T/T)$, when (7.4.16) does not hold, part (a) of Theorem 7.4.6 is established, for $s = 1$ and for the first statement of part (c). As explained, the remainder of part (a) is proved in the same way.

Turning to part (b), we show that

$$\hat{u}'\Gamma_{11}^{-1}\hat{u} = \frac{1}{T^2}\sum_1^h \left\{\sum_1^T \varepsilon(s)\varepsilon(s-j)\right\}^2 \{1 + o(1)\}.$$

This is proved by showing first that $\hat{u}'\Phi_h'\{(I_h + L_h\Gamma L_h')^{-1} - I_h\}\Phi_h\hat{u} = o(h/T)$, where Φ_h had $\phi(k-j)$ is row k, column j, $\phi(j) = 0$, $j < 0$, and L_h has $\phi(h - j + k)$ in row j, column k, $j = 1, \ldots, h$, $k = 1, 2, \ldots$. The proof is essentially as for Theorem 6.6.11 except that the spectral norm is used for $(I_h + L_h\Gamma L_h') - I_h$ instead of the uniform norm. In partitioning the matrix the row chosen is $\tau(T)$, $\tau(T)/\log T \to \infty$, for the second part of (b) of the theorem and we choose $\tau(T) =$ the integer part of $(\log T)^2$. For this second part we use the fact that $\hat{u}_k = O\{(\log T/T)^{1/2}\}$ uniformly and for the second part that, for $k \le c(\log T)^2$, $\hat{u}_k = O(Q_T)$. For details, see Hannan and Kavalieris (1984). Then it may be shown that

$$(7.4.33)\qquad \hat{u}'\Phi_h'\Phi_h\hat{u} = \frac{1}{T^2}\sum_1^h \left\{\sum_1^T \varepsilon(t)\varepsilon(t-j)\right\}^2 \{1 + o(1)\} + o(h/T).$$

Indeed, the left side is

$$\frac{1}{T^2}\sum_{s,t=1}^{T}\sum \varepsilon(s)\varepsilon(t)y(s,h)'\Phi_h'\Phi_h y(t,h),$$

and by essentially the same proof as for Theorem 7.4.3 and, using the properties of $\hat{u}_k$ just recalled above (7.4.33), that formula is established. Now we remove the terms

$$\frac{1}{T}\sum_1^h\sum_1^T \varepsilon(t)^2\varepsilon(t-j)^2 = \frac{1}{T^2}\sum_1^h\sum_1^T\{\varepsilon(t)^2-1\}\varepsilon(t-j)^2$$
$$+\frac{1}{T^2}\sum_1^h\sum_1^T\{\varepsilon(t-j)^2-1\}+\frac{h}{T}.$$

The first two terms are $o(h/T)$, the first term being shown to be of that order by the same argument, involving $\varepsilon(t)^2$ at t^a, $a<\frac{1}{2}$, and the choice of $\lambda_T = (\log T)^d/T$, $d>1$, as was used in the proof of Theorem 7.4.3. The second term is $o(h/T)$, by ergodicity. Next the terms

$$\frac{1}{T^2}\sum_1^h\sum_1^T \varepsilon(t)\varepsilon(t-j)^2\varepsilon(t-2j)$$

are shown to be negligible in the same way, so that (7.4.33) becomes

$$\left\{\frac{1}{T^2}\sum_{j=1}^{h}\sum\sum{}'\,\varepsilon(s)\varepsilon(s-j)\varepsilon(t)\varepsilon(t-j)\right\}(1+o(1))+\frac{h}{T}+o\left(\frac{h}{T}\right), \tag{7.4.34}$$

where the double sum is over $1\le s,\ t\le T,\ s\ne t,\ s\ne t-j,\ t\ne s-j$. The variance of the first factor in the first terms is $O(hT^{-2})$ since each summand has variance $O(T^2)$ and the covariances are $O(T)$, so that the variance is $T^{-4}\{hO(T^2)+h^2O(T)\}=O(hT^{-2})$. Now the first part of (b) is established since T/h_T by the first factor in the first part of (7.4.34) has variance $h_T^{-1}\to 0$.

The proof of the last part of (b) of the theorem is long and complex and will be omitted save for the observation that we may decompose

$$\sum_1^h\left[\left\{T^{-1/2}\sum_1^T\varepsilon(t)\varepsilon(t-j)\right\}^2-1\right]$$

as the sum of two martingales $S_1(T)$, $S_2(T)$, with differences, respectively,

$$\{\varepsilon(t)^2-1\}\sum_1^{h_t}\varepsilon(t-j)^2,\qquad 2\varepsilon(t)\sum_1^{h_t}\varepsilon(t-j)\sum_1^{t-1}\varepsilon(s)\varepsilon(s-j),$$

plus the "compensators" associated with the two martingales. We omit details and refer the reader to the reference cited. This completes the discussion of Theorem 7.4.6.

Proof of Theorem 7.4.7. Part (a) follows almost immediately from part (a) of Theorem 7.4.6. Thus, taking $s = 1$, for example,

$$\hat{\sigma}_h^2 = \frac{1}{T}\sum_t \hat{\varepsilon}_h(t)^2 = \frac{1}{T}\sum_t \varepsilon(t)^2 + \frac{1}{T}\sum_t \{\hat{\varepsilon}_h(t) - \varepsilon(t)\}^2 + \frac{2}{T}\sum_t \varepsilon(t)\{\hat{\varepsilon}_h(t) - \varepsilon(t)\}$$

$$(7.4.35) \qquad = \dot{\sigma}^2 + (\sigma_h^2 - \sigma^2)\{1 + o(1)\} + \sigma^2\frac{h}{T} + O(h\log T/T)$$

$$- 2\sigma^2\frac{h}{T} + O(h\log T/T)$$

so that

$$\log\hat{\sigma}_h^2 = \log\dot{\sigma}^2 + \{(\sigma_h^2 - \sigma^2)/\sigma^2\}\{1 + o(1)\} - \frac{h}{T} + O(h\log T/T)$$

using $\dot{\sigma}^2 \to \sigma^2$ a.s. Now part (a) follows immediately since $O(h\log T/T) = o\{(h(C_T - 1)/T\}$ for $C_T/\log T \to \infty$. However, it is part (b) that is important. Use the decomposition (7.4.35) and consider [see below (7.4.32)]

$$(7.4.36) \qquad \left\{h(C_T - 1)/T + \frac{\sigma_h^2 - \sigma^2}{\sigma^2}\right\}^{-1} T^{-2}\sum_1^h \left\{\sum_1^T \varepsilon(t)\varepsilon(t-j)\right\}^2.$$

If this can be shown to converge in probability to σ^2, the first two terms in (7.4.35) will provide a contribution

$$\frac{h(C_T - 1)}{T} + \frac{\sigma_h^2 - \sigma^2}{\sigma^2} + \log\dot{\sigma}^2.$$

Treating the last term on the right in (7.4.35) in the same way, the result is established. Now, following the proof of Theorem 7.4.6, we reduce (7.4.36) to a consideration of

$$f_T(h) = \left\{h(C_T - 1)/T + \frac{\sigma_h^2 - \sigma^2}{\sigma^2}\right\}^{-1} T^{-2}\sum_1^h \sum\sum{}' \varepsilon(s)\varepsilon(s-j)\varepsilon(t)\varepsilon(t-j),$$

where $\sum\sum'$ is defined below (7.4.34). Put $a_T^* = [h_T^*]^{1+\delta}$, $\delta > 0$, where $[h_T^*]$ is the integer part of h_T^* and h_T^* minimizes $h(C_T - 1)/T + [(\sigma_h^2 - \sigma^2)/\sigma^2]$. Put

$a_j = j^{1+\delta}$. Then

$$P\left\{\max_{1\le h\le H_T} |f_T(h)| > \varepsilon\right\} \le P\left\{\max_{1\le h\le a_T^*} |f_T(h)| > \varepsilon\right\} + \sum_{j>[h_T^*]} P\left\{\max_{a_j<h\le a_{j+1}} |f_T(h)| > \varepsilon\right\}. \tag{7.4.37}$$

Now we repeatedly use Menchoff's inequality (see Stout, 1974, p. 18). This deals with a sum $S_n = \sum_1^n X_i$, of orthogonal random variables and asserts that

$$E\left\{\max_{1\le n\le N} S_n^2\right\} \le \{(\log 4N)/\log 2\}^2 \sum_1^N E(X_i^2).$$

In our case the X_i are

$$\sum\sum{}' \varepsilon(s)\varepsilon(s-j)\varepsilon(t)\varepsilon(t-j),$$

which are not orthogonal. However, their covariance is $O(T)$ while their variance is $O(T^2)$, so that the sum of their variances is $O(T^2h)$ while the covariances sum to $O(Th^2)$ and $Th^2/(T^2h) = h/T$, so that a simple modification of the proof of Menchoff's inequality allows us to bound (7.4.37) by

$$c\frac{1}{(h_T^*)^2}(\log a_T^*)^2\frac{a_T}{\varepsilon^2} + c\sum_{j>[h_T^*]}\frac{1}{a_j^2}\{\log(a_{j+1}-a_j)\}^2 a_{j+1}/\varepsilon^2.$$

The first term converges to zero and the second is dominated by

$$c\sum_j \frac{(\log j)^2}{j^{1+\delta}} < \infty.$$

This completes the proof of Theorem 7.4.7.

Proof of Theorem 7.4.8. This is proved in Lewis and Reinsel (1985) with the $\varepsilon(t)$ serially independent. The modifications to achieve the result quoted here are minor.

(iv) Convergence Results for Maximum Likelihood Estimation

In Chapter 4, following a classical paradigm of statistics, the asymptotic properties of the maximum likelihood estimation have been investigated under the assumption that the data correspond to a system contained in the model class, or more precisely, that $\theta_0 \in \Theta^*$ holds. Here we consider the more general case, where the true system is not necessarily contained in the model class under consideration, in the sense that (k_0, l_0) may be of higher "order" or even not rational. This is an important generalization of the classical paradigm. The idea has been emphasized in a number of papers by Ljung (see, e.g., Ljung 1978). In a certain sense the idea is closely related to robustness.

The assumptions concerning the true system might be replaced by assumptions about the data-generating process $(y(t)', z(t)')$, that is, the stochastic process from which the data are sampled. Clearly, the concept of a true system, as well as that of a data-generating process, is an abstraction and therefore one wants to keep the corresponding assumptions as general as possible. To evaluate the properties of estimators, we introduce the set Θ_0 of all elements in the parameter space $\bar{\Theta}$ which give—in a certain sense—the best approximation to the true $\theta_0 = (k_0, l_0, \Sigma_0)$ within $\bar{\Theta}$. It is reasonable to generalize the concept of consistency in the sense that a sequence of estimators converges a.s. to the set Θ_0. Thereby we say that a sequence θ_T converges to a set Θ_0 if for every subsequence there is a sub-subsequence converging to a point in Θ_0. We use the notation $\lim \theta_T = \Theta_0$.

Now let us consider maximum likelihood estimation from this more general point of view. The standard assumptions of this section are as follows. The true system is of the form

$$y(t) = u(t) + v(t), \tag{7.4.38}$$

where $u(t)$ is stationary and linearly regular with Wold representation (1.3.2)

$$u(t) = \sum_{j=0}^{\infty} K_0(j)\varepsilon(t-j); \qquad \sum_{j=0}^{\infty} \|K_0(j)\|^2 < \infty \tag{7.4.39}$$

and

$$v(t) = \sum_{j=1}^{\infty} L_0(j)z(t-j); \qquad \operatorname{tr} \int_{-\pi}^{\pi} l_0\, dF_z(\omega) l_0^* < \infty. \tag{7.4.40}$$

In addition, we assume that

$$\Sigma_0 = E\varepsilon(t)\varepsilon(t)' > 0, \tag{7.4.41}$$

$K_0(0) = I$, asymptotic stationarity for $z(t)$ [see (4.1.1)], the persistent exciting condition (4.1.6) and (4.2.21) $Ez(s)\varepsilon(t)' = 0$. The true θ_0 then is written as (k_0, l_0, Σ_0) with

$$k_0(z) = \sum_{j=0}^{\infty} K_0(j)z^j; \qquad l_0(z) = \sum_{j=1}^{\infty} L_0(j)z^j.$$

For the parameter spaces Θ and $\hat{\Theta}$, respectively, the standard assumptions of this section are the standard assumptions of Section 4.2 and

$$\Theta^* \neq \varnothing, \tag{7.4.42}$$

where $\varnothing$ denotes the empty set. Condition (7.4.42) guarantees that

$$\inf_{\theta \in \Theta^*} L(\theta) < \infty$$

holds. In Chapter 4, (7.4.42) was fulfilled due to (4.2.2). Now (4.2.21), i.e.,

$$(\Gamma_z(i-j))_{i,j=1\ldots sp} > 0 \tag{7.4.43}$$

holds, where p is the smallest upper bound for the degrees of (a, b, d), for suitably chosen ARMAX systems corresponding to $\bar{U}$ [see (4.2.1)]. By exactly the same argument as in Section 4.2, we see that, for fixed T and for $\theta_n \to \theta$, $\theta_n \in \hat{\Theta}$, where $\theta = (k, l, \sigma(\Sigma))$ is such that Σ is singular or that k has a pole for $|z| = 1$, we have

$$L_T(\theta_n) \to \infty \qquad \text{for } n \to \infty.$$

The minimization of L_T can be performed over $\hat{\Theta}$ in order to obtain the MLE $\hat{\theta}_T$. Again, it is not apparent that L_T has a minimum over $\hat{\Theta}$ (Deistler and Pötscher 1984), but $L_T(\hat{\theta}_T)$ has a well-defined meaning as $\inf L_T(\theta)$, $\theta \in \hat{\Theta}$. From the proof of Lemma 4.2.1 we see that (ii) [where θ_0 satisfies the conditions of this section] and (iii) in this lemma remain true under the standard assumptions of the present section, since (4.2.11) remains valid under these assumptions. Thus $\inf L_T(\theta)$, $\theta \in \hat{\Theta}$ is finite.

Of course, the notion of best approximation of θ_0 by elements of $\bar{\Theta}$ has to be made precise. Clearly, if $\theta_0 \in \Theta^*$, the best approximation should be θ_0. From Lemma 4.2.1 it is clear that in this case θ_0 gives the unique minimum of L over $\bar{\Theta}$. Interpreting the asymptotic likelihood L as a measure of goodness of fit, we define the set Θ_0 of best approximating points for θ_0 within $\bar{\Theta}$ as

$$\Theta_0 = \left\{\theta \in \bar{\Theta} \,|\, L(\theta) = \inf_{\theta \in \bar{\Theta}} L(\theta)\right\}. \tag{7.4.44}$$

From the proof of the next theorem it will become clear that Θ_0 is always nonempty.

It is known that Θ_0 need not reduce to a single point. An example of this was given by Kabaila (1983) for the case of scalar moving-average processes (i.e., $s = 1$, $m = 0$, $p = 0$), where the model class are the moving-average processes of order smaller than or equal to 1 and the true process is $y(t) = (1 + b_0 z)(1 - b_0 z)\varepsilon_t$, $|b_0| < 1$. Then

$$L(\theta) = \log \sigma^2 + (2\pi\sigma^2)^{-1} \int_{-1}^{\pi} |1 + be^{i\omega}|^{-2} |1 + b_0 e^{i\omega}|^2 |1 - b_0 e^{i\omega}|^2 \, d\omega$$

has to be minimized with respect to b and σ^2. It is easy to see that a minimizing value $\hat{b}$ satisfies $|\hat{b}| < 1$ and that $-\hat{b}$ also minimizes.

Note that Lemma 4.2.2 does not depend at all on assumptions on the data-generating process. Also, Lemma 4.2.3 still holds under the standard assumptions of this section, as can be checked easily from its proof. We now state the following "generalized consistency" result (compare Theorem 4.2.1).

Theorem 7.4.10. *If the data-generating process satisfies (7.4.38)–(7.4.43), $\varepsilon(t)$ is independent of the $z(t)$ sequence and*

$$\lim T^{-1} \sum_{t=1}^{T-s} \varepsilon(t+s)\varepsilon(t)' = \delta_{s,0}\Sigma_0, \quad \text{a.s.}, \qquad \text{for } s \geq 0,$$
$$\inf_{\theta \in \overline{\Theta}} L(\theta) = \inf_{\theta \in \Theta^*} L(\theta) \tag{7.4.45}$$

then the maximum likelihood estimators $\hat{\theta}_T$ (over $\hat{\Theta}$) have the following property:

$$\lim_{T\to\infty} \hat{\theta}_T = \Theta_0, \quad \text{a.s.} \tag{7.4.46}$$

Proof. We reconsider the proof of Theorem 4.2.1. By essentially the same argument as in that proof, we obtain $\hat{\theta}_T \in \Theta_{c_1,c_2,c_3}$, a.s., for $T > T_0$, for suitably chosen T_0 and $c_1, c_2, c_3 > 0$. The only difference is that the last equality in (4.2.23) has to be replaced by

$$\inf_{\theta \in \Theta^*} L(\theta) \geq \log\det \Sigma_0 + s, \qquad \text{a.s.},$$

but this does not affect the argument. Since Θ_{c_1,c_2,c_3} is compact, every subsequence of $\hat{\theta}_T$ contains a subsubsequence (again denoted by $\hat{\theta}_T$), converging to a limit, θ say. Then, for a suitably chosen subsequence, in the same way as in the proof of Theorem 4.2.1, we obtain from (4.2.23) and (4.2.39)

$$\lim_{T\to\infty} \sup L_T(\hat{\theta}_T) \leq \inf_{\theta\in\Theta^*} L(\theta) = \inf_{\theta\in\overline{\Theta}} L(\theta) \leq \lim_{T\to\infty} \inf L_T(\hat{\theta}_T), \quad \text{a.s.} \tag{7.4.47}$$

Clearly, $\theta \in \Theta_0$, and thus $\lim \hat{\theta}_T = \Theta_0$ holds.

REMARK 1. Clearly, the theorem above, as a special case contains Theorem 4.2.1, where $\Theta_0 = \theta_0$ holds, by Lemma 4.2.1. A convergence result of the type in Theorem 7.4.10 was first derived by Ljung (see, e.g., Ljung 1978). Therein the additional assumption that Θ_0 contains only one element has been imposed. A theorem similar to the theorem above, for the ARMA case, but under more general assumptions on the data-generating process (e.g., allowing for linearly singular components) has been proved in Pötscher (1987).

REMARK 2. Of course, the fact that Θ_0 may contain more than one element causes some problems, for example with the central limit theorem. Under rather general assumptions the following "local uniqueness" result can be shown (Ploberger, 1982). Let S denote the set of all spectral measures (i.e., the measures corresponding to the spectral disbtribution function F) of the (asymptotically stationary) observed processes $(y(t)', z(t)')$, corresponding to $\hat{\Theta}$ (and F_z). Then there is a neighborhood of S in the weak topology of measures, such that for every true F_0 from this neighborhood, the set Θ_0 contains only one element. Central limit theorems for the case where the true system is not necessarily contained in the model class have been derived in Ljung and Caines (1979).

5. REAL-TIME CALCULATION

This topic has recently been surveyed extensively by Ljung and Söderström (1983) and Young (1984). It is a large subject and can be treated only briefly here. For further details the references may be consulted. It needs discussion, however, for it finally provides a main use for many of the techniques of this book and is very important in its own right. One purpose of such real-time calculation could be the (semi)optimal encoding of data for transmission, for example, over an international telephone channel, the linear model being fitted in order to form a basis for the encoding. In such a situation one cannot view the data being analyzed as generated by a stationary system, needless to say. See Section 7.3 for some discussion. Moreover, only recent data could conceivably be relevant to the current encoding, so that less recent data will have to be forgotten. In such a situation theorems based on stationarity have reduced but not insignificant meaning. Unless, for example, convergence of an algorithm can be established for a stationary regime, it can be doubted whether the algorithm will be useful, since it could not be expected reasonably to track a slowly evolving system. Of course, theorems of a more general nature will be desirable, but few, if any, are currently available.

Although we have headed this section "real-time calculation," nearly equivalent phrases are "on-line," "recursive," and "adaptive." The term "on-line" emphasizes the fact that calculations are done as the data come to hand, and since little time may be available for the calculations, these must be of the form of recursive adaptions of previously calculated values. Of course, calculations could be on-line, recursive, and adaptive without the real-time aspect being so important. However, in many cases that aspect is important, and great ingenuity has been used in designing such real-time algorithms.

Two things will be discussed here. The first is the on-line estimation of ARMA systems of fixed order. The second is the use of lattice (or ladder) forms of calculation so that all orders up to some maximum order can be examined. The lattice calculations are so called because of the appearance of the flow diagrams describing them. They seem very important and allow extensive calculations to be done in real time, with fast computers, even when data are being gathered at the rate of several kilohertz (i.e., several thousands of observations per second).

Because this account cannot hope to be complete, we shall restrict ourselves substantially to the scalar ARMA case. Extension to the ARMAX case, as far as calculations are concerned, is obvious and immediate. The basic lattice algorithms will be presented for the vector case and it will not be difficult to see how these might be used for the vector ARMAX case, although the details of that would be complicated in general.

(i) Recursive ARMA Calculation

We begin with a basic formula for a least-squares regression of one variable $w(s)$ on a vector $x(s)$, for $s = 1, 2, \ldots, t$. In later examples $w(t)$, $x(t)$, and θ_t will be identified with different quantities, but the basic calculation will always be the same, namely,

$$\begin{aligned} \theta_t &= \theta_{t-1} + P_t x(t) e(t), \qquad e(t) = w(t) - \theta_{t-1}' x(t) \\ P_t &= P_{t-1} - \{1 + x(t)' P_{t-1} x(t)\}^{-1} P_{t-1} x(t) x(t)' P_{t-1}. \end{aligned} \tag{7.5.1}$$

Here

$$P_t = \left\{\sum_1^t x(s)x(s)'\right\}^{-1}, \qquad \theta_t = P_t \sum_1^t w(s)x(s)$$

and the first line follows from

$$\begin{aligned} P_t^{-1}\theta_t = \sum_1^t w(s)x(s) &= P_{t-1}^{-1}\theta_{t-1} + x(t)w(t) \\ &= P_t^{-1}\theta_{t-1} + x(t)\{w(t) - x(t)'\theta_{t-1}\}, \end{aligned} \tag{7.5.2}$$

while the second line of (7.5.1), the matrix inversion lemma, is easily verified by multiplication on the right, say, by $P_t^{-1} = P_{t-1}^{-1} + x(t)x(t)'$. Initial values for

P_t, θ_t are needed. In a true regression situation one could calculate θ_{t_0}, P_{t_0} for some $t_0 \geq \dim x(t)$ and use these values to initiate at t_0. If we put $\theta_t = 0$, $P_t = cI$ for $t = 0$, these will be some distortion away from the true value of the regression vector, but if $P_t^{-1} \to \infty$, this effect will eventually vanish. *Note that we shall use θ for a true value and shall not write θ_0 for that.*

The vector θ_t is, of course, the solution to the problem of minimizing

$$\sum_1^t \{w(s) - \theta' x(s)\}^2,$$

but one will wish to replace this criterion by

$$\sum_1^t \alpha_t(s)\{w(s) - \theta' x(s)\}^2, \tag{7.5.3}$$

where $\alpha_t(s)$ forgets the past. The only really convenient formula, computationally, for $\alpha_t(s)$ appears to be

$$\alpha_t(s) = \alpha_s \prod_{s+1}^t \lambda_u, \qquad s < t; \qquad \alpha_t(t) = \alpha_t > 0. \tag{7.5.4}$$

We have in mind $0 < \lambda_u < 1$, so that as λ_u decreases, the past is more rapidly forgotten. We do not discuss here the choice of λ_u, α_s. This choice is an aspect of one of the most immanent of all antinomies, that between adequate damping and quick response. The smaller λ_u is, the more rapid is the forgetting and the quicker will be the response to change. However, there will also be a less damped response and a more erratic estimate of a stable system with a greater risk of wild oscillation, due, for example, to noise. However, as has been said, λ_u, which is called the forgetting factor, will be taken as given. From equations (7.5.2) it is easily seen that the first line of the following form of (7.5.1), beginning from (7.5.3), (7.5.4), holds:

$$\begin{aligned} &\theta_t = \theta_{t-1} + \alpha_t P_t x(t) e(t), \qquad e(t) = w(t) - x(t)'\theta_{t-1} \\ &P_t = \lambda_t^{-1}\left[P_{t-1} - \left\{\frac{\lambda_t}{\alpha_t} + x(t)'P_{t-1}x(t)\right\}^{-1} P_{t-1}x(t)x(t)'P_{t-1}\right]. \end{aligned} \tag{7.5.5}$$

The last line follows from

$$P_t^{-1} = \sum_1^t \alpha_t(s)x(s)x(s)' = \lambda_t P_{t-1}^{-1} + \alpha_t x(t)x(t)' = \lambda_t\left\{P_{t-1}^{-1} + \frac{\alpha_t}{\lambda_t}x(t)x(t)'\right\}.$$

Now using the matrix inversion lemma, the last line of (7.5.5) holds.

We now go on to give three examples of the use of (7.5.5) and then give a further example incorporating all three. In each case we have merely to identify $y(t)$, $x(t)$, and θ_t the forgetting procedure being taken as given.

AR. Here the underlying model is a scalar autoregression of order h. Then, using the notation of Section 5.6,

(7.5.6)

$$x(t)' = (-y(t-1),\ldots,-y(t-h)), \qquad \theta_t' = (\phi_t(1),\ldots,\phi_t(h)), \qquad w(t) = y(t).$$

If initiation was by calculation of the Toeplitz regression at $t_0 \geq h$, then θ_t' would be the Toeplitz estimate of $\theta' = (\phi(1),\ldots,\phi(h))$. However, other initiations might be preferable to avoid the bias in the Toeplitz calculation discussed in Sections 5.6 and 6.4. In any case, the asymptotic properties of θ_t are the same as for the off-line calculations, if $\alpha_t(s) \equiv 1$.

AML. Here the underlying model is the scalar ARMA model

$$y(t) = -\sum_1^p \alpha(j)y(t-j) + \sum_1^q \beta(j)\varepsilon(t-j) + \varepsilon(t) \tag{7.5.7}$$

viewed as a regression. Consequently,

$$\theta_t' = (\alpha_t(1),\alpha_t(2),\ldots,\alpha_t(p),\beta_t(1),\ldots,\beta_t(q)), \qquad w(t) = y(t). \tag{7.5.8}$$

However, since $\varepsilon(t)$ is not observed, it must be replaced by an estimate. One candidate is $e(t)$ from the algorithm itself, that is, to replace $\varepsilon(t-j)$, $j = 1,\ldots,q$, by $e(t-j)$ [see (7.5.5)], which has already been calculated at time t. A slightly better procedure is to use

$$\hat{\varepsilon}(t) = y(t) - \theta_t' x(t). \tag{7.5.9}$$

This can be done at time t since $\hat{\varepsilon}(t-j)$, $j > 0$, will be available and this procedure uses more-up-to-date information. Thus

$$x(t)' = (-y(t-1),\ldots,-y(t-p),\hat{\varepsilon}(t-1),\ldots,\hat{\varepsilon}(t-q)).$$

Here AML stands for "approximate maximum likelihood." The term "extended least squares" (ELS) is used in Ljung and Söderström (1983), no doubt because the method is in no sense maximum likelihood. Convergence of this algorithm has been widely discussed. Here we refer to Solo (1979). We introduce the "positive real condition" on the true MA transfer function, $b(z)$, namely,

$$\mathscr{R}\{b(e^{i\omega})^{-1}\} \geq \tfrac{1}{2}. \tag{7.5.10}$$

This is easily seen to be equivalent to

$$\left|\sum_1^q \beta(j)e^{ij\omega}\right| \leq 1$$

and is always satisfied for $q = 1$. Solo (1979) established the following

Theorem 7.5.1. *Let $y(t)$ be stationary and generated by (7.5.7) with $b(z) \neq 0$, $|z| \leq 1$ and satisfying (7.5.10). Let (4.1.15), (4.1.16) hold. If θ_t is obtained by AML with $\alpha_t(s) \equiv 1$, then $\theta_t \to \theta$ a.s.*

As indicated at the beginning of this section, this theorem has limited but by no means insignificant meaning. Although the condition (7.5.10) may not be necessary, there is no doubt that for some cases where it fails, the algorithm will not converge to θ. The near necessity of the condition is a disadvantage of AML. This procedure will also not produce an efficient estimate. No central limit theorem is available. Simulations and other considerations *suggest* that $t^{1/2}(\theta_t - \theta)$ will be asymptotically normal, under the conditions of Theorem 7.5.1, plus (4.1.20), with the same asymptotic distribution as would hold if the procedure II of Section 6.5 were used, for p, q known and using $\hat{\varepsilon}_h(t)$ with h determined by *BIC* (or *AIC*) at stage I. In Hannan and McDougall (1988) this normal distribution is evaluated. It is $D^{-1}XD^{-1}$, where

$$X = \frac{1}{2\pi}\int |k(e^{i\omega})|^2 x(e^{i\omega})x(e^{i\omega})^* \, d\omega,$$

$$x(z)' = (-z, -z^2, \ldots, -z^p, zk(z^{-1}), \ldots, z^q k(z^{-1}))$$

$$D = \frac{1}{2\pi}\int |k(e^{i\omega})|^2 b(e^{-i\omega})^{-1} x(e^{i\omega})x(e^{i\omega})^* \, d\omega.$$

However, that this is the limiting distribution of $t^{1/2}(\theta_t - \theta)$ is *not* established.

RML. We shall write down this algorithm and then justify it. It can be understood in relation to stage III of Section 6.5. The vector θ_t is, as in (7.5.8),

$$x(t)' = (-\eta(t-1), \ldots, -\eta(t-p), \xi(t-1), \ldots, \xi(t-q)),$$
$$w(t) = \eta(t) - \xi(t) + e(t) \tag{7.5.11}$$

$$\sum_0^q \beta_{t-1}(j)\{\eta(t-j), \xi(t-j)\} = (y(t), e(t)),$$
$$\sum_0^q \beta_{t-1}(j)e(t-j) = \sum_0^p \alpha_{t-1}(j)y(t-j). \tag{7.5.12}$$

Although $w(t)$ is as described, it does not occur explicitly in that form in the algorithm since

$$w(t) - \theta'_{t-1}x(t) = \eta(t) - \xi(t) + e(t) + \sum_1^p \alpha_{t-1}(j)\eta(t-j) - \sum_1^q \beta_{t-1}(j)\xi(t-j)$$

$$= \sum_0^p \alpha_{t-1}(j)\eta(t-j) = e(t),$$

if initiation is with $e(t) = y(t) = \eta(t) = \xi(t) = 0, t \leq 0$. Thus the algorithm may be rewritten as follows, since $w(t) - \theta'_{t-1}x(t) = e(t) = y(t) + \sum_1^p \alpha_{t-1}(j)y(t-j) - \sum_1^q \beta_{t-1}(j)e(t-j)$.

$$\theta_t = \theta_{t-1} + \alpha_t P_t x(t)e(t), \tag{7.5.13}$$
$$e(t) = y(t) - (-y(t-1), \ldots, -y(t-p), e(t-1), \ldots, e(t-q))\theta_{t-1}$$

with P_t given by (7.5.5) with $x(t)$ given by (7.5.11) and (7.5.12). Of course, in forming $x(t)$ one could replace $e(t)$ from (7.5.13) by a similar expression, with θ_{t-1} replaced by θ_t.

A discussion of the convergence of this is given in Hannan (1980b) and in Ljung and Söderström (1983). We shall not quote a theorem because such a theorem seems essentially to be complicated by the need for monitoring of $a_t(z)$, $b_t(z)$ [which are, of course, constructed from the $\alpha_t(j)$, $\beta_t(j)$] to decide whether their zeros lie in the stability region (i.e., outside if $|z| \leq 1$). If that condition fails, it seems, a projection procedure must be used wherein (7.5.12) must be constructed using a point θ^* in the stability region, in place of θ_{t-1}, until θ_t returns to the stability region. It seems that if this is not done, the algorithm may diverge. Again we refer to Ljung and Söderström (1983) for further details.

If the algorithm does converge, due to such monitoring and projection, which eventually will not be needed if the system is stationary and $\alpha_t(s) \equiv 1$, then $t^{1/2}(\theta_t - \theta)$ will, under the standard conditions (4.1.15), (4.1.16) and the existence of a moment higher than the fourth, be asymptotically normal with the covariance matrix for the off-line maximum likelihood estimator on Gaussian assumptions (see Solo, 1981). Again we do not explicitly state a theorem because of the complexity of the monitoring and projection requirement. Again this is a disadvantage of the RML algorithm.

AM. This procedure was suggested initially by Åström and Mayne (1982). It consists in running the three algorithms AR, AML, RML in parallel of with AR feeding into AML and AML into RML, so that the constructed variables in $x(t)$ for the AML part are got from the AR part, not the past of the AML, and the vector $x(t)$ for the RML part is got from the AML calculation.

Thus the calculation is as follows. AR is run as above. Now call $x^{(1)}(t)$, $\theta_t^{(1)}$ the vectors, defined in (7.5.6) for this calculation. Put

$$\hat{\varepsilon}^{(1)}(t) = y(t) - x^{(1)}(t)'\theta_t^{(1)}. \tag{7.5.14}$$

Alternatively, we could save computations by using $e(t)$ from (7.5.6) in place of (7.5.14). *It will now be convenient to call that* $e_h^{(1)}(t)$. Then AML is as in (7.5.8) but with $\hat{\varepsilon}(t)$ from (7.5.9) replaced by $\hat{\varepsilon}^{(1)}(t)$ from (7.5.14). Call $\theta_t^{(2)}$, $x^{(2)}(t)$ the new output and input vectors from this algorithm. Put

$$\hat{\varepsilon}^{(2)}(t) = y(t) - x^{(2)}(t)'\theta_t^{(2)}$$

and compute $x^{(3)}(t)$ from (7.5.11), (7.5.12) but with $\hat{\varepsilon}^{(2)}(t)$, $\theta_t^{(2)}$ replacing $e(t)$, θ_t. [Alternatively, we could reduce computation by using $e(t)$ from (7.5.13), with $x(t)$ from (7.5.14).] *It will now be convenient to call that* $e^{(2)}(t)$.

Call θ_t *the final output of the algorithm (i.e., of the RML part)*. For this algorithm to converge, clearly h will have to increase with t. This might be chosen at each t by *BIC* (or *AIC*). This requires us to construct $\sigma_h^2(t)$ for several values of t. A feasible method for doing that is described in Hannan, Kavalieris, and Mackisack (1986). However, we might, following the discussion in Section 7.3 [see (7.3.8) and above] equally well use [see following (7.5.14), for the definition]

$$\frac{1}{t}\sum_1^t e_h^{(1)}(s)^2, \tag{7.5.15}$$

which is trivially constructed, and use h_t that minimizes this. Of course, we now do not have a real-time algorithm, for as t increases, when (7.5.7) holds, then h_t will increase as $\log t$. Thus h in (7.5.15) will have to be restricted to be below some fixed maximum value $H < \infty$. Until t is quite large, this might not matter much, in practice. Moreover, in practice the past will be forgotten and this will also ensure that h_t does not get too large.

In Hannan, Kavalieris, and Mackisack (1986), the following is proved.

Theorem 7.5.2. *Let* $y(t)$ *be stationary and generated by (7.5.7) with* p, q *known, with (4.1.15), (4.1.16), (4.1.20) holding and* $a(z), b(z) \neq 0$, $|z| \leq 1$. *If AM is used, for* $h = h_t$ *determined by BIC and* $\alpha_t(s) \equiv 1$, *then* $\theta_t \to \theta$ *a.s. and* $t^{1/2}(\theta_t - \theta)$ *is asymptotically normal with the same covariance matrix as for the off-line maximum likelihood estimate.*

Again this theorem is relevant but does not cover the realistic situation. If $h_t \equiv h$, some chosen value, θ_t will still converge but not to θ_0. This in some ways is more relevant than this theorem quoted. A more relevant calculation again is one that allows p, q or more likely $p = q$, to be chosen from the data in the AML and RML stages. *This will then constitute a full on-line equivalent of the procedure of Section 6.5.* Such a calculation could be effected by the fast Givens transformation procedure of Hannan, Kavalieris, and Mackisack (1986). A preferable procedure will be based on the lattice calculations mentioned above and which we now discuss.

(ii) Lattice Calculations

Some references for this method are Friedlander (1982), Lee, Friedlander and Morf (1982), Lee, Morf, and Friedlander (1981), and Makhoul (1981). Earlier references to these techniques are given in Section 6.4. To introduce the method, let us consider (5.6.13) but now use t for the number of data points, in accordance with the notation earlier in this section. It is also necessary to indicate what data have been used in the calculation of the autoregressive coefficients, so that we write $\varepsilon_{h,t}(u)$ for the forward residual or forward prediction error at time u using data to time t to calculate the coefficients. Now

$$\varepsilon_{h,t}(u) = \sum_0^h \hat{\Phi}_{h,t}(j)y(u-j), \qquad r_{h,t}(u) = \sum_0^h \tilde{\Phi}_{h,t}(j)y(u-h+j),$$

so that, for the Toeplitz calculation, as in (6.4.1),

$$\begin{aligned} &\varepsilon_{h,t}(u) = \varepsilon_{h-1,t}(u) + \hat{\Phi}_{h,t}(h)r_{h-1,t}(u-1), \qquad &&\varepsilon_{0,t}(u) = r_{0,t}(u) = y(u) \\ &r_{h,t}(u) = r_{h-1,t}(u-1) + \tilde{\Phi}_{h,t}(h)\varepsilon_{h-1,t}(u), &&r_{h,t}(0) = 0. \end{aligned} \tag{7.5.16}$$

One might now aim at calculating just $\varepsilon_{h,t}(t)$, $r_{h,t}(t)$ and using these to calculate $\hat{\Phi}_{h,t}(h)$, $\tilde{\Phi}_{h,t}(h)$. As has been indicated at the beginning of this section, in many calculations only the $\varepsilon_{h,t}(t)$ and $r_{h,t}(t)$ are needed and not the predictor coefficients (autoregressive coefficients) $\hat{\Phi}_{h,t}(j)$, $j < h$. For example in the encoding problem discussed in Section 7.3, then, on Gaussian assumptions, to encode $y(t+1)$, given $y(1), \ldots, y(t)$, one needs only the mean and variance of the conditional distribution, which may be estimated by $m_h(t+1) = y(t+1) - \varepsilon_{h,t}(t+1)$, and

$$s_h(t)^2 = \frac{1}{t}\sum_0^{t-1} \varepsilon_{h,u}(u+1)\varepsilon_{h,u}(u+1)' = \left(1 - \frac{1}{t}\right)s_h(t-1)^2 + \frac{1}{t}\varepsilon_{h,t-1}(t)\varepsilon_{h,t-1}(t)'.$$

Moreover,

$$\begin{aligned} m_h(t+1) &= -\sum_1^h \hat{\Phi}_{h,t}(j)y(t+1-j) \\ &= m_{h-1}(t+1) - \hat{\Phi}_{h,t}(h)r_{h-1,t}(t), \qquad m_0(t+1) = 0 \end{aligned}$$

In other applications, for example, speech recognition, one may need the predictor coefficients only rarely.

However, the Toeplitz calculation is difficult to do in this manner as a real-time calculation. What one wishes to do is to complete the calculations for each t, for $h = 0, 1, \ldots, H$. The problem with the Toeplitz calculation is the time update, for as can be seen from (5.6.3) as T is increased, the last h rows of (5.6.3) change, as well as a new row being added. This is not so for the prewindowed calculation, which uses the data in (5.6.3) only down to the Tth row. It is to this type of calculation, therefore, that the lattice calculations correspond. This is all the more attractive because eliminating postwindowing also improves the quality of the estimates. We shall give below briefly derivation of these lattice calculations, which are well set out in Friedlander (1982). For the basic lattice calculation, without calculating the predictor coefficients, at each time point quantities

$$\varepsilon_{h,t} = \varepsilon_{h-1,t} - K^r_{h,t} r_{h,t-1}, \qquad r_{h,t} = r_{h-1,t-1} - (K^\varepsilon_{h,t})' \varepsilon_{h,t} \tag{7.5.17}$$

are calculated, these having the meaning corresponding to $\varepsilon_{h,t}(t)$, $r_{h,t}(t)$ above. The K^r_h, K^ε_h are called the backward and forward reflection coefficients. In (7.5.16) the quantity corresponding to $K^r_{h,t}$ is $-\hat{\Phi}_{h,t}(h)$. However, $K^r_{h,t}$ is called a backward reflection coefficient and $K^\varepsilon_{h,t}$ is the forward coefficient. We also have

$$K^r_{h,t} = \Delta_{h,t}(R^r_{h,t-1})^{-1}, \qquad K^\varepsilon_{h,t} = (R^\varepsilon_{h-1,t})^{-1}\Delta_{h,t}, \tag{7.5.18}$$

where $\Delta_{h,t}$ corresponds to Δ_h in (5.6.13) and $R^r_{h,t}$, $R^\varepsilon_{h,t}$ to $\tilde{\Sigma}_h$, $\hat{\Sigma}_h$ (except that there will be no division by t). The remaining calculations are therefore of $\Delta_{h,t}$, $R^r_{h,t}$, $R^\varepsilon_{h,t}$. These are of the form

$$\Delta_{h,t} = \Delta_{h,t-1} + \gamma^{-1}_{h-2,t-1}\varepsilon_{h-1,t}r'_{h-1,t-1} \tag{7.5.19}$$

$$\gamma_{h,t} = \gamma_{h-1,t} - r'_{h,t}(R^r_{h,t})^{-1}\varepsilon_{h,t} \tag{7.5.20}$$

$$R^\varepsilon_{h,t} = R^\varepsilon_{h-1,t} - K^r_{h,t}\Delta'_{h,t}, \qquad R^r_{h,t} = R^r_{h-1,t-1} - \Delta_{h,t}K^\varepsilon_{h,t}. \tag{7.5.21}$$

The calculations are initiated with

$$R^\varepsilon_{0,t} = R^r_{0,t} = R^\varepsilon_{0,t-1} + y(t)y(t)', \qquad \gamma_{-1,t} \equiv 1, \qquad \varepsilon_{0,t} = r_{0,t} = y(t). \tag{7.5.22}$$

It is evident that this calculation (7.5.17)–(7.5.22) involves a number of calculations that is linear in H, the maximum value of h. To obtain the predictor coefficients, a further $O(H^2)$ calculations must be effected. The derivation of these equations is not intrinsically difficult but involves some ingenuity. The basic technique is to represent the projection, P_{S+X}, onto the space spanned by the columns of $[S, X]$ as

(7.5.23) $$P_{S+X} = P_S + (I - P_S)X[X'(I - P_S)X]^{-1}X'(I - P_S).$$

For example, let

$$S = Y_{h,t} = \begin{bmatrix} 0 & 0 & \cdots & 0 \\ y(1)' & 0 & \cdots & 0 \\ y(2)' & y(1)' & \cdots & 0 \\ \vdots & \vdots & \cdots & \vdots \\ y(t-1)' & y(t-2)' & \cdots & y(t-h)' \end{bmatrix},$$

$$y_t = [y(1), \ldots, y(t)], \qquad y_t^h = [0, 0, \ldots, y(1), \ldots, y(t-h)].$$

Then $\varepsilon_{h,t} = y_t(I - P_S)\pi'$, $\pi = [0, \ldots, 1]$, as is obvious. Then $\varepsilon_{h+1,t}$ is obtained by taking $X' = [0, \ldots, 0, y_1, \ldots, y_{t-h-1}]$ and (7.5.17) results, since $y_t(I - P_S)X = \Delta_{h+1,t}$, fairly clearly, $X'(I - P_S)X = R^r_{h,t-1}$, again fairly clearly and $X'(I - P_S)\pi' = r_{h,t-1}$.

The time update depends on representing, for $X = \pi'$, $S = Y_{h,t}$,

(7.5.24) $$I - P_{S+X} = \begin{bmatrix} I - P_{Y_{h,t-1}} & 0 \\ 0 & 0 \end{bmatrix}.$$

This follows from the fact that the columns of $[Y_{h,t}\pi']$ span the same space as those of

$$\begin{bmatrix} y(1)' & 0 & \cdots & 0 \\ y(2)' & y(1) & \cdots & 0 \\ \vdots & \vdots & \cdots & \vdots \\ y(t-2)' & y(t-3) & \cdots & y'(t-h-1) \\ 0 & 0 & \cdots & 0 \\ 0 & 0 & \cdots & 1 \end{bmatrix},$$

so that, for this S, X,

$$I - P_{S+X} = I - \begin{bmatrix} P_{Y_{h,t-1}} & 0 \\ 0 & 1 \end{bmatrix},$$

as required. To see how this may be used, take $\Delta_{h+1,t}$, for example. This is $y_t(I - P_s)(y_t^h)'$ with $S = Y_{h,t}$. Then $\Delta_{h+1,t-1} = y_t(I - P_{S+X})(y_t^h)'$ with $X = \pi'$, from the immediately preceding result, (7.5.24). Thus, from (7.5.23),

$$\begin{aligned}\Delta_{h+1,t} &= \Delta_{h+1,t-1} + y_t(I - P_S)X[X'(I - P_S)X]^{-1}X'(I - P_S)(y_t^h)' \\ &= \Delta_{h+1,t-1} + \varepsilon_{h,t}r'_{h,t-1}\gamma_{h-2,t-1}^{-1},\end{aligned}$$

where $\{1 - \gamma_{h-2,t-1}\}$ is the bottom right-hand element of P_S and thus of the form

$$(y(t-1)', y(t-2)', \ldots, y(t-h)')(Y'_{h,t}Y_{h,t})^{-1}(y(t-1)', \ldots, y(t-h)')'.$$

Friedlander (1982) calls this "the likelihood variable." Other formulas in (7.5.17)–(7.5.22) may be obtained in the same way.

Another characteristic of many lattice calculations is normalization of the calculated quantities. This has been discussed in Section 6.4. See (6.4.12), for example. This normalization stabilizes the algorithm (see Section 6.4) but also results in fewer calculations if only the normalized forward and backward prediction errors are needed. Only the normalized reflection coefficients are now needed and there is only one of these, Ω_h, of Section 6.4.

Finally, the calculation may be (and usually will be) effected for the case where there is forgetting of the past, so that what is being minimized is

$$\sum \alpha_t(s)\varepsilon_\theta(s)^2, \qquad \varepsilon_\theta(t) = b^{-1}ay(t)$$

with $\alpha_t(s)$ given by (7.5.4). A standard case would be $\alpha_s = 1$, $0 < \lambda_u \equiv \lambda < 1$. The net effect of this forgetting on the lattice calculation is to require one additional multiplication. The number of data points effectively used to time t is now reasonably put at

$$n(t) = \sum_1^t \alpha_t(s) = \lambda_t n(t-1) + \alpha_t.$$

In the case where $\alpha_s \equiv 1$, $\lambda_u \equiv \lambda$, the only effect of forgetting on (7.5.17)–(7.5.22) is to replace $\Delta_{h,t-1}$ on the right side of (7.5.19) by $\lambda\Delta_{h,t-1}$ and to replace $R_{0,t-1}^\varepsilon$ by $\lambda R_{0,t-1}^\varepsilon$ on the right side in (7.5.22).

These methods may be used to calculate AML by taking $y(t)$ to be the vector process (or two channel input, in the jargon)

$$\begin{pmatrix} y(t) \\ \hat{\varepsilon}(t) \end{pmatrix}.$$

Then a criterion could also be used to determine p. We shall not give details here. For that matter RML or the algorithm AM could be effected via a lattice calculation. *However, now the order, p, will have to be prescribed because in these algorithms some inverse filtering takes place, for example,* $\eta(t) = b(z)^{-1}y(t)$. Because of the inverse nature the calculation of $\eta(t)$ must commence from a prescribed value for p. Of course, if back substitution is effected, all values of p up to some maximum value P can be considered, but then, whatever the method used, $O(P^2)$ calculations, in addition, must be done at each t. In some real-time situations this might be impossible unless P was very small. For example, in digital encoding of speech for transmission, even at the level of fidelity of a telephone, sampling will have to be at about 5 kHz. Even with a very fast computer, the number of calculations per time point is then very limited, even with some parallel processing, and the back substitution might have to be avoided. Of course, it is possible to produce a chip to effect the basic lattice calculation at each t, changing h to $h + 1$, these being set in series to effect the overall calculation. In that case, processing time would be greatly reduced.

These lattice calculations are certainly very important. There remains a large amount of theory to be completed in relation to the type of calculation they do.

References

Adamyan, V. M., Arov, D. F., and Krein, M. G. (1971). Analytic properties of Schmidt pairs for a Hankel operator and the generalised Schur-Takagi problem. *Math. USSR Sbornik* **15**, 31–73.

Adamyan, V. M., Arov, D. F., and Krein, M. G. (1978). Infinite Hankel block matrices and related extension problems. *Amer. Math. Soc. Translations* **111**, 133–156.

Akaike, H. (1969). Fitting autoregressive models for prediction. *Ann. Inst. Statist. Math.* **21**, 243–247.

Akaike, H. (1974). Stochastic theory of minimal realization. *IEEE Trans. Autom. Control* **AC-19**, 667–674.

Akaike, H. (1976). Canonical correlation analysis of time series and the use of an information criterion. In *System Identification: Advances and Case Studies* (eds. R. H. Mehra and D. G. Lainiotis), 27–96. New York, Academic Press.

Akaike, H. (1977). On entropy maximisation principle. In *Applications of Statistics* (ed. P. R. Krishnaiah), 27–41. Amsterdam, North-Holland.

An, Hong-Zhi (1982). Estimation of prediction error variance. *Stochastic Process. Appl.* **13**, 39–43.

An, Hong-Zhi, Chen, Zhao Guo, and Hannan, E. J. (1982). Autocorrelation, autoregression and autoregressive approximation. *Ann. Statist.* **10**, 926–936.

An, Hong-Zhi, Chen, Zhao Guo, and Hannan, E. J. (1983). A note on ARMA estimation. *J. Time Series Anal.* **4**, 9–17.

Anderson, T. W. (1958). *An Introduction to Multivariate Statistical Analysis*. New York, Wiley.

Anderson, T. W. (1971). *The Statistical Analysis of Time Series*. New York, Wiley.

Anderson, B. D. O., and Moore, J. B. (1979). *Optimal Filtering*. Englewood Cliffs, N.J., Prentice-Hall.

Åström, K. J., and Mayne, D. Q. (1982). A new algorithm for recursive estimation of controlled ARMA processes. In *Proceedings of the 6th IFAC Symposium on Identification and System Parameter Estimation* (eds. G. A. Bekey and G. W. Savidis). Oxford, Pergamon Press.

Auslander, L., and MacKenzie, R. E. (1977). *Introduction to Differentiable Manifolds*. New York, Dover.

Azencott, R., and Dacunha-Castelle, D. (1984). *Séries d'observations irrégulières.* Paris, Masson.

Barnett, S. (1972). *Matrices in Control Theory.* New York, Van Nostrand Reinhold.

Bartlett, M. S. (1955). *An Introduction to Stochastic Processes with Special Reference to Methods and Applications.* Cambridge, Cambridge University Press.

Bauer, F. L. (1955). Ein direktes Iterationsverfahren zur Hurwitz-Zerlegung eines Polynoms. *Arch. Elek. Ubertr.* **9**, 285–290.

Baxter, G. (1962). An asymptotic result for the finite predictor. *Math. Scand.* **10**, 137–144.

Baxter, G. (1963). A norm inequality for a "finite section" Wiener-Hopf equation. *Ill. J. Math.* **7**, 97–103.

Berman, S. (1971). Asymptotic independence of the number of high and low level crossings of stationary Gaussian processes. *Ann. Math. Statist.* **42**, 927–945.

Bhansali, R. J. (1980). Autoregressive and window estimates of the inverse correlation function. *Biometrika* **67**, 551–566.

Bhansali, R. J. (1983). The inverse partial correlation function of a time series and its applications. *J. Multivariate Anal.* **13**, 310–327.

Bhansali, R. J. (1986). The criterion autoregressive transfer function of Parzen. *J. Time Series Anal.* **7**, 79–104.

Billingsley, P. (1968). *Convergence of Probability Measures.* New York, Wiley.

Blomberg, H., and Ylinen, R. (1983). *Algebraic Theory for Multivariate Linear Systems.* London, Academic Press.

Bouaziz, M. (1983). Formes de Toeplitz de séries temporelles. *C.R. Acad. Sci.*, Ser. I, 1017–1019.

Bouaziz, M. (1984). Toeplitz forms of stationary time series. Research report, Université Paris-Sud.

Box, G. E. P., and Jenkins, G. M. (1970). *Time Series Analysis, Forecasting and Control,* San Fransisco, Holden-Day.

Brillinger, D. R. (1984). *The Collected Works of John Tukey,* Vol. II, *Time Series* (ed. D. R. Brillinger). Belmont, Colit., Wadsworth.

Burg, J. P. (1975). Maximum entropy spectral analysis. Ph.D. dissertation, Geophysics Dept., Stanford University, Stanford, California.

Byrnes, C. I., and Martin, C. F. (eds.) (1980). *Geometrical Methods for the Theory of Linear Systems.* Dordrecht, The Netherlands, D. Reidel.

Caines, P. E. (1978). Stationary linear and nonlinear identification and predictor set completeness. *IEEE Trans. Autom. Control* **AC-23**, 583–594.

Chow, Y. S., and Studden, W. J. (1969). Monotonicity of the variance under truncation and variations of Jensen's inequality. *Ann. Math. Statist.* **40**, 1106–1108.

Clark, J. M. C. (1976). The consistent selection of parameterization in system identification. In *Proceedings of the Joint Automatic Control Conference.* Purdue University, West Lafayette.

Cox, D. R., and Hinkley, D. V. (1974). *Theoretical Statistics.* London, Chapman & Hall.

Dahlhaus, R. (1983). Spectral analysis with tapered data. *J. Time Series Anal.* **4**, 163–176.

de Gooijer, J. G., Abraham, B., Gould, A., and Robinson, L. (1985). Methods for determining the order of an autoregressive-moving average process: a survey. *Internat. Statist. Rev.* **53**, 301–329.

Deistler, M. (1978). The structural identifiability of linear models with autocorrelated errors in the case of cross-equation restrictions. *J. Econometics* **8**, 23–31.

Deistler, M. (1983). The properties of the parameterization of ARMAX systems and their relevance for structural estimation. *Econometrica* **51**, 1187–1207.

Deistler, M. (1985). General structure and parameterization of ARMA and state-space systems and its relation to statistical problems. In *Handbook of Statistics*, Vol. 5 (eds. E. J. Hannan, P. R. Krishnaiah, and M. M. Rao), 257–277. Amsterdam, North-Holland.

Deistler, M., and Hannan, E. J. (1981). Some properties of the parameterization of ARMA systems with unknown order. *J. Multivariate Anal.* **11**, 474–484.

Deistler, M., and Pötscher, B. M. (1984). The behaviour of the likelihood function for ARMA models. *Adv. Appl. Probab.* **16**, 843–865.

Deistler, M., Dunsmuir, W., and Hannan, E. J. (1978). Vector linear time series models: corrections and extensions. *Adv. Appl. Probab.* **10**, 360–372.

Deistler, M., and Wang, Liqun (1987). The common structure of parameterizations for linear systems. To appear.

Denham, M. J. (1974). Canonical forms for the identification of multivariable linear systems. *IEEE Trans. Autom. Control* **AC-19**, 646–656.

Dickinson, B. W., Kailath, T., and Morf, M. (1974). Canonical matrix fraction and state-space descriptions for determinstic and stochastic linear systems. *IEEE Trans. Autom. Control* **AC-19**, 656–667.

Dunsmuir, W. (1976). Estimation for vector linear time series models. Thesis submitted to the Australian National University, Canberra.

Dunsmuir, W., and Hannan, E. J. (1976). Vector linear time series models. *Adv. Appl. Probab.* **8**, 339–364.

Durbin, J. (1960). The fitting of time series models. *Rev. Internat. Inst. Statist.* **28**, 233–244.

Dym, H., and McKean, H. P. (1976). *Gaussian Processes, Function Theory and the Inverse Spectral Problem.* New York, Academic Press.

Findley, D. F. (1985). On the unbiasedness property of *AIC* for exact or approximating linear stochastic time series models. *J. Time Series Anal.* **6**, 229–252.

Fisher, R. A. (1944). *Statistical Methods for Research Workers*, 9th ed. Edinburgh, Oliver and Boyd.

Forney, G. D. (1975). Minimal bases of rational vector spaces, with applications to multivariable linear systems. *SIAM J. Control* **13**, 493–520.

Franke, J. (1985). A Levinson-Durbin recursion for autoregressive-moving average processes. *Biometrika* **72**, 573–581.

Franke, J., Gasser, Th., and Steinberg, H. (1985). Fitting autoregressive processes to EEG time series: an empirical comparison of estimates of the order. *IEEE Trans. Acoustics Speech Signal Process.* **ASSP-33**, 143–150.

Friedlander, B. (1982). Lattice filters for adaptive processing. *Proc. IEEE* **20**, 830–867.

Fuller, W., Hasza, D. P., and Goebel, J. J. (1981). Estimation of the parameters of a stochastic difference equation. *Ann. Statist.* **9**, 531–543.

Gihman, I. I., and Skorohod, A. V. (1974). *The Theory of Stochastic Processes I.* Berlin, Springer-Verlag.

Glover, K. (1984). All optimal Hankel-norm approximations of linear multivariable systems and their L^∞ error bounds. *Internat. J. Control* **39**, 1115–1193.

Glover, K., and Willems, J. C. (1974). Parametrizations of linear dynamical systems: canonical forms and identifiability. *IEEE Trans. Autom. Control* **AC-19**, 640–645.

Godolphin, E. J. (1976). On the Cramér-Wold factorization. *Biometrika* **63**, 367–380.

Godolphin, E. J. (1984). A direct representation for the large-sample maximum likelihood estimator of a Gaussian autoregressive-moving average process. *Biometrika* **71**, 281–289.

Gray, H. L., Kelley, G. D., and McIntire, D. D. (1978). A new approach to ARMA modelling. *Communications Statist.* **B7**, 1–77.

Green, M. (1987). Allpass matrices, the positive real lemma and unit canonical correlations between future and past. *J. Multivariate Anal.* (to appear).

Grenander, U., and Szegö, G. (1958). *Toeplitz Forms and Their Applications.* Berkeley. Calif., University of California Press.

Guidorzi, R. (1975). Canonical structures in the identification of multivariable systems. *Automatica* **11**, 361–374.

Guidorzi, R. (1981). Invariants and canonical forms for systems structural and parametric identification. *Automatica* **17**, 177–133.

Hall, P., and Heyde, C. C. (1980). *Martingale Limit Theory and Its Application.* New York, Academic Press.

Hannan, E. J. (1969a). The identification of vector mixed autoregressive-moving average systems. *Biometrika* **56**, 223–225.

Hannan, E. J. (1969b). The estimation of mixed autoregressive moving average systems. *Biometrika* **56**, 579–594.

Hannan, E. J. (1970). *Multiple Time Series.* New York, Wiley.

Hannan, E. J. (1971a). Nonlinear time series regression. *J. Appl. Probab.* **8**, 767–780.

Hannan, E. J. (1971b). The identification problem for multiple equation systems with moving average errors. *Econometrica* **39**, 751–765.

Hannan, E. J. (1973). The asymptotic theory of linear time series models. *J. Appl. Probab.* **10**, 130–145.

Hannan, E. J. (1974). The uniform convergence of autocovariances. *Ann. Statist.* **2**, 803–806.

Hannan, E. J. (1978). Rates of convergence for time series regression. *Adv. Appl. Probab.* **10**, 740–743.

Hannan, E. J. (1980a). The estimation of the order of an ARMA process. *Ann. Statist.* **8**, 1071–1081.

Hannan, E. J. (1980b). Recursive estimation based on ARMA models. *Ann. Statist.* **8**, 762–777.

Hannan, E. J. (1981). Estimating the dimension of a linear system. *J. Multivariate Anal.* **11**, 459–473.

Hannan, E. J. (1982). Testing for autocorrelation and Akaike's criterion. In *Essays in Statistical Science* (eds. J. M. Gani and E. J. Hannan), 403–412. Sheffield, England, Applied Probability Trust.

Hannan, E. J., and Heyde, C. C. (1972). On limit theorems for quadratic functions of discrete time series. *Ann. Math. Statist.* **43**, 2058–2066.

Hannan, E. J., and Kavalieris, L. (1983). The convergence of autocorrelations and autoregressions. *Aust. J. Statist.* **25**, 287–297.

Hannan, E. J., and Kavalieris, L. (1984). Multivariate linear time series models. *Adv. Appl. Probab.* **16**, 492–561.

Hannan, E. J., and Kavalieris, L. (1986). Regression, autoregression models. *J. Time Series Anal.* **7**, 27–50.

Hannan, E. J., and McDougall, A. J. (1988). Regression procedures for ARMA estimation. *J. Amer. Statist. Assoc.* (to appear).

Hannan, E. J., and Poskitt, D. S. (1987). Unit canonical correlation between future and past. *Ann. Statist.* (to appear).

Hannan, E. J., and Rissanen, J. (1982). Recursive estimation of ARMA order. *Biometrika* **69**, 81–94.

Hannan, E. J., Dunsmuir, W., and Deistler, M. (1980). Estimation of vector ARMAX models. *J. Multivariate Anal.* **10**, 275–295.

Hannan, E. J., Kavalieris, L., and Mackisack, M. (1986). Recursive estimation of linear systems. *Biometrika* **73**, 119–134.

Hardy, G. H., Littlewood, J. E., and Pólya, G. (1964). *Inequalities*, 3rd ed. Cambridge, Cambridge University Press.

Hartman, P., and Wintner, A. (1941). On the law of the iterated logarithm. *Amer. J. Math.* **63**, 169–176.

Hazewinkel, M. (1979). On identification and the geometry of the space of linear systems. *Lecture Notes in Control and Information Sciences*, Vol. 16, 401–415. Berlin, Springer-Verlag.

Hazewinkel, M., and Kalman, R. E. (1976). Invariants, canonical forms and moduli for linear, constant, finite dimensional, dynamical systems. In *Lecture Notes in Economics and Mathematical Systems*, Vol. 131 (eds. G. Marchesini and S. K. Mitter), 48–60. Berlin, Springer-Verlag.

Hosoya, Y. (1984). Information criteria and tests for time series models. In *Time Series Analysis: Theory and Practice*, Vol. 5 (ed. O. D. Anderson), 39–52, Amsterdam, North-Holland.

Ibragimov, I. A., and Linnik, Y. V. (1971). *Independent and Stationary Sequences of Random Variables.* Groningen, The Netherlands, Wolters-Noordhoff.

Jazwinski, A. H. (1970). *Stochastic Processes and Filtering Theory.* New York, Academic Press.

Jennrich, R. I. (1969). Asymptotic properties of non-linear least squares estimators. *Ann. Math. Statist.* **40**, 633–643.

Jonckheere, E. A., and Helton, J. W. (1985). Power spectrum reduction by optimal Hankel norm approximation of the phase of the outer spectral factor. *IEEE Trans. Autom. Control* **AC-30**, 1192–1201.

Kabaila, P. (1983). Parameter values of ARMA models minimising the one-step-ahead prediction error when the true system is not in the model set. *J. Applied Prob.* **20**, 405–408.

Kailath, T. (1980). *Linear Systems.* Englewood Cliffs, N.J., Prentice-Hall.

Kalman, R. E. (1960). A new approach to linear filtering and prediction problems. *Trans. Amer. Soc. Mech. Eng., J. Basic Engineering* **82**, 35–45.

Kalman, R. E. (1963). Mathematical description of linear dynamical systems. *SIAM J. Control* **1**, 152–192.

Kalman, R. E. (1965). Irreducible realizations and the degree of a rational matrix. *SIAM J. Appl. Math.* **13**, 520–544.

Kalman, R. E. (1974). Algebraic geometric description of the class of linear systems of constant dimension. *8th Annual Princeton Conference on Information Sciences and Systems*, Princeton, N.J.

Kalman, R. E., and Bucy, R. S. (1961). New results in linear filtering and prediction theory. *Trans. Amer. Soc. Mech. Eng., J. Basic Engineering* **83**, 95–108.

Kalman, R. E., Falb, P. L., and Arbib, M. A. (1969). *Topics in Mathematical System Theory.* New York, McGraw-Hill.

Kavalieris, L. (1984). Statistical problems in linear systems. Thesis submitted to the Australian National University, Canberra.

Kohn, R. (1979). Asymptotic estimation and hypothesis testing results for vector linear time series models. *Econometrica* **47**, 1005–1030.

Lee, D. T. L., Friedlander, B., and Morf, M. (1982). Recursive ladder algorithms for ARMA modelling. *IEEE Trans. Autom. Control* **AC-27**, 753–764.

Lee, D. T. L., Morf, M., and Friedlander, B. (1981). Recursive least squares ladder estimation algorithms. *IEEE Trans. Acoustics Speech Signal Process.* **ASSP-29**, 627–641.

Lewis, R., and Reinsel, G. C. (1985). Prediction of multivariate autoregressive time series by autoregressive model fitting. *J. Multivariate Anal.* **16**, 393–411.

Ljung, L. (1978). Convergence analysis of parametric identification methods. *IEEE Trans. Autom. Control* **AC-23**, 770–783.

Ljung, L., and Caines, P. E. (1979). Asymptotic normality of prediction error estimation for approximate system models. *Stochastics*, **3**, 89–100.

Ljung, L., and Rissanen, J. (1976). On canonical forms, parameter identifiability and the concept of complexity. *Proceedings of the 4th IFAC Symposium on Identification and Systems Parameter Estimation*, Tbilisi, USSR.

Ljung, L., and Söderström, T. (1983). *Theory and Practice of Recursive Identification.* Cambridge, Mass., MIT Press.

Luenberger, D. G. (1967). Canonical forms for linear multivariable systems. *IEEE Trans. Autom. Control* **AC-12**, 290–293.

MacDuffee, C. C. (1956). *The Theory of Matrices.* New York, Chelsea.

Macleish, D. L. (1977). On the invariance principle for non stationary mixingales. *Ann. Probab.* **5**, 616–621.

Makhoul, J. (1981). Lattice methods in spectral estimation. In *Applied Time Series Analysis*, Vol. II (ed. D. F. Findley), 301–326. New York, Academic Press.

Mallows, C. L. (1973). Some comments on C_p. *Technometrics* **15**, 661–675.

Monden, Y., Yamada, M., and Arimoto, S. (1982). Fast algorithms for identification of an ARX model and its order determination. *IEEE Trans. Acoustics Speech Signal Process.* **ASSP-30**, 350–398.

Moore, B. C. (1981) Principal component analysis in linear systems: controllability, observability and model reduction. *IEEE Trans. Automatic Control*, AC-26, 17–27.

Morf, M., Vieira, A., and Kailath, T. (1978). Covariance characterization of partial autocorrelation matrices. *Ann. Statist.* **6**, 643–645.

Nehari, Z. (1957). On bounded bilinear forms. *Ann. Math.* **65**, 153–162.

Neveu, J. (1975). *Discrete-Parameter Martingales.* Amsterdam, North-Holland.

Newman, M. (1972). *Integral Matrices.* New York, Academic Press.

Nicholls, D. F. (1976). The efficient estimation of vector linear time series models. *Biometrika* **63**, 381–390.

Ogata, Y. (1980). Maximum likelihood estimation of incorrect Markov models and the derivation of *AIC*. *J. Appl. Probab.* **17**, 59–72.

Parzen, E. (1974). Some recent advances in time series modelling. *IEEE Trans. Autom. Control* **AC-19**, 723–730.

Paulsen, J., and Tjøstheim, D. (1985). On the estimation of residual variance and order in autoregressive time series. *J. Roy. Statist. Soc.* **47**, 216–228.

Ploberger, W. (1982). Slight misspecifications of linear systems. In *Operations Research in Progress* (eds. G. Feichtinger and P. Kall), 413–424. Dordrecht, The Netherlands, D. Reidel.

Popov, V. M. (1969). Some properties of the control systems with irreducible matrix-transfer functions. In *Seminar on Differential Equations and Dynamical Systems II, Lecture Notes in Mathematics*, No. 144, 169–180. Berlin, Springer-Verlag.

Popov, V. M. (1972). Invariant description of linear time-invariant controllable systems. *SIAM J. Control* **10**, 252–264.

Poskitt, D. S., and Tremayne, A. R. (1980). Testing the specification of a fitted autoregressive-moving average model. *Biometrika* **67**, 359–363.

Poskitt, D. S., and Tremayne, A. R. (1981). An approach to testing linear time series models. *Ann. Statist.* **9**, 974–986.

Poskitt, D. S., and Tremayne, A. R. (1982). Diagnostic tests for multiple time series models. *Ann. Statist.* **10**, 114–120.

Pötscher, B. M. (1983). Order estimation in ARMA-models by Lagrangian multiplier tests. *Ann. Statist.* **11**, 827–885.

Pötscher, B. M. (1985). The behaviour of the Lagrangian multiplier test in testing the orders of an ARMA model. *Metrika*, **32**, 129–150.

Pötscher, B. M. (1987). Convergence results for maximum likelihood type estimators in multivariable ARMA models. *J. Multivariate Anal.* **21**, 29–52.

Rissanen, J. (1974). Basis of invariants and canonical forms for linear dynamic systems. *Automatica* **10**, 175–182.

Rissanen, J. (1978). Modeling by shortest data description. *Automatica* **14**, 465–471.

Rissanen, J. (1983). Universal prior for parameters and estimation by minimum description length. *Ann. Statist.* **11**, 416–431.

Rissanen, J. (1986). Stochastic complexity and modeling. *Ann. Statist.* **14**, 1080–1100.

Rissanen, J., and Caines, P. E. (1979). The strong consistency of maximum likelihood estimators for ARMA processes. *Ann. Statist.* **7**, 297–315.

Rosenbrock, H. H. (1970). *State-Space and Multivariable Theory*. New York, Wiley.

Rozanov, Yu. A. (1967). *Stationary Random Processes*. San Fransisco, Holden-Day.

Rudin, W. (1973). *Functional Analysis*. New York, McGraw-Hill.

Saikonnen, P. (1987). Asymptotic properties of some preliminary estimators for autoregressive-moving average models. *J. Time Series Anal.* (To appear.)

Schwarz, G. (1978). Estimating the dimension of a model. *Ann. Statist.* **6**, 461–464.

Shibata, R. (1976). Selection of the order of an autoregressive model by Akaike's information criterion. *Biometrika* **63**, 117–126.

Shibata, R. (1980). Asymptotically efficient selection of the order of the model for estimating parameters of a linear process. *Ann. Statist.* **8**, 147–164.

Shibata, R. (1981). An optimal autoregressive spectral estimate. *Ann. Statist.* **9**, 300–306.

Shimizu, R. (1978). Entropy maximization principle and selection of the order of an autoregressive Gaussian process. *Ann. Inst. Stat. Math.* **30**, 263–270.

Shumway, R. H., and Stoffer, D. S. (1982). An approach to time series smoothing and forecasting using the EM algorithm. J. Time Series Anal. **3**, 253–264.

Solo, V. (1979). The convergence of AML. *IEEE Trans. Autom. Control* **AC-24**, 958–962.

Solo, V. (1981). The second order properties of a time series recursion. *Ann. Statist.* **9**, 307–317.

Solo, V. (1984). The exact likelihood for a multivariate ARMA model. *J. Multivariate Anal.* **15**, 164–173.

Solo, V. (1986). Topics in advanced time series analysis. In *Lectures in Probability and Statistics* (eds. G. del Pino and R. Rebodedo). Berlin, Springer-Verlag.

Son, L. H., and Anderson, B. D. O. (1971). Design of Kalman filters using signal model output statistics. *Proc. IEEE* **120**, 312–318.

Spitzer, F. (1964). *Principles of Random Walk*. Princeton, N.J., D. Van Nostrand.

Spliid, H. (1983). A fast estimation method for vector autoregressive-moving average model with exogenous variables. *J. Amer. Statist. Assoc.* **78**, 843–849.

Stout, W. F. (1970). The Hartman-Wintner law of the iterated logarithm for martingales. *Ann. Math. Statist.* **41**, 2158–2160.

Stout, W. F. (1974). *Almost Sure Convergence*. New York, Academic Press.

Tiao, G. C., and Tsay, R. S. (1983). Consistency properties of least squares estimates of autoregressive parameters in ARMA models. *Ann. Statist.* **11**, 856–871.

Tjøstheim, P., and Paulsen, J. (1983). Bias of some commonly used time series estimates. *Biometrika* **70**, 389–400.

Tong, H. (1979). A note on a local equivalence of two recent approaches to autoregressive order determination. *Internat. J. Control* **29**, 441–446.

Tsay, R. S., and Tiao, G. C. (1984). Consistent estimates of autoregressive parameters and extended sample autocorrelation function for stationary and nonstationary ARMA models. *J. Amer. Statist. Assoc.* **79**, 84–96.

Tuan, Pham-Dinh (1978). Estimation of parameters in the ARMA model when the characteristic polynomial of the MA operator has a unit root. *Ann. Statist.* **6**, 1369–1383.

van Overbeek, A. J. M. (1982). On line structure selection for the identification of multivariable systems. Linköping, Department of Electrical Engineering, Linköping University.

van Overbeek, A. J. M., and Ljung, L. (1982). On line structure selection for multivariable state space models. *Automatica* **18**, 529–544.

Veres, S. (1987). Asymptotic distributions of likelihood ratios and overparameterized ARMA processes. *J. Time Series Anal.* **8**, 345–348.

Walker, A. M. (1962). Large-sample estimation of parameters for autoregressive processes with moving-average residuals. *Biometrika* **49**, 117–131.

Walker, A. M. (1964). Asymptotic properties of least squares estimates of parameters of the spectrum of a stationary non-deterministic time series. *J. Aust. Math. Soc.* **4**, 363–384.

Whittle, P. (1951). *Hypothesis Testing and Time Series Analysis*. Uppsala, Almqvist & Wiksell.

Whittle, P. (1963). On the fitting of multivariate autoregression and the approximate canonical factorisation of a spectral density matrix. *Biometrika* **50**, 129–134.

Wiener, N. (1949). *Extrapolation, Interpolation and Smoothing of Stationary Time Series*. New York, Wiley.

Willems, J. C. (1986). From time series to linear system. *Automatica*, **22**, 561–580.

Wilson, G. T. (1969). Factorization of the covariance generating function of a pure moving average process. *SIAM J. Number. Anal.* **6**, 1–8.

Wolovich, W. A. (1974). *Linear Multivariable Systems.* New York, Springer-Verlag.

Young, P. C. (1984). *Recursive Estimation and Time Series Analysis.* Berlin, Springer-Verlag.

Author Index

Subject Index

Applied Probability and Statistics (Continued)

HOCHBERG and TAMHANE • Multiple Comparison Procedures
HOEL • Elementary Statistics, *Fourth Edition*
HOEL and JESSEN • Basic Statistics for Business and Economics, *Third Edition*
HOGG and KLUGMAN • Loss Distributions
HOLLANDER and WOLFE • Nonparametric Statistical Methods
IMAN and CONOVER • Modern Business Statistics
JESSEN • Statistical Survey Techniques
JOHNSON • Multivariate Statistical Simulation
JOHNSON and KOTZ • Distributions in Statistics
Discrete Distributions
Continuous Univariate Distributions—1
Continuous Univariate Distributions—2
Continuous Multivariate Distributions
JUDGE, GRIFFITHS, HILL, LÜTKEPOHL and LEE • The Theory and Practice of Econometrics, *Second Edition*
KALBFLEISCH and PRENTICE • The Statistical Analysis of Failure Time Data
KISH • Statistical Design for Research
KISH • Survey Sampling
KUH, NEESE, and HOLLINGER • Structural Sensitivity in Econometric Models
KEENEY and RAIFFA • Decisions with Multiple Objectives
LAWLESS • Statistical Models and Methods for Lifetime Data
LEAMER • Specification Searches: Ad Hoc Inference with Nonexperimental Data
LEBART, MORINEAU, and WARWICK • Multivariate Descriptive Statistical Analysis: Correspondence Analysis and Related Techniques for Large Matrices
LINHART and ZUCCHINI • Model Selection
LITTLE and RUBIN • Statistical Analysis with Missing Data
McNEIL • Interactive Data Analysis
MAGNUS and NEUDECKER • Matrix Differential Calculus with Applications
MAINDONALD • Statistical Computation
MALLOWS • Design, Data, and Analysis by Some Friends of Cuthbert Daniel
MANN, SCHAFER and SINGPURWALLA • Methods for Statistical Analysis of Reliability and Life Data
MARTZ and WALLER • Bayesian Reliability Analysis
MIKÉ and STANLEY • Statistics in Medical Research: Methods and Issues with Applications in Cancer Research
MILLER • Beyond ANOVA, Basics of Applied Statistics
MILLER • Survival Analysis
MILLER, EFRON, BROWN, and MOSES • Biostatistics Casebook
MONTGOMERY and PECK • Introduction to Linear Regression Analysis
NELSON • Applied Life Data Analysis
OSBORNE • Finite Algorithms in Optimization and Data Analysis
OTNES and ENOCHSON • Applied Time Series Analysis: Volume I, Basic Techniques
OTNES and ENOCHSON • Digital Time Series Analysis
PANKRATZ • Forecasting with Univariate Box-Jenkins Models: Concepts and Cases
PLATEK, RAO, SARNDAL and SINGH • Small Area Statistics: An International Symposium
POLLOCK • The Algebra of Econometrics
RAO and MITRA • Generalized Inverse of Matrices and Its Applications
RÉNYI • A Diary on Information Theory
RIPLEY • Spatial Statistics
RIPLEY • Stochastic Simulation